Crop Authenticity in Organic Horticultural Production: Recent Studies and Applications

Crop Authenticity in Organic Horticultural Production: Recent Studies and Applications

Guest Editors

Simona Fabroni
Francesco Montemurro
Luana Bontempo

MDPI

Basel • Beijing • Wuhan • Barcelona • Belgrade • Novi Sad • Cluj • Manchester

Guest Editors
Simona Fabroni
Research Center for Olive, Fruit and Citrus Crops
Council for Agricultural Research and Economics (CREA)
Acireale
Italy

Francesco Montemurro
Research Center for Agriculture and Environment
Council for Agricultural Research and Economics (CREA)
Bari
Italy

Luana Bontempo
Traceability Unit
Edmund Mach Foundation
San Michele all'Adige
Italy

Editorial Office
MDPI AG
Grosspeteranlage 5
4052 Basel, Switzerland

This is a reprint of the Special Issue, published open access by the journal *Horticulturae* (ISSN 2311-7524), freely accessible at: https://www.mdpi.com/journal/horticulturae/special_issues/8AX4D84I73.

For citation purposes, cite each article independently as indicated on the article page online and as indicated below:

Lastname, A.A.; Lastname, B.B. Article Title. *Journal Name* **Year**, *Volume Number*, Page Range.

ISBN 978-3-7258-4857-7 (Hbk)
ISBN 978-3-7258-4858-4 (PDF)
https://doi.org/10.3390/books978-3-7258-4858-4

Cover image courtesy of Francesco Montemurro

Contents

horticulturae

MDPI

Editorial

Crop Authenticity in Organic Horticulture: Insights from Recent Research and Practical Applications

Simona Fabroni [1,*], Luana Bontempo [2] and Francesco Montemurro [3]

[1] Council for Agricultural Research and Economics (CREA), Research Center for Olive, Fruit and Citrus Crops, Corso Savoia 190, 95024 Acireale, Italy
[2] Traceability Unit, Edmund Mach Foundation, Via E. Mach 1, 38098 San Michele All'Adige, Italy; luana.bontempo@fmach.it
[3] Council for Agricultural Research and Economics (CREA), Research Center for Agriculture and Environment, Via Celso Ulpiani, 5, 70125 Bari, Italy; francesco.montemurro@crea.gov.it
* Correspondence: simona.fabroni@crea.gov.it

1. Introduction

For more than two decades, organic management has been associated with beneficial effects on human and soil health, while being environmentally sustainable due to the removal of/reduction in pesticides used as plant protection agents and chemical fertilizers. Therefore, organic agriculture implies that the use of mineral chemical synthetic fertilizers is harshly prohibited, thus involving the application of organic fertilizers and alternative methods to improve plant nutrition. Accordingly, the existing knowledge gaps can be first identified as the optimization and validation of innovative and safe agronomic practices to improve organic crop yield, at the same time preserving and enhancing their quality. A recent comprehensive review on the impact of the most relevant agronomic practices on soil chemical and biological properties have highlighted that, if properly carried out, sustainable agronomic management can enhance soil fertility, nutrient cycling, and agricultural productivity while mitigating environmental consequences [1].

As far as the transition towards organic management is concerned, detailed reports on the effects of the application of organic fertilizers, the introduction of cover crops, and on the set-up of appropriate varieties specifically developed for organic farming in the specific environments are needed to help farmers make the right management choices. Through a meta-analysis approach, it has been recently demonstrated that the application of organic fertilizers can significantly improve the yield, growth parameters, and quality of Chinese cabbage in China, while increasing soil nutrients [2]. Regarding the need for organic farming focused breeding programs, Rempelos et al. [3] have recently compared the performance of modern varieties developed for conventional management with older and new varieties suitable for organic farming. They also reviewed evidence from wheat variety trials and factorial field experiments supporting the definition of trait combinations that should be targeted to improve crop yield.

In addition to that, one of the most important gaps to be filled for redefining the value of organic agriculture is related to the lack of a reliable and rigorous traceability system for organic productions. In fact, for more than two decades, several studies have been proposed targeting this specific need. Very recently, Giannioti et al. [4] reviewed the isotopic-based methodologies developed for organic food authentication, highlighting that, for this purpose, the stable isotope ratio of nitrogen (N) has been defined as the most reliable organic marker. They also reviewed the challenges of this kind of approach, concluding that innovative multivariate approaches should be validated to overcome these limitations related to organic crop authenticity.

Finally, taking into account the increasing interest towards consumer health and safety, a relevant issue to be addressed is also related to health risk assessment. Indeed, Hurtado-Barroso et al. [5] concluded that large long-term intervention studies are needed to determine whether an organic diet is healthier than a diet including conventionally grown food products.

This Special Issue of *Horticulturae* has provided new insights from recent research and practical applications supporting the transition to the sustainable organic crop management while providing relevant validated tools for organic crop authenticity.

2. Innovative Agronomic Practices to Improve Organic Crop Yield

A study by Kilic et al. [contribution 1] has proposed a novel approach for organic strawberry fertilization based on joint vermicompost and microbial complementary nutrition, providing promising results for enhancing organic strawberry production. Another study by Foughar et al. [contribution 2] has compared black soldier fly and mealworm larvae frass with another organic amendment in two growth cycles of lettuce. They have provided evidence that mealworm larvae frass exhibits plant biostimulant effects, boosting productivity on lettuce, beyond nutrient release and improved soil properties. Useful insights for the application of mixed formulations containing soybean meal and peanut grain residue for the increase in organic button mushroom yield have been provided by Cardoso Caitano et al. [contribution 3]. In addition, Guerra et al. [contribution 4] implemented a novel technique for the management of crop load in conference pear orchards, validating an effective preharvest organic strategy for mitigating postharvest quality losses during storage, thereby promoting overall orchard sustainability. A novel sustainable phytosanitary management strategy based on the use of fungal extracts as plant growth promoters with biocontrol capabilities has been validated against *Pseudomonas syringae* in organic tomato plants under greenhouse conditions by Garcia-Latorre et al. [contribution 5]. A review article by Ierna et al. [contribution 6] on crop nutrition and soil fertility management in organic potato production systems highlighted the most relevant agronomic practices needed to effectively maintain good soil fertility and satisfy the nutritional needs of the crop. They also provided to growers an overview of the current innovative tools to reduce the yield gap between organic and conventional potato growing systems.

3. Evidence Supporting the Sustainable Organic Crop Management

A research study by Mineață et al. [contribution 7] has shown that significant quality differences increased between the organically grown cherry fruits depending on the ripening phase and cultivar, enabling growers to appropriately choose the better agronomic settings to be applied. In this context, Tagiakas et al. [contribution 8] have demonstrated that selected inbred lines originated from tomato landraces developed under low-input conditions were better suited to organic farming systems. Moreover, they also ensured high-quality products with satisfactory yield and low requirements, offering an innovative pathway toward sustainable agriculture. In addition, it has been demonstrated that organic fertilization exhibited a performance similar to that of a conventional one, in terms of plant growth indexes, physiological parameters, and antioxidant levels, in wild edible species (purslane and sow thistle) cultivation under field conditions [contribution 9]. Roussos et al. [contribution 10] have demonstrated that a multisite two-year evaluation of the quality characteristics of olive oils produced under organic and conventional cultivation systems highlighted no significant variations. However, sustainable management practices leading to healthy fruit production and oil extraction may significantly influence the quality attributes of the oil produced by applying different fertilization and plant protection programs.

4. Traceability Systems for Organic Productions Authenticity

Based on the multivariate results of their long-term study on 'Navelina' oranges and 'Clemenules' mandarins under organic and conventional management, Domínguez-Gento et al. [contribution 11] have proposed a method that allows the successful differentiation of citrus fruits from organic and conventional production as a tool to predict the production system of citrus fruits based on easy-to-measure parameters. While stable isotopes have emerged as a powerful tool for this purpose, Muñoz-Redondo et al. [contribution 12] have isotopically characterized fertilizers, soil conditioners, and micronutrient fertilizers commonly used in intensive organic agriculture in the largest organic production region in the world (Almería, Spain). They demonstrated that some plant-based fertilizers, protein hydrolysed fertilizers, and chelated nutrients presented a wide range of variability in their $\delta^{15}N$ values, including some very low or even negative values, more similar to those of conventional fertilizers. These results have highlighted the challenges of authenticating organic foods in agriculture when using only nitrogen isotope analysis. With an attempt to overcome these limitations, Fabroni et al. [contribution 13] have proposed an integrated approach to discriminate organic from conventional products based on different isotopic signatures, due to the different nitrogenous sources, combined with the qualitative profile of the crops, which are significantly affected by the different agronomic treatments. This approach is intended to be coupled to the application of chemometric multivariate analysis on quality and nutraceutical parameters combined with isotopic data. As a case study, Montemurro et al. [contribution 14] have validated the aforementioned integrated approach for the reliable classification of organically and conventionally grown fennel productions obtained by applying different treatments at diverse percentages of mineral fertilizers, even introducing N-fixing plant (alfalfa) as an agroecological service crop in the intercropping fennel.

5. Health Risk Assessment

Elbagory et al. [contribution 15] have provided valuable insights into the biomonitoring of potentially toxic element contamination in widely consumed summer fruits of Northern India and the subsequent health risk assessment. Moreover, based on the environmental issue linked to the presence of high copper concentrations in soils as a consequence of its use for vegetable productions, Schmitt et al. [contribution 16] have compared Cu levels in soils from different agronomic systems on 25 sites in the Rio Grande do Sul region in Brazil. They showed that the Cu concentration in soils cultivated with vegetables increased in comparison to native soil as a result of agricultural management.

6. Conclusions

The Special Issue "Crop Authenticity in Organic Horticultural Production: Recent Studies and Applications" has been aimed at providing growers with new effectual elements for the improvement of organic crops' growth, yield, quality, safety, and authenticity. Further research should extend its application to a wider range of horticultural and/or fruit crops to confirm and expand upon the current findings.

Author Contributions: Writing—original draft preparation, S.F.; logical conception, S.F.; writing—review and editing, S.F., F.M. and L.B. All authors have read and agreed to the published version of the manuscript.

Conflicts of Interest: The authors declare no conflicts of interest.

List of Contributions:

1. Kilic, N.; Dasgan, H.Y.; Gruda, N.S. A Novel Approach for Organic Strawberry Cultivation: Vermicompost-Based Fertilization and Microbial Complementary Nutrition. *Horticulturae* **2023**, *9*, 642. https://doi.org/10.3390/horticulturae9060642.
2. Foughar, M.; Arrobas, M.; Rodrigues, M.Â. Mealworm Larvae Frass Exhibits a Plant Biostimulant Effect on Lettuce, Boosting Productivity beyond Just Nutrient Release or Improved Soil Properties. *Horticulturae* **2024**, *10*, 711. https://doi.org/10.3390/horticulturae10070711.
3. Caitano, C.E.C.; Vieira Júnior, W.G.; Alves, L.d.S.; Teixeira, P.A.G.; de Paula, L.C.; Zied, D.C. Applicability Analysis of Peanut Addition to Button Mushroom Substrate Supplement Formulation. *Horticulturae* **2024**, *10*, 1088. https://doi.org/10.3390/horticulturae10101088.
4. Guerra, M.; Álvarez-Taboada, F.; Marabel, V.; Felices, A.M.; Rodríguez-González, Á.; Casquero, P.A. Organic Agricultural Practice: Crop Load Management Enhancing Quality and Storability of High-Russet Pears. *Horticulturae* **2024**, *10*, 686. https://doi.org/10.3390/horticulturae10070686.
5. García-Latorre, C.; Rodrigo, S.; Santamaria, O. Biological Control of Pseudomonas syringae in Tomato Using Filtrates and Extracts Produced by Alternaria leptinellae. *Horticulturae* **2024**, *10*, 334. https://doi.org/10.3390/horticulturae10040334.
6. Ierna, A.; Distefano, M. Crop Nutrition and Soil Fertility Management in Organic Potato Production Systems. *Horticulturae* **2024**, *10*, 886. https://doi.org/10.3390/horticulturae10080886.
7. Mineață, I.; Murariu, O.C.; Sîrbu, S.; Tallarita, A.V.; Caruso, G.; Jităreanu, C.D. Effects of Ripening Phase and Cultivar under Sustainable Management on Fruit Quality and Antioxidants of Sweet Cherry. *Horticulturae* **2024**, *10*, 720. https://doi.org/10.3390/horticulturae10070720.
8. Tagiakas, R.; Vardaki, E.; Katsanoulas, E.; Kostoula, S.; Mavromatis, A.; Xynias, I.; Avdikos, I.D. Exploring the Benefits of Greek Tomato Landraces for Development of High Yield and Quality Hybrids in Organic Farming Systems. *Horticulturae* **2025**, *11*, 296. https://doi.org/10.3390/horticulturae11030296.
9. Hajisolomou, E.; Neofytou, G.; Petropoulos, S.A.; Tzortzakis, N. The Application of Conventional and Organic Fertilizers During Wild Edible Species Cultivation: A Case Study of Purslane and Common Sowthistle. *Horticulturae* **2024**, *10*, 1222. https://doi.org/10.3390/horticulturae10111222.
10. Roussos, P.A.; Karyda, A.-G.; Kaibalis, V.; Zontanos, K. Olive Oil Quality Produced Under Conventional and Organic Farming Systems in a Multisite Two-Year Evaluation in Greece. *Horticulturae* **2025**, *11*, 130. https://doi.org/10.3390/horticulturae11020130.
11. Domínguez-Gento, A.; Di Giorgi, R.; García-Martínez, M.D.; Raigón, M.D. Effects of Organic and Conventional Cultivation on Composition and Characterization of Two Citrus Varieties 'Navelina' Orange and 'Clemenules' Mandarin Fruits in a Long-Term Study. *Horticulturae* **2023**, *9*, 721. https://doi.org/10.3390/horticulturae9060721.
12. Muñoz-Redondo, J.M.; Cuevas, F.J.; Montenegro, J.C.; Ordóñez-Díaz, J.L.; Moreno-Rojas, J.M. An Updated Isotopic Database of Fertilizers Used in Intensive Organic Farming: A Case Study on Protein Hydrolyzed Derivatives and Chelated Nutrients. *Horticulturae* **2024**, *10*, 523. https://doi.org/10.3390/horticulturae10050523.
13. Fabroni, S.; Bontempo, L.; Campanelli, G.; Canali, S.; Montemurro, F. Innovative Tools for the Nitrogen Fertilization Traceability of Organic Farming Products. *Horticulturae* **2023**, *9*, 723. https://doi.org/10.3390/horticulturae9060723.
14. Montemurro, F.; Diacono, M.; Fiore, A.; Bontempo, L.; Roncone, A.; Amenta, M.; Ballistreri, G.; Timpanaro, N.; Torrisi, B.; Fabroni, S. Innovative Tools for Nitrogen Fertilization Traceability in Organic Farming Products: A Fennel Case Study. *Horticulturae* **2025**, *11*, 329. https://doi.org/10.3390/horticulturae11030329.
15. Elbagory, M.; Abd El-Aziz, M.A.; Omara, A.E.-D.; Abou Fayssal, S.; Kumar, V. Assessment of Potentially Toxic Elements in Four Melon Fruit Varieties Grown in the Ganges and Yamuna River Basin. *Horticulturae* **2025**, *11*, 216. https://doi.org/10.3390/horticulturae11020216.
16. Schmitt, O.J.; Andriolo, J.L.; Tiecher, T.L.; De Conti, L.; Lourenzi, C.R.; Vieira, R.C.B.; Kaiser, D.R.; Marchezan, C.; Corrêa, A.M.; Silva, I.C.B.; et al. Cu Concentration in Vegetable-Cultivated Soil: Comparison Between Open Fields and Protected Environments in Brazil. *Horticulturae* **2024**, *10*, 1183. https://doi.org/10.3390/horticulturae10111183.

References

1. Al-Shammary, A.A.G.; Al-Shihmani, L.S.S.; Fernández-Gálvez, J.; Caballero-Calvo, A. Optimizing sustainable agriculture: A comprehensive review of agronomic practices and their impacts on soil attributes. *J. Environ. Manag.* **2024**, *364*, 121487. [CrossRef] [PubMed]
2. Zhang, Y.; Zhang, J.; Zhang, J.; Li, H.; Li, C.; Wang, X. Effects of the Application of Organic Fertilizers on the Yield, Quality, and Soil Properties of Open-Field Chinese Cabbage (*Brassica rapa* spp. pekinensis) in China: A Meta-Analysis. *Agronomy* **2024**, *14*, 2555. [CrossRef]
3. Rempelos, L.; Wang, J.; Sufar, E.K.; Almuayrifi, M.S.B.; Knutt, D.; Leifert, H.; Leifert, A.; Wilkinson, A.; Shotton, P.; Hasanaliyeva, G.; et al. Breeding Bread-Making Wheat Varieties for Organic Farming Systems: The Need to Target Productivity, Robustness, Resource Use Efficiency and Grain Quality Traits. *Foods* **2023**, *12*, 1209. [CrossRef] [PubMed]
4. Giannioti, Z.; Ogrinc, N.; Suman, M.; Camin, F.; Bontempo, L. Isotope ratio mass spectrometry (IRMS) methods for distinguishing organic from conventional food products: A review. *TrAC Trends Anal. Chem.* **2024**, *170*, 117476. [CrossRef]
5. Hurtado-Barroso, S.; Tresserra-Rimbau, A.; Vallverdú-Queralt, A.; Lamuela-Raventós, R.M. Organic food and the impact on human health. *Crit. Rev. Food Sci. Nutr.* **2019**, *59*, 704–714. [CrossRef] [PubMed]

horticulturae

MDPI

Article

A Novel Approach for Organic Strawberry Cultivation: Vermicompost-Based Fertilization and Microbial Complementary Nutrition

Neslihan Kilic [1], Hayriye Yildiz Dasgan [2,*] and Nazim S. Gruda [3]

1 Department of Organic Farming Management, Kadirli Faculty of Applied Sciences, University of Osmaniye Korkut Ata, Osmaniye 80000, Turkey; neslihankilic@osmaniye.edu.tr
2 Department of Horticulture, Faculty of Agriculture, University of Cukurova, Adana 01330, Turkey
3 Institute of Plant Sciences and Resource Conservation, Division of Horticultural Sciences, University of Bonn, D-53113 Bonn, Germany; nazim.gruda@hu-berlin.de
* Correspondence: dasgan@cu.edu.tr

Abstract: This study investigated the effects of vermicompost fertilization with complementary microbial nutrition on the plant growth, yield, and fruit quality of the organically grown strawberry "Monterey" cultivar. Along with vermicompost, five different microbial fertilizers containing plant-growth-promoting rhizobacteria (PGPR) and arbuscular mycorrhizal fungi (AMF) were used as complementary nutrition. Here, we examined plant growth parameters, strawberry yield, fruit weight, pH, total soluble solids, and acidity in fruit and leaf mineral nutrient concentrations. Vermicompost-based fertilization with PGPR and AMF improved plant growth, yield, and fruit quality. The highest total yield (216.75 g per plant^{-1}) and heaviest fruits with an average of 18.11 g were obtained from the vermicompost-based fertilization with PGPR containing complementary fertilization. This included *Bacillus amyloliquefaciens*, *Bacillus pumilus*, *Bacillus subtilis*, *Bacillus licheniformis*, *Bacillus megaterium*, *Trichoderma harzianum*, and *Trichoderma konigii*. This treatment also resulted in the best ratio of total soluble solids to acidity (18.74), pH (3.95), and mineral nutrient concentrations in leaves. The novel approach with vermicompost-based fertilization and complementary microbial nutrition improves organic strawberries' growth, yield, and fruit quality. These results are promising for enhancing organic strawberry production.

Keywords: *Fragaria* × *ananassa* Duch.; vermicompost; bacteria; mycorrhiza; yield; fruit quality

1. Introduction

The strawberry (*Fragaria* × *ananassa* Duch.), which can be grown in different ecological conditions, is one of the most popular berry fruits globally. The strawberry is an excellent source of natural antioxidants, such as carotenoids, phenolics, vitamins, anthocyanins, and flavonoids. It has a very high capacity to scavenge free radicals that can prevent certain types of cancer, cardiovascular diseases, inflammation, obesity, and type II diabetes. Additionally, strawberries are a good vitamin C source and are in high demand by consumers due to their sensory and nutritional properties [1–3]. The strawberry is available for fresh consumption, processed, or frozen. The recoup of strawberry investments in the first year, especially when other fruits are unavailable in the market, provides a decent source of income for the producers [4].

The plantation area and production of strawberries have increased substantially in the last five years. According to FAO data, 377,112 ha of worldwide strawberry production area in 2015 expanded to 389,665 ha in 2021. Meanwhile, the annual production amount of 8,221,263 tons in 2015 increased to 9,175,384 tons in 2021. Among all the countries producing strawberries, Turkey ranks third worldwide with 669,195 tons of strawberry production [5]. However, organic strawberry production is very limited. The excessive utilization of chemical fertilizers and pesticides has contributed to soil quality and ecological stability degradation, leading to human health concerns. As a response, there has been a growing trend towards organic farming practices and a corresponding increase in demand for organic fruit products. Pesticide use is intense during conventional strawberry cultivation. Producing organic strawberries is valuable as it provides a safe, chemical-free, and health-protective food source that is particularly appealing to children. However, one of the significant challenges in organic farming is ensuring adequate plant nutrition. As a result, numerous studies have been conducted to enhance the efficacy of various organic sources, such as farm manure, vermicompost, and different biofertilizers for use in organic strawberry cultivation [6–9].

Consumers have shown increasing interest in the crop's nutritional quality [10–12]. According to Dasgan et al. [13–15], biofertilizers could improve crops' antioxidant, vitamin, and mineral contents.

In recent years, the use of vermicompost has increased in horticulture as a growing substrate component [16,17] or as a fertilizer with an increasing momentum [18–22]. It has been acknowledged for its elevated levels of NPK, humic contents, micronutrients, actinomycetes, and beneficial soil microbes, which enhance soil's physical, chemical, and biological characteristics. Earthworms facilitate the biodegradation of organic waste materials through enzymatic action and decomposer microorganisms. In addition to enhancing the composting process, earthworms disinfect and detoxify the product [18]. The resulting improvements in plant growth, development, and productivity lead to a rise in strawberry yield, alongside essential elements, vitamins, and enzymes attributable to vermicompost application [19–22].

In addition, biofertilizers, such as plant-growth-promoting rhizobacteria (PGPR) and arbuscular mycorrhizal fungi (AMF), have gained popularity in horticulture due to their positive impact on plant nutrition [13–15]. These biofertilizers promote plant growth through various mechanisms, including nitrogen fixation; phosphorus solubilization; the production of plant hormones, such as indole acetic acid and cytokinins; and the promotion of gibberellin synthesis within the plant [23,24]. They are also reported to provide organic acids and enzymes, such as 1-aminocyclopropane-1-carboxylate-deaminase, chitinase, and glucanase, to increase water and nutrient uptakes and enhance systemic resistance against diseases [25–28].

The novelty of this study lies in the utilization of "vermicompost-based fertilization and microbial complementary (top-dress) nutrition" as an organic approach to plant nutrition. In organic strawberry cultivation, using mineral chemical synthetic fertilizers is strictly prohibited. Consequently, alternative methods must be employed to enhance plant nutrition. To ensure accessibility for strawberry farmers, PGPR and AMF biofertilizers were chosen from commercially available preparations. While previous studies have explored the individual use of vermicompost, PGPR, and AMF, these investigations were not explicitly focused on organic farming. Our study, however, synergistically combines vermicompost and microbial fertilizers to determine the most suitable combination for organic strawberry nutrition.

This study was carried out to improve plant nutrition in organic strawberry production to increase yields and enrich product quality. We aimed to investigate combinations of different strains of bacteria and mycorrhizae. It was investigated whether vermicompost can promote organic strawberry production alone or in combination with PGPR and AMF. We investigated the effects of vermicompost as a basic fertilizer mixed into the soil in combination with top fertilization by PGPR and AMF treatments on the "Monterey" strawberry variety.

2. Material and Methods

2.1. Trial Area, Plant Material, and Treatments

This study was carried out in the experimental area of Osmaniye Korkut Ata University in 2019–2020. This region has a Mediterranean climate, and its height from the sea is 121 m.

A medium-neutral day variety, the "Monterey", was used as a strawberry plant material with outstanding flavour and a distinct sweet taste. The fruits are large, firm, slightly carved, and dark red. "Monterey" fruits are used for both fresh markets and processing. The variety is sensitive to mildew, and its earliness and plant structure are comparatively strong. The strawberry crop phenotypic characteristics can vary according to growing and environmental conditions [22].

The vermicompost used in the study, which was produced by a company called Ekosol, was of two types: The first is in a granular form with the name of Ekosolfarm®. It is 100% organic vermicompost obtained from the red California culture worms. It contains 35% organic matter, 1.2% total nitrogen (N), 1% organic nitrogen, and 20% total humic and fulvic acids. The second, a liquid vermicompost fertilizer, contains 7% organic matter, 1% total nitrogen, and 6% total humic fulvic acid (https://www.ekosol.net/urunler_menu, accessed on 5 January 2023). The granular vermicompost as a basic fertilizer was mixed into the soil before seedling transplanting. Liquid vermicompost with the same trade name, Ekosolfarm®, was also applied every week periodically with drip irrigation. Moreover, we examined several biofertilizers as top fertilizers containing PGPR and AMF microorganisms. The biofertilizer with the trade name RhizoFill® included three PGPR species: *Bacillus subtilis, Bacillus megaterium,* and *Pseudomonas fluorescens*, at a concentration of 1 × 109 cfu/mL. The biofertilizer Subtima® included only *Bacillus subtilis* at a concentration of 1 × 109 cfu/mL. The biofertilizer Fontera microzone® contained nine PGPR species at a concentration of 5 × 108 cfu/g, including *Azosprillium brasilense, Azotobacter vinelandii, Rhizobium trifollii, Pseudomonas fluorescens, Bacillus subtilis, Bacillus licheniformis, Azotobacter chroococcum, Bacillus amyloliquefaciens,* and *Bacillus mucilaginosus*. The biofertilizer Endo Roots Soluble (ERS)® contained nine AMF species at a concentration of 1 × 104 cfu/g, including *Glomus intraradices, Glomus aggregatum, Glomus mosseae, Glomus clarum, Glomus monosporum, Glomus deserticola, Glomus brasilianum, Glomus etunicatum,* and *Gigaspora margarita*. Finally, the biofertilizer Bontera® contained five PGPR and two AMF species at a concentration of 1.2 × 108 cfu/mL, including *Bacillus amyloliquefaciens, Bacillus pumilus, Bacillus subtilis, Bacillus licheniformis, Bacillus megaterium, Trichoderma harzianum,* and *Trichoderma konigii*. In this study, the treatments are abbreviated as follows:

T_1 (control with no fertilizer);
T_2 (Ekosolfarm);
T_3 (Ekosolfarm + RhizoFill);
T_4 (Ekosolfarm + Subtima);
T_5 (Ekosolfarm + Fontera microzone);
T_6 (Ekosolfarm + Endo Roots Soluble—ERS);
T_7 (Ekosolfarm + Bontera).

A schedule was established for watering the plants with liquid vermicompost and other microbial fertilizers via drip irrigation. Precisely, the plants were watered with the liquid vermicompost and the microbial fertilizers at a rate of 1 mL L^{-1} every 15 days via drip irrigation, beginning one week after planting [14].

2.2. Soil Analyses, Cultural Practices, and Measurements of Plant Parameters

Soil analysis was made prior to the trial. The application area had a loamy soil structure with a pH of 7.9 (Table 1).

The soil was covered with black polyethylene mulch. Except for the control (T_1) in the other six applications (T_2, T_3, T_4, T_5, T_6, T_7), 30 g of granular vermicompost Ekosalfarm was applied per plant by mixing it with the soil during the seedling planting. The AMF-containing T_6 treatment was used once at 0.2 g of ERS for each seedling by mixing it into the soil during the seedling planting. Fresh strawberry seedlings were planted in 30 × 30 cm using the triangle planting method in November 2019. The experiment was established in a randomized blocks experimental design with 4 replications and 20 plants in each replication (Figure 1). A low polyethene tunnel was established to ensure the seedlings' survival during winter. The tunnels protected the plants from harsh weather conditions until March. The gradual opening of the tunnels ensured that the plants were gradually exposed to the external environment, avoiding shock to the plants' systems. The experiment was concluded at the end of May.

Table 1. Soil properties of the trial area.

Soil Properties	Value
Texture	loamy
pH	7.9
Saltiness (%)	0.05
Lime (%)	37.13
Organic matter (%)	0.51
P_2O_5 (g m^{-2})	2.97
K_2O (g m^{-2})	37.77
Ca (%)	0.7033
Mg (%)	0.0358
Na (%)	0.0056
Fe (mg kg^{-1})	1.99
Cu (mg kg^{-1})	1.16
Mn (mg kg^{-1})	0.94
Zn (mg kg^{-1})	0.28

Figure 1. General images of seedling planting and low tunnels in organic strawberry cultivation experiment.

Plant vegetative growth parameters were measured at the end of the experiment using ten plants for each replication. The primary stem diameter, most extended root length and thickness, leaf area, and shoot and root dry matter per plant were determined. At the end of the experiment, the whole plant leaf area was measured by the Digimizer software version 5.3.5 in three randomly selected plants in each replication. In addition, the root length, root thickness, stem diameter, root, and stem dry matter content were determined in three randomly selected plants from each replication of the treatments. Their fresh weights were recorded initially to assess the root and stem dry matter. Then, they were dried in an oven at 65 °C until a constant weight was reached, and the dry matter ratios were calculated [14].

The strawberry harvest started in March and continued until the end of May when the weather warmed up. The fruit yield was recorded in the weekly strawberry harvests. In the middle of the harvest period in April, 20 fruits at the fully ripened stage were randomly taken from each replication for pomological analysis. The fruit weight, pH, total soluble solids (TSSs), and titrable acidity were measured according to Saygı [22]. The pH of the fruit was measured from 100 mL of strawberry juice using a digital pH meter (WTW pH 3110, Weilheim, Germany). The total soluble solids (TSSs) were measured with a temperature-compensated digital refractometer (Atago PR-101, Tokyo, Japan) at 25 °C and the results were expressed as % °Brix. The titratable acidity (TA) was measured via potentiometric titration against 0.1 N NaOH up to a pH of 8.1. All the determinations were made in triplicate and the results were expressed as g of citric acid equivalent per 100 g of fresh weight.

2.3. Determination of Leaf Minerals by Atomic Absorption Spectrophotometry

In April, leaf samples, 15 fully mature leaves of 10 plants per replicate, were collected for mineral nutrient analysis. The leaves were dried in a forced-air oven at 65 °C for 48 h and ground through a 40-mesh sieve for elemental analysis [14]. The samples were dry-ashed in a muffle furnace at 550 °C for six hours. After the samples became ash, 3.3% of HCl acid was added to the ashes, and the obtained solution was filtered using blue band filter papers. Potassium (K), calcium (Ca), magnesium (Mg), iron (Fe), manganese (Mn), zinc (Zn), and copper (Cu) concentrations were determined using an atomic absorption spectrophotometer [29].

2.4. Determination of Leaf Phosphorus

The dry-ashed, furnaced, and dissolved leaf samples (as mentioned above) were reacted with Barton's solution. The phosphorus (P) concentration was determined at a wavelength of 430 nm in the spectrophotometer [14,29].

2.5. Determination of Leaf Total Nitrogen

A 0.2 g ground sample was digested with 5 mL of concentrated H_2SO_4 at 380 °C using a selenium tablet in the combustion unit of a Kjeldahl apparatus for 1 h until the color turned pale. Then, distillation was performed with 28% NaOH according to a standard Kjeldahl protocol and then titration was performed with 0.01 N H_2SO_4. The total nitrogen (N) of the leaf was calculated with the amount of H_2SO_4 consumed in the titration [14,29].

2.6. Statistical Analyses

Data were subjected to the MSTAT-C software for the statistical analysis. The LSD (Least Significant Difference) test was used to determine the differences between the means at $p \leq 0.05$.

3. Results and Discussion

3.1. Plant Growth Parameters

The synergistic effect of vermicompost and microbial fertilizers on plant growth can be attributed to several key parameters. The application of vermicompost, PGPR, and AMF leads to elevated levels of essential plant growth regulators, such as cytokinins, auxins, and humic acids. These bioactive compounds play a significant role in enhancing the plant height and leaf area, as well as the dry weights of shoots and roots [20]. The increased availability of nutrients facilitated by these organic amendments stimulates the activity of enzymes involved in chlorophyll synthesis and photosynthesis. Consequently, this biochemical activation contributes to the overall improvement in plant growth [20].

The effects of the applications on the root length, root thickness, root and stem dry matter ratio, and stem diameter were statistically significant (Table 2). The longest root was in T_7 (27.49 cm), followed by the T_5 (26.28 cm) and T_3 (25.73 cm) treatments, respectively. The root lengths of all the biofertilizer treatments were higher than the control (T_1). Ciylez and Esitken [30] reported a root length between 17.50 and 27.16 cm, and the biofertilizers generally increased vegetative plant growth.

Table 2. The effects of treatments on root length, thickness, root and stem dry matter, and stem diameter.

Treatments	The Longest Root (cm)	The Thickest Root (mm)	Root Dry Matter (%)	Shoot Dry Matter (%)	Stem Diameter (mm)
T_1	22.06 c	0.89 e	33.29 c	20.97 d	11.71 c
T_2	24.84 b	0.94 d	40.56 b	27.75 c	13.04 b
T_3	25.73 ab	1.04 c	42.98 ab	31.54 bc	13.03 b
T_4	24.52 b	1.06 c	40.03 b	32.12 b	13.20 ab
T_5	26.28 ab	1.12 b	45.75 a	35.62 b	13.27 ab
T_6	25.08 b	1.11 b	43.30 ab	35.79 b	13.45 ab
T_7	27.49 a	1.22 a	45.81 a	44.35 a	14.06 a
$LSD_{0.05}$	1.88	0.03	3.98	4.36	0.94

The means, which differ significantly at the 5% level according to the LSD test, are shown with different letters. (T_1) no fertilizer (control); (T_2) Ekosolfarm; (T_3) Ekosolfarm + RhizoFill; (T_4) Ekosolfarm + Subtima; (T_5) Ekosolfarm + Fontera microzone; (T_6) Ekosolfarm + Endo Roots Soluble; (T_7) Ekosolfarm + Bontera.

When the root thickness data were examined, it was determined that the highest value, similar to the root length, was in the T_7 (1.22 mm) treatment, and the lowest was 0.89 mm in the control (T_1). The stem diameters of all the treatments were higher than the control. In addition, we observed that the plants with the thickest stems provided higher yields, as was the case in the T_7 application (14.06 mm). Similarly, Balci [31] reported a 13 mm stem diameter in an organically grown strawberry plant.

The highest amount of dry matter ratio in roots was revealed in the T_7 (45.81%) and T_5 (45.75%) treatments and the lowest in the control (T_1) (33.29%). In this study, the highest stem dry matter percentage was observed in T_7 (44.35%), followed by T_6 (35.79%) and T_5 (35.62%). The stem dry matter ratios of all treatments were higher than the control. Burgut et al. [32] stated that the highest dry matter ratios were 25% and 12.5% in roots and stems, respectively, in organically fertigated strawberry plants.

Shoot, root, and leaf growth characteristics were significantly affected by the applications and these effects are probably due to the combined effects of vermicompost and biofertilizers. Additionally, factors such as biofertilizer content, dosage, application time, and soil characteristics could be inter-related aspects.

The effect of the treatments on the leaf area was statistically significant. The maximum leaf area was measured in T_7 (667.03 cm^2 per $plant^{-1}$) followed by T_5 with 620.55 cm^2 per $plant^{-1}$ (Figure 2). Ipek et al. [33] reported that the highest leaf area value was 465.9 cm^2 using PGPR in conventional growing, while Khalil and Agah [34] stated that the maximum leaf area was 529.66 cm^2 by the application of 50% mineral fertilizer + *Trichoderma* and *Bacillus* strains. In our study, the increase in leaf area in the vermicompost with PGPR and AMF may have enabled the plant to efficiently uptake more water and nutrients from the soil. This, in turn, may have increased the number of fruits, weight, and yield due to better root, stem, and leaf development and higher photosynthesis. Seema et al. [35] noted that the increase in leaf area might be due to plant-growth-promoting rhizobacteria's ability to produce plant growth hormones, better root development, better displacement, and an efficient use of water and nutrients.

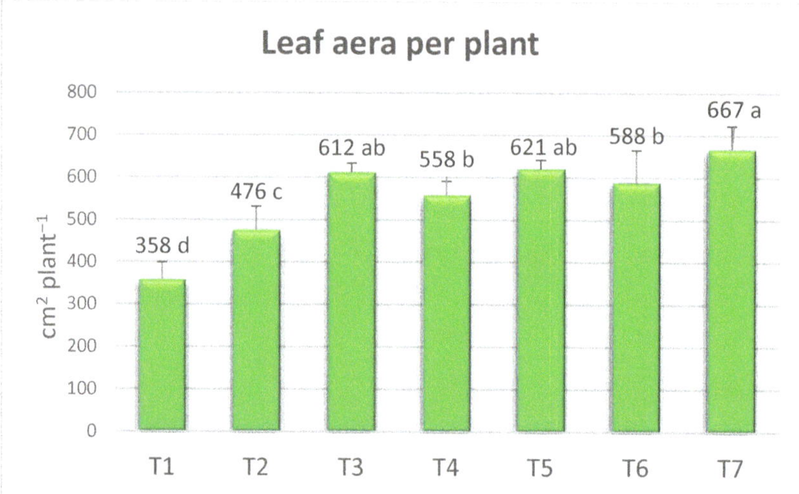

Figure 2. Effects of the treatment on leaf area of organically grown strawberry plants. The means, which differ significantly at the 5% level according to the LSD test, are shown with different letters. (T_1) no fertilizer (control); (T_2) Ekosolfarm; (T_3) Ekosolfarm + RhizoFill; (T_4) Ekosolfarm + Subtima; (T_5) Ekosolfarm + Fontera microzone; (T_6) Ekosolfarm + Endo Roots Soluble (Mycorrhiza); (T_7) Ekosolfarm + Bontera.

3.2. *Total Yield and Average Fruit Weight per Plant*

The differences between the treatments in terms of total yield per plant were statistically significant. As seen in Figure 3, the highest total yield was promoted in the T_7 treatment with 216.75 g per plant^{-1} and the lowest yield was obtained in the control (T_1) treatment with 108.36 g per plant^{-1}. The yield increase in T_7 (Ekosolfarm + Bontera) was 100.03% compared to no fertilizer (T_1) and 29.09% compared to vermicompost alone (T_2). Compared to vermicompost alone (T_2), the yield was higher in other treatments of the combined use of the vermicompost with PGPR and AMF (T_3, T_4, T_5, T_6, T_7). Therefore, the growth in yield may be due to the fact that vermicompost, PGPR, and AMF facilitate the plant's nutrient uptake. Imriz et al. [25] and Seema et al. [35] have reported that the application of biofertilizers containing auxins and cytokinins can potentially influence various physiological processes of plants, such as cell division, flower and root development, and leaf growth. Moreover, biofertilizers are believed to enhance the uptake and utilization of water and nutrients in plants, ultimately leading to increased fruit and flower production and higher yields. Similarly, previous studies have reported that the application of vermicompost and Azotobacter in the "Sweet Charlie" strawberry cultivar resulted in the highest yield of 124.46 g per plant^{-1} [6], while the combined application of vermicompost, poultry manure, and Azotobacter induced a yield of 144.77 g per plant^{-1} [8]. The "Chandler" strawberry cultivar produced the highest yield of 185.08 g per plant^{-1} in the treatment where organic manure and plant-growth-promoting rhizobacteria (PGPR), such as *Azotobacter chroococcum*, and *Pseudomonas fluorescens*, were applied in combination [9]. Moreover, we observed findings similar to those of the researchers mentioned above.

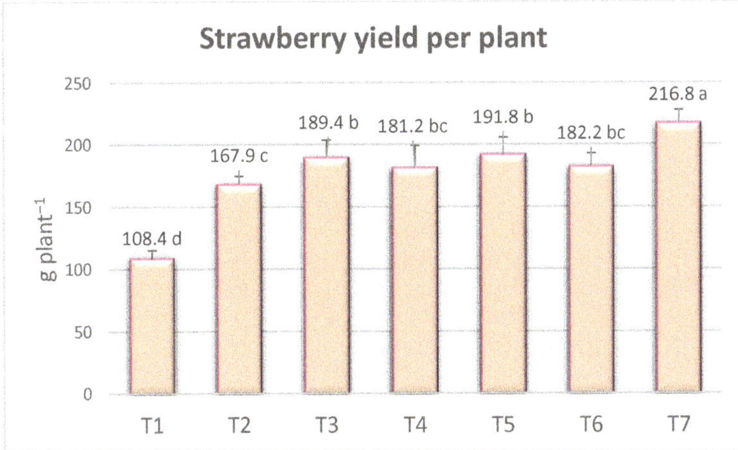

Figure 3. The effect of treatments on strawberry yield per plant. The means, which differ significantly at the 5% level according to the LSD test, are shown with different letters. (T_1) no fertilizer (control); (T_2) Ekosolfarm; (T_3) Ekosolfarm + RhizoFill; (T_4) Ekosolfarm + Subtima; (T_5) Ekosolfarm + Fontera microzone; (T_6) Ekosolfarm + Endo Roots Soluble (Mycorrhiza); (T_7) Ekosolfarm + Bontera.

The differences in the fruit weight ratios of the treatments were statistically significant (Figure 4). All treatments were more effective on fruit weights than the control (T_1). Enormous fruits were observed in the T_7 treatment (18.11 g) and minor fruits in the control (T_1) (15.51 g). In the T_7 treatment, the fruit weight increased by 16.76% compared to no fertilizer (T_1) and 7.41% compared to vermicompost alone (T_2). Seema et al. [35] reported that the maximum fruit weight in the PGPR biofertilizer treatments in the Chandler strawberry variety was 14.62 g. Negi et al. [9] reported that the highest fruit weight in organic manure with PGPR treatments in the same variety was 13.30 g. Although fruit size is a variety-specific feature, environmental conditions, applied fertilizers, and treatment methods can be effective.

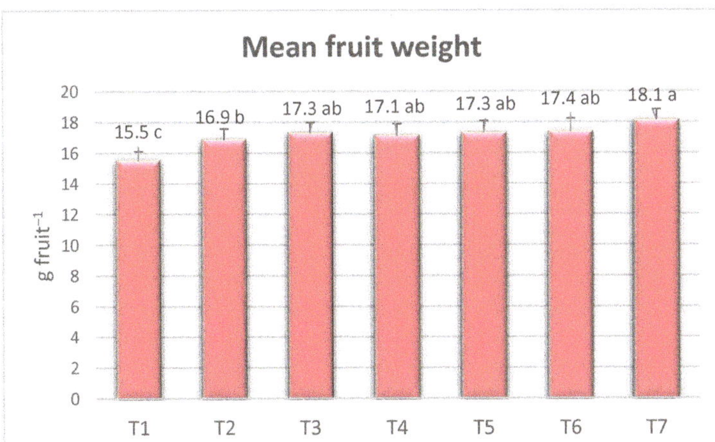

Figure 4. The effect of the treatments on fruit weight. The means, which differ significantly at the 5% level according to the LSD test, are shown with different letters. (T_1) no fertilizer (control); (T_2) Ekosolfarm; (T_3) Ekosolfarm + RhizoFill; (T_4) Ekosolfarm + Subtima; (T_5) Ekosolfarm + Fontera microzone; (T_6) Ekosolfarm + Endo Roots Soluble (Mycorrhiza); (T_7) Ekosolfarm + Bontera.

Additionally, researchers have suggested that the increase in fruit size may be attributed to the beneficial microorganisms present in vermicompost. The enhancement of the nutrient uptake may have promoted the production of phytohormones, particularly gibberellins, leading to increased nutrient and water absorption by the plants [8,13,15,36].

Fruit growth can be characterized by two distinct stages: cell division and cell enlargement. The initial phase involves cell division, followed by a subsequent process of cell volume increase. The influence of gibberellin on fruit enlargement primarily occurs during the rapid growth period in the first phase of fruit development. At this stage, the presence of gibberellin facilitates the transition from cell division to cell enlargement and also facilitates the mobilization, transportation, and accumulation of nutrients within the fruit [8,24]. Additionally, gibberellin induces an increase in auxin content, which enhances nutrient absorption by the fruit and promotes growth in overall fruit size [8,24]. Moreover, we observed findings similar to those of the researchers mentioned above. The extra beneficial microorganisms and organic matter in the vermicompost content increased the yields and fruit sizes of the strawberry plants.

The strawberry yield results obtained were dependent on the composition, application methods, and dosages of the vermicompost and biofertilizers used, as well as the soil and climatic characteristics of the experimental site.

3.3. Fruit pH, Total Soluble Solids (TSSs), Titrable Acid, and the Ratio of TSSs/Acid

Among all the treatments, T_7 provided the highest pH (3.95) in strawberry fruits. The lowest fruit juice pH was observed in the control treatment (T_1) at 3.83, followed by the T_2 treatment (3.87). The differences in pH results were statistically significant (Table 3). Jain et al. [37] stated that the pH of strawberries varied between 2.63 and 3.90, while Sener and Duran [38] reported that the pH values were between 3.59 and 3.81. Strawberry pH values may also vary according to the region's climate, the plant's genetic material, and fertilizer [22].

Table 3. The effect of treatment on pH, TSSs, acidity, and TSSs/acid ratio of the strawberry fruit.

Treatments	pH	TSSs (%)	TA (%)	TSSs/TA
T_1	3.83 e	9.48 e	0.67 a	14.27 e
T_2	3.87 d	9.82 d	0.61 b	16.17 d
T_3	3.92 b	10.62 a	0.62 b	17.24 bc
T_4	3.90 bc	10.07 cd	0.61 b	16.54 cd
T_5	3.91 b	10.23 bc	0.57 c	18.07 ab
T_6	3.89 cd	10.06 cd	0.60 b	16.86 cd
T_7	3.95 a	10.42 ab	0.56 c	18.74 a
$LSD_{0.05}$	0.02	0.29	0.03	0.91

The means, which differ significantly at the 5% level according to the LSD test, are shown with different letters. (T_1) no fertilizer (control); (T_2) Ekosolfarm; (T_3) Ekosolfarm + RhizoFill; (T_4) Ekosolfarm + Subtima; (T_5) Ekosolfarm + Fontera microzone; (T_6) Ekosolfarm + Endo Roots Soluble (Mycorrhiza); (T_7) Ekosolfarm + Bontera; TSSs: total soluble solids; TA: titratable acidity.

The effects of vermicompost alone or combined with the PGPR and AMF on TSSs were statistically significant (Table 3). T_3 showed the highest TSSs, 10.62%, followed by T_7 at 10.42%, while the lowest TSSs were measured in the control treatment (T_1), 9.48%. Similarly, Negi et al. [9] stated that the highest TSSs value in the "Chandler" strawberry cultivar was observed in organic manure with a PGPR treatment, 10.47%, and the lowest in the control treatment, 7.60%. The researchers additionally stated that the observed rise in TSSs levels could be attributed to the fast metabolic conversion of the manure, vermicompost, and biofertilizer combination into soluble starch and pectin compounds. Furthermore, they suggested that the transition of sugar from leaves to developing fruits might result from converting polysaccharides into simpler sugars. Table 3 shows that the treatments significantly affected the titrable acidity of the strawberries. For instance, the lowest acidity was obtained in T_7, 0.56%, and T_5, 0.57%, and the highest in the control (T_1), 0.67%. Singh et al. [39] observed that the acidity values were between 0.35 and 0.65%, while

Anuradha et al. [40] reported that the values varied between 0.73 and 0.84%. TSSs/acidity ratios determined strawberries' taste and were statistically significant in our study.

Strawberry consumers prefer fruits characterized by a high sugar content, specifically total soluble solids (TSSs), and low acidity levels. The augmentation of TSSs in strawberry fruits through the activity of plant-growth-promoting Rhizobacteria (PGPR) and arbuscular mycorrhizal fungi (AMF) during cultivation is likely attributed to the provision of nutrients by these microorganisms. This nutrient supply enhances the vigor of strawberry plants, leading to an increase in leaf area (Figure 2) and a subsequent boost in the synthesis of photo-assimilates, resulting in an elevated rate of photosynthesis [41]. In addition, the high mobility of photosynthetic products from leaves to developing fruits contributes to the rise in TSSs levels within the fruits [41]. However, the lower TSSs values observed in the control group (T_1) and the vermicompost alone treatment (T_2) can be attributed to the limited availability of metabolites primarily allocated for vegetative development, leaving only a tiny portion for fruit development [41]. This limitation in photosynthetic capacity and nutrient uptake may account for the reduced TSSs in these treatments.

The titratable acidity reached its maximum level in the control group, which received no fertilizer. The application of vermicompost alone or a combined inoculation of PGPR and AMF reduced titratable acidity. This decline in acidity can be attributed to the presence of nitrogen, potassium, calcium, magnesium, and micronutrients, as changes in their availability directly influence fruit quality, particularly titratable acidity, which tends to increase when the availability of these nutrients decreases (Tables 3 and 4) [41]. In our experiment, a significant increase in nutrient availability (Tables 4 and 5) was observed in plots treated with vermicompost alone, especially in those treated with combined PGPR and AMF. Fruit pH is inversely related to acidity; therefore, higher acidity corresponds to lower pH values.

Table 4. The effect of the treatments on the macroelement concentrations of strawberry leaves (%).

Treatments	N	P	K	Ca	Mg
T_1	2.63 e	0.14 d	0.91 c	1.61 c	0.19 c
T_2	3.00 d	0.33 c	1.33 b	1.98 b	0.25 b
T_3	3.24 cd	0.68 a	1.37 ab	2.07 b	0.25 b
T_4	3.19 cd	0.34 c	1.34 b	2.03 b	0.25 b
T_5	3.68 b	0.43 b	1.38 ab	2.37 a	0.26 b
T_6	3.30 c	0.48 b	1.41 ab	2.04 b	0.25 b
T_7	3.97 a	0.70 a	1.46 a	2.47 a	0.30 a
$LSD_{0.05}$	0.25	0.07	0.11	0.23	0.02

The means, which differ significantly at the 5% level according to the LSD test, are shown with different letters. (T_1) no fertilizer (control); (T_2) Ekosolfarm; (T_3) Ekosolfarm + RhizoFill; (T_4) Ekosolfarm + Subtima; (T_5) Ekosolfarm + Fontera microzone; (T_6) Ekosolfarm + Endo Roots Soluble (Mycorrhiza); (T_7) Ekosolfarm + Bontera.

Table 5. The effect of treatments on microelement concentrations of strawberry leaves (ppm).

Treatments	Fe	Zn	Mn	Cu
T_1	52.00 e	27.00 d	34.50 c	3.00 d
T_2	73.30 d	32.75 c	52.50 b	4.00 c
T_3	110.00 a	51.56 a	58.00 ab	5.67 a
T_4	85.00 bc	42.67 b	61.75 a	5.00 b
T_5	92.67 b	42.33 b	57.25 ab	5.00 b
T_6	84.00 c	40.56 b	58.25 ab	4.00 c
T_7	84.11 c	50.67 a	63.00 a	4.00 c
$LSD_{0.05}$	8.28	5.11	8.86	0.67

The means, which differ significantly at the 5% level according to the LSD test, are shown with different letters. (T_1) no fertilizer (control); (T_2) Ekosolfarm; (T_3) Ekosolfarm + RhizoFill; (T_4) Ekosolfarm + Subtima; (T_5) Ekosolfarm + Fontera microzone; (T_6) Ekosolfarm + Endo Roots Soluble (Mycorrhiza); (T_7) Ekosolfarm + Bontera.

The TSSs/acidity ranged between 14.27 and 18.74, with the highest TSSs/acidity ratio in the T_7 treatment (Ekosolfarm + Bontera), 18.74 (Figure 3). These confirm the values from the literature. Kumar et al. [6] reported that the TSSs/acidity ratio in the "Sweet Charlie" cultivar varied between 12.84 and 17.05 in applications of vermicompost and biofertilizer treatments. Similarly, Kumar et al. [36] found that the TSSs/acidity ratio in the "Chandler" cultivar ranged from 11.3 to 14.0.

3.4. Plant Nutrient Analyses in Leaves

One crucial finding of this study demonstrates that relying solely on vermicompost-based fertilization is inferior, and the use of complementary top dressing with biofertilization seems to be necessary for superior plant nutrition, especially in T_7 conditions. The concentrations of macro- and micronutrients in strawberry leaves could be influenced by the combination of vermicompost and the biofertilizers, their application ratios and timing, and the soil properties employed in the experiment. The effects of the vermicompost alone or in combination with PGPR and AMF on the leaves' total nitrogen, phosphorus, potassium, calcium, and magnesium concentrations were statistically significant (Table 4). The leaf nitrogen concentration varied between 2.63% and 3.97%, and the highest nitrogen among all the treatments was revealed in T_7 (3.97%). Nitrogen concentration increases in T_2–T_7 applications ranged from 14% to 51% compared to the control (T_1). Jones et al. [42] indicated that the nitrogen adequacy level in the strawberry leaf was between 2.5 and 4.00%. Therefore, the nitrogen concentration of all treatments was within the adequacy limits. The increase in nitrogen may be due to the fixation of air nitrogen by the PGPR, as well as their facilitation and uptake of the nutrient from the soil [25].

Meanwhile, Joshi et al. [19] observed the highest nitrogen content in the leaves in a vermicompost + *Azotobacter* treatment (3.31%). He reported that the reasons for the increase were biological nitrogen fixation, enzyme complex production, and the dissolved nutrient form, which was readily available to be used. Tomic et al. [43] reported that the leaf nitrogen varied between 1.94 and 2.03%, while Negi et al. [9] noted that the leaf nitrogen content of the Chandler strawberry cultivar was between 0.81 and 2.67% in a manure and biofertilizer treatment.

In this study, the leaf phosphorus concentration was between 0.14 and 0.70%, and the highest P was in T_7 (0.70%) and T_3 (0.68%) treatments, while the lowest P was in the control (T_1) with 0.14%. Phosphorus concentration increases in T_2–T_7 treatments ranged from 135% to 400% compared to the control (T_1). Jones et al. [42] stated that the phosphorus sufficiency level in leaves was between 0.25 and 1.00%. Although the phosphorus content in the leaves was within the appropriate limits in vermicompost alone or in combination with PGPR and AMF, it was insufficient in the control treatment (T_1). Accordingly, the increase in the phosphorus in the leaves may be due to the phosphorus solubility feature of the biofertilizers. Similarly, Imriz et al. [25] stated that organic acids are essential in the mineralization of organic phosphorus in the soil as it is equally critical to make it readily available for use by the plant. They also noted that *Bacillus subtilis, Bacillus polymyxa, B megatarium, Pseudomonas striata, P. rathonia,* and *Rhizobium leguminosarum* are among the bacteria that can dissolve phosphorus by producing organic acids. Ipek et al. [33] reported that leaf phosphorus values were between 0.27 and 0.41% with PGPR application. Basu et al. [27] indicated that an essential feature of biofertilizers is the solubilization and mineralization of phosphorus, which converts it into a usable form for plants.

We also observed that the leaf potassium was between 0.91–1.46% in our trial. The potassium levels of T_7 (1.46%) were higher than in other treatments. Potassium concentration increases in T_2–T_7 applications ranged from 46% to 60% compared to the control (T_1). Jones et al. [42] pointed out that the potassium sufficiency level in the leaves is between 1.30 and 3.00%. While the potassium was sufficient in vermicompost alone and combined with PGPR and AMF treatments, it was insufficient in the control (T_1).

The leaf calcium levels observed in the treatments were between 1.61 and 2.47%, while the highest calcium concentrations were in T_7 and T_5 (2.47% and 2.37%, respectively) and the lowest was 1.61% in the control (T_1). Calcium concentration increases in T_2–T_7 applications ranged from 23% to 53% compared to the control (T_1). Beer and Singh [44] reported the highest calcium in vermicompost with a phosphate solubilizing biofertilizer treatment (2.50%) and the lowest in a control treatment (1.95%). Jones et al. [42] that the calcium sufficiency level in leaves was between 1.00 and 2.50%. Likewise, the calcium contents of the treatments of the current study are within the sufficiency limits.

The total magnesium concentrations in the leaves ranged from 0.19 to 0.30%, and the highest magnesium was 0.30% in the T_7 treatment. Magnesium concentration increases in T_2–T_7 applications ranged from 32% to 58% compared to the control (T_1). Jones et al. [42] stated that the magnesium sufficiency level in leaves was between 0.25 and 1.00%. In this study, while the amount of magnesium was sufficient in vermicompost alone and combined with PGPR and AMF treatments, it was insufficient in the control (T_1).

Table 5 shows that the iron, zinc, manganese, and copper concentrations in leaves are statistically significant. The highest iron concentration was detected in the T_3 treatment, 110 ppm, and the lowest in the control (T_1) at 52 ppm. Iron concentration increases in T_2–T_7 treatments ranged from 41% to 62% compared to the control (T_1). Jones et al. [42] reported that leaf iron in strawberries was sufficient between 50 and 200 ppm, and Ipek et al. [33] stated that it was between 46.1 and 85.7 ppm in their study with PGPR.

The zinc concentration in the leaves was between 27 and 51.56 ppm. The highest zinc concentrations were observed in T_3 (51.56 ppm) and T_7 (50.67 ppm), while the lowest zinc was in the control treatment (T_1) (27 ppm). Increases in zinc concentration in T_2–T_7 applications ranged from 21% to 88% compared to the control (T_1). Jones et al. [42] pointed out that the zinc sufficiency level in leaves was between 20 ppm and 200 ppm.

The highest manganese levels were in the T_7 and T_4 treatments (63.00–61.75 ppm, respectively), and the lowest Mn was in the control (T_1), 34.50 ppm. Manganese concentration increases in T_2–T_7 treatments ranged from 34% to 83% compared to the control (T_1). Jones et al. [42] stated that the manganese sufficiency level in leaves was between 50 and 200 ppm. The manganese concentration was observed to be sufficient in vermicompost alone and combined with PGPR and AMF, yet it was insufficient in T_1 (control).

The copper concentration ranged between 3.00 and 5.67 ppm, with the highest value in T_3 (5.67 ppm). Copper concentration increases in T_2–T_7 treatments ranged from 33% to 89% compared to the control (T_1). Jones et al. [42] reported that the copper sufficiency level in the leaf is between 6 and 50 ppm. Accordingly, the concentrations of copper observed in the current study were insufficient and should be pointed out for further studies.

The contributions of PC1 and PC2 to variability were 85.3% and 8.3%, respectively (Figure 5). The principal component analysis (PCA) based on an analysis of twenty-one variables has scattered the T_1–T_7 treatments. The control (T_1) and the vermicompost alone (T_2) treatments disintegrated and dissociated from the complementary biofertilizer PGPR and AMF applications. The complementary biofertilizer PGPR and AMF applications have been grouped and affiliated in the scatterplot.

Heat maps provide a visual tool for comprehending numerical values by presenting them in a graphical format. This approach enables the simultaneous depiction of multiple data points and their inter-relationships within a single image, employing a color-coded system [15]. In Figure 6, the collective impact of vermicompost-based fertilization and microbial complementary nutrition on plant growth, yield, fruit quality, and plant nutrition parameters can be observed effectively through heat maps. Furthermore, heat maps can be utilized literally, such as illustrating "hot and cold" areas (e.g., from red to blue) on a map. Among the treatments examined, "T_7: Ekosolfarm + Bontera" exhibited the highest prevalence of warm red tones, signifying favorable outcomes. Conversely, applying "T_1: No-fertilizer control" resulted in predominantly cold, dark-blue hues, indicating fewer desirable results. The relationships between the remaining treatments (T_2, T_3, T_4, T_5, and T_6) and the examined parameters are demonstrated in Figure 6, based on the intensity of red and blue shades. Consequently, the effectiveness of vermicompost-based fertilization and microbial complementary nutrition is visually categorized and interpreted.

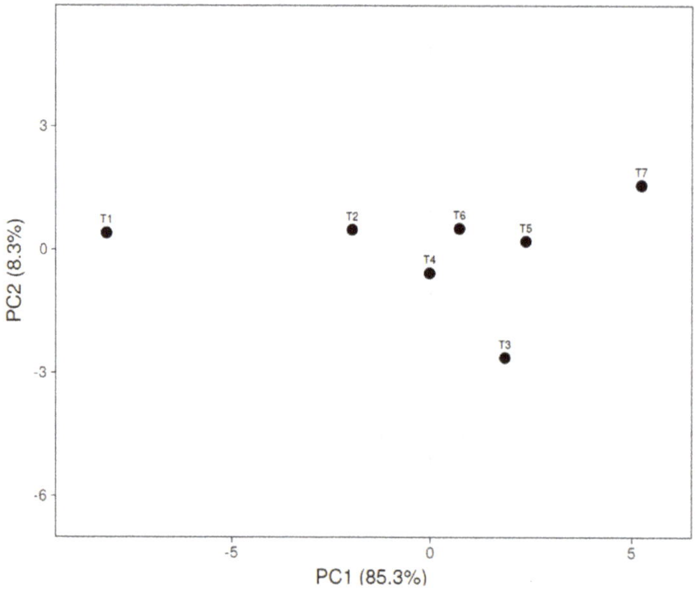

Figure 5. The principal component analysis of the vermicompost-based fertilization and microbial complementary nutrition. (T_1) no fertilizer (control); (T_2) Ekosolfarm; (T_3) Ekosolfarm + RhizoFill; (T_4) Ekosolfarm + Subtima; (T_5) Ekosolfarm + Fontera microzone; (T_6) Ekosolfarm + Endo Roots Soluble (Mycorrhiza); (T_7) Ekosolfarm + Bontera.

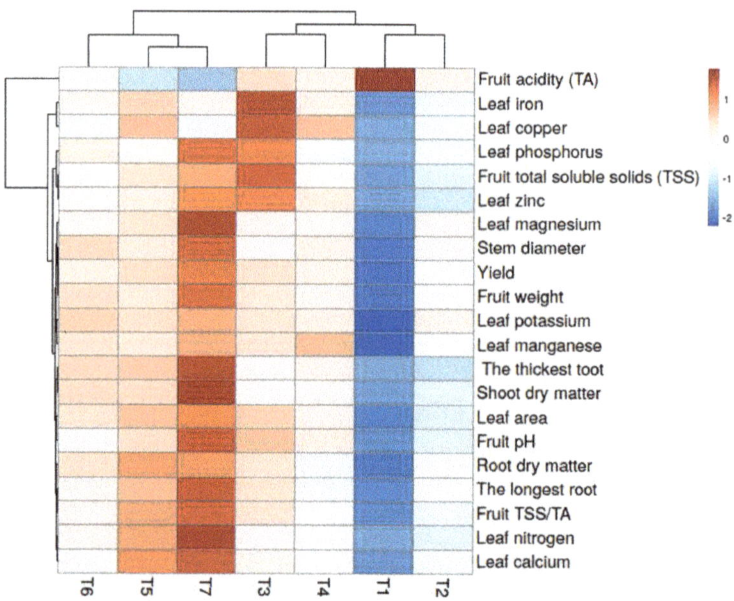

Figure 6. Heat map of the vermicompost-based fertilization and microbial complementary nutrition. (T_1) no fertilizer (control); (T_2) Ekosolfarm; (T_3) Ekosolfarm + RhizoFill; (T_4) Ekosolfarm + Subtima; (T_5) Ekosolfarm + Fontera microzone; (T_6) Ekosolfarm + Endo Roots Soluble (Mycorrhiza); (T_7) Ekosolfarm + Bontera.

4. Conclusions

In conclusion, this study demonstrates that combining vermicompost-based fertilization and complementary microbial nutrition is a practical approach to improving organic strawberries' growth, yield, plant nutrition, and fruit quality. Treatment T_7 (Ekosolfarm and Bontera) showed the most promising results among the biofertilizer combinations tested. These findings have significant implications for sustainable agriculture and organic farming. They offer a practical and sustainable solution to meet the high demand for organic strawberries while increasing producers' income. These results are promising for enhancing organic strawberry production. The outcomes of this study depend on several factors, including the species of PGPR and AMF present in the microbial fertilizers, the concentration of viable microorganisms, and the application rates and timing of the biofertilizers, as well as the physical and chemical properties of the experimental soil in which the strawberry plants were organically grown.

Author Contributions: All the authors contributed to this research. N.K. and H.Y.D. designed the experiment. Conceptualization; data curation; formal analysis; investigation; resources; funding acquisition, N.K. and H.Y.D. Supervision and writing—review and editing, H.Y.D., N.K. and N.S.G. All authors have read and agreed to the published version of the manuscript.

Funding: This research received no external funding.

Data Availability Statement: The data presented in this study are available in the article.

Conflicts of Interest: The authors declare no conflict of interest.

References

1. Kumra, R.; Saravanan, R.S.; Bakshi, P.; Kumar, A.; Singh, M.; Kumar, V. Influence of plant growth regulators on Strawberry: A review. *Int. J. Chem. Stud.* **2018**, *6*, 1236–1239.
2. Morais, M.C.; Mucha, Â.; Ferreira, H.; Gonçalves, B.; Bacelar, E.; Marques, G. Comparative study of plant growth-promoting bacteria on the physiology, growth and fruit quality of strawberry. *J. Sci. Food Agric.* **2019**, *99*, 5341–5349. [CrossRef]
3. Almli, L.V.; Asioli, D.; Rocha, C. Organic consumer ghoices for nutrient labels on dried strawberries among different health attitude segments in Norway, Romania and Turkey. *Nutrients* **2019**, *11*, 2951. [CrossRef]
4. Turemis, N.; Ağaoğlu, Y.S. Berry Fruits, Chapter II (Strawberry). In *Education Publications 1*; Ağaoğlu, S., Gerçekcioglu, R., Eds.; Tomurcuk Bag Ltd. Sti.: Ankara, Turkey, 2013; pp. 55–100.
5. Food and Agriculture Organization (FAO). FAOSTAT (Statistics). 2023. Available online: http://www.fao.org/faostat/en/#data/QC (accessed on 3 January 2023).
6. Kumar, N.; Ram, R.B.; Mishra, P.K. Response of vermicompost and Azotobacter on growth and yield of 'Sweet Charlie' strawberry. *Int. J. Agri. Sci. Res.* **2015**, *5*, 13–20.
7. Srivastav, A.; Singh, B.K.; Pandey, R.; Singh, K.; Singh, V. Effect of organic manures and bio-fertilizers on vegetative growth and yield of strawberry cv. chandler. *J. Pharmacogn. Phytochem.* **2018**, *7*, 2841–2844.
8. Soni, S.; Kanawjia, A.; Chaurasiya, R.; Chauhan, P.S.; Kumar, R.; Dubey, S. Effect of organic manure and biofertilizers on growth, yield and quality of strawberry (*Fragaria × ananassa* Duch) cv. Sweet Charlie. *J. Pharmacogn. Phytochem.* **2018**, *2*, 128–132.
9. Negi, Y.K.; Sajwan, P.; Uniyal, S.; Mishra, A.C. Enhancement in yield and nutritive qualities of strawberry fruits by the application of organic manures and biofertilizers. *Sci. Hortic.* **2021**, *283*, 110038. [CrossRef]
10. Gruda, N. Impact of environmental factors on product quality of greenhouse vegetables for fresh consumption. *Crit. Rev. Plant Sci.* **2005**, *24*, 227–247. [CrossRef]
11. Gruda, N. Do soilless culture systems have an influence on product quality of vegetables? *J. Appl. Bot. Food Qual.* **2009**, *82*, 141–147. Available online: https://ojs.openagrar.de/index.php/JABFQ/article/view/2093 (accessed on 6 May 2009).
12. Gruda, N.; Savvas, D.; Youssuf, R.; Colla, G. Impacts of genetic material and current technologies on product quality of selected greenhouse vegetables–A review. *Eur. J. Hortic. Sci.* **2018**, *83*, 319–328. [CrossRef]
13. Dasgan, H.Y.; Aldiyab, A.; Elgudayem, F.; İkiz, B.; Gruda, N.S. Effect of biofertilizers on leaf yield, nitrate amount, mineral content and antioxidants of basil (*Ocimum basilicum* L.) in a floating culture. *Sci. Rep.* **2022**, *12*, 20917. [CrossRef] [PubMed]
14. Dasgan, H.Y.; Yilmaz, M.; Dere, S.; Ikiz, B.; Gruda, N.S. Bio-Fertilizers Reduced the Need for Mineral Fertilizers in Soilless-Grown Capia Pepper. *Horticulturae* **2023**, *9*, 188. [CrossRef]
15. Dasgan, H.Y.; Kacmaz, S.; Arpaci, B.B.; İkiz, B.; Gruda, N.S. Biofertilizers Improve the Leaf Quality of Hydroponically Grown Baby Spinach (*Spinacia oleracea* L.). *Agronomy* **2023**, *13*, 575. [CrossRef]
16. Gruda, N.S. Increasing Sustainability of Growing Media Constituents and Stand-Alone Substrates in Soilless Culture Systems. *Agronomy* **2019**, *9*, 298. [CrossRef]

17. Abul-Soud, M.; Emam, M.; El-Rahman, N. The Potential Use of Vermicompost in Soilless Culture for Producing Strawberry. *Int. J. Plant Soil Sci.* **2015**, *8*, 1–15. [CrossRef]
18. Singh, S.; Sinha, R.K. Vermicomposting of organic wastes by earthworms: Making wealth from waste by converting 'garbage into gold' for farmers (Chapter 7). In *Advanced Organic Waste Management*; Hussain, C., Hait, S., Eds.; Elsevier: Amsterdam, The Netherlands, 2022; pp. 93–120. ISBN 9780323857925. [CrossRef]
19. Joshi, R.; Singh, J.; Vig, A.P. Vermicompost as an effective organic fertilizer and biocontrol agent: Effect on growth, yield and quality of plants. *Rev. Environ. Sci. Biotechnol.* **2015**, *14*, 137–159. [CrossRef]
20. Zuo, Y.; Zhang, J.; Zhao, R.; Dai, H.; Zhang, Z. Application of vermicompost improves strawberry growth and quality through increased photosynthesis rate, free radical scavenging and soil enzymatic activity. *Sci. Hortic.* **2018**, *233*, 132–140. [CrossRef]
21. Čabilovski, R.; Manojlović, M.S.; Popović, B.M.; Radojčin, M.T.; Magazin, N.; Petković, K.; Kovačević, D.; Lakićević, M.D. Vermicompost and Vermicompost Leachate Application in Strawberry Production: Impact on Yield and Fruit Quality. *Horticulturae* **2023**, *9*, 337. [CrossRef]
22. Sayğı, H. Effects of Organic Fertilizer Application on Strawberry (*Fragaria vesca* L.) Cultivation. *Agronomy* **2022**, *12*, 1233. [CrossRef]
23. Baum, C.; El-Tohamy, W.A.; Gruda, N. Increasing the productivity and product quality of vegetable crops using arbuscular mycorrhizal fungi: A review. *Sci. Hortic.* **2015**, *187*, 131–141. [CrossRef]
24. El-Tohamy, W.A.; El-Abagy, H.M.; El-Greadly, N.H.M.; Gruda, N. Hormonal changes, growth and yield of tomato plants in response to chemical and bio-fertilization application in sandy soils. *J. Appl. Bot. Food Qual.* **2009**, *82*, 179–182.
25. Imriz, G.; Ozdemir, F.; Topal, I.; Ercan, B.; Tas, M.N.; Yakısır, E.; Okur, O. Rhizobacteria (PGPRs) promoting plant growth in plant production and their mechanisms of action. *Electron. J. Microbiol.* **2014**, *12*, 1–19.
26. Nadeem, S.M.; Ahmad, M.; Zahir, Z.A.; Javaid, A.; Ashraf, M. The role of mycorrhizae and plant growth promoting rhizobacteria (PGPR) in improving crop productivity under stressful environments. *Biotechnol. Adv.* **2014**, *32*, 429–448. [CrossRef] [PubMed]
27. Basu, S.; Rabar, R.; Negi, S. Towards a better greener future—An alternative strategy using biofertilizers. I: Plant growth promoting bacteria. *Plant Gene* **2017**, *12*, 43–49. [CrossRef]
28. Emmanuel, C.O.; Babalola, O.O. Productivitiy and quality of horticultural crops through co-inoculation of arbuscular mycorrhizal fungi and plant promoting bacteria. *Microbiol. Res.* **2020**, *239*, 126569. [CrossRef]
29. Jones, J.B. *Laboratory Guide for Conducting Soil Tests and Plant Analysis*; CRC Press: Boca Raton, FL, USA, 2001; 384p.
30. Ciylez, S.; Esitken, A. The Effect of Mycorrhiza and PGPR application on strawberry of growth. *Selcuk. J. Agric. Food Sci.* **2018**, *32*, 361–365. [CrossRef]
31. Balci, G. The Effects of Different Organic Waste on Yield and Quality of Organically Grown Strawberry. Ph.D. Thesis, Ondokuz Mayıs University, Samsun, Turkey, 2012; 132p.
32. Burgut, A.; Saygı, H.; Turemis, N.F.; Buyukyel, S.; Aktas, B.; Ekinci, K. The effects of different compost applications on organic strawberry seedling production. *Acta Hortic.* **2020**, *1286*, 33–38. [CrossRef]
33. Ipek, M.; Pirlak, L.; Esitken, A.; Figen Dönmez, M.; Turan, M.; Sahin, F. Plant growth-promoting rhizobacteria (PGPR) increase yield, growth and nutrition of strawberry under high-calcareous soil conditions. *J. Plant Nutr.* **2014**, *37*, 990–1001. [CrossRef]
34. Khalil, N.H.; Agah, R.J. Effect of chemical, organic and bio fertilization on growth and yield of strawberry plant. *Int. J. Adv. Chem. Eng. Biol. Sci.* **2017**, *4*, 5.
35. Seema, K.; Mehta, K.; Singh, N. Studies on the effect of plant growth promoting rhizobacteria (PGPR) on growth, physiological parameters, yield and fruit quality of strawberry cv. Chandler. *J. Pharmacogn. Phytochem.* **2018**, *7*, 383–387.
36. Kumar, P.; Sharma, N.; Sharma, S.; Gupta, R. Rizosphere stochiometry, fruit yield, quality attibutes and growth response to PGPR transplant amendments in strawberry (*Fragaria* × *ananassa* Duch.) growing on solarized soils. *Sci. Hortic.* **2020**, *265*, 109215. [CrossRef]
37. Jain, N.; Mani, A.; Kumari, S.; Kasera, S.; Bahadur, V. Influence of INM on yield, quality, shelf life and economics of cultivation of strawberry (*Fragaria* × *ananassa* Duch.) cv. Sweet Charlie. *J. Pharm. Phytochem.* **2017**, *6*, 1178–1181.
38. Sener, S.; Duran, C.N. Effects of some microbial plant growth and development regulators on vegetatıve development and fruit quality of strawberry under greenhouse conditions. *Euroasia J. Math. Eng. Nat. Med. Sci.* **2020**, *8*, 31–38.
39. Singh, D.; Kumar, S.; Verma, R.; Maurya, R.; Shukla, A. Effect of organic manure and bio-fertilizers on quality parameters of strawberry (*Fragaria* × *ananassa* Duch.) cv. Chandler. *J. Pharma. Phyto.* **2018**, *7*, 1227–1228.
40. Anuradha; Goyal, R.K.; Sindhu, S.; Godara, A.K. Effect of PGPR on strawberry cultivation under greenhouse conditions. *Indian J. Hort.* **2019**, *76*, 400–404. [CrossRef]
41. Kumar, S.; Kundu, M.; Das, A.; Rakshit, R.; Siddiqui, M.W.; Rani, R. Substitution of mineral fertilizers with biofertilizer: An alternate to improve the growth, yield and functional biochemical properties of strawberry (*Fragaria* × *ananassa* Duch.) cv. Camarosa. *J. Plant Nutr.* **2019**, *42*, 1818–1837. [CrossRef]
42. Jones, J.R.; Wolf, B.; Mills, H.A. *Plant Analysis Handbook*; Micro Macro Publishing, Inc.: Athens, GA, USA, 1991; p. 213. ISBN 13: 9781878148001.

43. Tomic, J.M.; Milivojevic, J.M.; Pesakovic, M.I. The response to bacterial inoculation is cultivar-related in strawberries. *Turk. J. Agric. For.* **2015**, *39*, 332–341. [CrossRef]
44. Beer, K.; Singh, A.K. Effect of vermicompost and biofertilizers on strawberry: Chlorophyll and nutrients concentration in leaves. *Ann. Plant Soil Res.* **2015**, *17*, 211–214.

horticulturae

MDPI

Article

Mealworm Larvae Frass Exhibits a Plant Biostimulant Effect on Lettuce, Boosting Productivity beyond Just Nutrient Release or Improved Soil Properties

Meroua Foughar, Margarida Arrobas and Manuel Ângelo Rodrigues *

Centro de Investigação de Montanha (CIMO), Laboratório para a Sustentabilidade e Tecnologia em Regiões de Montanha (SusTEC), Instituto Politécnico de Bragança, Campus de Santa Apolónia, 5300-253 Bragança, Portugal; foughar.maroua@gmail.com (M.F.); marrobas@ipb.pt (M.A.)

* Correspondence: angelor@ipb.pt

Abstract: There is a need for alternatives or complements to synthetic fertilizers to enhance agricultural sustainability. Applying organic amendments can play a significant role in this. Insect droppings show high potential, though studies evaluating their agronomic value have only recently begun to emerge. This study compared black soldier fly (*Hermetia illucens* L.) and mealworm (*Tenebrio molitor* L.) larvae frass with another organic amendment (Nutrimais) derived from composting forestry, agro-industrial, and domestic waste. The experiment also included ammonium nitrate at two rates [the same as the organic amendments, 50 kg ha^{-1} nitrogen (N) (FullR), and half that rate (HalfR)] and an unfertilized control. The study spanned two growth cycles of lettuce (*Lactuca sativa* L.) grown in pots, followed by unfertilized oats (*Avena sativa* L.) to assess the residual effects of the fertilizing treatments. Mealworm larvae frass mineralized rapidly, with an apparent N recovery of 37.4% over the two lettuce growth cycles, indicating its high availability to soil heterotrophic microorganisms. The average dry matter yield (DMY) of lettuce was the highest among all treatments (12.8 and 9.8 g plant^{-1} in the first and second lettuce cycles), even compared to the FullR treatment (12.2 and 7.8 g plant^{-1}), though without significant differences. Although mealworm larvae frass exhibited a high mineralization rate, the DMY cannot be attributed solely to N supply, as plants in the FullR treatment showed better N nutritional status. Mealworm larvae frass provided strong evidence of a plant biostimulant effect, not explained by the variables measured in this study. Black soldier fly larvae frass exhibited typical behavior of a moderately reactive organic amendment, while Nutrimais showed low reactivity, with a near-neutral mineralization/immobilization balance. The results suggest mealworm larvae frass is recommended for early maturing vegetable crops, whereas Nutrimais appears more suitable for perennial crops with low short-term nutrient requirements.

Keywords: black soldier fly frass; *Hermetia illucens*; *Tenebrio molitor*; insect droppings; manuring effect

1. Introduction

The global population is projected to continue increasing in the coming decades. According to the latest United Nations report [1], estimates suggest there will be 8.5 billion people by 2030 and 9.8 billion by 2050. This upward trend in world population presents significant challenges in terms of meeting food demand and ensuring sustainable agriculture practices. Achieving a balance between environmentally friendly farming methods and high-quality yields is paramount in the 21st century [2]. While intensive farming may offer improved income, it tends to be unsustainable and detrimental to the environment [3]. Reliance on mineral fertilizers, pesticides, and intensive tillage practices may have adverse effects on soil quality [4,5], exacerbating climate change and rendering traditional agricultural systems susceptible to extreme weather events [6].

Organic amendments have historically supplied essential nutrients to crops, but their usage declined with the mechanization of agricultural practices and the availability of affordable chemical fertilizers [7]. Chemical fertilizers, containing vital NPK nutrients, rapidly enhance soil fertility and support plant growth [8]. However, their long-term impacts can lead to soil pollution, reduced fertility, and imbalances in soil nutrients [9]. Organic matter plays a vital role in soil fertility and overall soil functioning, influencing its physical, chemical, and biological properties [10,11]. Excessive application of chemical fertilizers, in turn, may lead to a reduction in soil organic matter, thereby negatively impacting general soil properties [12,13].

Intensive agricultural practices without organic amendments result in reduced soil carbon (C) content, microbial biomass, enzymatic activities, species diversity, and increased soil salinity [14]. Establishing a soil conservation baseline entails implementing conservation agriculture techniques and enriching soils with organic matter. Common sources of organic matter for soil enrichment may include farmyard manures [13,15,16], municipal solid waste [17,18] and sewage sludge [19,20]. The use of high-quality compost is crucial for its safe and effective utilization in soil [20,21].

Scholars worldwide are currently researching sustainable waste sources and decomposition methods [22]. Insects and their excreta have shown promising potential for use in agriculture, particularly as soil amendments [23]. The byproduct of insect bioconversion, known as "frass", contains significant amounts of nutrients and can be marketed as fertilizing materials, offering additional income for breeding operations [24]. With the expanding insect industry globally, frass production is expected to increase, driven by waste recycling and circular economy policies [25,26]. Furthermore, there is an ongoing debate regarding the effectiveness of organic agriculture in ensuring food security [27]. Farmers have the option to adopt conservation practices, and their decisions are influenced by factors such as community pressure, financial incentives, and information availability [28].

Therefore, additional research is necessary to establish the best conditions for utilizing insect frass and to scale up this industry significantly [29]. This includes studying the effects of insect byproducts on soil properties and plant growth. To address current knowledge gaps, our study began examining the effect of two commercial products resulting from black soldier fly (*Hermetia illucens* L.) larvae frass and mealworm (*Tenebrio molitor* L.) larvae frass on lettuce (*Lactuca sativa* L.) growth, plant nutritional status, and soil properties. For comparison, the study included a commercially available organic compost that is widely used by farmers. This compost is produced from forestry, agro-industrial, and domestic waste through composting. It is sold as a pelletized and dehydrated material and is permitted in organic farming in Portugal. Additionally, the study included two treatments with mineral N fertilization at equivalent and half rates of the N provided by organic amendments, as well as a control treatment without fertilization. In this way, through multiple comparisons, we aim to understand how to position these new organic amendments in crop recommendation systems.

2. Materials and Methods

2.1. Experimental Conditions

A pot experiment was conducted at the experimental farm of Instituto Politécnico de Bragança located in Northeastern Portugal (41.797254° N–6.762279° W). The experiment involved two growing cycles of lettuce (cv. Summer Wonder) and a single growing cycle of oats (*Avena sativa* L., cv. Boa Fé). Lettuce was cultivated during the first cycle, spanning from May to June, and again in the second cycle, which took place from August to October 2023. Oats were grown from October 2023 to January 2024.

The region's climate falls under the Mediterranean type, classified as Csb according to the Köppen classification, characterized by mild, dry summers [30]. Weather data collected during the experimental period are presented in Figure 1.

The soil used in this study was collected from the surface layer (0–0.20 m) of a fallow plot. It is classified as a Regosol of coluvial origin, presenting a sandy-clay-loam texture (soil separates of 228.7, 241.0, and 530.3 g kg^{-1} of clay, silt, and sand, respectively). Soil organic C content was 11.0 g kg^{-1} and the pH (H_2O) was 6.5. Extractable phosphorus (P) and potassium (K) levels were 133.5 and 191.7 mg kg^{-1}, P_2O_5 and K_2O, respectively.

Exchangeable calcium (Ca), magnesium (Mg), and cation exchange capacity (CEC) were 9.6, 4.7, and 15.3 $cmol_c\ kg^{-1}$, respectively.

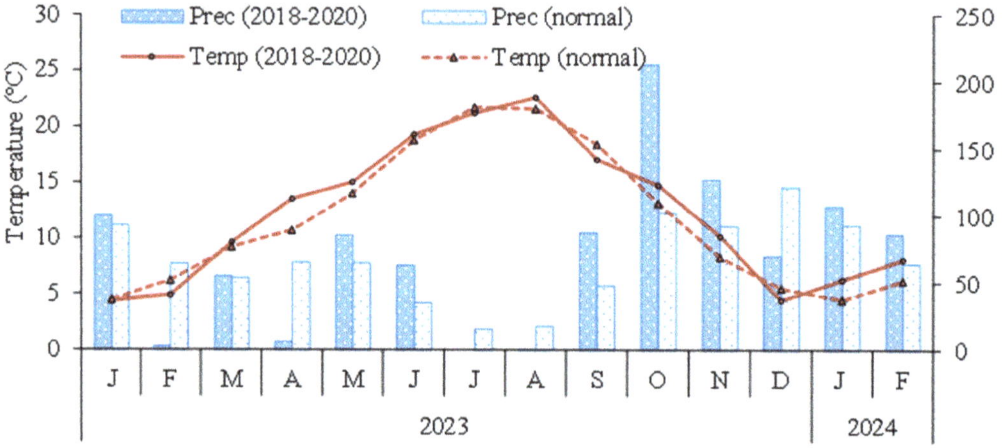

Figure 1. Climatological normal data for the region (1981–2010) and monthly average air temperature and precipitation during the experimental period.

2.2. Experimental Design and Fertilizing Materials

The first growing cycle of the lettuce experiment followed a completely randomized design, comprising six treatments and six replicates (six pots, one lettuce per pot). The treatments consisted of three commercial organic amendments: (1) Nutrimais® (Gondomar, Portugal) (Nutri), an organic amendment derived from composting forestry, agro-industrial, and domestic waste, available in pelletized and dehydrated form, permitted for organic farming in Portugal; (2) black soldier fly larvae frass (BSFly), a pelletized organic amendment produced from insect droppings by Entogreen® (Santarém, Portugal); (3) mealworm larvae frass (Mealw), also a natural pelletized organic amendment from insect droppings and plant debris, manufactured by ThunderFoods® (Santarém, Portugal); (4) ammonium nitrate (27% N, 50% NH_4^+, 50% NO_3^-), applied at the same N rate (FullR) as the three organic amendments; (5) ammonium nitrate, applied at half the N rate (HalfR) of the organic amendments; and (6) a non-fertilized control (Cont). Table 1 presents the main properties of the three organic amendments used in this study.

The treatments Nutri, BSFly, Mealw, and FullR were applied at a rate of 50 kg N ha^{-1}. To calculate the quantities of amendments and fertilizer for each plot, a standard planting density of 140,000 plants ha^{-1} was used. Thus, each pot received 1/140,000 of the amendments corresponding to one hectare. Considering the moisture content and N concentration of the organic amendments, they were applied at rates of 19.93, 15.33, and 22.72 g pot^{-1} for BSFly, Mealw, and Nutri. Ammonium nitrate was applied at rates of 1.59 g pot^{-1} for FullR and 0.79 g pot^{-1} for HalfR treatments. The application of ammonium nitrate at full and half rates was split into two applications: at transplanting and in the middle of the growing season. Each pot received 3 kg of dried (40 °C) and sieved (2 mm) soil, 50 g of perlite to ensure aeration and the corresponding quantity of amendments and fertilizer. The first growing cycle ran from 3 May to 15 June.

After completing the first lettuce cycle, the experiment was divided into two, forming two groups of pots with three replicates each (three pots, one lettuce per pot). In one group, the fertilization from the first cycle was repeated to assess the continued effect of the fertilizer treatments on lettuce growth and nutrition. In the other group, no fertilizers were applied. Since organic amendments release nutrients gradually through the mineralization process, the aim of this second group of pots was to evaluate the residual effect of fertilization from the first cycle. This second lettuce growth cycle ran from 25 August to 2 October 2023.

Table 1. Selected properties (mean ± standard deviation, n = 3) of the organic amendments used in this study: black soldier fly larvae frass (BSFly), mealworm larvae frass (Mealw) and Nutrimais (Nutri).

	BSFly	Mealw	Nutri
Moisture content (%)	25.3 ± 0.72	7.4 ± 0.05	8.0 ± 0.19
pH (H_2O)	7.6 ± 0.21	6.5 ± 0.31	8.1 ± 0.15
El. conduct. (mS cm^{-1})	9.9 ± 0.46	7.3 ± 0.56	9.7 ± 0.45
Carbon (g kg^{-1})	499.5 ± 2.30	517.7 ± 0.64	419.0 ± 3.90
Nitrogen (g kg^{-1})	26.8 ± 0.06	31.3 ± 0.51	25.3 ± 1.00
Carbon/nitrogen ratio	18.6 ± 0.11	16.5 ± 0.26	16.5 ± 1.14
Phosphorus (g kg^{-1})	11.7 ± 0.66	10.7 ± 0.18	4.8 ± 0.13
Potassium (g kg^{-1})	31.2 ± 0.89	18.4 ± 2.11	17.3 ± 0.16
Calcium (g kg^{-1})	3.1 ± 0.02	3.8 ± 0.07	75.9 ± 3.12
Magnesium (g kg^{-1})	4.9 ± 0.06	6.5 ± 0.28	3.2 ± 0.07
Boron (mg kg^{-1})	32.3 ± 0.35	26.0 ± 0.61	8.3 ± 0.65
Iron (mg kg^{-1})	761.9 ±31.11	422.9 ± 12.18	5230.3 ± 212.42
Manganese (mg kg^{-1})	87.4 ± 1.50	155.7 ± 5.67	133.7 ± 2.96
Zinc (mg kg^{-1})	108.8 ± 0.94	107.5 ± 1.82	103.7 ± 2.60
Copper (mg kg^{-1})	19.6 ± 0.36	14.0 ± 0.21	36.5 ± 0.10

After completing the second lettuce growth cycle, oats, a species with high N requirements capable of thriving during the autumn–winter period, were sown. Oats were planted in both groups of pots with 20 seeds per pot, later adjusted to 15 plants per pot after germination. The oats were not fertilized and were intended to evaluate the residual effect of the fertilization applied to the lettuce. The oat cropping cycle began on 13 October 2023 and concluded on 18 January 2024, when all plants exhibited chlorosis and stunted growth, at the stem elongation phase (with node 2 at least 2 cm above node 1) [31]. This approach aimed to gather detailed information on the short and long-term effects of organic fertilization. Mineral fertilization was included in the experiment as a positive control to assess the nutrient use efficiency of organic amendments compared to the more standardized use of mineral fertilizers.

2.3. Conducting the Experiment

Seedlings at the phenological growth stage 13, 3rd true leaf unfolded [31] were used, having been previously grown in commercial organic substrate. The pots were placed outdoors on a wooden platform with lateral protection to prevent direct sunlight and minimize pot heating. Regularly, the pots were rotated to ensure equivalent radiation exposure conditions. Weeds were promptly removed after germination.

The pots were watered as needed, ensuring that the lettuce plants remained well hydrated. In this type of experiment, the same amount of water is not applied to all pots because different fertilizations induce varying growth and transpiration rates, necessitating constant adjustment of water quantities per pot. Plates were placed under the pots to prevent nutrient leaching.

2.4. Sampling and Laboratory Analyses

The lettuces and oats were harvested by cutting them at ground level and then divided into several segments to facilitate the drying process. Once cut, the plants were oven-dried at 70 °C until reaching constant weight. Thereafter, the entire plant was ground through a 1 mm mesh. Subsequently, tissue samples were digested using nitric acid. Elemental tissue analysis was conducted using the Kjeldahl method for N, colorimetry for boron (B) and P, flame emission spectrometry for K, and atomic absorption spectrophotometry for Ca, Mg, copper (Cu), iron (Fe), zinc (Zn), and manganese (Mn). For a detailed description of these methodologies, the reader is referred to Temminghoff and Houba [32]. Nitrate concentration in lettuce tissues was determined by UV-vis spectrophotometry in a water extract (dry matter/water, 1:50 m/v) [33].

After the conclusion of the second lettuce growing cycle, soil samples were collected from the pots immediately before they were used for the oat experiment. The soil samples were oven-dried at 40 °C, sieved again through a 2 mm mesh, and analyzed for various parameters including organic C (determined using the wet digestion Walkley-Black method), soil pH (measured in water and KCl solutions at a soil-to-solution ratio of 1:2.5 using potentiometry), exchangeable bases (extracted with ammonium acetate at pH 7.0 and determined by atomic absorption spectrometry), exchangeable acidity (extracted with KCl and quantified through titration with sodium hydroxide), P and K (analyzed using the Egnér–Riehm method), and B (extracted with hot water and determined using the azomethine-H method). For detailed information on these analytical procedures, please refer to Van Reeuwijk [34]. The other micronutrients (Cu, Fe, Zn, and Mn) were extracted using diethylenetriaminepentaacetic acid (DTPA) buffered at pH 7.3 and determined by atomic absorption spectrometry, following the procedures of FAO [35]. The analyses conducted on the initial soil samples (Table 1) also followed these same procedures. Additionally, soil separates were determined in these initial soil samples using the Robison pipette method.

2.5. Data Analysis

The data were subjected to analysis of variance (one-way ANOVA) after being tested for normality and variance homogeneity using the Shapiro–Wilk and Levene's tests, respectively. Whenever the ANOVA revealed significant differences among treatments, means were separated using the Tukey HSD multiple comparison test ($\alpha = 0.05$).

Apparent N recovery (ANR) was estimated by the following equation: ANR (%) = 100 × (N recovered by plants in fertilized treatments − N recovered by plants in the control)/N applied as fertilizer.

3. Results

3.1. Lettuce Dry Matter Yield

Lettuce dry matter yield (DMY) varied significantly between treatments in the first lettuce growing cycle (Figure 2). Mealw, FullR, and HalfR gave significantly higher values than the other treatments, while Nutri and Cont displayed significantly lower values. It should also be noted that the highest average value was obtained in the Mealw treatment (12.8 g plant^{-1}), even higher than the value recorded in the FullR treatment (12.2 g plant^{-1}). DMY in the BSFly treatment was lower than in the Mealw, FullR, and HalfR treatments, but higher than in the Cont and Nutri treatments. It is also worth noting that the average DMY in the Nutri treatment (5.1 g plant^{-1}) was lower than in the Cont treatment (6.7 g plant^{-1}).

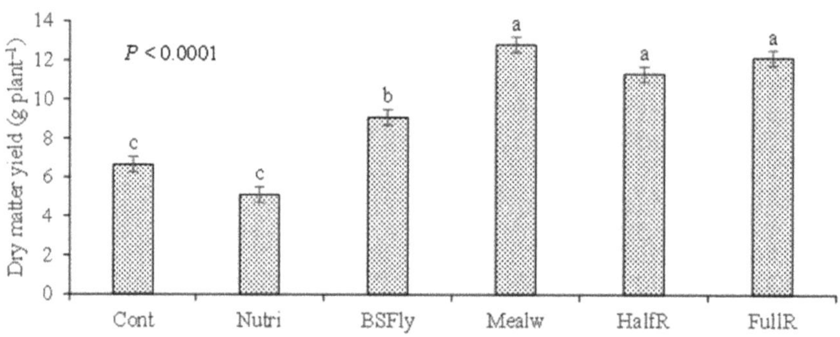

Figure 2. Dry matter yield of lettuce across fertilization treatments [control (Cont), Nutrimais (Nutri), black soldier fly larvae frass (BSFly), mealworm larvae frass (Mealw), half the rate ammonium nitrate (HalfR), full rate of ammonium nitrate (FullR)] in the first growing cycle. Means with the same letter are not significantly different (Tukey's HSD test, $\alpha = 0.05$). Vertical bars represent standard errors.

The pots subjected to the second rate of organic amendments and fertilizers (Figure 3a) exhibited results consistent with those observed during the initial growing cycle. However, the mean value of the Mealw treatment (9.8 g plant^{-1}) stood out, being significantly higher than all others except for the FullR treatment (7.8 g plant^{-1}). The BSFly treatment exhibited a lower mean value (5.7 g plant^{-1}), although not significantly different from that of the HalfR treatment (6.9 g plant^{-1}). The Nutri treatment maintained a very low mean value (2.7 g plant^{-1}), comparable to the Cont treatment (3.1 g plant^{-1}), and lower than that of the BSFly treatment.

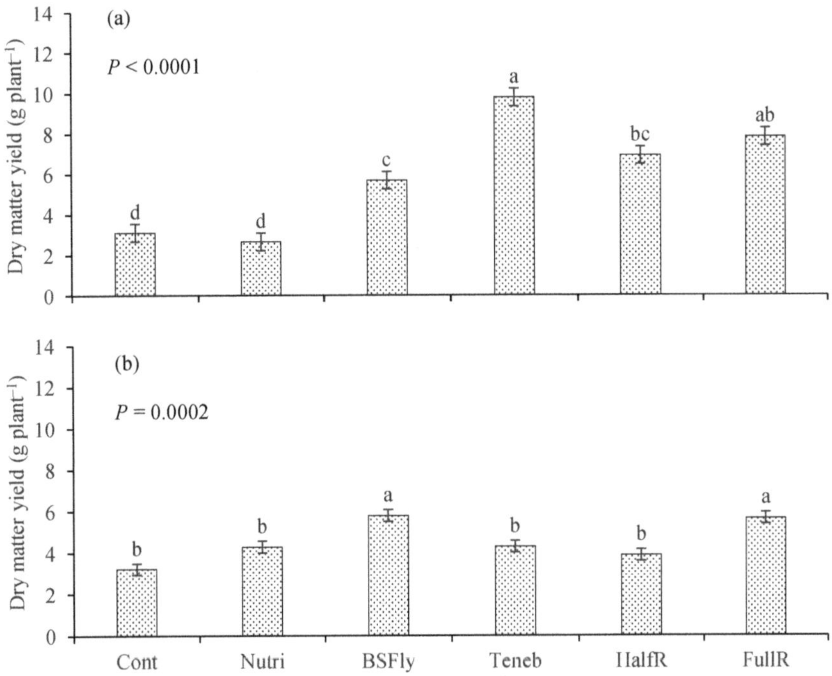

Figure 3. Dry matter yield of lettuce across fertilization treatments [control (Cont), Nutrimais (Nutri), black soldier fly larvae frass (BSFly), mealworm larvae frass (Mealw), half the rate ammonium nitrate (HalfR), full rate of ammonium nitrate (FullR)] in the second growing cycle, in pots that received (**a**) and did not receive (**b**) the second dose of organic amendments and fertilizers. Means with the same letter are not significantly different (Tukey's HSD test, $\alpha = 0.05$). Vertical bars represent standard errors.

In the pots that did not receive the second dose of organic amendments and fertilizers, where plant growth depended on the residual fertility from the treatments of the first lettuce cycle, the BSFy treatment stood out (5.8 g plant^{-1}), with no significant differences compared to the FullR treatment (5.7 g plant^{-1}), but significant differences were observed with all other treatments (Figure 3b). Nutri (4.3 g plant^{-1}) and Mealw (4.3 g plant^{-1}) displayed slightly higher mean values, but without significant differences compared to the Cont treatment (3.2 g plant^{-1}).

3.2. Plant Nitrogen Nutritional Status

The N concentration in the tissues of lettuces grown during the first growing cycle followed a slightly different pattern from DMY (Figure 4). The highest values were observed in the FullR treatment (33.0 g plant^{-1}), followed by HalfR (26.5 g kg^{-1}), and Mealw (19.6 g kg^{-1}). The mean values observed in the BSFly (15.2 g kg^{-1}) and Nutri (14.4 g kg^{-1}) treatments did not differ significantly from those observed in the Cont treatment (13.6 g kg^{-1}).

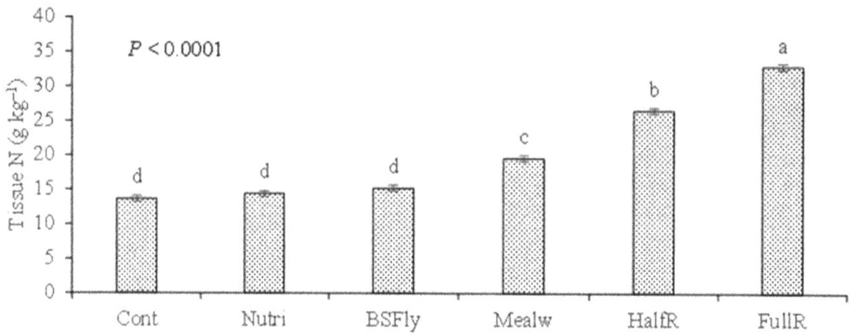

Figure 4. Nitrogen (N) concentration in lettuce tissues across fertilization treatments [control (Cont), Nutrimais (Nutri), black soldier fly larvae frass (BSFly), mealworm larvae frass (Mealw), half the rate ammonium nitrate (HalfR), full rate of ammonium nitrate (FullR)] in the first growing cycle. Means with the same letter are not significantly different (Tukey's HSD test, α = 0. 05). Vertical bars represent standard errors.

The response pattern observed in the first growing cycle of lettuce was repeated in the second cycle in pots that received the second rate of fertilizers (Figure 5a). Treatments receiving mineral fertilizer resulted in higher N concentrations in the tissues compared to other treatments. In pots not fertilized for the second growing cycle of lettuce (Figure 5b), the significant effect of the FullR treatment over others was maintained, but no significant differences were recorded among the others.

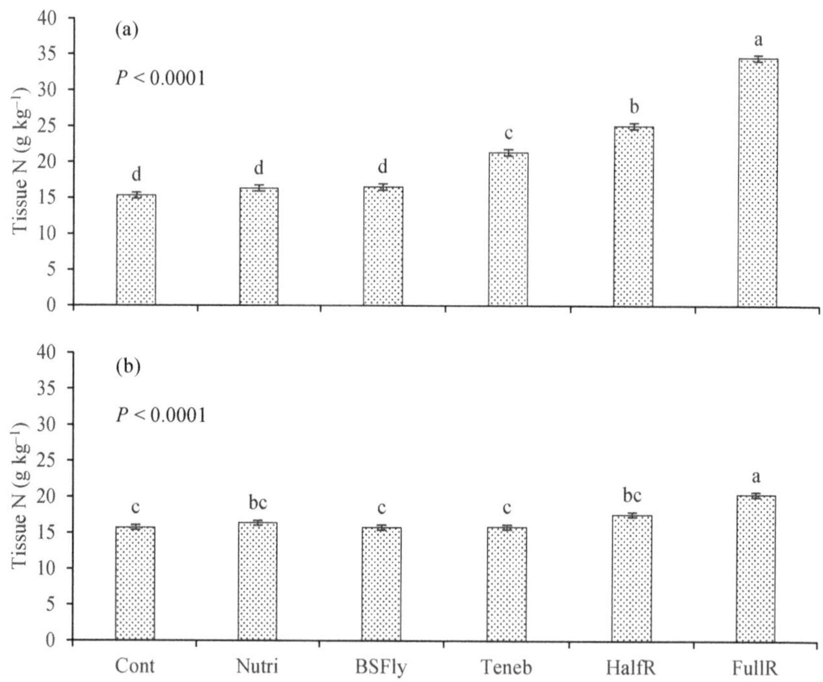

Figure 5. Nitrogen (N) concentration in lettuce tissues across fertilization treatments [control (Cont), Nutrimais (Nutri), black soldier fly larvae frass (BSFly), mealworm larvae frass (Mealw), half the rate ammonium nitrate (HalfR), full rate of ammonium nitrate (FullR)] in the second growing cycle, in pots that received (**a**) and did not receive (**b**) the second dose of organic amendments and fertilizers. Means with the same letter are not significantly different (Tukey's HSD test, α = 0.05). Vertical bars represent standard errors.

The nitrate concentration in the tissues did not vary as markedly with the fertilizer treatments as the N concentration in the tissues (Figure 6). In the first growth cycle of lettuce, significant differences occurred only between the FullR treatment (10,917 mg kg^{-1}) and the Cont treatment (8975 mg kg^{-1}). No significant differences were observed between the other fertilized treatments and the control treatment.

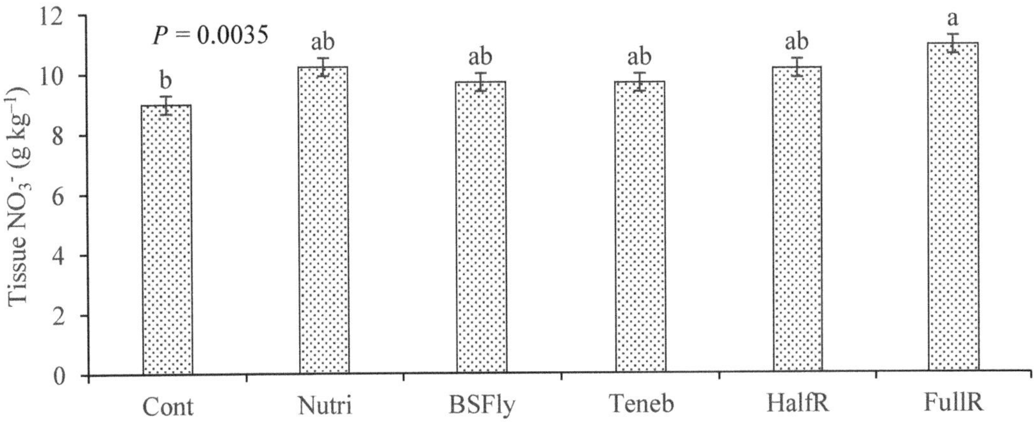

Figure 6. Nitrate (NO_3^-) concentration in lettuce tissues across fertilization treatments [control (Cont), Nutrimais (Nutri), black soldier fly larvae frass (BSFly), mealworm larvae frass (Mealw), half the rate ammonium nitrate (HalfR), full rate of ammonium nitrate (FullR)] in the first growing cycle. Means with the same letter are not significantly different (Tukey's HSD test, $\alpha = 0.05$). Vertical bars represent standard errors.

In the second growing cycle of lettuce, in pots that received the second rate of fertilizers, the Mealw and BSFly treatments resulted in nitrate tissue values equivalent to those of treatments with mineral fertilizer and significantly higher than those of the Nutri and Cont treatments (Figure 7a). Once again, Nutri resulted in a value equivalent to that of the control. In the second growing cycle of lettuce, in pots that did not receive a second dose of fertilizers, a greater gradient between treatments was observed in tissue nitrate levels (Figure 7b). The mean values of the FullR and Mealw treatments were significantly higher than those of the control.

The ANR in the first growing cycle of lettuce was very high in the HalfR treatment (96.6%), followed by the FullR treatment (71.9%) (Table 2). Among organic amendments, the mean value of Mealw (37.3%) was higher than that of BSFly (11.1%), being negative in the Nutri treatment (−4.0%). In pots where the second dose of fertilizer was not applied, the values increased comparatively to those of the first growing cycle in all treatments. The mean value exceeded 100% (105.9%) in the HalfR treatment and was slightly above 0% (0.6%) in the Nutri treatment. In pots that received the second dose of fertilizer, organic amendments maintained ANR values similar to those of the first lettuce cycle, while they decreased in treatments with mineral N application. Considering both growing cycles and pots that received both rates of fertilizers, the ANR values followed the decreasing order HalfR (78.3%), FullR (62%), Mealw (37.4%), BSFly (10.9%), and Nutri (−2.5%).

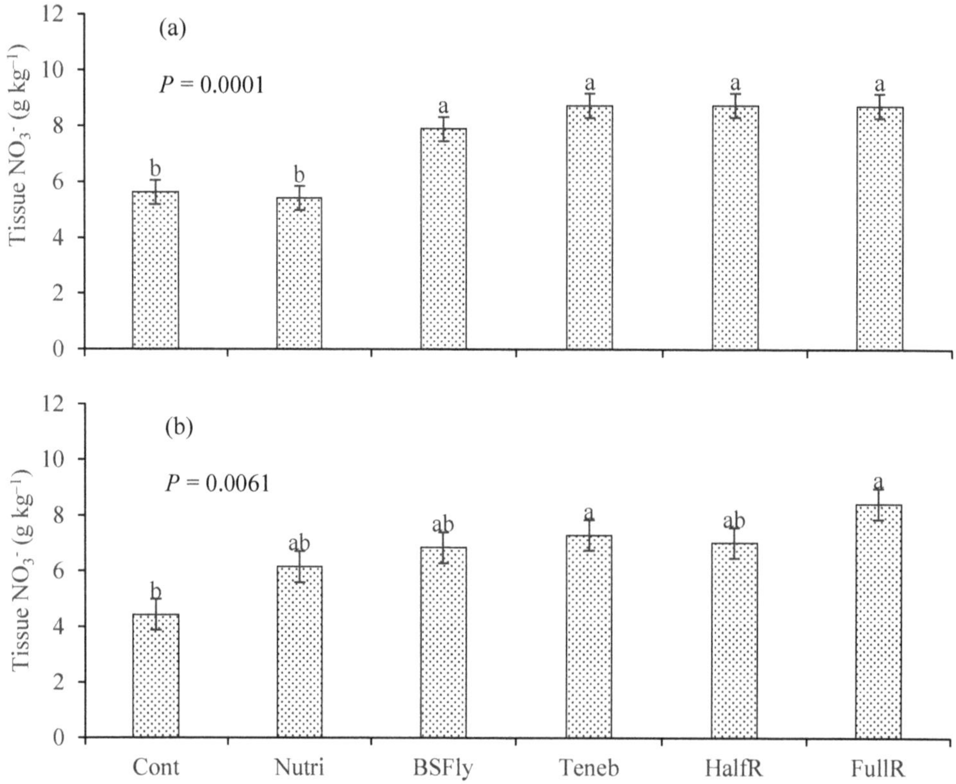

Figure 7. Nitrate (NO_3^-) concentration in lettuce tissues across fertilization treatments [control (Cont), Nutrimais (Nutri), black soldier fly larvae frass (BSFly), mealworm larvae frass (Mealw), half the rate ammonium nitrate (HalfR), full rate of ammonium nitrate (FullR)] in the second growing cycle, in pots that received (**a**) and did not receive (**b**) the second dose of organic amendments and fertilizers. Means with the same letter are not significantly different (Tukey's HSD test, $\alpha = 0.05$). Vertical bars represent standard errors.

Table 2. Apparent nitrogen (N) recovery by lettuces across fertilization treatments [Nutrimais (Nutri), black soldier fly larvae frass (BSFly), mealworm larvae frass (Mealw), half the rate ammonium nitrate (HalfR), full rate of ammonium nitrate (FullR)] in the first and second growing cycle (GC) in pots that received (with) and did not receive (without) the second dose of organic amendments and fertilizers.

	Apparent N Recovery (%)			
	1st GC	**2nd CG (without)**	**2nd GC (with)**	**Total (1st + 2nd with)**
Nutri	−4.0	0.6	−1.0	−2.5
BSFly	11.1	20.5	10.8	10.9
Mealw	37.3	41.3	37.6	37.4
HalfR	96.6	105.9	59.0	78.3
FullR	71.9	87.2	52.0	62.0

Apparent N recovery (%) = 100 × [(N recovered in the fertilized treatments − N recovered in the control)/N applied as a fertilizer].

3.3. Concentration of Other Macro and Micronutrients in the Tissues

The concentration of P in lettuce tissues varied significantly between treatments in the first growing cycle, with the highest average value found in the Mealw treatment (3.7 g kg^{-1}), followed by BSFly (3.2 g kg^{-1}) (Table 3). The average values of the other treatments did not vary significantly. The Mealw treatment also exhibited the highest average concentration of K in lettuce tissues (56.4 g kg^{-1}), which was significantly higher than that of the Cont treatment (43.8 g kg^{-1}). The Nutri treatment showed significantly higher Ca levels in tissues (8.3 g kg^{-1}) than those of FullR (6.7 g kg^{-1}). FullR showed the highest values of Mg (2.9 g kg^{-1}), while BSFly (2.3 g kg^{-1}) and Cont (2.3 g kg^{-1}) showed the lowest ones. Tissue B concentration did not vary between treatments. The Fe levels in tissues appeared lower in treatments associated with higher DMY, a trend apparently followed by Cu values. The Mn and Zn values seem to show a tendency to be higher in treatments with mineral fertilization.

Table 3. Concentration of macro and micronutrients in lettuce tissues across fertilization treatments [control (Cont), Nutrimais (Nutri), black soldier fly larvae frass (BSFly), mealworm larvae frass (Mealw), half the rate ammonium nitrate (HalfR), full rate of ammonium nitrate (FullR)] in the first growing cycle.

	P	K	Ca	Mg	B	Fe	Mn	Zn	Cu
		g kg^{-1}					mg kg^{-1}		
Cont	2.6 c	43.8 b	6.9 ab	2.3 b	35.5 a	2161.9 a	88.4 bc	76.1 ab	32.7 a
Nutri	2.7 c	53.5 ab	8.3 a	2.7 ab	32.8 a	2062.8 a	73.8 c	58.4 b	29.4 ab
BSFly	3.2 b	54.2 ab	7.2 ab	2.3 b	34.3 a	1889.1 a	86.0 c	67.7 ab	33.8 a
Mealw	3.7 a	56.4 a	7.3 ab	2.4 ab	34.6 a	1407.5 b	82.1 c	71.9 ab	28.3 ab
HalfR	2.6 c	44.1 ab	7.2 ab	2.5 ab	33.9 a	1133.2 b	103.2 ab	80.0 ab	25.6 b
FullR	2.6 c	45.0 ab	6.7 b	2.9 a	31.3 a	1115.1 b	107.0 a	87.7 a	23.9 b
Prob	<0.0001	0.0077	0.0453	0.0115	0.1314	<0.0001	<0.0001	0.0118	0.0002

In columns, means followed by the same letter are not significantly different by Tukeys' HSD test ($\alpha = 0.05$).

The concentration of P in lettuce tissues in the second growth cycle, and in pots that received the second dose of fertilizer, did not vary significantly between treatments (Table 4). K values were lower in the FullR treatment compared to treatments with organic amendments and the control. Ca values in the tissues were again high in the Nutri treatment, this time accompanied by those of the control treatment. Mg, B, Mn, and Zn levels in lettuce tissues did not vary significantly with the treatments. Fe levels in the tissues showed the same trend of being lower in the more productive treatments, and Cu levels did not show consistency with the values observed in the first growing cycle of lettuce.

Table 4. Concentration of macro and micronutrients in lettuce tissues across fertilization treatments [control (Cont), Nutrimais (Nutri), black soldier fly larvae frass (BSFly), mealworm larvae frass (Mealw), half the rate ammonium nitrate (HalfR), full rate of ammonium nitrate (FullR)] in the second growing cycle in pots that received the second dose of organic amendments and fertilizers.

	P	K	Ca	Mg	B	Fe	Mn	Zn	Cu
		g kg^{-1}					mg kg^{-1}		
Cont	1.3 a	49.5 a	8.9 a	3.8 a	27.8 a	1332.5 ab	70.5 a	57.2 a	11.0 ab
Nutri	1.0 a	50.7 a	9.5 a	4.1 a	26.1 a	1487.8 a	58.8 a	70.4 a	14.8 a
BSFly	1.1 a	49.9 a	7.5 b	3.2 a	27.9 a	660.2 ab	62.1 a	79.1 a	10.0 b
Mealw	1.0 a	50.2 a	7.2 b	3.5 a	27.7 a	816.9 ab	84.1 a	68.9 a	11.4 ab
HalfR	0.9 a	41.4 ab	7.5 b	4.0 a	24.2 a	733.4 ab	80.6 a	76.8 a	12.5 ab
FullR	1.3 a	39.6 b	6.7 b	3.8 a	22.8 a	570.9 b	87.6 a	79.3 a	12.2 ab
Prob	0.3630	0.0300	<0.0001	0.2630	0.0554	0.0171	0.1772	0.4862	0.0476

In columns, means followed by the same letter are not significantly different by Tukey's HSD test ($\alpha = 0.05$).

In the pots that did not receive the second dose of fertilizers, no significant differences occurred in the concentration of macronutrients in lettuce tissues (Table 5). In this group of pots, significant differences in tissue B levels appeared for the first time, with the lowest mean values observed in the FullR treatment. No significant differences were observed among treatments for metallic micronutrients.

Table 5. Concentration of macro and micronutrients in lettuce tissues across fertilization treatments [control (Cont), Nutrimais (Nutri), black soldier fly larvae frass (BSFly), mealworm larvae frass (Mealw), half the rate ammonium nitrate (HalfR), full rate of ammonium nitrate (FullR)] in the second growing cycle in pots that did not receive the second dose of organic amendments and fertilizers.

	P	K	Ca	Mg	B	Fe	Mn	Zn	Cu
	g kg^{-1}				mg kg^{-1}				
Cont	1.2 a	47.8 a	9.1 a	3.9 a	27.2 ab	1117.0 a	69.0 a	63.1 a	11.6 a
Nutri	1.1 a	47.4 a	7.9 a	3.3 a	27.6 a	1058.4 a	53.5 a	58.0 a	10.8 a
BSFly	1.2 a	42.2 a	7.3 a	3.3 a	27.2 a	691.2 a	54.5 a	54.3 a	10.0 a
Mealw	0.9 a	38.9 a	8.3 a	3.5 a	28.3 ab	983.7 a	74.3 a	54.7 a	13.3 a
HalfR	1.1 a	48.2 a	9.1 a	4.0 a	28.7a	986.0 a	78.0 a	75.0 a	11.4 a
FullR	1.1 a	44.2 a	7.0 a	3.4 a	23.8 b	661.5 a	87.2 a	71.9 a	10.6 a
Prob	0.8482	0.1440	0.0686	0.0604	0.0119	0.2245	0.0443	0.4587	0.3428

In columns, means followed by the same letter are not significantly different by Tukey's HSD test ($\alpha = 0.05$).

3.4. Residual Effect of Treatments on Inorganic Nitrogen in Soil and Oat Crop

The nitrate, ammonia, and total inorganic N levels in the soil at the end of the second growing cycle of lettuce were the highest in the FullR treatment, with significant differences compared to all other treatments in ammonia and total inorganic N levels (Table 6). In pots that did not receive the second dose of fertilizer, elevated values were also observed in some treatments with organic amendments, such as Mealw and Nutri. The control values were typically the lowest.

Table 6. Nitrogen (N) in the form of nitrate (NO_3^-) and ammonium (NH_4^+) in the soil at the end of the second lettuce growing cycle across fertilization treatments [control (Cont), Nutrimais (Nutri), black soldier fly larvae frass (BSFly), mealworm larvae frass (Mealw), half the rate ammonium nitrate (HalfR), full rate of ammonium nitrate (FullR)] in pots that received and did not receive the second dose of organic amendments and fertilizers.

	Two Fertilizer Applications			Fertilizers Applied Once		
	NO_3^--N	NH_4^+-N mg kg^{-1}	Inorg-N	NO_3^--N	NH_4^+-N mg kg^{-1}	Inorg-N
Cont	2.2 b	0.3 b	2.5 b	2.3 bc	0.5 c	2.8 b
Nutri	3.9 ab	1.5 b	5.4 b	3.8 ab	1.6 bc	5.4 ab
BSFly	1.7 b	1.8 b	3.4 b	1.4 c	2.9 ab	4.3 b
Mealw	4.1 ab	3.4 b	7.4 b	4.3 a	3.8 a	8.1 a
HalfR	3.4 ab	3.0 b	6.4 b	2.6 abc	1.2 bc	3.8 b
FullR	6.4 a	10.1 a	16.5 a	3.0 abc	2.7 ab	5.7 ab
Prob	0.0039	0.0001	<0.0001	0.0035	0.0013	0.0011

In columns, means followed by the same letter are not significantly different by Tukey's HSD test ($\alpha = 0.05$).

After the second lettuce cycle, oats were grown in pots without receiving further fertilization. In pots that received the second dose of fertilizer, DMY was highest in the BSFly treatment (3.1 g pot^{-1}), followed by Mealw (2.8 g pot^{-1}) and FullR (2.6 g pot^{-1}) (Table 7). The Cont (1.2 g pot^{-1}) and Nutri (1.4 g pot^{-1}) treatments recorded the lowest values. The trend in DMY results in pots that did not receive the second dose persisted, with higher values in the BSFly treatment (2.0 g pot^{-1}) and significantly different from the Cont (1.3 g pot^{-1}) and Nutri (1.3 g pot^{-1}) treatments. No significant differences were observed in N concentration in the tissues, whether in pots that received or did not receive the second dose of fertilizer. However, the amount of N taken up by the plant followed the trend of DMY, with the highest mean values observed in the BSFly treatment, although no significant differences between treatments were observed in pots that did not receive the second dose of fertilizer.

Table 7. Dry matter yield (DMY), tissue N and N recovery of oats across fertilization treatments [control (Cont), Nutrimais (Nutri), black soldier fly larvae frass (BSFly), mealworm larvae frass (Mealw), half the rate ammonium nitrate (HalfR), full rate of ammonium nitrate (FullR)] in pots that received and did not receive the second dose of organic amendments and fertilizers.

	Two Fertilizer Applications			Fertilizers Applied Once		
	DMY g pot^{-1}	Tissue N g kg^{-1}	N Recovery mg pot^{-1}	DMY g pot^{-1}	Tissue N g kg^{-1}	N Recovery mg pot^{-1}
Cont	1.2 c	17.0 a	21.4 c	1.3 b	18.6 a	26.0 a
Nutri	1.4 c	16.1 a	22.2 c	1.3 b	17.3 a	22.0 a
BSFly	3.1 a	15.7 a	48.1 a	2.0 a	17.2 a	34.3 a
Mealw	2.8 ab	16.5 a	46.2ab	1.8 ab	15.5 a	28.3 a
HalfR	2.2 b	15.9 a	35.1 b	1.4 ab	17.7 a	25.2 a
FullR	2.6 ab	14.2 a	37.3 ab	1.5 ab	16.8 a	25.0 a
Prob	<0.0001	0.0571	<0.0001	0.0318	0.1232	0.3163

In columns, means followed by the same letter are not significantly different by Tukey's HSD test ($\alpha = 0.05$).

3.5. Soil Properties

Soil organic C in the pots that received the second rate of fertilizers varied significantly between treatments (Table 8). The higher average values were found in the Mealw treatment (11.5 g kg^{-1}), while the lower ones were observed in the Cont treatment (8.8 g kg^{-1}). Soil pH did not vary significantly between treatments. Extractable P was particularly high in the BSFly and Mealw treatments and low in the HalfR and FullR treatments. Soil K levels were also particularly low in the HalfR and FullR treatments. Exchangeable Ca did not vary significantly between treatments, nor did the CEC, although significant differences were observed among treatments for soil Mg levels, which were particularly high in the BSFly treatment and low in the FullR treatment. Cation micronutrients did not vary significantly between treatments.

Table 8. Soil properties at the end of the second growing cycle of lettuce across fertilization treatments [control (Cont), Nutrimais (Nutri), black soldier fly larvae frass (BSFly), mealworm larvae frass (Mealw), half the rate ammonium nitrate (HalfR), full rate of ammonium nitrate (FullR)] in pots that received the second dose of organic amendments and fertilizers.

	Cont	Nutr	BSFly	Mealw	HalfR	FullR	Prob
Organic C (g kg^{-1})	8.8 b	9.9 ab	11.4 a	11.5 a	8.4 b	10.6 ab	0.0337
pH (H_2O)	6.6 a	6.7 a	6.5 a	6.7 a	6.7 a	6.5 a	0.1076
Extract P (mg kg^{-1})	124.5 bc	148.8 bc	216.6 a	193.4 a	89.8 c	97.4 c	<0.0001
Extract K (mg kg^{-1})	128.3 ab	123.7 b	151.7 a	110.7 b	83.3 c	75.3 c	<0.0001
Exchang Ca (cmol$_c$ kg^{-1})	16.0 a	17.0 a	16.8 a	15.9 a	16.4 a	14.8 a	0.2424
Exchang Mg (cmol$_c$ kg^{-1})	5.3 bc	5.3 bc	6.4 a	5.8 ab	5.3 bc	4.8 c	0.0023
CEC (cmol$_c$ kg^{-1})	21.9 a	23.2 a	24.0 a	22.4 a	22.4 a	20.1 a	0.0858
Extract Fe (mg kg^{-1})	123.0 a	106.4 a	107.1 a	100.3 a	108.2 a	107.7 a	0.1963
Extract Zn (mg kg^{-1})	4.1 a	4.6 a	5.1 a	5.3 a	4.0 a	4.2 a	0.2432
Extract Cu (mg kg^{-1})	71.0 a	67.7 a	69.5 a	67.0 a	73.2 a	75.3 a	0.8506
Extract Mn (mg kg^{-1})	176.3 a	144.2 a	153.6 a	146.0 a	159.5 a	154.0 a	0.0781

In rows, means followed by the same letter are not significantly different by the Tukey HSD test ($\alpha = 0.05$).

In pots that did not receive the second dose of fertilizer, soil organic C varied similarly between treatments as in pots that received the second dose, although with generally lower values, even in treatments that did not receive organic amendments (Table 9). Soil pH also did not vary significantly between treatments. P and K values followed the trend observed in pots that received the second dose of fertilizers, although P levels were generally lower now. Exchangeable Ca and Mg, as well as CEC, showed a very similar pattern to what was observed in pots that received the second dose of fertilizer. Among cationic micronutrients, no significant differences occurred between treatments.

Table 9. Soil properties at the end of second growing cycle of lettuce across fertilization treatments [control (Cont), Nutrimais (Nutri), black soldier fly larvae frass (BSFly), mealworm larvae frass (Mealw), half the rate ammonium nitrate (HalfR), full rate of ammonium nitrate (FullR)] in pots that did not receive the second dose of organic amendments and fertilizers.

	Cont	Nutri	BSFly	Mealw	HalfR	FullR	Prob
Organic C (g kg^{-1})	8.2 c	11.2 a	11.3 a	10.5 ab	8.6 bc	9.4 abc	0.0014
pH (H_2O)	6.5 a	6.7 a	6.7 a	6.7 a	6.5 a	6.5 a	0.0531
Extract P (mg kg^{-1})	96.7 b	153.9 a	166.2 a	163.4 a	90.1 b	90.3 b	<0.0001
Extract K (mg kg^{-1})	104.0 b	128.0 a	115.3 b	115.3 b	91.7 c	82.0 c	<0.0001
Exchang Ca (cmol$_+$ kg^{-1})	15.8 a	17.1 a	16.1 a	15.9 a	14.8 a	15.4 a	0.1119
Exchang Mg (cmol$_+$ kg^{-1})	5.3 ab	5.3 ab	5.8 a	5.5 ab	5.0 b	4.9 b	0.0147
CEC (cmol$_c$ kg^{-1})	21.7 a	23.2 a	22.6 a	22.0 a	20.5 a	20.9 a	0.1178
Extract Fe (mg kg^{-1})	117.8 a	118.0 a	109.0 a	94.3 a	108.7 a	104.7 a	0.2596
Extract Zn (mg kg^{-1})	5.0 a	5.1 a	5.0 a	4.9 a	3.9 a	4.1 a	0.3151
Extract Cu (mg kg^{-1})	71.2 a	73.0 a	76.0 a	69.5 a	69.0 a	70.4 a	0.8828
Extract Mn (mg kg^{-1})	175.5 a	162.0 a	160.4 a	137.8 a	160.6 a	151.0 a	0.2789

In rows, means followed by the same letter are not significantly different by the Tukey HSD test ($\alpha = 0.05$).

4. Discussion

4.1. Mealworm Larvae Frass

Mealworm larvae frass (Mealw) resulted in the highest average DMY among all treatments in the first lettuce growing cycle, significantly surpassing the BSFly, Nutri, and Cont treatments (Figure 2). This trend of high values persisted in the second lettuce cycle upon reapplication of fertilizers, with DMY significantly exceeding that of the other treatments, except for the FullR treatment (Figure 3a). N concentration in the tissues was notably higher in the Mealw treatment compared to treatments with other organic amendments. Often considered the most crucial nutrient in crop fertilization programs, N has an undeniable effect on crop growth [13,36,37], and this study is no exception.

The ANR in the Mealw treatment was very high for an organic amendment, reaching 37.4% over the two growing cycles that received fertilization. This result indicates a high mineralization rate of the organic substrate. Houben et al. [38] reported a significant effect of mealworm larvae frass application on ryegrass grown in pots, also attributing the result to the rapid appearance of nutrients in the soil in readily available forms for plants. The mineralization rate of organic substrates in the soil largely depends on their N content and/or the C/N ratio. The balance between biological immobilization and mineralization is typically found for N values between 15 and 20 g kg^{-1} and C/N ratios between 20 and 30 [11,39]. The N content and C/N ratio of mealworm larvae frass were 31.3 g kg^{-1} and 16.5, respectively. Although these values may suggest net mineralization in the short term, they would be unlikely to predict such high mineralization rates, given results reported in previous studies using other organic amendments [16,18,40,41], especially considering the short growing cycle of lettuce.

However, the availability of N alone cannot account for the good performance of lettuces subjected to the Mealw treatment. Lettuces that received ammonium nitrate, particularly those in the FullR treatment, exhibited higher N concentrations in tissues compared to those in the Mealw treatment. Moreover, the N concentration in the tissues of the Mealw treatment is still below the sufficiency range for lettuce [42], indicating that the FullR treatment did not supply excess N. The ANR in treatments with mineral N reached notably high percentages, especially when compared to typical field trial values, which usually fall below 50% [41,43,44]. This also suggests that excess N was not used. The N rate used in the FullR treatment, amounting to 50 kg ha^{-1}, is not excessive when considering the common N requirements for lettuce [45].

Lettuces in the Mealw treatment exhibited high tissue P levels, particularly during the first growing cycle. This result suggests that P might have played a significant role in the DMY of the lettuce, considering the well-documented importance of P in crop fertilization across various regions worldwide [46,47]. Additionally, with organic amendments, the simultaneous availability of multiple nutrients can create positive synergies in root system development, thereby benefiting plant growth [48]. However, some of the more productive treatments, FullR and HalfR, showed lower tissue P concentrations, diminishing the relevance of P in lettuce growth. Given the concentrations of other nutrients in the tissues, there seems to be no justification to assert that the role of any specific nutrient was significant in explaining the high DMY in the Mealw treatment. Therefore, other factors unrelated to tissue nutrient concentrations must be identified to explain the DMY results.

Organic amendments can positively affect plant growth in several ways, particularly by improving the physical, chemical, and biological properties of the soil [18,49,50]. Some studies have demonstrated that the positive effects on plants achieved by applying organic amendments cannot be replicated by using only mineral fertilizers. This beneficial action of organic amendments is sometimes referred to as the "manuring effect", a term used to describe measurable growth benefits that may be difficult to attribute to a specific cause [11,16,51]. By testing various organic materials produced by mealworm larvae, Poveda et al. [52] suggested that some microorganisms present in the excreta act as plant growth promoters, which explains the benefits they observed in the growth of chard (*Beta vulgaris* L., var. Cicla) and bean (*Phaseolus vulgaris* L.) seedlings. The authors also found that each type of excrement had a different potential to promote plant growth and stress resistance, depending on the diets of the insects. In another study, although positive effects on lettuce growth were recorded when using mealworm frass, Fuertes-Mendizábal et al. [53] found that the use of the product at a dose above 2.5% of the peat substrate resulted in a negative effect on lettuce plant growth, particularly on root development, indicating that the applied dose may be relevant in the management of this organic amendment.

In the study presented here, the Mealw treatment appears to have a positive effect on plant growth that extends beyond its nutrient content and its impact on the nutritional status of the plants. The result suggests a clear manuring or biostimulant effect. Considering the use of soil as substrate and a fertilizer rate similar to field cultivation of lettuce, no negative effects were observed. Thus, only further studies can determine whether this result was incidental or consistent and attributable to the insect or its diet. On the other hand, a promising aspect to note is that this study used a commercially available product, suggesting that it may maintain the stability of its properties and agronomic behavior in the soil.

4.2. Black Soldier Fly Frass

Black soldier fly frass initially presented a composition not very different from that of mealworm larvae frass (Table 1). The composition of the black soldier fly frass used in this study is also comparable to organic materials of the same origin reported in other studies [24]. However, the lettuce DMY was significantly lower than that of the Mealw treatment in the cycles where fertilizers were applied. The N concentration in the tissues was also lower (Figures 4 and 5), as well as the ANR (Table 2), which averaged 10.9% over the two growing cycles where fertilizers were applied. The N content (26.8 g kg^{-1}) and the C/N ratio (18.6) of this material also indicate a high likelihood of short-term net mineralization, since the C/N ratio value that leads to a balance between mineralization and biological immobilization ranges from 20 to 30, as previously mentioned [11,39].

The BSFly treatment increased lettuce DMY compared to the control treatment. Menino et al. [54], using a black soldier fly frass organic amendment, also observed an increase in ryegrass production in a greenhouse experiment compared to the control treatment. Similar results were found by Chavez et al. [55] in a pot experiment with arugula (*Eruca sativa* L.) and lettuce. However, in practice, the results of the BSFly treatment were not significantly different from those reported worldwide with other organic amendments [16,40,41]. Carroll et al. [56], for instance, using black soldier fly frass, reported cereal seedling growth at a level equivalent to that observed with pelletized chicken manure. The productivity loss in the BSFly treatment compared to the FullR treatments was likely due to the reduced availability of N in the soil, as the nutrient levels in the tissues were much lower in the BSFly treatment and below the lower limit of the sufficiency range for lettuce [42]. N availability in the soil is crucial for cell growth and expansion, particularly for the formation of proteins and other organic molecules [57]. Unlike the Mealw treatment, the BSFly treatment did not appear to have any evident effect beyond what is typically observed with high-quality organic amendments.

4.3. Nutrimais

The lettuce DMY in the Nutri treatment remained consistently at the level of the Cont treatment in both lettuce cycles, regardless of whether fertilization was repeated or not. The N levels in the tissues were always low, and not significantly different from those in the Cont treatment. On the other hand, the nitrate levels in the tissues, another indicator of N availability in the soil [20,58,59], tended to be lower than those in the other fertilized treatments. There did not appear to be significant differences between the Nutri treatment and the other treatments regarding the concentration of nutrients other than N in the tissues. Thus, the lack of available N for the plants was likely the main cause of the low productivity of the lettuces subjected to this treatment, considering that the N concentration in the tissues was well below the lower limit of the sufficiency range as reported by Bryson et al. [42].

The ANR by lettuces subjected to the Nutri treatment was globally negative, including when considering the two cropping cycles that received fertilization (Table 2). This indicates that biological immobilization occurred, meaning that the lettuces in this treatment had less available N than those in the Cont treatment. Biological immobilization resulting from the application of an organic amendment tends to occur when the amendment has a very low N concentration or a very high C/N ratio [11,39]. Under these conditions, energy is available for the heterotrophic communities in the soil (reflected in the C of the substrate), but there is little N (essential for the formation of amino acids and proteins). This leads to competition between heterotrophic microorganisms and plants for the available mineral N in the soil [11,39]. Although this organic substrate initially had a composition not very different from the others (C/N ratio, 16.5), it proved to be unreactive in the soil, not releasing nutrients to the plants. Previous studies using this organic amendment have already indicated its low reactivity [60,61]. Perhaps because it is a highly dehydrated and pelletized material, it becomes biologically inactive when applied to the soil. This aspect, combined with the short duration of the lettuce growing cycle, resulted in a situation where its effect on crop productivity was practically negligible. Objectively, it seems to be an organic amendment unsuitable for early maturing horticultural crops that require a rapid supply of nutrients.

4.4. Residual Effect of Fertilization and Soil Properties

The effect of lettuce fertilization on unfertilized oats highlighted the role of the BSFly treatment, which showed higher average values for DMY and N recovery both when fertilizers were applied in both lettuce growing cycles and when applied in only one of them (Table 7). This may indicate that, although Mealw had a higher short-term effect, over time, BSFly also released its nutrients, increasing consistency in the medium term. This result emphasizes the general knowledge about organic amendments and the care that should be taken in their management. Although N content and the C/N ratio have been used as the main indicators of the balance in the mineralization/immobilization process and nutrient release for crops [11,39], the process is highly unpredictable. Therefore, it is important to have the best information about organic amendments for more effective management in agricultural fields [41,51].

The effect of treatments on soil properties, considering the situations of two or just one application, demonstrated a tendency for an increase in soil organic matter in the Nutri, BSFly, and Mealw treatments compared to the unfertilized control. However, the FullR treatment consistently exhibited higher average values compared to the control, although without significant differences. This study's limited duration prevents providing fully consolidated results on soil organic matter dynamics. Nevertheless, it seemed evident that the application of organic amendments increased soil organic matter content through direct C input, a phenomenon frequently observed following such applications [13,18,49,54]. However, mineral fertilization can also sometimes increase soil organic matter content [61,62]. Increased primary productivity results in more organic substrate being deposited into the soil through roots and associated mycorrhizal fungi [63,64]. In this study, comparing organic amendments is complex because some (e.g., Mealw) may have undergone more mineralization, potentially reducing soil organic C content. On the other hand, the stimulation of lettuce growth was greater, contributing to higher C deposition in the soil through root system development. With the Nutri treatment, for example, the opposite may have occurred, with more original C remaining in the soil and a lesser contribution from the lettuce root system.

Due to the short duration of the experiments, only a few other soil properties varied significantly with the treatments. Extractable P and K values were the most dynamic variables, reflecting higher values in treatments with organic amendments, where significant amounts of nutrients were applied to the soil, and lower values in mineral treatments, where nutrients were not applied and there may have been greater nutrient removal by lettuce. Similar findings were observed in studies by Menino et al. [54] utilizing black soldier fly frass, demonstrating an elevation in soil nutrient levels corresponding to the main components of the organic amendment.

5. Conclusions

Mealworm larvae frass mineralized very rapidly, releasing N to lettuce at a much faster rate than its C/N ratio would predict. This suggests that the action of mealworm larvae produces an organic substrate that is very easily degraded by soil heterotrophic microorganisms. Consequently, it becomes a highly valuable organic amendment for application in early maturing annual crops that require a rapid release of nutrients. It should also be noted that these fertilizers compete freely in the market. Due to their origin and market acceptance, they are likely to be expensive and used mainly for high-value horticultural crops. However, the high N release does not fully explain its effect on DMY, which tended to be even higher than that of the treatment that received the highest rate of mineral fertilizer (FullR) and resulted in plants with higher tissue N concentrations. A manuring effect was observed that cannot be achieved solely with mineral fertilizer. The results suggest the possibility of compounds within the mealworm larvae frass composition exerting a plant biostimulant effect, capable of enhancing productivity beyond nutrient release and its impact on soil properties. In future research, efforts should be made to assess the persistence of the biostimulant effect and elucidate its underlying mechanisms.

Black soldier fly frass exhibited the typical behavior of a conventional high-quality organic fertilizer, with just over 10% of the N released during the short lettuce growing cycle. It did not appear as reactive in the soil as mealworm larvae frass, nor did it demonstrate any apparent manuring or plant biostimulant effects.

Despite its low C/N ratio, Nutrimais showed little reactivity, with a nearly negligible N mineralization/immobilization balance during the short lettuce growing cycle. This result suggests that this fertilizer should preferably be used in perennial crops that do not rely on immediate nutrient release following the application of the organic amendment.

Author Contributions: M.F. investigation, writing—review and editing.: M.A.: funding acquisition, methodology, supervision, review and editing. M.Â.R.: conceptualisation, funding acquisition, project administration, data curation, writing—review and editing. All authors have read and agreed to the published version of the manuscript.

Funding: The authors are grateful to the Foundation for Science and Technology (FCT, Portugal) for financial support from national funds FCT/MCTES, to CIMO (UIDB/AGR/00690/2020) and SusTEC (LA/P/0007/2020). They also thank ENTOGREEN® company for providing the insect frass.

Data Availability Statement: The data presented in this study are available on request from the corresponding author. The data are not publicly available due to privacy.

Conflicts of Interest: The authors declare no conflicts of interest.

References

1. United Nations, Department of Economic and Social Affairs, Population Division. *World Population Prospects: The 2017 Revision, Key Findings and Advance Tables (ESA/P/WP/248)*; United Nations: New York, NY, USA, 2017.
2. Terfa, N.G. Role of black soldier fly (*Hermetia illucens*) larvae frass bio-fertilizer on vegetable growth and sustainable farming in Sub-Saharan Africa. *Rev. Agric. Sci.* **2021**, *9*, 92–102. [CrossRef] [PubMed]
3. Pretty, J.; Toulmin, C.; Williams, S. Sustainable intensification in African agriculture. *Int. J. Agric. Sustain.* **2011**, *9*, 5–24. [CrossRef]
4. Scotti, R.; Dascoli, R.; Gonzalez Caceres, M.; Bonanomi, G.; Sultana, S.; Cozzolino, L.; Rao, M.A. Combined use of compost and wood scraps to increase carbon stock and improve soil quality in intensive farming systems. *Eur. J. Soil Sci.* **2015**, *66*, 463–475. [CrossRef]
5. Kopittke, P.M.; Menzies, N.W.; Wang, P.; McKenna, B.A.; Lombi, E. Soil and the intensification of agriculture for global food security. *Environ. Int.* **2019**, *232*, 105078. [CrossRef] [PubMed]
6. Fahad, S.; Ullah, A.; Ali, U.; Ali, E.; Saud, S.; Hakeem, K.R.; Arif, M. Drought tolerance in plants: Role of phytohormones and scavenging system of ROS. In *Plant Tolerance to Environmental Stress*; CRC Press: Boca Raton, FL, USA, 2019; pp. 103–114.
7. Gong, W.; Yan, X.; Wang, J.; Hu, T.; Gong, Y. Long-term manure and fertilizer effects on soil organic matter fractions and microbes under a wheat–maize cropping system in northern China. *Geoderma* **2009**, *149*, 318–324. [CrossRef]
8. Hakeem, K.R.; Dar, G.H.; Mehmood, M.A.; Bhat, R.A. *Microbiota and Biofertilizers: A Sustainable Continuum for Plant and Soil Health*; Springer International Publishing: Cham, Switzerland, 2021.
9. Savci, S. Investigation of effect of chemical fertilizers on environment. *Apcbee Procedia* **2012**, *1*, 287–292. [CrossRef]
10. Huber, G.; Schaub, C. La fertilité des sols: L'importance de la matière organique. In *Agriculture et Terroir*; Chambre d'Agriculture Bas Rhin: Schiltigheim, France, 2011.
11. Weil, R.R.; Brady, N.C. *The Nature and Properties of Soils*; Pearson Education Limited: Edinburgh, UK, 2017.
12. Azizi, S.; Thomas, T.; Rao, S. Effect of different levels of chemical fertilizers on soil physicochemical properties of inceptisols. *Int. J. Multidisc. Res. Dev.* **2016**, *3*, 29–32.
13. Lopes, J.I.; Gonçalves, A.; Brito, C.; Martins, S.; Pinto, L.; Moutinho-Pereira, J.; Raimundo, S.; Arrobas, M.; Rodrigues, M.A.; Correia, C.M. Inorganic fertilization at high n rate increased olive yield of a rainfed orchard but reduced soil organic matter in comparison to three organic amendments. *Agronomy* **2021**, *11*, 2172. [CrossRef]
14. Bonanomi, G.; Ascoli, R.; Antignani, V.; Capodilupo, M.; Cozzolino, L.; Marzaioli, R.; Puopolo, G.; Rutigliano, F.A.; Scelza, R.; Scotti, R.; et al. Assessing soil quality under intensive cultivation and tree orchards in Southern Italy. *Appl. Soil Ecol.* **2011**, *47*, 187–194. [CrossRef]
15. Goss, M.J.; Tubeileh, A.; Goorahoo, D. A review of the use of organic amendments and the risk to human health. *Adv. Agron.* **2013**, *120*, 275–379. [CrossRef]
16. Rodrigues, M.A.; Ladeira, L.C.; Arrobas, M. Azotobacter-enriched organic manures to increase nitrogen fixation and crop productivity. *Eur. J. Agron.* **2018**, *93*, 88–94. [CrossRef]
17. Nanda, S.; Berruti, F. Municipal solid waste management and landfilling technologies: A review. *Environ. Chem. Lett.* **2021**, *19*, 1433–1456. [CrossRef]
18. Arrobas, M.; Carvalho, J.T.N.; Raimundo, S.; Poggere, G.; Rodrigues, M.A. The safe use of compost derived from municipal solid waste depends on its composition and conditions of application. *Soil Use Manag.* **2022**, *38*, 917–928. [CrossRef]
19. Dhanker, R.; Chaudhary, S.; Goyal, S.; Garg, V.K. Influence of urban sewage sludge amendment on agricultural soil parameters. *Environ. Technol. Innov.* **2021**, *23*, 101642. [CrossRef]
20. Arrobas, M.; Meneses, R.; Gusmão, A.G.; da Silva, J.M.; Correia, C.M.; Rodrigues, M.Â. Nitrogen-rich sewage sludge mineralized quickly, improving lettuce nutrition and yield, with reduced risk of heavy metal contamination of soil and plant tissues. *Agronomy* **2024**, *14*, 924. [CrossRef]
21. Afonso, S.; Pereira, E.; Arrobas, M.; Rodrigues, M.A. Recycling nutrient-rich hop leaves by composting with wheat straw and farmyard manure in suitable mixtures. *J. Environ. Manag.* **2021**, *284*, 112105. [CrossRef]
22. Policastro, G.; Cesaro, A. Composting of organic solid waste of municipal origin: The role of research in enhancing its sustainability. *Int. J. Environ. Res. Public Health* **2023**, *20*, 312. [CrossRef]
23. Beesigamukama, D.; Mochoge, B.; Korir, N.; Musyoka, M.W.; Fiaboe, K.; Nakimbugwe, D.; Tanga, C.M. Nitrogen fertilizer equivalence of black soldier fly frass fertilizer and synchrony of nitrogen mineralization for maize production. *Agronomy* **2020**, *10*, 1395. [CrossRef]
24. Beesigamukama, D.; Subramanian, S.; Tanga, C.M. Nutrient quality and maturity status of frass fertilizer from nine edible insects. *Sci. Rep.* **2022**, *12*, 7182. [CrossRef]

25. International Platform of Insects for Food and Feed (IPIFF). *IPIFF Contribution Paper on the Application of Insect Frass as Fertilising Product in Agriculture*; IPIFF: Brussels, Belgium, 2019.
26. EC (European Commission). *Circular Economy Action Plan: The European Green Deal*; European Union: Brussels, Belgium, 2020.
27. Pickett, J.A. Food security: Intensification of agriculture is essential, for which current tools must be defended and new sustainable technologies invented. *Food Energy Secur.* **2013**, *2*, 167–173. [CrossRef]
28. Liu, T.; Bruins, R.J.F.; Heberling, M.T. Factors Influencing Farmers' Adoption of Best Management Practices: A Review and Synthesis. *Sustainability* **2018**, *10*, 432. [CrossRef]
29. Dzepe, D.; Mbenda, T.K.; Ngassa, G.; Mube, H.; Chia, S.Y.; Aoudou, Y.; Djouaka, R. Application of black soldier fly frass, *Hermetia illucens* (Diptera: Stratiomyidae) as sustainable organic fertilizer for lettuce, *Lactuca sativa* production. *Open J. Appl. Sci.* **2022**, *12*, 1632–1648.
30. Instituto Português do Mar e da Atmosfera (IPMA). Normais Climatológicas. 2024. Available online: https://www.ipma.pt/pt/oclima/normais.clima/ (accessed on 15 April 2024).
31. Meier, U. *Growth Stages of Mono and Dicotyledonous Plants*; Federal Biological Research Centre for Agriculture and Forestry: Berlin, Germany, 2018.
32. Temminghoff, E.E.; Houba, V.J. *Plant Analysis Procedures*, 2nd ed.; Kluwer Academic Publishers: London, UK, 2004. [CrossRef]
33. Baird, R.B.; Eaton, A.D.; Rice, E.W. Nitrate by ultraviolet spectrophotometric method. In *Standard Methods for the Examination of Water and Wastewater*; American Public Health Association, American Water Works Association, Water Environment Federation: Washington, DC, USA, 2017.
34. Van Reeuwijk, L.P. *Procedures for Soil Analysis*, 6th ed.; Technical Paper 9; ISRIC; FAO: Rome, Italy, 2002.
35. FAO. *Standard Operating Procedure for Soil Available Micronutrients (Cu, Fe, Mn, Zn) and Heavy Metals (Ni, Pb, Cd), DTPA Extraction Method*; FAO: Rome, Italy, 2022; Available online: https://www.fao.org/3/cc0048en/cc0048en.pdf (accessed on 9 April 2024).
36. Yahbi, M.; Nabloussi, A.; Maataoui, A.; El Alami, N.; Boutagayout, A.; Daoui, K. Effects of nitrogen rates on yield, yield components, and other related attributes of different rapeseed (*Brassica napus* L.) varieties. *OCL* **2022**, *29*, 8. [CrossRef]
37. Aghabeygi, M.; Dönmez, C. Estimating yield response functions to nitrogen for annual crops in Iran. *Agronomy* **2024**, *14*, 436. [CrossRef]
38. Houben, D.; Daoulas, G.; Faucon, M.P.; Dulaurent, A.M. Potential use of mealworm frass as a fertilizer: Impact on crop growth and soil properties. *Sci. Rep.* **2020**, *10*, 4659. [CrossRef] [PubMed]
39. Myrold, D.D.; Bottomley, P.Y. Nitrogen mineralization and immobilization. In *Nitrogen in Agricultural Systems*; Schepers, J., Raun, W.R., Eds.; Agronomy Monograph No. 49; ASA, CSSA, SSSA: Madison, WI, USA, 2008; pp. 157–172.
40. Cao, Y.; Zhao, F.; Zhang, Z.; Zhu, T.; Xiao, H. Biotic and abiotic nitrogen immobilization in soil incorporated with crop residue. *Soil Tillage Res.* **2020**, *202*, 104664. [CrossRef]
41. Xie, J.; Shi, X.; Zhang, Y.; Wan, Y.; Hu, Q.; Zhang, Y.; Wang, J.; He, X.; Evgenia, B. Improved nitrogen use efficiency, carbon sequestration and reduced environmental contamination under a gradient of manure application. *Soil Tillage Res.* **2022**, *220*, 105386. [CrossRef]
42. Bryson, G.M.; Mills, H.A.; Sasseville, D.N.; Jones, J.J., Jr.; Barker, A.V. *Plant Analysis Handbook II: A Guide to Sampling, Preparation, Analysis, Interpretation and Use of Results of Agronomic and Horticultural Crop Plant Tissue*; Micro-Macro Publishing, Inc.: Athens, GA, USA, 2014.
43. Havlin, J.L.; Beaton, J.D.; Tisdale, S.L.; Nelson, W.L. *Soil Fertility and Fertilizers: An Introduction to Nutrient Management*, 8th ed.; Pearson, Inc.: Chennai, India, 2017.
44. Rodrigues, M.Â.; Torres, L.N.D.; Damo, L.; Raimundo, S.; Sartor, L.; Cassol, L.C.; Arrobas, M. Nitrogen use efficiency and crop yield in four successive crops following application of biochar and zeolites. *J. Soil Sci. Plant Nutr.* **2021**, *21*, 1053–1065. [CrossRef]
45. Almeida, D. *Manual de Culturas Hortícolas*, 2nd ed.; Editorial Presença: Queluz de Baixo, Portugal, 2006; Volume 1.
46. Li, G.; Huang, G.; Li, H.; van Ittersum, M.K.; Leffelaar, P.A.; Zhang, F. Identifying potential strategies in the key sectors of China's food chain to implement sustainable phosphorus management: A review. *Nutr. Cycl. Agroecosyst.* **2016**, *104*, 341–359. [CrossRef]
47. Tian, D.; Li, Z.; O'Connor, D.; Shen, Z. The need to prioritize sustainable phosphate-based fertilizers. *Soil Use Manag.* **2020**, *36*, 351–354. [CrossRef]
48. Du, M.; Zhang, W.; Gao, J.; Liu, M.; Zhou, Y.; He, D.; Zhao, Y.; Liu, S. Improvement of root characteristics due to nitrogen, phosphorus, and potassium interactions increases rice (*Oryza sativa* L.) yield and nitrogen use efficiency. *Agronomy* **2022**, *12*, 23. [CrossRef]
49. Su, L.; Bai, T.; Qin, X.; Yu, H.; Wu, G.; Zhao, Q.; Tan, L. Organic manure induced soil food web of microbes and nematodes drive soil organic matter under jackfruit planting. *Appl. Soil Ecol.* **2021**, *166*, 103994. [CrossRef]
50. Hua, W.; Luo, P.; An, N.; Cai, F.; Zhang, S.; Chen, K.; Yang, J.; Han, X. Manure application increased crop yields by promoting nitrogen use efficiency in the soils of 40-year soybean-maize rotation. *Sci. Rep.* **2020**, *10*, 14882. [CrossRef] [PubMed]
51. Dimande, P.; Arrobas, M.; Rodrigues, M.Â. Effect of bat guano and biochar on okra yield and some soil properties. *Horticulturae* **2023**, *9*, 728. [CrossRef]
52. Poveda, J.; Jiménez-Gómez, A.; Saati-Santamaría, Z.; Usategui-Martín, R.; Rivas, R.; García-Fraile, P. Mealworm frass as a potential biofertilizer and abiotic stress tolerance-inductor in plants. *Appl. Soil Ecol.* **2019**, *142*, 110–122. [CrossRef]
53. Fuertes-Mendizábal, T.; Salcedo, I.; Huérfano, X.; Riga, P.; Estavillo, J.M.; Ávila Blanco, D.; Duñabeitia, M.K. Mealworm frass as a potential organic fertilizer in synergy with pgp-based biostimulant for lettuce plants. *Agronomy* **2023**, *13*, 1258. [CrossRef]

54. Menino, R.; Felizes, F.; Castelo-Branco, M.A.; Fareleira, P.; Moreira, O.; Nunes, R.; Murta, D. Agricultural value of black soldier fly larvae frass as organic fertilizer on ryegrass. *Heliyon* **2021**, *7*, e05855. [CrossRef]
55. Chavez, M.Y.; Uchanski, M.; Tomberlin, J.K. Impacts of black soldier fly, *Hermetia illucens*, larval frass on lettuce and arugula production. *Front. Sustain. Food Syst.* **2024**, *8*, 1399932. [CrossRef]
56. Carroll, A.; Fitzpatrick, M.; Hodge, S. The effects of two organic soil amendments, biochar and insect frass fertilizer, on shoot growth of cereal seedlings. *Plants* **2023**, *12*, 1071. [CrossRef]
57. Hawkesford, M.J.; Cakmak, I.; Coskun, D.; De Kok, L.J.; Lambers, H.; Schjoerring, J.K.; White, P.J. Functions of macronutrients. In *Marschner's Mineral Nutrition of Higher Plants*, 4th ed.; Rengel, Z., Cakmak, I., White, P.J., Eds.; Academic Press: Cambridge, MA, USA; Elsevier: London, UK, 2023; pp. 201–281. [CrossRef]
58. Conversa, G.; Bonasia, A.; Lazzizera, C.; La Rotonda, P.; Elia, A. Reduction of nitrate content in baby-leaf lettuce and *Cichorium endivia* through the soilless cultivation system, electrical conductivity and management of nutrient solution. *Front. Plant Sci.* **2021**, *12*, 645671. [CrossRef]
59. Luetic, S.; Knezovic, Z.; Jurcic, K.; Majic, Z.; Tripkovic, K.; Sutlovic, D. Leafy vegetable nitrite and nitrate content: Potential health effects. *Foods* **2023**, *12*, 1655. [CrossRef]
60. Arrobas, M.; de Almeida, S.F.; Raimundo, S.; da Silva Domingues, L.; Rodrigues, M.Â. Leonardites rich in humic and fulvic acids had little effect on tissue elemental composition and dry matter yield in pot-grown olive cuttings. *Soil Syst.* **2022**, *6*, 7. [CrossRef]
61. Arrobas, M.; Andrade, M.; Raimundo, S.; Mazaro, S.M.; Rodrigues, M.A. Lettuce response to the application of two commercial leonardites and their effect on soil properties in a growing medium with nitrogen as the main limiting factor. *J. Plant Nutr.* **2023**, *46*, 4280–4294. [CrossRef]
62. Silva, E.; Arrobas, M.; Gonçalves, A.; Martins, S.; Raimundo, S.; Pinto, L.; Brito, C.; Moutinho-Pereira, J.; Correia, C.M.; Rodrigues, M.A. A controlled-release fertilizer improved soil fertility but not olive tree performance. *Nutr. Cycl. Agroecosyst.* **2021**, *120*, 1–15. [CrossRef]
63. Godbold, D.L.; Hoosbeek, M.R.; Lukac, M.; Cotrufo, M.F.; Janssens, I.A.; Ceulemans, R.; Polle, A.; Velthorst, E.J.; Scarascia-Mugnozza, G.; Angelis, P.; et al. Mycorrhizal hyphal turnover as a dominant process for carbon input into soil organic matter. *Plant Soil* **2006**, *281*, 15–24. [CrossRef]
64. Högberg, M.N.; Skyllberg, U.; Högberg, P.; Knicker, H. Does ectomycorrhiza have a universal key role in the formation of soil organic matter in boreal forests? *Soil Biol. Biochem.* **2020**, *140*, 107635. [CrossRef]

horticulturae

Article

Applicability Analysis of Peanut Addition to Button Mushroom Substrate Supplement Formulation

Cinthia Elen Cardoso Caitano [1], Wagner Gonçalves Vieira Júnior [1], Lucas da Silva Alves [1], Pedro Afonso Gomes Teixeira [1], Laura Cristina de Paula [2] and Diego Cunha Zied [2,*]

1 Programa de Pós-Graduação em Microbiologia Agropecuária, Faculdade de Ciências Agrárias e Veterinárias, Universidade Estadual Paulista (UNESP), Jaboticabal 14884-900, Brazil; cinthia.cardoso@unesp.br (C.E.C.C.); vieira.jr@unesp.br (W.G.V.J.); silva.alves@unesp.br (L.d.S.A.); afonso.gomes@unesp.br (P.A.G.T.)

2 Departamento de Produção Vegetal, Faculdade de Ciências Agrárias e Tecnológicas, Universidade Estadual Paulista (UNESP), Dracena 17915-899, Brazil; lc.paula@unesp.br

* Correspondence: diego.zied@unesp.br

Abstract: This study evaluated the agronomic response of *Agaricus bisporus* strains supplemented with various formulations containing soybean meal and peanut grain residue, which has traces of aflatoxins. Six supplement formulations were developed, starting with 100% soybean meal and no peanuts and gradually increasing peanut inclusion by 10% while proportionally reducing soybean meal until reaching a 50% soybean and 50% peanut ratio. The substrate was produced using the traditional method and supplemented at two points: during inoculation and when adding the cover layer. The strains ABI 22/01 and ABI 22/02 were utilized. After the supplementation and incubation periods in a controlled environment (19 ± 2 °C and 85 ± 5% humidity), the fungus was cultivated. Three production cycles were conducted, evaluating yield, weight, number of mushrooms, biological efficiency, and precocity. The concentrations of aflatoxins in the supplements and mushrooms were determined, along with the nutritional characterization of the substrate and supplements. The inclusion of up to 30% peanuts in the supplement formulation was beneficial for yield, particularly for ABI 22/01. The formulation with 80% soybean meal and 20% peanuts resulted in a 53% yield increase compared to the control. Traces of aflatoxin were found in the supplement but not in the mushrooms.

Keywords: *Agaricus bisporus*; yield; aflatoxins; supplementation; strains

1. Introduction

Edible mushrooms have been increasingly produced on an industrial scale in recent years [1], driven by the growing global population and the demand for nutritionally rich and healthy foods [2]. Among the cultivated species, *Agaricus bisporus* stands out as one of the most significant mushrooms, and it is highly valued for its nutritional properties and culinary uses [3]. The cultivation of this fungus requires, initially, the preparation of a composted substrate, followed by pasteurization [4]. This production process involves the use of biomass as a base substrate, which is often composed of agricultural residues such as straw from grasses, rice, wheat, and others. A critical factor in substrate formulation is the carbon-to-nitrogen (C/N) ratio, which determines the balance of carbon and nitrogen available for the fungus's development [5].

In this context, substrate supplementation emerges as a technology aimed at providing the fungus with an additional source of nutrients, primarily nitrogen, phosphorus, and other essential micronutrients, thereby promoting optimal conditions for its growth and development. Various approaches to supplementation are available for the cultivation of *A. bisporus*, including the addition of supplements during compost inoculation and the application of supplements after fungal colonization but before the addition of the casing layer [6,7].

The wide genetic variability among different mushroom strains leads to distinct behaviors, which are influenced by factors such as environmental conditions and the cultivation region [8]. This genetic diversity can affect crucial aspects, such as productivity, disease resistance, harvest time, and mushroom quality, making the selection of the appropriate strain a strategic factor for optimizing the results and adapting cultivation to specific local conditions [8–10].

There is a wide variety of commercial supplements available for mushroom production; however, most of these products are costly [11] and present logistical challenges, particularly in regions with low availability and production. In this context, the scientific community has been exploring alternatives for developing supplements using regional by-products that can achieve equivalent or superior results in terms of fungal productivity. Various by-products, such as soybean meal, cottonseed meal, pistachio meal, and bean meal, can be employed as supplements [8,10]. Moreover, the reuse of agricultural waste for supplementation has proven to be a promising technique, promoting the sustainable management of residues [12].

Peanut (*Arachis hypogaea*) is a crop of significant relevance to the food sector and is widely cultivated in tropical and subtropical regions [13]. However, its production generates a considerable amount of agricultural waste [14,15]. Among these residues are the grains, which, when contaminated by fungi, contain aflatoxins—substances toxic to humans—primarily produced by Aspergillus parasiticus and Aspergillus flavus, rendering the grains unsuitable for consumption due to health risks [16]. Conversely, mushrooms possess enzymatic complexes capable of degrading a wide range of molecules [17–20]. Furthermore, mushrooms require an adequate protein source for their development [21], which can be provided by supplementing these grains. Subsequently, the mushrooms can be utilized for human consumption [22] if they do not contain traces of toxins.

In this context, the objective of this study is to investigate the use of peanut grains contaminated with aflatoxins as a protein source for the supplementation of *A. bisporus* cultivation. This study aims to evaluate the final quality of the produced mushrooms and ensure their safety for human consumption.

2. Materials and Methods

The experiment was conducted at the facilities of the Mushroom Studies Center (CECOG) located at the Faculty of Agricultural and Technological Sciences (UNESP), Dracena, SP, Brazil.

2.1. Strains

The *Agaricus bisporus* strains used are currently deposited in the public culture collection of CECOG under codes ABI 22/01 and ABI 22/02. The inoculum used was produced following the methodology of [8], involving a primary and secondary matrix cultivated on Petri dishes containing a Potato Dextrose Agar (PDA) medium followed by propagation in sterilized, autoclaved sorghum grains with the addition of 2% calcite.

2.2. Substrate and Supplements

The substrate was produced by the company Compobras®, which is located in the city of Castro (Paraná—Brazil), following the composting stages, including Phase I, which involved turning for oxygen renewal (18 days), and Phase II (6 days) with pasteurization and conditioning of the substrate. Chemical analysis of the substrate was conducted according to the Manual of Official Analytical Methods for Fertilizers and Amendments, revealing a substrate C/N ratio of 21:1 and a nitrogen content of 2.3%.

The supplements were formulated using peanuts and soybean meal. Six formulations were obtained, denoted by abbreviations representing their composition, with 'S' for soybean and 'P' for peanut according to the percentage of materials in the composition as follows: 100S with 100% soybean meal, 90S10P with 90% soybean meal and 10% peanut, 80S20P with 80% soybean meal and 20% peanut, 70S30P with 70% soybean meal and 30%

peanut, 60S40P with 60% soybean meal and 40% peanut, and finally, formulation 50S50P with 50% soybean meal and 50% peanut. All formulations were supplemented with 2% calcite to adjust the pH. The commercial mushroom supplement Promycel Gold® (Amycel, Ittervoort, The Netherlands) was also used as a treatment.

For the preparation of the supplements, the peanuts were ground in a blender and then sieved to obtain fragments smaller than 3 mm. Disinfection of the formulated supplements was carried out using formaldehyde with a concentration of 6000 ppm, following the methodology of [23]. The supplements were sealed in polypropylene bags together with the formalin solution and, after 24 h, opened for evaporation of the product. The supplements were subjected to analysis according to the Manual of Official Analytical Methods for Fertilizers and Amendments to verify the composition of macro- and micronutrients. Table 1 presents the composition of the substrate and supplements used.

Table 1. Chemical analysis of the substrate and supplements for the production of *Agaricus bisporus*.

Materials	N	P_2O_5	K_2O	Ca	Mg	S	M.O.	Rel. C/N	Na	Cu	Fe	Mn	Zn	pH
	% (Natural)								mg/kg (Natural)					Natural
Substrate	2.34	0.91	2.88	3.61	0.44	0.86	89	21/1	1899	120	1908	137	118	7.3
Supplements														
100s	5.6	0.98	2	0.7	0.4	0.2	66	7	100	11	163	27	40	6
90s10p	5.7	1.14	2.4	0.8	0.5	0.2	80	7	124	9	184	25	41	6.6
80s20p	5.5	0.96	1.9	0.7	0.4	0.1	76	7	105	8	432	25	37	6.2
70s30p	6.1	1.05	2.1	0.7	0.4	0.1	83	7	109	9	475	25	37	6
60s40p	5.6	0.86	1.8	0.6	0.3	0	86	7	110	7	200	22	38	6
50s50p	5.7	0.96	1.9	0.7	0.4	0.1	86	8	110	7	200	22	39	6.1
Promycel	6.8	0.34	1.4	0.6	0.2	0.3	87	6	na	11	197	18	67	5.7

The letters S and P in the supplement nomenclature indicate soybean and peanut, respectively. The numbers next to each letter represent the percentage of each product in the supplement formulation.

2.3. Cultivation

The inoculation process was conducted in a controlled environment (greenhouse) using a 1% inoculum. Each plastic box (40 × 30 × 30 cm), containing 5 kg of compost, received 50 g of inoculum. A portion of the boxes was supplemented according to the treatment described above, while the remaining boxes received no supplementation. The compost was then incubated for 15 days at 25 ± 2 °C and 85 ± 5% relative humidity. After complete colonization, the experimental units that had not been supplemented during inoculation were supplemented. For this, the compost was fragmented, and the supplement was added and homogenized, after which all experimental units were covered with a casing layer. A covering layer consisting of peat with a thickness of 4cm was added above the compost.

Subsequently, the experimental units continued under the temperature and humidity regime for the colonization of the casing layer. Seven days after applying the casing layer, the primordia induction process began. This involved gradually lowering the temperature from 26 ± 2 °C to 19 ± 2 °C, with a reduction of one degree per day. The humidity was kept constant at 85 ± 5%.

2.4. Analyses of Aflatoxins

Formulated supplements with any percentage of peanuts and the resulting mushrooms from cultivation with these supplements were analyzed for aflatoxins. The samples were dried and ground. Supplement samples were collected prior to mixing with compost and mushrooms during the first and second flushes.

An adapted methodology from Kamal et al. [24] was used for the analysis, conducted in collaboration with Cooperativa Casul (Junqueirópolis, SP, Brazil). The Aflatest kit was utilized for the analysis, which consists of affinity columns and a revealing solution for reading in a fluorometer.

The materials were homogenized in a blender with deionized water, methanol, and sodium chloride. They were filtered, and the content was diluted in a 1:1 ratio. The liquid was subjected to a second vacuum filtration before passing through the affinity column. The obtained content was supplemented with 1ml of the revealing solution and read in a VICAM Series-4 Fluorometer.

2.5. Variables Analyzed

The mushrooms were harvested over three production flushes at their optimal harvest point, just before the veil ruptured. Based on the data for mushroom mass and quantity, the values for average mass, mushroom count, yield, biological efficiency, and precocity were determined [25].

Yield was calculated by dividing the weight of the harvested mushrooms by the fresh weight of the substrate and multiplying by 100 to obtain the result in percentage. Biological efficiency was calculated by dividing the weight of the harvested mushrooms by the dry weight of the substrate and multiplying by 100 to obtain the result in percentage. The average weight of the mushrooms is calculated by dividing the weight of the mushrooms by the number of mushrooms harvested. Precocity was obtained by separating the harvesting time into two equal periods and dividing the harvested mass of the period by the total harvested weight. The number of mushrooms was determined by counting the mushrooms during cultivation.

2.6. Statistical Analysis

Two experiments were conducted with two different strains. The experiments were carried out in a double factorial design (supplement vs. supplementation time), totaling 16 treatments each with 5 repetitions per treatment. The repetitions consisted of boxes with 5 kg of substrate for the cultivation of button mushrooms. The obtained data were subjected to analysis of variance (ANOVA) and a Scott–Knott mean comparison test at 5% probability using SISVAR 5.6 software [26].

A Principal Component Analysis (PCA) was conducted between the composition parameters of the supplements and the production parameters of the cultivation. The PCA was performed using R software version 4.3.3 (Auckland, New Zealand) [27].

3. Results

For the productivity data of strain ABI 22/01 (Figure 1), the best supplements were those with up to 30% peanut in the supplement composition (100S, 90S10P, 80S20P, and 70S30P) and Promycel added at the time of inoculation. This was also observed in the second and third harvesting cycles.

Supplement 80S20P, which showed the highest productivity result (31%), also performed better in each cycle whenever added to the substrate at the time of inoculation. When added to the substrate after its colonization, the supplements did not show any statistically significant differences in yield within each individual flush. However, when considering the total productivity (sum of all flushes), it was found that supplements 90s10p, 70s30p, and 50s50p stood out with higher average yields.

The data on the average weight and number of mushrooms (Table 2) showed differences between application times only for the 80S20P supplement, with greater weight mushrooms observed when the supplement was added at the time of applying the casing layer. Among the supplements, greater weight mushrooms were seen with 60S40P, 50S50P, and Promycel Gold® when added during inoculation, which also resulted in fewer mushrooms (10.28 g, 11.75g, and 10.02g, and 83, 90.4, and 120.4 mushrooms, respectively). The precocity of the ABI 22/01 strain showed no statistical differences for any of the factors evaluated.

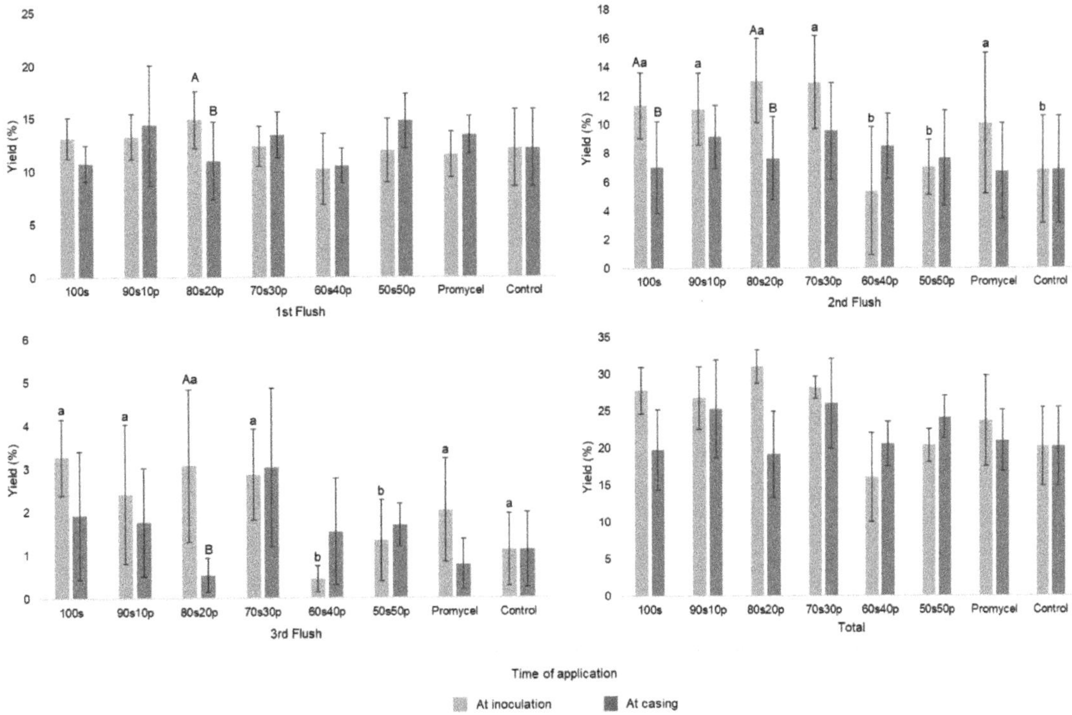

Figure 1. Productivity of strain ABI 22/01 in the first, second, third flow, and total supplemented at inoculation and casing. Different lowercase letters within the same flow indicate a difference between the supplements according to the Scott–Knott statistical test at 5% probability. Different uppercase letters indicate statistical differences between the moments of supplement application according to the Scott–Knott statistical test at 5% probability. The absence of letters indicates no statistical difference.

Table 2. Biological efficiency (%), number of mushrooms (unit), average mushroom weight (g), and precocity (%) of the ABI 22/01 strain supplemented at the time of inoculation and the time of addition of the covering layer.

Time of Application	Supplement							
	100S	**90S10P ***	**80S20P**	**70S30P**	**60S40P**	**50S50P**	**Promycel**	**Control**
				Biological Efficiency, %				
At inoculation	85.15 ± 9.76 Aa	82.19 ± 13.02 a	95.09 ± 6.84 Aa	86.36 ± 4.80 a	49.20 ± 18.36 b	62.34 ± 6.68 b	72.60 ± 18.84 a	61.92 ± 16.22 b
At casing	60.56 ± 16.61 Bb	77.58 ± 20.24 a	58.82 ± 17.73 Bb	79.85 ± 18.57 a	63.03 ± 9.22 b	74.02 ± 8.88 a	64.34 ± 12.67 b	61.92 ± 16.22b
				Weight of mushrooms, g				
At inoculation	8.59 ± 1.32 b	9.15 ± 1.08 b	7.44 ± 1.17 Ab	7.73 ± 0.99 b	10.28 ± 1.86 a	11.75 ± 2.90 a	10.02 ± 1.07 a	7.82 ± 1.46 b
At casing	10.95 ± 2.49 b	9.40 ± 1.35 b	14.36 ± 3.92 Ba	9.68 ± 1.68 b	10.90 ± 1.54 b	10.23 ± 1.39 b	10.92 ± 1.64 b	7.82 ± 1.46 b
				Number of mushrooms, un				
At inoculation	166.8 ± 44.64 Aa	147.8 ± 29.26 a	213 ± 42.63 Aa	184 ± 23.20 a	83 ± 44.05 b	90.4 ± 32.28 b	120.4 ± 40.43 b	129.6 ± 24.18 b
At casing	99.6 ± 52.79 Bb	138.4 ± 47.20 a	70 ± 22.04 Bb	141.2 ± 54.81 a	97.6 ± 30.98 b	120.8 ± 29.28 a	98.8 ± 27.95 b	129.6 ± 24.18 a
				Precocity, %				
At inoculation	69.51 ± 9.68	72.93 ± 17.01	70.05 ± 14.57	77.8 ± 8.16	89 ± 8.99	78.51 ± 11.16	86.53 ± 6.20	80.18 ± 15.03
At casing	75.96 ± 15.72	84.52 ± 10.41	79.65 ± 11.96	87.27 ± 7.97	86.49 ± 3.89	78.08 ± 6.00	91.49 ± 8.88	80.18 ± 15.03

Averages ± standard deviation. Different lowercase letters in the same row indicate a difference between the supplements according to the Scott–Knott statistical test at a 5% probability level. Different uppercase letters indicate statistical differences between the application moments of the supplementation according to the Scott–Knott statistical test at a 5% probability level. The absence of letters indicates no statistical difference. * The letters S and P in the supplement nomenclature indicate soybean and peanut, respectively. The numbers next to each letter represent the percentage of each product in the supplement formulation.

The productivity of strain ABI 22/02 (Figure 2) was only affected by supplementation in the second cycle, where supplements 70S30P, 60S40P, and Promycel added at the time of inoculation, as well as the control, stood out (5.41%, 4.82%, 7.55%, and 5.38%, respectively). When supplemented together with the covering layer, this strain was more responsive, with all treatments performing better than the control, except for supplement 90S10P.

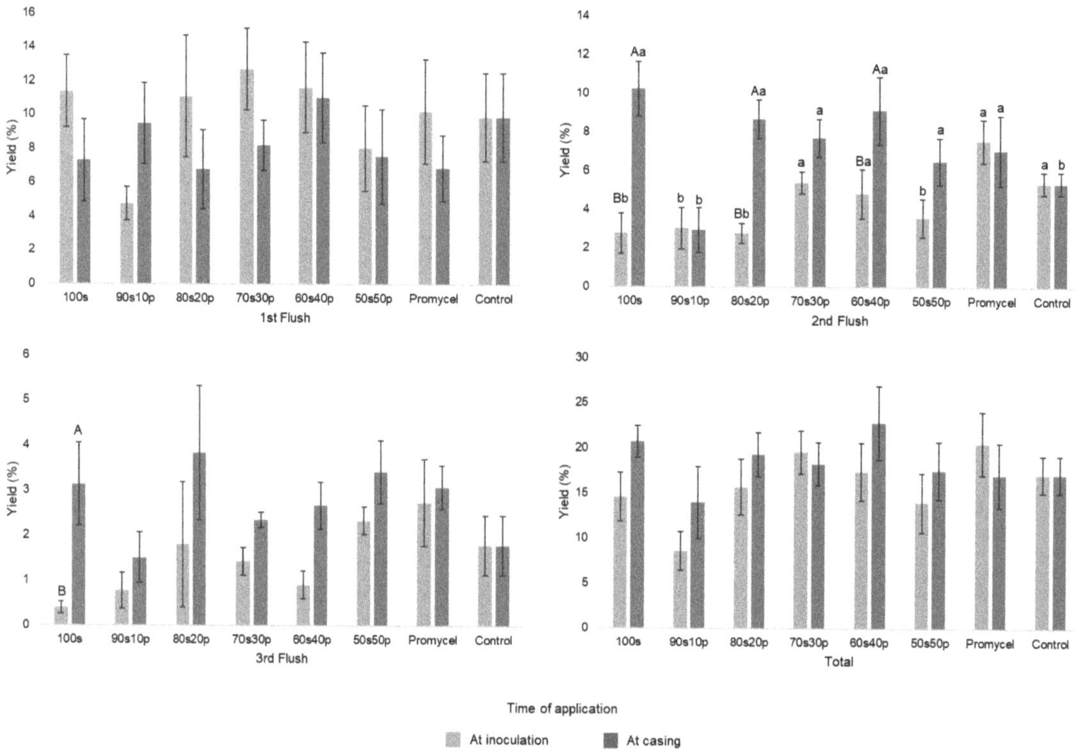

Figure 2. Productivity of strain ABI 22/02 in the first, second, third flow, and total supplemented at the moment of inoculation and at the moment of adding the covering layer. Different lowercase letters within the same flow indicate a difference between the supplements according to the Scott–Knott statistical test at 5% probability. Different uppercase letters indicate statistical differences between the moments of supplement application according to the Scott–Knott statistical test at 5% probability. The absence of letters indicates no statistical difference.

The ABI 22/02 strain demonstrated a positive response to supplementation, as observed in the analysis of the average mushroom mass (Table 3). Specifically, the 90S10P supplement formulation, applied at the time of inoculation, resulted in an average mushroom mass of 23.58 g, indicating superior efficacy compared to other supplementation practices. Additionally, the addition of the supplement at the time of inoculation also positively influenced the earliness of cultivation, leading to faster mushroom development. These findings suggest that synchronizing supplementation with the inoculation process may be a crucial factor in maximizing productive performance, optimizing not only the mushroom mass but also accelerating the time to harvest. Such practice can be particularly useful in commercial operations, where production time and uniformity of the final product are critical factors.

Table 3. Biological efficiency (%), number of mushrooms (unit), average mushroom weight (g), and precocity (%) of the ABI 22/02 strain supplemented at the time of inoculation and at the time of addition of the covering layer.

Time of Application	Supplement							
	100S	90S10P *	80S20P	70S30P	60S40P	50S50P	Promycel	Control
				Biological Efficiency, %				
At inoculation	44.63 ± 18.61	26.24 ± 14.59	48 ± 21.13	59.9 ± 16.27	53.23 ± 22.03	42.74 ± 20.63	62.88 ± 24.03	52.3 ± 14.40
At casing	63.39 ± 11.87	42.66 ± 27.62	59.1 ± 16.49	55.88 ± 16.58	69.89 ± 27.98	53.63 ± 21.95	52.05 ± 24.65	52.3 ± 14.40
				Weight of mushrooms, g				
At inoculation	13.25 ± 1.67 b	23.58 ± 14.47 Aa	12.77 ± 2.36 b	12.12 ± 2.51 b	12.87 ± 0.68 b	16.79 ± 5.30b	13.67 ± 3.98b	15.63 ± 3.25 b
At casing	14.7 ± 2.92	15.37 ± 3.97 B	16.16 ± 5.18	15.96 ± 2.29	13.48 ± 2.17	14.46 ± 2.37	11.35 ± 1.07	15.63 ± 3.25
				Number of mushrooms, un				
At inoculation	57.4 ± 26.28	34.2 ± 13.14	68.6 ± 40.82	85.8 ± 33.70	68.4 ± 30.24	60 ± 33.49	84.4 ± 44.93	54.4 ± 29.58
At casing	73.2 ± 23.30	57.6 ± 45.09	64.6 ± 27.96	61.4 ± 26.96	91.8 ± 54.09	64 ± 35.64	77.2 ± 35.64	54.4 ± 29.58
				Precocity, %				
At inoculation	61.51 ± 22.06Aa	39.54 ± 31.08 Ba	54.72 ± 33.51 Aa	47.02 ± 16.51 a	53.94 ± 11.54 Aa	20.12 ± 36.68 b	23.39 ± 11.96 b	43.99 ± 18.49 a
At casing	27.44 ± 22.38 Bc	68.33 ± 12.23 Aa	20.26 ± 9.66 Bc	28.84 ± 5.35 c	28.76 ± 4.04 Bc	26.51 ± 19.04 c	16.87 ± 12.28 c	43.99 ± 18.49 b

Averages ± standard deviation. Different lowercase letters in the same line indicate a difference between the supplements according to the Scott–Knott statistical test at 5% probability. Different uppercase letters indicate statistical differences between the moments of supplement application according to the Scott–Knott statistical test at 5% probability. The absence of letters indicates no statistical difference. * The letters S and P in the nomenclature of the supplements indicate soybean and peanut, respectively. The numbers next to each letter represent the percentage of each product in the supplement formulation.

The results of the Principal Component Analysis (PCA) revealed that Principal Component 1 (PC1) accounted for the majority of the variation in the data, totaling 53.65% of the total variance. Principal Component 2 (PC2) represented 24.97%, thus raising the representativeness of the analysis to 78.62% (Figure 3). The correlation between the principal components and the analyzed variables showed that PC1 was influenced by the nutritional data of the supplements, while PC2 was influenced by the agronomic data.

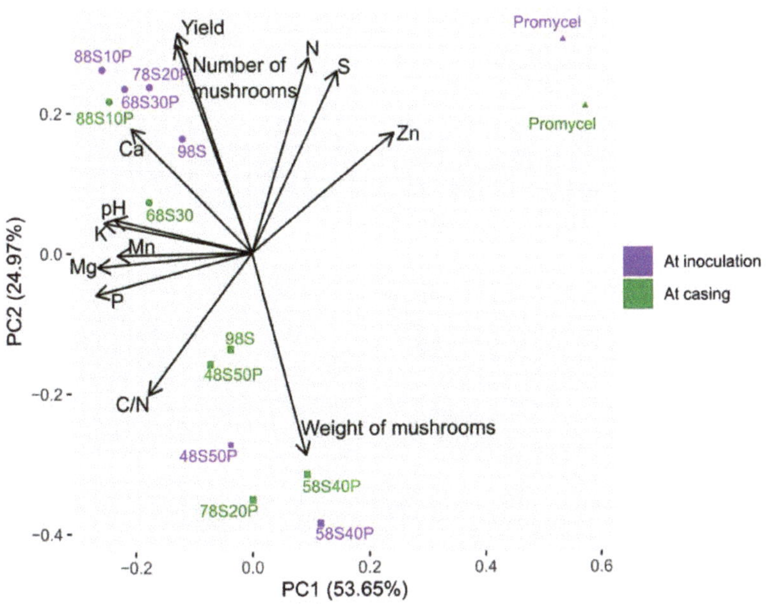

Figure 3. PCA biplot of the productivity parameters of strain ABI 22/01 and nutrient content of the supplements added at the time of inoculation and at the time of addition of the covering layer. The geometric shapes represent the clusters formed by the analysis.

The Principal Component Analysis allowed the categorization of treatments into three distinct groups for each evaluated strain. The first grouping, related to strain ABI 22/01, was primarily influenced by productivity and the number of mushrooms. This group included supplements 100S, 90S10P, 80S20P, and 70S30P, which were added at the time of inoculation, as well as supplements 90S10P and 70S30P, which were incorporated during the application of the substrate covering layer. The second group is more associated with the C/N ratio of the supplements and the average mushroom mass. This grouping encompassed supplements 60S40P and 50S50P, which were added at the time of inoculation, and supplements 100S, 80S20P, 60S40P, and 50S50P, which were applied during the addition of the substrate covering layer.

Lastly, the third group was composed of the Promycel supplement, in both application times, which was primarily influenced by nitrogen, zinc, and sulfur content.

For the ABI 22/02 strain, Principal Component Analysis (PCA) explained 80.69% of the total variation observed in the data. The First Principal Component (PC1) accounted for 57.58% of the variation, while the Second Principal Component (PC2) represented 23.11% (Figure 4). These results indicate that the majority of the variability in the data can be attributed to the first two principal components, providing a detailed insight into the main factors influencing the characteristics of the ABI 22/02 strain.

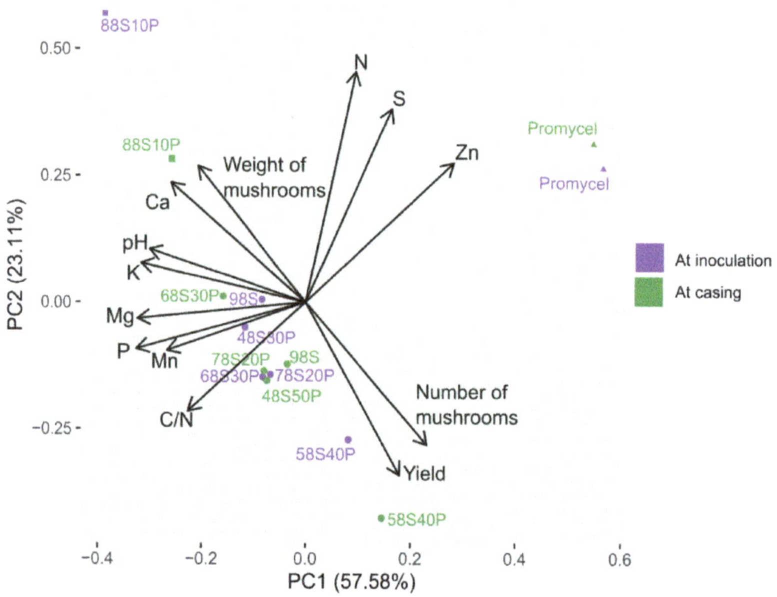

Figure 4. PCA biplot of the productive parameters of strain ABI 22/02 and nutrient content of supplements added at inoculation and at the addition of the cover layer. The geometric shapes represent the clusters formed by the analysis.

The treatments were also grouped into three groups. The first group, composed of supplement 90S10P at both application times, was particularly influenced by the average mushroom mass and calcium content in the supplement. The second group, characterized by the presence of the Promycel supplement at both application times, was influenced by the levels of zinc, sulfur, and nitrogen in the supplements used. Finally, the third group encompassed the remaining treatments that were shown to be influenced by nutrients and productivity.

All materials analyzed for aflatoxins achieved values lower than those accepted in Brazil by the National Health Surveillance Agency (ANVISA): 16 of 20ppb. These data infer that the aflatoxin was either eliminated from the medium or not accumulated by the mushroom (Table 4).

Table 4. Aflatoxin content (ppb) is presented in the supplements and the cultivated mushrooms.

Formulation	Supplement	Mushroom (Supplementation during Inoculation)	Mushroom (Supplementation in the Casing Layer)
90S10P	1.50	ND *	ND
80S20P	2.00	ND	ND
70S30P	2.00	ND	ND
60S40P	1.50	ND	ND
50S50P	1.20	ND	ND

* ND = Not Detected.

4. Discussion

Although supplementation techniques are widely used in several countries, Brazilian mushroom growers do not find alternatives in the local market for implementing this practice. Therefore, it is necessary to import commercial supplements used abroad.

Supplements can provide proteins, lipids, carboxylic acids, and minerals, each acting specifically in cultivation [6]. Soybean is one of the most studied materials as a protein source for mushroom cultivation [7,28]. One of the reasons for this is the high production and availability, especially in North and South America, which contains the three main producers of this legume: Brazil, the United States, and Argentina [29], but this extends worldwide. In Brazil, a variety of products are used for mushroom supplementation; however, due to the country's vast territorial extent, the availability and utilization of each material may be limited. Among the available resources are peanut residues, soybean meal, cottonseed meal, corn meal [30], commercial supplements [11], acerola residues, and specialty grains [8], among others.

However, soybean is widely used in other areas, such as animal feed production and human consumption. Considering this perspective, peanuts stand out as a rich source of lipids, comprising, on average, about 49% of their composition [31]. This characteristic positions it as a valuable option for supplementing cultivation substrates, expanding the range of choices, and contributing to efficiency and cost savings in the mushroom cultivation process. Thus, it was possible to achieve similar productivity levels by adding peanuts to the supplement formulation, with proportions of up to 30%. Previous studies, such as the one conducted by [8], demonstrated the significant potential of this oilseed in the production of *Pleurotus ostreatus* mushrooms. In their research, supplementation with 80% peanut grains resulted in a 32% increase in productivity compared to the control group. These results indicate that peanuts can be an effective source of nutrients in mushroom cultivation, enhancing crop yields without compromising the quality of the final product.

Studies related to *A. bisporus* report productivities ranging, on average, from 15% to 23% [32,33]. This variation can be attributed to a series of factors, including genetic differences in the strains used, an inadequate substrate [34–36], and the nutritional properties of the substrate employed [37]. Such factors play a crucial role in explaining the diversity of productivities obtained in this study, which ranged from 8.55% to 20.5% for the ABI 22/02 strain and from 16.04% to 31% for the ABI 22/01 strain. This wide range of productivity demonstrates the complexity of *A. bisporus* cultivation and the importance of considering various aspects to optimize performance in production.

Although some numerical variations were observed, particularly in total productivity, supplementation with the ABI 22/02 strain did not produce results that justified the use of this practice. The data indicate that this strain does not respond positively to the added supplements, resulting in no significant increase in production. On the other hand, the ABI 22/01 strain demonstrated a substantial increase in productivity, achieving up to a 53% increment when subjected to the same supplementation practice. Therefore, the choice of strain is a critical factor in optimizing productivity in commercial mushroom cultivation, making it essential to consider the genetic responsiveness of strains to management interventions, such as nutritional supplementation.

The inverse relationship between the average mushroom mass and the number of mushrooms has been documented in previous studies [29,38], and this relationship was also evident in the present study, particularly in the context of the ABI 22/01 strain. This study found that when the treatments resulted in larger mushrooms, there was a corresponding reduction in the number of mushrooms produced, while the opposite also occurred; an increase in the quantity of mushrooms was observed when the average mushroom mass decreased, reflecting the interdependence of these two variables.

In general, larger mushrooms (in terms of weight) tend to be more acceptable to producers, primarily due to the reduced demand for labor. Harvesting a smaller number of units is quicker, resulting in lower labor costs, which are one of the main components of production costs [39]. However, this preference may vary according to the intended use of the mushrooms; for example, mushrooms destined for preservation are preferably smaller [40,41].

Aflatoxins are present in various foods, such as peanuts, powdered milk, and cereals, such as corn [42]. Kamal et al. [24] found aflatoxin levels in canned mushrooms of the genus *Agaricus*; however, fresh mushrooms were free from toxins. White rot fungi can even be allies in the degradation of aflatoxins. In studies conducted by Branà et al. [43,44], the degradation of aflatoxins by mycelium of *Pleurotus eryngii* in liquid and solid media and by laccase and Mn-peroxidase enzymes isolated from an exhausted mushroom substrate was higher than 90%.

It is important to emphasize that mushroom production requires an integrated approach, involving not only nutrients, centesimal elements, and substrate microbiota but also the covering layer. Fat, protein, and carbohydrates present in the substrate reflect on the quality of the mushrooms produced and productivity [10,33]; these elements can be increased through supplementation. The microbiological factors of the compost can be related to productivity, disease control, and other factors [45,46].

The application of the supplementation technique is directly related to increased yields, particularly concerning the control treatment of this study. Furthermore, financial factors must be considered prior to the implementation of this technique, assessing whether the increase in productivity is sufficient to cover the costs of supplement production and operational expenses, including potential investments in infrastructure and increased labor.

5. Conclusions

The incorporation of peanuts into the supplement formulation for *Agaricus bisporus* cultivation can increase productivity by up to 53%, depending on the strain's responsiveness to supplementation. The greatest efficacy is observed when combined with the inoculation practice, without the need to alter the cultivation process. The nutrients in the supplements, such as lipids and proteins, have a limited direct contribution to the harvest outcomes. This effect may be related to the composition and balance of macronutrients and micronutrients, which influence absorption and metabolism by the mycelium. Further studies are needed to better understand these interactions and optimize supplement formulations to maximize productive efficiency.

Author Contributions: Conceptualization, D.C.Z.; data curation, C.E.C.C.; formal analysis, C.E.C.C.; investigation, C.E.C.C., W.G.V.J., L.d.S.A., P.A.G.T. and L.C.d.P.; methodology, C.E.C.C. and D.C.Z.; resources, D.C.Z.; software, C.E.C.C. and W.G.V.J.; supervision, D.C.Z.; visualization, D.C.Z.; writing—original draft, C.E.C.C.; writing—review and editing, W.G.V.J. and D.C.Z. All authors have read and agreed to the published version of the manuscript.

Funding: This research was funded by Coordenação de Aperfeiçoamento de Pessoal de Nível Superior—Brasil (CAPES)—Funding Code 001 and the São Paulo Research Foundation (FAPESP) (grant number 2020/13230-8).

Data Availability Statement: The raw data supporting the conclusions of this article will be made available by the authors upon request.

Conflicts of Interest: The authors declare no conflicts of interest.

References

1. Dedousi, M.; Melanouri, E.-M.; Karayannis, D.; Kaminarides, E.-I.; Diamantopoulou, P. Utilização de Substratos Usados e Resíduos Do Cultivo de Cogumelos Para Produzir Novas Culturas de *Pleurotus ostreatus*, *Pleurotus eryngii* e *Agaricus bisporus*. *Carbon Resour. Convers.* **2024**, *7*, 100196. [CrossRef]
2. Usman, M.; Murtaza, G.; Ditta, A. Nutritional, Medicinal, and Cosmetic Value of Bioactive Compounds in Button Mushroom (*Agaricus bisporus*): A Review. *Appl. Sci.* **2021**, *11*, 5943. [CrossRef]
3. Wasser, S.P.; Akavia, E. Regulatory issues of mushrooms as functional foods and dietary supplements: Safety and efficacy. In *Mushrooms as Functional Foods*; Cheung, P.C.K., Ed.; John Wiley & Sons, Ltd.: Hoboken, NJ, USA, 2008; pp. 199–226.
4. Pardo-Gimenez, A.; Pardo-Gonzalez, J.E.; Cunha-Zied, D. Supplementation of High Nitrogen *Agaricus* Compost: Yield and Mushroom Quality. *J. Agric. Sci. Technol.* **2017**, *19*, 1589–1601.
5. Buth, J. Compost as a Food Base for *Agaricus bisporus*. In *Edible and Medicinal Mushrooms*; Zied, D.C., Pardo-Giménez, A., Eds.; John Wiley & Sons, Ltd.: Hoboken, NJ, USA, 2017; pp. 129–147.
6. Burton, K.; Noble, R.; Rogers, S.; Wilson, J. *Understanding Mushroom Nutri-Tion: Project Aimed at Improving Yield, Substrate Efficiency and Utilisation and Flavor*; M056 Final Report; Agriculture and Horticulture Development Board: Warwickshire, UK, 2015; 54p.
7. Carrasco, J.; Zied, D.C.; Pardo, J.E.; Preston, G.M.; Pardo-Giménez, A. Supplementation in Mushroom Crops and Its Impact on Yield and Quality. *AMB Express* **2018**, *8*, 146. [CrossRef]
8. Zied, D.C.; Caitano, C.E.C.; Pardo-Gimenez, A.; Dias, E.S.; Zeraik, M.L.; Pardo, J.E. Using of Appropriated Strains in the Practice of Compost Supplementation for *Agaricus subrufescens* Production. *Front. Sustain. Food Syst.* **2018**, *2*, 26. [CrossRef]
9. Santos, R.C.; Freire, R.M.M.; de Lima, L.M. *O Agronegócio Do Amendoim No Brasil*, 2nd ed.; Bases de Dados da Pesquisa Agropecuária: Brasilia, Brazil, 2013.
10. Pardo-Giménez, A.; Catalán, L.; Carrasco, J.; Álvarez-Ortí, M.; Zied, D.; Pardo, J. Effect of Supplementing Crop Substrate with Defatted Pistachio Meal on *Agaricus bisporus* and *Pleurotus ostreatus* Production. *J. Sci. Food Agric.* **2016**, *96*, 3838–3845. [CrossRef]
11. Ferrari, A.B.S.; Azevedo de Oliveira, G.; Mannochio Russo, H.; de Carvalho Bertozo, L.; da Silva Bolzani, V.; Zied, D.C.; Farias Ximenes, V.; Zeraik, M.L. *Pleurotus ostreatus* and *Agaricus subrufescens*: Investigation of Chemical Composition and Antioxidant Properties of These Mushrooms Cultivated with Different Handmade and Commercial Supplements. *Int. J. Food Sci. Technol.* **2021**, *56*, 452–460. [CrossRef]
12. Otieno, O.D.; Mulaa, F.J.; Obiero, G.; Midiwo, J. Utilization of Fruit Waste Substrates in Mushroom Production and Manipulation of Chemical Composition. *Biocatal. Agric. Biotechnol.* **2022**, *39*, 102250. [CrossRef]
13. Putra, N.R.; Rizkiyah, D.N.; Che Yunus, M.A.; Abdul Aziz, A.H.; Md Yasir, A.S.H.; Irianto, I.; Jumakir, J.; Waluyo, W.; Suparwoto, S.; Qomariyah, L. Valorization of Peanut Skin as Agricultural Waste Using Various Extraction Methods: A Review. *Molecules* **2023**, *28*, 4325. [CrossRef]
14. Mohammadi Torkashvand, A.; Alidoust, M.; Mahboub Khomami, A. The Reuse of Peanut Organic Wastes as a Growth Medium for Ornamental Plants. *Int. J. Recycl. Org. Waste Agric.* **2015**, *4*, 85–94. [CrossRef]
15. Zhao, X.; Chen, J.; Du, F. Potential Use of Peanut By-Products in Food Processing: A Review. *J. Food Sci. Technol.* **2012**, *49*, 521–529. [CrossRef] [PubMed]
16. Eskola, M.; Kos, G.; Elliott, C.T.; Hajšlová, J.; Mayar, S.; Krska, R. Worldwide Contamination of Food-Crops with Mycotoxins: Validity of the Widely Cited 'FAO Estimate' of 25%. *Crit. Rev. Food Sci. Nutr.* **2020**, *60*, 2773–2789. [CrossRef] [PubMed]
17. Agrawal, N.; Verma, P.; Shahi, S.K. Degradation of Polycyclic Aromatic Hydrocarbons (*Phenanthrene* and *Pyrene*) by the Ligninolytic Fungi *Ganoderma lucidum* Isolated from the Hardwood Stump. *Bioresour. Bioprocess.* **2018**, *5*, 11. [CrossRef]
18. Bankole, P.O.; Adekunle, A.A.; Jeon, B.-H.; Govindwar, S.P. Novel Cobiomass Degradation of NSAIDs by Two Wood Rot Fungi, *Ganoderma applanatum* and *Laetiporus sulphureus*: Ligninolytic Enzymes Induction, Isotherm and Kinetic Studies. *Ecotoxicol. Environ. Saf.* **2020**, *203*, 110997. [CrossRef] [PubMed]
19. Hanafiah, Z.M.; Wan Mohtar, W.H.M.; Hasan, H.A.; Jensen, H.S.; Klaus, A.; Sharil, S.; Wan-Mohtar, W.A.A.Q.I. Ability of *Ganoderma lucidum* Mycelial Pellets to Remove Ammonia and Organic Matter from Domestic Wastewater. *Int. J. Environ. Sci. Technol.* **2022**, *19*, 7307–7320. [CrossRef]
20. Yadav, A.; Rene, E.R.; Kanti Mandal, M.; Kumar Dubey, K. Biodegradation of Cyclophosphamide and Etoposide by White Rot Fungi and Their Degradation Kinetics. *Bioresour. Technol.* **2022**, *346*, 126355. [CrossRef]
21. Noble, R.; Thai, M.; Kertesz, M.A. Nitrogen Balance and Supply in Australasian Mushroom Composts. *Appl. Microbiol. Biotechnol.* **2024**, *108*, 151. [CrossRef]
22. Hong, J.Y.; Kim, M.K.; Yang, N. Mushroom Consumption and Cardiometabolic Health Outcomes in the General Population: A Systematic Review. *Nutr. Res. Pract.* **2024**, *18*, 165–179. [CrossRef]
23. Pardo-Giménez, A.; Zied, D.C.; Álvarez-Ortí, M.; Rubio, M.; Pardo, J.E. Effect of Supplementing Compost with Grapeseed Meal on Production. *J. Sci. Food Agric.* **2012**, *92*, 1665–1671. [CrossRef]

24. Kamal, A.; Khair, A.; Dawlatana, M.; Hassan, M.; Begum, F.; Rahim, M. Evaluation of Aflatoxins and Pesticide Residues in Fresh and Different Processed Mushrooms. *Bangladesh J. Sci. Ind. Res.* **2009**, *44*, 193–198. [CrossRef]
25. Pardo-Giménez, A.; Pardo-González, J.E.; Zied, D.C. Evaluation of Harvested Mushrooms and Viability of *Agaricus bisporus* Growth Using Casing Materials Made from Spent Mushroom Substrate. *Int. J. Food Sci. Technol.* **2011**, *46*, 787–792. [CrossRef]
26. Ferreira, D.F. SISVAR: Um Programa Para Análises e Ensino de Estatística. *Rev. Cient. Symp.* **2008**, *6*, 36–41.
27. R: The R Project for Statistical Computing. Available online: https://www.r-project.org/ (accessed on 12 August 2024).
28. El-Shemy, H. Soybean and Nutrition, 1st ed.IntechOpen: London, UK, 2011.
29. FAO—Food and Agriculture Organizations of the United Nations. FAO Data Base. Available online: https://www.fas.usda.gov/data/world-agricultural-production (accessed on 12 August 2024).
30. Zied, D.C.; Prado, E.P.; Dias, E.S.; Pardo, J.E.; Pardo-Gimenez, A. Use of Peanut Waste for Oyster Mushroom Substrate Supplementation—Oyster Mushroom and Peanut Waste. *Braz. J. Microbiol.* **2019**, *50*, 1021–1029. [CrossRef]
31. Davis, J.P.; Dean, L.L. Peanut composition, flavor and nutrition. In *Peanuts*; Stalker, H.T., Wilson, R.F., Eds.; AOCS Press: Washington, DC, USA, 2016; pp. 289–345.
32. Caitano, C.E.C.; Iossi, M.R.; Pardo-Giménez, A.; Vieira Júnior, W.G.; Dias, E.S.; Zied, D.C. Design of a Useful Diagrammatic Scale for the Quantification of *Lecanicillium fungicola* Disease in *Agaricus bisporus* Cultivation. *Curr. Microbiol.* **2020**, *77*, 4037–4044. [CrossRef] [PubMed]
33. Zied, D.C.; Nunes, J.S.; Nicolini, V.F.; Gimenez, A.P.; Rinker, D.L.; Dias, E.S. Tolerance to *Lecanicillium fungicola* and Yield of *Agaricus bisporus* Strains Used in Brazil. *Sci. Hortic.* **2015**, *190*, 117–122. [CrossRef]
34. Harfi, T.; Alireza, M.-A.; Farzad, R.; Fariborz, Z.-N. Induced Mutation in *Agaricus bisporus* by Gamma Ray to Improve Genetic Variability, Degradation Enzyme Activity, and Yield. *Int. J. Radiat. Biol.* **2021**, *97*, 1020–1031. [CrossRef] [PubMed]
35. Foulongne-Oriol, M.; Rodier, A.; Rousseau, T.; Savoie, J.-M. Quantitative Trait Locus Mapping of Yield-Related Components and Oligogenic Control of the Cap Color of the Button Mushroom, *Agaricus bisporus*. *Appl. Environ. Microbiol.* **2012**, *78*, 2422–2434. [CrossRef]
36. Zied, D.C.; Pardo, J.E.; Noble, R.; Pardo-Giménez, A. Efficiency of Mushrooms for Food Production—Fundamental Strategic Decision-Making. *J. Food Compos. Anal.* **2024**, *125*, 105734. [CrossRef]
37. Wang, Q.; Xiao, T.; Juan, J.; Qian, W.; Zhang, J.; Chen, H.; Shen, X.; Huang, J. Lignocellulose Degradation Efficiency of *Agaricus bisporus* Strains Grown on Wheat Straw-Based Compost. *J. Agric. Food Chem.* **2023**, *71*, 10607–10615. [CrossRef]
38. Kaur, M.; Kaur, S.; Devi, R.; Kapoor, S. Wheat Straw and Maize Stalks Based Compost for Cultivation of *Agaricus bisporus*. *Pharma Innov. J.* **2019**, *8*, 393–397.
39. Li, C.; Xu, S. Edible Mushroom Industry in China: Current State and Perspectives. *Appl. Microbiol. Biotechnol.* **2022**, *106*, 3949–3955. [CrossRef] [PubMed]
40. Dawadi, E.; Magar, P.B.; Bhandari, S.; Subedi, S.; Shrestha, S.; Shrestha, J. Nutritional and Post-Harvest Quality Preservation of Mushrooms: A Review. *Heliyon* **2022**, *8*, e12093. [CrossRef] [PubMed]
41. Verma, A.K.; Shivani, P.C.S.; Kumar, M.; Rani, N. Processing of Mushrooms: A Viable Option to Sustain the Growing Population of the Developing Countries. *Int. J. Chem. Stud.* **2020**, *8*, 1416–1423. [CrossRef]
42. Kwiatkowski, A.; Alves, A.P.F. Importância da detecção e do controle de aflatoxinas em alimentos. *SaBios* **2007**, *2*, 45–54.
43. Branà, M.T.; Cimmarusti, M.T.; Haidukowski, M.; Logrieco, A.F.; Altomare, C. Bioremediation of Aflatoxin B1-Contaminated Maize by King Oyster Mushroom (*Pleurotus eryngii*). *PLoS ONE* **2017**, *12*, e0182574. [CrossRef]
44. Branà, M.T.; Sergio, L.; Haidukowski, M.; Logrieco, A.F.; Altomare, C. Degradation of Aflatoxin B1 by a Sustainable Enzymatic Extract from Spent Mushroom Substrate of *Pleurotus eryngii*. *Toxins* **2020**, *12*, 49. [CrossRef]
45. Suwannarach, N.; Kumla, J.; Zhao, Y.; Kakumyan, P. Impact of Cultivation Substrate and Microbial Community on Improving Mushroom Productivity: A Review. *Biology* **2022**, *11*, 569. [CrossRef]
46. Carrasco, J.; Tello, M.L.; de Toro, M.; Tkacz, A.; Poole, P.; Pérez-Clavijo, M.; Preston, G. Casing Microbiome Dynamics during Button Mushroom Cultivation: Implications for Dry and Wet Bubble Diseases. *Microbiology* **2019**, *165*, 611–624. [CrossRef]

horticulturae

MDPI

Article

Organic Agricultural Practice: Crop Load Management Enhancing Quality and Storability of High-Russet Pears

Marcos Guerra [1], Flor Álvarez-Taboada [2], Verónica Marabel [3], Amanda M. Felices [3], Álvaro Rodríguez-González [4] and Pedro A. Casquero [4,*]

1 Grupo Universitario de Investigación en Ingeniería y Agricultura Sostenible (GUIIAS), Escuela de Ingeniería Agraria y Forestal (EIAF), Campus de Ponferrada, Universidad de León, Avenida de Astorga s/n, 24401 Ponferrada, Spain; mgues@unileon.es
2 DRACONES, Escuela de Ingeniería Agraria y Forestal (EIAF) (Campus de Ponferrada), Universidad de León, Av. Astorga s/n, 24400 Ponferrada, Spain; flor.alvarez@unileon.es
3 Escuela de Ingeniería Agraria y Forestal (EIAF) (Campus de Ponferrada), Universidad de León, Av. Astorga s/n, 24400 Ponferrada, Spain
4 Grupo Universitario de Investigación en Ingeniería y Agricultura Sostenible (GUIIAS), Instituto de Medio Ambiente Recursos Naturales y Biodiversidad (INMARENBIO), Escuela de Ingeniería Agraria y Forestal (EIAF), Universidad de León, Avenida de Portugal 41, 24071 León, Spain; alrog@unileon.es
* Correspondence: pacasl@unileon.es

Abstract: The variability of crop load in Conference pears significantly influences fruit quality and postharvest losses. This study aimed to investigate and implement a novel technique for managing crop load in Conference pear orchards, with a focus on contributing to sustainable orchard management practices. High-russet pear cv. Conference fruit was harvested from trees categorized into two groups based on yield efficiency. A two-way ANOVA was used to test the effects of crop load level and year on yield properties and fruit quality parameters. Results demonstrated that low crop loads exerted a positive influence on fruit quality, resulting in increased fresh-market yield, enhanced fruit-mass parameters, and improved firmness and acidity. Additionally, fruit from low crop load levels exhibited a heightened level of characteristic skin russeting, a desirable attribute that enhances product appeal, compared to fruit from high crop load levels. Thus, crop load management in Conference pear orchards emerged as an effective preharvest organic strategy for mitigating postharvest quality losses during storage, thereby promoting overall orchard sustainability.

Keywords: conference pear; cold storage; fruit quality; organic strategy; skin russeting

1. Introduction

The Conference pear (*Pyrus communis* L.) holds a prominent position in Spain and throughout the European Union [1]. This cultivar is distinguished by its Guarantee Mark (Pera Conferencia del Bierzo), which certifies compliance with quality and origin standards under Law 17/2001, 7 December 2001. Typically, russet on pear fruit skin diminishes its commercial value [2]. However, intriguingly, the Conference pear's Guarantee Mark highlights a notable feature: a pronounced degree of russeting on its skin, enhancing its value compared to pears from other regions in Spain. A substantial portion of the Conference fruit surface is covered by russeting, that is, cork cells, which exfoliated and revealed deeper layers [3]. Russeting occurs during the fruit's rapid growth phase when micro-cracks develop in the epidermis, leading to the formation of a waterproofing periderm to regulate water loss through the skin [4]. While russeting is genetically determined, environmental factors and cultivation practices significantly influence its formation [5]. Contrary to earlier beliefs that russeting detracted from fruit quality and market value, it is now considered a desirable trait, enhancing fruit quality and providing protection against environmental stress [3,5–7].

The Conference pear, characterized by its abundant russeting, is a prolific and high-yielding cultivar [8]. However, within an orchard, crop load in Conference pears may vary annually [9] due to its classification as a biennial bearing cultivar, yielding heavily in 'on'-years and minimally in 'off'-years [10]. Additionally, adverse weather conditions during pear blossom time can lead to low yields (less than 1.5 kg/tree) [11,12]. Apart from biennial bearing and climatic influences, seasonal crop load is determined by carbohydrate availability, rootstock, and flower- and fruit-thinning practices [13].

Variability in crop load profoundly impact fruit quality and tree physiology [13,14], with significant economic implications due to their effects on orchard productivity and fruit quality [15]. The value of harvested Conference pears, like other cultivars, heavily depends on fruit size distribution [16]. Harvesting pears with diameters of 65 mm or larger is economically advantageous as they fetch higher returns compared to smaller fruit sizes [9]. Market prices for Conference fruit with diameters exceeding 60 mm are twice those of smaller-sized fruit (<55 mm) [17]. Thus, limiting crop load to a specific number of fruits per tree enhances fruit quality and economic returns [18]. To achieve optimal productivity and adjust crop load to enhance fruit quality and economic returns [18,19], thinning treatments are employed. These treatments aim to reduce the crop load of the tree, increase mean fruit weight and size, and improve fruit quality [10,20,21].

Historically, orchards focused on maximizing yield through excessive fertilizer and pesticide use, often neglecting fruit quality [18]. However, recent years have witnessed a shift towards organic agricultural practices, optimizing input efficiency and minimizing environmental impacts [22,23]. According to European Union regulations CEE/2092/91 and CEE/1840/99, organic agricultural practices are defined as an organic production management system that promotes and enhances biodiversity, biological cycles, and soil biological activity [24], being recognized as a sustainable agricultural practice [25]. Organic agricultural practices include the addition of nutrient amendments that enhance soil fertility and health and the exclusion of mineral inorganic fertilizers [24]. Thus, research suggests that incorporating organic fertilization practices, which involve replacing chemical fertilizers with organic alternatives, is crucial for enhancing crop growth, soil fertility, and sustainability compared to relying solely on chemical fertilizers [26]. This approach also protects the local environment, making it compatible with environmentally sustainable development [27]. In addition to biofertilizers, other organic agricultural practices employed include: the use of biopesticides, biocontrol agents, preparation of farmyard manure, and trash composting [28]. According to [29], organic control management ranges from the use of host resistance, rootstock, and scion cultivar selection to the planting location of trees, the application of biological control agents, soil amendments, cultural management, postharvest treatments, disease modelling, and forecasting [29]. Regarding cultural management, there is a growing need to replace chemical thinners, particularly for pear blossom thinning, as no approved chemicals are available in Europe [30]. Organic crop load management techniques, such as mechanical or hand thinning, offer environmentally friendly alternatives, aligning with the principles of organic horticulture. They provide a low-environmental-impact method for crop load management and are a good fit for organic orchards [31]. Such practices gain significance for premium fruits, such as the Conference pear, awarded with a Guarantee Mark, as customers are willing to pay premium prices for organically produced high-quality fruit [32].

The impact of crop load and management techniques on pear yield and fruit weight is well documented [33]. However, there is limited information available for managing crop load in European pear cultivars, and even fewer studies focus on the impact of crop load on pear fruit quality [31]. Moreover, the comprehensive effects of crop load management, implemented as an organic agricultural practice, on overall fruit quality, both at harvest and after cold storage, particularly in high-russet Conference pears expressed as kg/TCSA (Trunk Cross-Sectional Area), remain understudied. Furthermore, little is known about the influence of crop load on the distinctive russeting characteristic of Conference pears. Therefore, this study aims to investigate the effects of organic agricultural practices, specifically crop load management, on fruit quality properties, including skin russet, in high-russet Conference pears.

2. Materials and Methods

2.1. Plant Material and Tree Selection

The experiment was conducted over the 2017 (year 1) and 2018 (year 2) growing seasons as a means of replication over time. Pear cv. Conference from 19-year-old trees on BA-29 rootstock, planted on 3.5 × 1.5 m spacing and trained as central leaders, were harvested in accordance with commercial harvesting standards based on days after full bloom and colour. The trees were sourced from a commercial orchard located in Cabañas Raras, El Bierzo Valley (León, Castilla y León, Spain) (Lat. 42°37′ N, Long. −6°38′ W, elevation 573 m). The orchard has 15 rows planted in an east–west orientation and includes three pear cultivars: Conference as the main cultivar and Buena Luisa and Passe Crassanne as pollinators. Orchard and nutritional management practices remained consistent between years and were those commonly used by the growers. The orchard was irrigated by gravity during the dry season. Dormant pruning was carried out each preceding winter season following recommendations for this cultivar [34]. Soil characteristics were consistent across both years, characterized by loam texture, neutral pH, and medium organic matter content. However, average annual rainfall increased from 681 mm in year 1 to 787 mm in year 2. Initially, 60 trees were randomly selected within the orchard in both years.

Tree selection for each crop load level changed slightly depending on the year, since climatic conditions or biennial bearing greatly affected the crop load of the trees during the second year of the study. Additionally, different trees were used for sample collection each year because the crop load of each tree varied across years.

In year 1, when climatic conditions or biennial bearing did not greatly affect the crop load of the trees, 300 suitable trees were selected out of the 826 trees in the orchard, focusing particularly on those with a sufficient crop load level to study the effect of thinning. From this selection, 60 trees were randomly chosen to conduct the study. Once the 60 trees were selected, it was necessary to conduct hand thinning on 30 of these trees to obtain a wide range of crop loads suitable for studying their impact on fruit quality.

In year 2, all the fruit trees in the orchard were examined to determine their estimated production levels. Subsequently, a random selection of 30 trees with a low crop load and a random selection of 30 trees with a high crop load were made, from which fruit samples were later collected. In this year, thinning was unnecessary due to reduced crop loads resulting from rainfall during flowering.

2.2. Crop Load Management and Experimental Design

During the first year, trees received two fruit thinning levels: no thinning and a light hand thinning carried out 75 days after full bloom (leaving an average of 75% of the fruit compared to trees with no thinning). Thinning aimed to achieve 2–3 fruits per cluster, with the number of fruits retained within the cluster depending on the fruit size (when the fruits within a cluster were large, three fruits per cluster were retained, whereas when the fruits were small, only two fruits per cluster were left) (Figure 1). Besides the advantages that the hand thinning method has over chemical methods, such as being more environmentally friendly and applicable in organic farming, this thinning method was chosen because it facilitates the retention of the largest and healthiest fruits within the cluster while removing the smaller and damaged ones. Thinned fruits averaging a diameter of 22.8 mm and a weight of 10.1 g.

Tree vigour was evaluated annually in October, marking the end of the growing season, and expressed as TCSA measured 20 cm above the graft [35].

Organic management in this study comprised two key strategies: firstly, the adoption of environmentally friendly hand thinning when necessary, and secondly, the sorting of trees based on yield efficiency. These organic practices were implemented to effectively address the variability of crop load levels.

(**a**) (**b**)

Figure 1. Illustration of hand thinning performed in Conference pear trees: (**a**) cluster before thinning; (**b**) cluster after thinning.

The experiments were conducted using a completely randomized design with twenty replicates, focusing solely on the factor of crop load level.

2.3. Skin Russeting Quantification

The quantification of skin russeting was conducted through digital image processing and classification. The sample consisted of three pears randomly chosen per tree in the first year, each fruit representing one of three size classes: large, medium, and small. In the second year, the sample consisted of two pears per tree, representing large and medium/small sizes. Overall, the russetted area (%) was assessed in 180 pears during the first year and 120 pears during the second year. Three close-up photographs, using a Canon Powershot A3300 IS 16 MgP digital camera with a 28 mm focal lens and optical image stabilizer, were taken per pear to cover the entire surface area. A total of 900 images were taken and processed, with 540 in the first year and 360 in the second year.

An object-based image classification approach was applied to each image, defining two classes: green (non-russetted area) and brown (russetted area). The multiresolution segmentation algorithm in eCOGNITION® 8.9 (Trimble Germany GmbH, Munich, Germany, 2021). [36] was used to merge pixels and obtain homogeneous non-overlapping polygons, which were then classified using the nearest neighbour algorithm. The accuracy assessment of the classification was assessed using an independent sample of validation points [37]. A total of 19 out of the 900 images were randomly selected. Within each of these 19 images, a systematic sampling design (1 cm × 1 cm) was applied to collect the validation points (one point corresponded to one object resulting from the segmentation). Overall, 1375 points belonging to the green class and 2108 to the brown class were used for validation, resulting in an overall accuracy of 98%, indicating that the method is sufficiently reliable to calculate russeting for all samples.

Once the classification method was validated, each classified image was exported as a raster file. The number of pixels assigned to the class 'russetted area' was calculated, along with the total number of pixels representing the pear (including both 'non-russetted' and 'russetted' areas). The russetted area of the pear (%) was determined by dividing the number of 'russetted skin' pixels by the total number of 'pear' pixels in the three images per pear. The mean russetted area (%) per tree was then calculated as the mean value of russeting obtained for the three pears in the first year and two pears in the second year. To account for pear size in the estimation of russeting per tree, a weighed mean of russeting was calculated (Equation (1)).

$$\text{Russeting Weighted mean} = \frac{\sum_{i=1}^{n} w_i x_i}{\sum_{i=1}^{n} \omega_i} \quad (1)$$

where:

russeting weighted mean: russeting weighted mean per tree (%)
n: number of pears in the sample (3 in year 1, and 2 in year 2)
x: russetted area of the pear (%)
w: weight of the pear (g)

2.4. Yield and Fruit Quality Evaluations

Yield efficiency, total number of fruits per tree, and yield per tree (kg) were analysed. The collected fruits were subdivided into five classes according to their sizes based on a size card in steps of 5 mm. Yield parameters were then calculated, including the number of fruits per size class and the weight of fruits per size class (<55, 55/60, 60/65, 65/70, and >70 mm). Fresh-market yield for each tree was determined based on fruit with a maximum cheek diameter equal to or higher than 60 mm. The financial return from fruits with a diameter higher than 60 mm is twice that of smaller-sized fruits (less than 55 mm) [38]. Mean fruit mass was calculated by dividing the total yield per tree (kg) by the total number of fruits per tree.

In both seasons, samples consisting of 20 commercial fruits, representative in size from the total fruit of the tree, were selected for quality evaluation. Then, once in the lab, samples were divided into two sub-samples of 10 fruits each, one for quality evaluation at harvest and the other for postharvest quality evaluation. The latter were stored in plastic containers for 70 days in standard cold storage at 1.5 °C, 95% RH. To simulate commercial storage conditions, the fruit was placed on plastic trays enclosed with paper.

Weight loss (%) was determined by measuring the total fresh weight of the 10-fruit samples both at harvest and after storage. Fruit quality evaluation included skin ground colour, firmness, starch index (SI) (only at harvest), total soluble solids (TSS), and titratable acidity (TA) assessments. For each fruit, skin ground colour was determined at three points in the area free from russet with a colorimeter (Minolta, CR-200, Ahrensburg, Germany), expressed in L*, a*, b* system and converted to hue angle (h°), where h° = tan − 1 b*/a*. If the background colour was obscured by russetted areas covering the entire fruit, colour analysis for that particular fruit was omitted to prevent measurement errors. Three determinations of the colour parameters were made along the equatorial axis of each fruit. Flesh firmness was measured using a penetrometer (Effegi TR Turoni & C., Forlì, Italy) mounted on a hand-operated press fitted with a 7.9 mm diameter plunger. Measurements were taken at two equidistant points on the equatorial axis of the fruit. The iodine–starch index was evaluated by dipping an equatorial slide of the fruit into iodine–iodize solution (20 g IK + 10 g I2 + 1 L H_2O) for 1 min and comparing the colour pattern with a reference chart, Ctifl-Eurofru Code, using a scale of 1 (fully stained, all starch) to 10 (no staining, no starch).

The 10-fruit sub-samples were divided into two 5-fruit groups. Juice extraction was performed by homogenizing the fruit flesh in a blender to determine TSS and TA. The TSS (%) of the juice was measured with a digital refractometer (Atago, DR-A1, Tokyo, Japan), while TA was determined by titrating 10 mL of juice with NaOH 0.1 N up to pH 8.1. Acidity of pear fruit was reported as a percentage of malic acid concentration.

2.5. Incidence of Physiological Disorders

For stored samples, the incidence of fruit affected by disorders such as shrivelling or rot incidence was recorded. Core breakdown was visually scored on fruit cut through the equator.

2.6. Statistical Analysis

A two-way analysis of variance (ANOVA) was used to test the effects of crop load level and year on yield properties and fruit quality parameters, followed by a Fisher's LSD post-hoc test to compare means. The significance level for the analyses was set at 5% ($p \leq 0.05$). All statistical analyses were carried out using SPSS software v. 29.0 (SPSS Inc., Chicago, IL, USA).

3. Results

3.1. Yield Efficiency and Yield Properties

Each year, all fruits from every experimental tree were harvested, counted, and weighed. Yield efficiency was then calculated as the ratio of the total cumulative yield of each tree (g) per TCSA (cm^2). Subsequently, each year, the 60 trees, out of the 826 trees on the orchard, were ranked according to yield efficiency, forming two categories of crop load: a high crop load level, consisting of the 20 trees with the highest yield efficiency (high TCSA), and a low crop level, consisting of the 20 trees with the lowest yield efficiency (low TCSA) (Table 1). The 20 trees in each crop load level were selected as replicates, while the 20 remaining trees in the middle were excluded from the analysis. Each sample tree in the orchard was labelled with a unique code. The number of fruits per tree in the orchard was 329.9 in year 1 and 53.1 in year 2.

Table 1. Yield efficiency (g cm^{-2}).

Crop Load Level	Year 1			Year 2		
	Minimum	Mean	Maximum	Minimum	Mean	Maximum
Low	168.06	288.77	365.93	34.58	81.10	119.85
High	510.03	651.46	919.60	219.59	332.95	760.51
Mean		470.12			207.02	

The mean yield efficiency varied between the two years, decreasing from 470.12 g cm^{-2} of TCSA in year 1 to 207.02 g cm^{-2} of TCSA in year 2 (Table 1). Consequently, the crop load of pear cv. Conference, measured in kg per tree, varied considerably from high (27.92 kg per tree in year 1) to low (13.57 kg per tree in year 2). Yield efficiency (g cm^{-2}) ranged from 168.06 (representing the tree with the lowest crop load in year 1) to 919.60 (indicating the tree with the highest crop load in the same year). In year 2, the range was from 34.58 (lowest) to 760.51 (highest). When considering crop load levels, during the first season, yield efficiency ranged from 168.06 to 365.93 g cm^{-2} in low crop load trees and from 510.03 to 919.60 g cm^{-2} in high crop load trees. In the second season, yield efficiency varied from 34.58 to 119.85 g cm^{-2} in low crop load trees and from 219.59 to 760.51 g cm^{-2} in high crop load trees. The mean yield efficiency changed from 288.77 g cm^{-2} of TCSA (low crop load) to 651.46 g cm^{-2} of TCSA (high crop load) in year 1 and from 81.10 g cm^{-2} of TCSA (low crop load) to 332.95 g cm^{-2} of TCSA (high crop load) in year 2.

Trees with low crop load produced larger fruits, characterized by higher percentages in the upper size category (>70) and lower percentages in the lower size categories (<55, 55–60). These trees also yielded a higher percentage of fresh-market produce and had greater fruit mass (Table 2). Fresh-market yield varied from 63.01% to 70.35%, and fruit mass varied from 219.22 g to 237.16 g between crop loads, whereas fresh-market yield varied from 38.42% to 94.94%, and fruit mass varied from 170.55 g to 285.83 g between years. The two-way ANOVA test revealed statistically significant interactions between crop load and year for the 60–65 mm and 65–70 mm size classes.

Table 2. Means for yield properties.

Factor		<55 mm (%)	55–60 mm (%)	60–65 mm (%)	65–70 mm (%)	>70 mm (%)	Fresh-Market Yield (%)	Fruit Mass (g)
Crop load	Low	19.26 b [1]	10.39 b	11.60 a	13.04 a	45.72 a	70.35 a	237.16 a
	High	24.55 a	12.44 a	11.92 a	12.08 a	39.01 b	63.01 b	219.22 b
Year	1	41.51 a	20.07 a	17.67 a	12.96 a	7.80 b	38.42 b	170.55 b
	2	2.30 b	2.76 b	5.85 b	12.16 a	76.93 a	94.94 a	285.83 a

Table 2. *Cont.*

Factor	<55 mm (%)	55–60 mm (%)	60–65 mm (%)	65–70 mm (%)	>70 mm (%)	Fresh-Market Yield (%)	Fruit Mass (g)
Main effects [2]							
Crop load	*	*	ns	ns	**	**	*
Year	***	***	***	ns	***	***	***
Interaction							
Crop load × Year	ns	ns	*	*	ns	ns	ns

[1] For each yield property, means with different letters within the same factor are significantly different (p value ≤ 0.05, according to the two-way ANOVA). [2] p value: ns—not significantly different; * $p < 0.05$; ** $p < 0.01$; *** $p < 0.001$.

3.2. Quality Properties

Skin colour and TSS were not influenced by crop load either at harvest or after storage. However, significant differences in firmness, starch index (analysed only at harvest), and TA were found between crop loads (Table 3).

Table 3. Means for fruit quality properties.

Storage Stage	Factor		h°	Firmness (N)	SI (1–10)	TSS (%)	TA (%)
Harvest	Crop load	Low	115.00 a [1]	52.18 a	5.02 b	12.64 a	0.217 a
		High	114.84 a	48.81 b	5.78 a	12.62 a	0.201 b
	Year	1	114.26 b	51.05 a	5.93 a	12.06 b	0.203 b
		2	115.58 a	49.94 a	4.87 b	13.20 a	0.215 a
Main effects [2]							
Crop load			ns	***	**	ns	**
Year			***	ns	***	***	*
Interaction							
Crop load × Year			ns	ns	*	ns	ns
Postharvest	Crop load	Low	106.83 a	21.89 a	-	13.78 a	0.180 a
		High	106.12 a	18.74 b	-	13.88 a	0.159 b
	Year	1	106.74 a	24.35 a	-	13.20 b	0.171 a
		2	106.22 a	16.29 b	-	14.46 a	0.169 a
Main effects							
Crop load			ns	*	-	ns	***
Year			ns	***	-	***	ns
Interaction							
Crop load × Year			ns	ns	-	ns	ns

[1] For each quality parameter, means with different letters within the same factor are significantly different (p value ≤ 0.05, according to the two-way ANOVA). [2] p value: ns—not significantly different; * $p < 0.05$; ** $p < 0.01$; *** $p < 0.001$. h°, hue angle; SI, starch index; TSS, total soluble solids; TA, titratable acidity.

Fruit from low crop load trees exhibited greater hardness compared to those from high crop load trees. This trend was consistent both at harvest and after storage. Differences in fruit firmness between years were only observed at the end of storage. Fruit coming from low crop load trees had a lower SI value compared to fruit from high crop load trees, while fruit harvested in year 2 had lower SI values than fruit harvested in year 1. TSS was the only parameter affected by year during both storage stages, with TSS being higher in year 2 compared to year 1. Higher TA values were observed in fruit coming from low crop loads compared to fruit from high crop loads. Differences in TA between years were observed only at harvest. Significant interactions between crop load and year were observed only for SI at harvest.

The russetted area was influenced by crop load, with russeting being more pronounced in fruit from low crop load trees (Table 4). The mean russetted area was higher in low crop load trees compared to high crop load trees (47.22% and 38.22%, respectively). Similarly, the weighed mean russetted area was higher in low crop load trees compared to high crop load trees (47.51% and 39.03%, respectively). Skin russeting was also affected by year, with both russeting parameters being more pronounced in year 2 than in year 1. Interactions between crop load and year for russeting parameters were not significant.

Table 4. Means for fruit russeting parameters.

Factor		Weighted Mean	Mean
Crop load	Low	47.51 a [1]	47.22 a
	High	39.03 b	38.22 b
Year	1	39.58 b	38.20 b
	2	46.96 a	47.24 a
Main effects [2]			
Crop load		**	**
Year		*	**
Interaction			
Crop load × Year		ns	ns

[1] For each russeting parameter, means with different letters within the same factor are significantly different (p value ≤ 0.05, according to the two-way ANOVA). [2] p value: ns—not significantly different; * $p < 0.05$; ** $p < 0.01$.

No significant differences were found for disorders or weight loss between crop load levels.

4. Discussion

The highest yield efficiency observed for the high crop load aligns closely with the yield observed by a previous study on pear cv. Conference, which recorded a yield of 690 g cm^{-2} [39]. When considering the mean yield efficiency in terms of kg/ha, both low and high crop load levels in year 1 were consistent with high yield levels, as defined by [1] for Conference pear trees (above 38t per ha). However, in year 2, only the mean yield efficiency from the high crop load level could meet this criterion. It is important to note that crop load, as measured by yield efficiency, is considered low in Conference pear due to its high TCSA and low productivity [39].

The notable difference in yield efficiencies between the two years could be attributed to several factors. One major factor is the impact of rainfall during flowering in the second season, which likely led to a reduction in crop load. Additionally, the phenomenon of biennial bearing may have influenced yield efficiency. The authors of [40] highlighted the importance of crop load management in order to avoid biennial bearing and regulate fruit size and quality in apple cultivars.

Despite the pear trees in this orchard being winter-pruned not only during the two years of the experiment but also in previous years, and the well-established understanding of how management practices such as pruning are utilized to restore the balance between vegetative and reproductive shoots in alternate bearing varieties [41], thereby limiting the detrimental effect of alternate bearing [42], there are certain cultivars for which alternate bearing is less affected by pruning and training [43], such as the Conference pear variety.

Environmental factors can moderate fruit levels through suboptimal conditions for flower induction or by reducing the number of fruits, while growers can intervene by providing optimal conditions for the tree through practices such as pruning [41]. However, growers are often hesitant to remove flowers and fruit from trees as unexpected climatic events, such as those experienced during the second year, can further reduce crop load after grower intervention.

The differences in production between the two years likely contributed to the larger discrepancies in fresh-market yield between years (56.52%). Additionally, better fruit quality, particularly in terms of fruit size, was achieved with low crop load. The tendency of fruit trees to produce heavy crop loads often results in the production of numerous small fruits that are unsuitable for fresh market sale [44]. Therefore, when comparing both years, fruit quality in terms of fruit size was superior during the year with a lower yield efficiency, which was the second season. Variations in fruit weight between years has been observed in other studies of pear cv. Conference [45]. The interaction between crop load and year on intermediate size classes, specifically the 60–65 mm and 65–70 mm classes, is likely influenced by the significant difference in fruit mass between the two years. In year 2, when the average fruit mass is significantly higher compared to year 1, the majority of the fruit falls into the >70 mm size class, especially in low crop load level. This leaves a smaller percentage of fruit for the other size classes, including 60–65 mm and 65–70 mm. Conversely, the smaller fruit size in year 1 resulted in higher percentages of fruit in the intermediate size classes under low crop load.

The natural tendency of fruit trees to produce heavy crop loads causes the sugars produced in the leaves (sources) to be distributed over too many fruits (sinks) [44]. In contrast, thinning, as a preharvest practice that decreases crop load, can enhance fruit quality in Conference pears, particularly fruit size, by improving the source–sink relationship and increasing the allocation of photoassimilates to the remaining fruit [33]. Moreover, [23] studies have found that the eating quality of fruit such as Deveci pears was better in larger fruit, as lower fruit size led to reduced juiciness and crispness. Lower crop loads have also been associated with enhanced fruit quality in terms of sugar content, as there is less competition for soluble carbohydrates between plant organs [23], as well as improved firmness [22,33,40,46] and even acidity in apples [15]. The authors of [47] found that low crop load reduced the photosynthetic performance of Honeycrisp apple trees, possibly due to the increase in leaf fructose, glucose, and sorbitol concentrations. This suggested that the pathways underlying the distribution and concentration of these sugars could be sensitive to changes in crop load treatments, highlighting the key role of crop load in the allocation of photoassimilates. In our experiment, these changes between crop load levels were clear for firmness and TA, while TSS remained invariable regardless of the crop load.

After 70 days of storage, firmness in all cases exceeded the optimum level recommended for acceptable consumption in pear cv. Conference, which is typically between 10.8 and 12.8 N according to [48], or within the range of 10–30 N as suggested by [49]. However, the potential storability of fruit from trees with low crop loads would be extended compared to fruit from trees with high crop loads, especially in years like year 2, where the firmness value in postharvest was close to the minimum recommended level for consumption. Hence, the storability of fruit from trees with low crop load would be greater compared to fruit from trees with high crop loads, particularly in terms of mid-term storage. Firmness during storage was also dependent on the year. This could be attributed to the fact that, as explained in the Materials and Methods Section, the samples of fruit selected for quality evaluation were representative in size of the total fruit from each tree. Since the majority of the sample by weight (and even more in terms of the number of pieces) consisted of small fruits in the first year, this could have had a large impact on firmness during postharvest, which was contrary to the effect of the factor under study (crop load). Johnston et al. [50] noted that larger-sized apple fruits harvested at advanced maturity usually soften more rapidly than smaller fruits.

TSS exhibited notable variation across years, making it the quality property with the greatest change from year to year, whereas the crop load factor did not affect TSS. Specifically, TSS varied from 12.06% (at harvest) in year 1, the year with higher yield efficiency, to 13.20% (at harvest) in year 2, the year with lower yield efficiency. These findings align with previous research [10], which highlights that the individual sugar content in pear cv. Conference is highly dependent on the year rather than on treatments such as chemical thinning. Similarly to the effect observed with firmness, the variation in fruit size between years could have had influenced TSS, as lower TSS was found in year 1 (the year with a higher proportion of small fruits) compared to year 2. Furthermore, disparities in TSS between years could also be attributed to the average solar radiation over a 4-week period [51].

Conference pear is considered a low-acidity cultivar, with a TA level around 0.2–0.3% [35]. In our experiment, the TA levels in fruit from trees with low crop loads were consistently above the threshold representing ripe Conference pears (0.16%) [52]. However, fruit from trees with high crop loads showed TA values after storage that were significantly lower than those for fruit from trees with low crop loads, and, in fact, they were lower than that threshold. Similarly, the SI values were consistent with the firmness and TA values. Fruit from low crop levels were less ripe as their starch content was higher (on a scale of 1 (all starch) to 10 (no starch)), suggesting they could withstand longer storage periods. However, while crop load affected SI in year 1, with lower values observed in low crop load compared to high crop load, significant differences between crop load levels could not be found in year 2. Thus, the effect of crop load on SI was contingent upon the year, indicating that it cannot be definitively stated that crop load affects SI. Ultimately, except for SI, it could be inferred that regardless of the season, fruit coming from low crop loads would have better behaviour during postharvest as its storability would be increased compared to fruit coming from high crop loads, which will be riper and therefore would have lower storability.

As previously mentioned, russeting in pear cv. Conference is considered a significant quality attribute of the fruit. Research has shown that differential growth rates between regions in pear cv. Conference contributed to variations in russeting [53]. The russetted area for Conference pears grown in El Bierzo was found to be higher than in other areas where Conference pear are cultivated [54]. Thus, [1] estimated that the russetted area in Conference pears grown in Dąbrowice (Poland) was minimal, mainly around the calyx, averaging 1.9 and 2.2 on a scale from 1 (no russeting) to 5 (russeting > 75% of fruit skin surface), indicating that the percentage of russeting was around 20%, significantly lower than the percentage observed in our work.

According to [53], russeting in Conference pear increases when the relative growth rates of the surface area exceeded 0.03 per day. This finding could explain the differences observed in russetted area between crop loads. Larger fruit, such as those coming from low crop loads, compared to those from high crop loads, would have higher growth rates, resulting in a higher percentage of russetted area. Therefore, in addition to location, the higher percentage of russetted area found in fruit from low crop loads will further enhance the quality of this cultivar, making it more desirable. This aspect contributed to achieving organic management practices, as it eliminated the need for farmers to use chemical products such as thinners to increase the percentage of russetted area in the pear fruit to enhance its appearance [18].

It could be asserted that thinning to manage the crop load of Conference pear trees will not be necessary every year. Therefore, only years with high crop loads will necessitate hand thinning as an organic preharvest technique to optimize fruit quality and enhance environmental sustainability. Conversely, in years with low crop loads, it would be more productive to concentrate on identifying or classifying fruit according to yield efficiency to improve the quality and storability of the different batches picked. This approach could be implemented by farmers to contribute to organic management.

5. Conclusions

Despite the significant influence of annual variations on fruit quality, particularly evident in cultivars susceptible to biennial bearing, our study underscores a consistent finding: maintaining a low crop load in pear cv. Conference reliably enhances fruit quality. Specifically, it improves fruit size, firmness, and acidity, and increases mid-term storability. Additionally, our findings highlight the superiority of fruit from low crop load levels in terms of russetted area, a key factor in consumer preference for this variety, where russeting importantly contributes to its commercial value; hence, crop load management should be the key area of focus for increasing russetted area in high-russetted cultivars such as Conference pears. Considering horticultural management, the implementation of organic agricultural practices such as crop load management not only optimizes orchard productivity but also fosters organic practices by minimizing postharvest quality losses in Conference pears. Procedures for classifying fruit trees according to yield efficiency should be implemented to improve the quality and storability of the different harvested batches.

Author Contributions: Conceptualization, M.G. and F.Á.-T.; methodology, M.G. and F.Á.-T.; software, M.G., F.Á.-T., A.M.F. and V.M.; validation, M.G., F.Á.-T., A.M.F. and V.M.; formal analysis, M.G., F.Á.-T., A.M.F. and V.M.; investigation, M.G., F.Á.-T., A.M.F. and V.M.; data curation, all authors; writing—original draft preparation, M.G. and F.Á.-T.; writing—review and editing, M.G., F.Á.-T., A.M.F., V.M., Á.R.-G. and P.A.C.; visualization, M.G. and F.Á.-T.; supervision, M.G., F.Á.-T. and P.A.C.; funding acquisition, P.A.C. All authors have read and agreed to the published version of the manuscript.

Funding: This research was funded by the research project 'Adaptación de los modelos predictivos para el control del Stemphylium vesicarium en plantaciones de frutales de Cefrubierzo' (2024/00027/001), Universidad de León-Cefrubierzo.

Data Availability Statement: The original contributions presented in the study are included in the article, further inquiries can be directed to the corresponding author.

Acknowledgments: We express our gratitude to the research program of the Universidad de León for their funding support. Additionally, we extend our appreciation to the 'Asociación Berciana de Agricultores' and to Aquilino Guerra for his collaboration in the development of the field assays, for their professional cultivation of the experimental fruit trees, and for granting permission for sampling in their orchard.

Conflicts of Interest: The authors declare no conflicts of interest.

References

1. Wójcik, P.; Skorupińska, A.; Filipczak, J. Impacts of preharvest fall sprays of calcium chloride at high rates on quality and 'Conference' pear storability. *Sci. Hortic.* **2014**, *168*, 51–57. [CrossRef]
2. Kim, Y.; Oh, S.; Han, H.; Kim, D. QTL analysis and CAPS marker development linked with russet in pear (*Pyrus* spp.). *Plants* **2022**, *11*, 3196. [CrossRef] [PubMed]
3. Konarska, A. The relationship between the morphology and structure and the quality of fruits of two pear cultivars (*Pyrus communis* L.) during their development and maturation. *Sci. World J.* **2013**, *2013*, 846796. [CrossRef] [PubMed]
4. Khanal, B.P.; Grimm, E.; Knoche, M. Russeting in apple and pear: A plastic periderm replaces a stiff cuticle. *AoB PLANTS* **2013**, *5*, pls048. [CrossRef] [PubMed]
5. Falginella, L.; Cipriani, G.; Monte, C.; Gregori, R.; Testolin, R.; Velasco, R.; Troggio, M.; Tartarini, S. A major QTL controlling apple skin russeting maps on the linkage group 12 of 'Renetta Grigia di Torriana'. *BMC Plant Biol.* **2015**, *15*, 150. [CrossRef] [PubMed]
6. Inoue, E.; Kasumi, M.; Sakuma, F.; Anzai, H.; Amano, K.; Hara, H. Identification of RAPD marker linked to fruit skin color in Japanese pear (*Pyrus pyrifolia* Nakai). *Sci. Hortic.* **2006**, *107*, 254–258. [CrossRef]
7. Jia, B.; Cheng, Z.; Wang, Q.; Zhang, S.; Heng, W.; Zhu, L. Characterization of the composition and gene expression involved the shikimate pathway in the exocarp of 'Dangshansuli' pear and its russet mutant. *Hortic. Environ. Biotechnol.* **2021**, *62*, 125–134. [CrossRef]
8. Sansavini, S. Pear fruiting-branch models related to yield control and pruning. *Acta Hortic.* **2002**, *596*, 627–633. [CrossRef]
9. Maas, F.; van der Steeg, P. Crop load regulation in 'Conference' pears. *Acta Hortic.* **2011**, *909*, 369–379. [CrossRef]
10. Hudina, M.; Štampar, F. Influence of NAA thinning on yield of pear (*Pyrus communis* L.) cvs. 'Williams', 'Conference' and 'Harrow sweet'. In Proceedings of the 21st Scientific-Expert Conference of Agriculture and Food Industry, Sarajevo, Bosnia and Herzegovina, 29 September–2 October 2010; pp. 223–230.
11. Pfeiffer, B.; Eis, B.; Zimmer, J.; Fieger-Metag, N. Optimizing crop loading of apples and pears—Results 2004–2006 (foliar fertilizers, thinning). In Proceedings of the Ecofruit-13th International Conference on Cultivation Technique and Phytopathological Problems in Organic Fruit-Growing, Weinsberg, Germany, 18–20 February 2008; pp. 324–329.
12. Wójcik, P. Quality and 'Conference' pear storability as influenced by preharvest sprays of calcium chloride. *J. Plant Nutr.* **2012**, *35*, 1970–1983. [CrossRef]
13. Wünsche, J.N.; Ferguson, I.B. Crop load interactions in apple. *Hortic. Rev.* **2005**, *31*, 231–290.
14. Guerra, M.; Casquero, P. Post-harvest quality of 'Green Gage' European plum in integrated production: Effects of year and fruit maturity. *J. Hortic. Sci. Biotechnol.* **2010**, *85*, 66–70. [CrossRef]

15. Greene, D. Increasing the flowering of bearing apple trees with PGRs. *HortScience* **2008**, *43*, 1058.
16. Maas, F.M.; Kanne, H.J.; van der Steeg, P.A.H. Chemical thinning of 'Conference' pears. *Acta Hortic.* **2010**, *884*, 293–304. [CrossRef]
17. Janssens, P.; Deckers, T.; Elsen, F.; Elsen, A.; Schoofs, H.; Verjans, W.; Vandendriessche, H. Sensitivity of root pruned 'Conference' pear to water deficit in a temperate climate. *Agric. Water Manag.* **2011**, *99*, 58–66. [CrossRef]
18. Suo, G.-D.; Xie, Y.-S.; Zhang, Y.; Cai, M.-Y.; Wang, X.-S.; Chuai, J.-F. Crop load management (CLM) for sustainable apple production in China. *Sci. Hortic.* **2016**, *211*, 213–219. [CrossRef]
19. Lee, J.-Y.; Lee, S.-M.; Lee, M.-J.; Han, S.-Y.; Jung, H.-W.; Lee, Y.-H. Crop load adjustment based on branch vigor for producing uniform fruit in young apple trees. *Hortic. J.* **2015**, *84*, 202–213. [CrossRef]
20. Bound, S.A. Optimising crop load and fruit quality of 'Packham's Triumph' pear with ammonium thiosulfate, ethephon and 6-benzyladenine. *Sci. Hortic.* **2015**, *192*, 187–196. [CrossRef]
21. Kacal, E. Crop load regulation with chemical thinners in Deveci pear (*Pyrus communis* L.). *Appl. Ecol. Environ. Sci.* **2018**, *16*, 7203–7212. [CrossRef]
22. Kılıç, O.; Boz, I.; Eryılmaz, G.A. Comparison of conventional and good agricultural practices farms: A socio-economic and technical perspective. *J. Clean. Prod.* **2020**, *258*, 120666. [CrossRef]
23. Guerra, M.; Sanz, M.Á.; Rodríguez-González, Á.; Casquero, P.A. Effect of sustainable preharvest and postharvest techniques on quality and storability of high-acidity 'Reinette du Canada' apple. *Horticulturae* **2022**, *8*, 86. [CrossRef]
24. Winter, C.K.; Davis, S.F. Organic foods. *J. Food Sci.* **2006**, *71*, R117–R124. [CrossRef]
25. Knight, K.W.; Newman, S. Organic agriculture as environmental reform: A cross-national investigation. *Soc. Natur. Resour.* **2013**, *26*, 369–385. [CrossRef]
26. Tong, Y.; Wang, Z.; Gong, D.; Huang, C.; Ma, X.; Ma, X.; Yuan, F.; Fu, S.; Feng, C. Enhancing soil fertility and elevating pecan fruit quality through combined chemical and organic fertilization practices. *Horticulturae* **2024**, *10*, 25. [CrossRef]
27. Ananthi, T.; Vennila, C. Influence of organic manures and inorganic fertilizers on growth and yield of fodder maize (*Zea mays* L.) grown in north eastern zone of Tamil Nadu. *Curr. J. Appl. Sci. Technol.* **2021**, *40*, 70–78. [CrossRef]
28. Elakkiya, S.; Karthikeyan, C. An analytical study on training needs of farmers on organic farming. *Int. J. Farm Sci.* **2020**, *10*, 40–44. [CrossRef]
29. Shuttleworth, L.A. Alternative disease management strategies for organic apple production in the United Kingdom. *CABI Agric. Biosci.* **2021**, *2*, 34. [CrossRef]
30. Seehuber, C.; Damerow, L.; Kunz, A.; Blanke, M.M. Mechanical thinning of 'Lucas' and 'Conference' pear improves fruit quality. *Acta Hortic.* **2015**, *1094*, 289–295. [CrossRef]
31. Bound, S.A. Managing Crop Load in European Pear (*Pyrus communis* L.)—A Review. *Agriculture* **2021**, *11*, 637. [CrossRef]
32. Guerra, M.; Sanz, M.Á.; Rodríguez-González, Á.; Casquero, P.A. Summer pruning, an eco-friendly approach to controlling bitter pit and preserving sensory quality in highly vigorous apple cv. 'Reinette du Canada'. *Agriculture* **2021**, *11*, 1081. [CrossRef]
33. Lopez, G.; Larrigaudière, C.; Girona, J.; Behboudian, M.H.; Marsal, J. Fruit thinning in 'Conference' pear grown under deficit irrigation: Implications for fruit quality at harvest and after cold storage. *Sci. Hortic.* **2011**, *129*, 64–70. [CrossRef]
34. Guerra, A.; Guerra, M. *Evolución de Fruticultura y Poda de Árboles Frutales*, 2nd ed.; Consejería de Agricultura y Ganadería: Valladolid, Spain, 2009; pp. 127–136.
35. Lepsis, J.; Blanke, M.M. The trunk cross-section area as a basis for fruit yield modelling in intensive apple orchards. *Acta Hortic.* **2006**, *707*, 231–235. [CrossRef]
36. Jakovljevic, G.; Miro, G.; Alvarez-Taboada, F. A deep learning model for automatic plastic mapping using Unmanned Aerial Vehicle (UAV) data. *Remote Sens.* **2020**, *12*, 1515. [CrossRef]
37. Congalton, R.G.; Green, K. *Assessing the Accuracy of Remotely Sensed Data*, 2nd ed.; CRC Press: Boca Raton, FL, USA, 2009.
38. Janssens, P.; Elsen, F.; Elsen, A.; Deckers, T.; Vandendriessche, H.; Verstraeten, W.W.; Coppin, P.; Sase, S.; DeMeloAbreu, J.P. Adapted soil water balance model for irrigation scheduling in 'Conference' pear orchards. *Acta Hortic.* **2011**, *919*, 39–46. [CrossRef]
39. Milošević, T.; Milošević, N.; Mašković, P. Do the rootstocks determine tree growth, productivity and fruit quality of pears, which grow on typical heavy and acidic soil? *Erwerbs-Obstbau* **2015**, *57*, 125–134. [CrossRef]
40. Sidhu, R.S.; Hunt, I.; Bound, S.A.; Swarts, N.D. Crop load, fruit quality and mineral nutrition as predictors of fruit softening and internal flesh browning in modern firm fleshed apple cultivars. *Sci. Hortic.* **2024**, *330*, 113035. [CrossRef]
41. Smith, H.M.; Samach, A. Constraints to obtaining consistent annual yields in perennial tree crops. I: Heavy fruit load dominates over vegetative growth. *Plant Sci.* **2013**, *207*, 158–167. [CrossRef]
42. Bussi, C.; Genard, M. Thinning and pruning to overcome alternate bearing in peach trees. *Eur. J. Hortic. Sci.* **2014**, *79*, 313–317.
43. Lauri, P.E.; Grappadelli, L.C. Tree architecture, flowering and fruiting—Thoughts on training, pruning and ecophysiology. *Acta Hortic.* **2014**, *1058*, 291–298. [CrossRef]
44. Wouters, N.; De Ketelaere, B.; Deckers, T.; De Baerdemaeker, J.; Saeys, W. Multispectral detection of floral buds for automated thinning of pear. *Comput. Electron. Agric.* **2015**, *113*, 93–103. [CrossRef]
45. Lace, B.; Lacis, G. Evaluation of pear (*Pyrus communis* L.) cultivars in Latvia. *Hortic. Sci.* **2015**, *42*, 107–113. [CrossRef]
46. Link, H. Significance of flower and fruit thinning on fruit quality. *Plant Growth Regul.* **2000**, *31*, 17–26. [CrossRef]
47. Baldassi, C.; Berim, A.; Roeder, S.; Losciale, P.; Serra, S.; Gang, D.R.; Musacchi, S. Rootstock and crop load effects on 'honeycrisp' photosynthetic performance and carbohydrate accumulation. *Plants* **2023**, *12*, 4035. [CrossRef] [PubMed]

48. Plocharski, W.J.; Konopacka, D. The relation between mechanical and sensory parameters of apples and pears. *Acta Hortic.* **1999**, *485*, 309–317. [CrossRef]
49. Torregrosa, L.; Echeverria, G.; Illa, J.; Giné-Bordonaba, J. Ripening behaviour and consumer acceptance of 'Conference' pears during shelf life after long term DCA-storage. *Postharvest Biol. Technol.* **2019**, *155*, 94–101. [CrossRef]
50. Johnston, J.W.; Hewett, E.W.; Hertog, M.L.A.T.M.; Harker, R. Harvest date and fruit size affect postharvest softening of apple fruit. *J. Hortic. Sci. Biotechnol.* **2002**, *77*, 355–360. [CrossRef]
51. Iwanami, H.; Moriya-Tanaka, Y.; Hanada, T.; Baba, T.; Sakamoto, D. Meteorological and tree-management factors related to soluble solids content of apple fruit and crop load management for producing high soluble solids content fruit in high-density planted 'Fuji'. *Sci. Hortic.* **2023**, *310*, 111755. [CrossRef]
52. Colás-Medà, P.; Abadias, M.; Alegre, I.; Usall, J.; Viñas, I. Effect of ripeness stage during processing on Listeria monocytogenes growth on fresh-cut 'Conference' pears. *Food Microbiol.* **2015**, *49*, 116–122. [CrossRef]
53. Scharwies, J.D.; Grimm, E.; Knoche, M. Russeting and relative growth rate are positively related in 'Conference' and 'Condo' pear. *HortScience* **2014**, *49*, 746–749. [CrossRef]
54. Guerra, M.; Casquero, P.A.; Valenciano, J.B. Evolución de madurez de pera Conferencia mediante diferentes técnicas de conservación frigorífica. In *Innovaciones Fisiológicas y Tecnológicas de la Maduración y Post-Recolección de frutas y Hortalizas*; Valero, D., Serrano, M., Eds.; Limencop S.L.: Elche, Spain, 2006; pp. 37–40.

horticulturae

Article

Biological Control of *Pseudomonas syringae* in Tomato Using Filtrates and Extracts Produced by *Alternaria leptinellae*

Carlos García-Latorre [1], Sara Rodrigo [2] and Oscar Santamaria [3,*]

[1] School of Agricultural Engineering, University of Extremadura, Avda. Adolfo Suárez s/n, 06007 Badajoz, Spain; cgarcialn@unex.es
[2] Institute of Dehesa Research (INDEHESA), University of Extremadura, Avda. Elvas s/n, Campus de Badajoz, 06006 Badajoz, Spain; saramoro@unex.es
[3] Department of Plant Production and Forest Resources, Sustainable Forest Management Research Institute (iuFOR), University of Valladolid, Avda. Madrid 57, 34004 Palencia, Spain
* Correspondence: oscar.santamaria@uva.es

Abstract: Endophytic fungi offer promising alternatives for sustainable plant disease management strategies, often through the production of bioactive secondary metabolites. This study investigated the biocontrol potential of filtrates and extracts, produced under controlled conditions, from *Alternaria leptinellae* E138 against *Pseudomonas syringae* in tomato plants under greenhouse conditions. To understand the main mechanisms involved in biocontrol, the direct inhibition of bacterial growth and disruption of quorum sensing activity caused by metabolites were studied in vitro, as well as indirect mechanisms, such as their capacity to produce phytohormone-like substances, nutrient mobilization, and antioxidant activity, which can enhance plant growth and fitness. Moreover, a mass spectrometry analysis was used to tentatively identify the secondary metabolites present in the extract with antimicrobial properties, which could explain the biocontrol effects observed. Mycopriming assays, involving the direct treatment of tomato seeds with the fungal *A. leptinellae* E138 extracts, produced increased germination rates and seedling vigor in tomato seeds. As another treatment, postemergence application of the extracts in greenhouse conditions significantly improved plant health and resulted in a 41% decrease in disease severity. Overall, this study underscores the potential of *A. leptinellae* E138 extract as a plant growth promoter with biocontrol capabilities, offering promising avenues for sustainable plant disease management.

Keywords: fungal endophytes; biological control; plant growth promotion; metabolites; sustainable agriculture; *Alternaria*; *Pseudomonas*

1. Introduction

Tomato (*Solanum lycopersicum* L.) is a significant agricultural crop globally, serving as a vital source of food, income, and exports across many regions, including Europe, East Africa, and Indonesia [1,2]. Its economic importance is underscored by global production of 189.1 million t and cultivation on 5.1 million ha in 2021 [3], due to its widespread consumption and utilization in the food industry [4]. However, the tomato crop faces numerous challenges, including diseases, pest infestations, and environmental stressors. Consequently, comprehensive research and management strategies are required to ensure sustainable tomato production, thereby safeguarding food security and economic stability [5,6]. Among the challenges faced, tomato is notably susceptible to the pathogen *Pseudomonas syringae* pathovar tomato, the causal agent of the bacterial speck (BS) disease, which has been associated with substantial crop losses in regions like Europe and Mediterranean Africa [7]. BS disease manifests as small, dark lesions with a speck-like appearance on tomato leaves, stems, and fruit [8], often merging to cause extensive tissue damage [9]. In severe cases, infected fruit can become unmarketable due to their low quality and unsightly appearance [10]. Thus, the economic impact caused by BS of tomato can be profound, resulting in substantial yield losses and rendering the affected fruit unsuitable for both fresh consumption and processing.

Conventional disease control methods primarily involve synthetic chemicals, posing environmental hazards when applied indiscriminately. More novel and environmentally friendly strategies to reduce the impact of this pathogen include the implementation of varietal resistance or epidemiological modeling. Notably, the utilization of fungal endophytes as biocontrol agents (BCAs) to protect tomato plants against pathogens, such as *P. syringae,* has garnered attention due to their potential to enhance plant defense mechanisms and promote growth. Fungal endophytes, residing within plant tissues without causing apparent harm to the host [11], play a crucial role in enhancing plant tolerance to biotic and abiotic stresses, including pathogen defense [12–14]. For instance, Amruthesh et al. [15] demonstrated that fungal endophytic seed treatment induced disease resistance against early blight in tomatoes. Similarly, Constantin et al. [16] found that endophyte-mediated resistance (EMR) against Fusarium wilt disease in tomatoes was independent of certain defense signaling pathways, indicating the potential of endophytic fungi in enhancing plant resistance.

However, the effective use of fungal endophytes as BCAs requires a comprehensive understanding of the endophyte–host relationship, endophyte–pathogen interactions, and the influence of environmental factors on these dynamics [17,18]. The nature of the endophyte-host relationship may vary across tomato cultivars, highlighting the importance of considering host specificity in endophyte selection for biocontrol purposes [19]. Furthermore, the adaptability of endophytes to the edaphoclimatic conditions of the application site compared to those of their isolation is crucial [20]. These factors may affect the efficacy and suitability of direct application of these endophytes for disease control. The utilization of the secondary metabolites produced by the endophyte, rather than the living organism, could provide greater efficacy and mitigate such challenges, as these metabolites have often been identified as the primary agents responsible for biocontrol activity [21,22], and may not be so influenced by host specificity or edaphoclimatic conditions. The application of these bioactive compounds produced in vitro has shown to be promising not only in biocontrol activity but also in promoting various plant growth traits such as germination, vigor, and chlorophyll content, highlighting their potential to enhance overall plant health [23]. Additionally, fungal filtrates have been found to be effective in controlling certain tomato crop pathogens, such as *Alternaria solani* [24]. These compounds also offer the advantage of generally being non-toxic and posing minimal environmental risk, though further evaluation is necessary [25]. Furthermore, they can be easily and inexpensively produced, making them a cost-effective solution for tomato farmers.

In the specific case of *P. syringae,* its biocontrol using fungal endophytes has been demonstrated in various plant species [26,27], including the use of their fungal filtrates [28], which promote plant growth and stimulate plant resistance. Hence, the hypotheses of this study were (1) that the biocontrol activity of a fungal endophyte species (if any) is mainly attributed to the secondary metabolites it produces, and (2) that the filtrates/extracts produced in vitro by this endophyte, which may contain such metabolites, could protect plants against pathogens when applied externally. To test these hypotheses, the main goal of the present study was to assess whether filtrates or extracts of *Alternaria leptinellae,* a fungal endophyte previously isolated from healthy plants and demonstrated to possess antimicrobial activity, could effectively control *P. syringae* upon application to tomato seeds and seedlings. Furthermore, to gain insight into the mechanisms underlying potential protection, several traits of the endophyte and its filtrates/extracts that could directly or indirectly affect plant protection—such as direct antagonism with bacteria, phytohormone production, antioxidant activity, or nutrient solubilization activity—were also evaluated in vitro.

2. Materials and Methods

2.1. Fungal and Plant Material

The fungal endophyte used in this study was previously isolated from healthy leaves of *Ornithopus compressus* plants growing in the *dehesas* of Extremadura, southwestern Spain. The fungus, designated with the internal code E138, was identified as *Embellisia leptinellae* E.G. *Simmons & C.F. Hill* (Thom), the basionym of *Alternaria leptinellae*. Initial identification involved a morphological assessment of its reproductive structures, followed by a molecular characterization through comparison of its ITS region sequence with entries in two databases, GenBank (www.NCBI.nlm.nih.gov) and UNITE (https://unite.ut.ee), employing a BLAST search method [29]. Consequently, the isolate was assigned the Genbank accession number KP698337 [30]. Selection criteria for this species included its high frequency of isolation from the original plant host and promising bioactivity observed in previous assays [31,32].

In vitro antagonism tests were conducted using two bacterial strains, irrespective of their pathogenicity: Gram-positive *Bacillus subtilis* NCTC 8236 and Gram-negative *Pseudomonas syringae* pv. tomato NCPPB 1464, to assess the range of activity of the extract. The same *P. syringae* strain was also utilized for the in-planta experiments. The *Chromobacterium violaceum* strain NCTC 9757 for the quorum sensing test was provided by the Spanish Type Culture Collection (CECT), along with *B. subtilis* NCTC 8236 and *P. syringae* NCPPB 1464 pv. tomato. Bacterial strains were cultivated on EBS medium (in g per L: casein peptone, 5; D-glucose, 5; meat extract, 1; yeast extract, 1; 50 mM Hepes, 11.9) at 30 °C in the dark for 24 h to obtain sufficient inocula. Colony-forming unit (CFU) concentrations were calculated using a Neubauer chamber. For the greenhouse experiments, commercial seeds of *Solanum lycopersicum* (Cultivar Marmande) were utilized.

2.2. Filtrate and Extract Obtention

Two 5 mm agar discs were excised from an actively growing colony (7-day-old) of the endophyte and placed into each of three 500 mL Erlenmeyer flasks containing 250 mL of yeast malt broth (YMB, in g per L: yeast extract: 6; malt extract: 10; D-glucose: 6; pH adjusted to 6.3). The flasks were then incubated in a thermoshaker (Orbital Shaker Incubator COMECTA 1102, Barcelona, Spain) at 23 °C and 140 rpm. Two days after the total consumption of glucose in the medium, the fungal culture was filtered using sterile paper discs (pore = 0.2 μm) to separate the mycelium from the liquid filtrate containing the secondary metabolites [33]. A portion of this filtrate was subjected to an extraction process following the protocol described by Halecker et al. [34], mixing the filtrate with an equal volume of ethyl acetate and shaking vigorously for 2 min. The resulting mixture was then poured into a separatory funnel to allow the two phases to separate. To remove any water residues, a small amount of sodium sulfate was added to the organic phase, which was subsequently filtered through 0.16 mm filters (MN 615 $\frac{1}{4}$). The resulting sample was evaporated using a rotary evaporator to remove the ethyl acetate (Hei-Vap ML/G1, Thermo Fisher Scientific, Waltham, MA, USA). The solid residue was then resuspended in methanol for the in vitro tests, in dimethyl sulfoxide (DMSO, Panreac, Barcelona, Spain) for the quorum sensing activity tests, and in water for the in-planta tests.

2.3. *Effect of the Fungal Filtrate on the Bacterial Growth In Vitro (Disc Diffusion Assay)*

The antimicrobial activity was initially assessed using the disc diffusion method [35]. For this assay, plates (3 replicates per bacterial species) were divided into quadrants. Then, 18 mL of potato dextrose agar (PDA, VWR Chemicals, Pennsylvania, PA, USA, 39 g L^{-1}) was poured into each plate. Before solidification, 2 mL per plate of bacterial solution was added to achieve a concentration of 5×10^4 CFU in the total 20 mL and gently shaken for homogenization. After solidification, a sterile 6 mm diameter filter paper containing 20 μL of the filtrate was placed in the center of each plate. Penicillin, at a concentration of 1.5 mg mL^{-1}, was used as the positive control, while sterile YMB served as the negative control. The plates were then incubated at 30 °C in the dark. In each quadrant, the radius of inhibition produced by the filtrate was measured every 12 h up to 72 h. Based on these measurements, the effective radius of action of the filtrate, where no bacterial growth occurred, was calculated as the average across each quadrant.

2.4. Effect of the Extract on the Quorum Sensing Activity

The efficacy of the *A. leptinellae* 138 extract in inhibiting bacterial quorum sensing activity was tested based on its ability to prevent the production of violacein by *Chromobacterium violaceum* NCTC 9757. The agar diffusion method was employed for this purpose [36,37]. Initially, 50 µL of the organism *C. violaceum* NCTC 9757, cultivated overnight at 30 °C with agitation, were added at a concentration of 10^8 CFU to 5 mL of liquid Luria-Bertani (LB, in g per L: casein peptone, 5; yeast extract, 5; NaCl, 5; agar, 15; pH adjusted to 7.22) agar at 40 °C. Before solidification, 15 mL of the mixture was poured into Petri dishes. Once the medium solidified, 6 mm diameter wells were made and filled with 250 µg of the extract dissolved in 500 µL of DMSO. The plates were then incubated in the dark at 30 °C for 12 h. DMSO was used as the negative control, and 20 µg of the compound (Z-)-4-bromo-5-(bromomethylene)-2(5H)-furanone (Sigma Aldrich, San Luis, MO, USA), a known quorum sensing inhibitor [38], was used as the positive control. Inhibition, indicated by an opaque halo around the disc, was estimated using the equation R2–R1 (in mm), where R1 corresponds to the radius of inhibition of bacterial growth (transparent zone), and R2 includes R1 + the zone where violacein production was inhibited but not the microorganism growth (opaque white zone).

2.5. In Vitro Minimum Inhibitory Concentration of the Extract

The minimum inhibitory concentration (MIC) of the fungal extract against *Bacillus subtilis* NCTC 8236 and *Pseudomonas syringae* NCPPB 1464 was determined using a broth microdilution assay in 96-well plates, following the protocol described by Halecker et al. [34]. A freshly prepared solution (6.7×10^5 CFU mL^{-1}) of each bacterial species (130 µL) was added to each well, followed by 20 µL (300 µg mL^{-1}) of the extract in the first row. Subsequently, a 50% serial dilution was performed for the subsequent rows. To ensure the validity of the test, a positive control of 1.5 mg mL^{-1} of penicillin and a negative control of 20 µL of methanol were included. The plates were then incubated at 30 °C and 600 rpm for 48 h before the results were evaluated. At the end of this period, growth inhibition was qualitatively assessed based on the turbidity or clarity of the culture medium. The test was performed in triplicate.

2.6. In Planta Assays

To assess the efficacy of the extract in protecting tomato plants against *Pseudomonas syringae* NCPPB 1464, two greenhouse experiments were conducted: (i) a mycopriming assay based on the application of the fungal extract to seeds, and (ii) a post-emergence assay, where extracts were applied to seedlings. Before both experiments, tomato seeds were surface-disinfected by immersion in 70% ethanol for one minute, followed by subsequent immersion in a 2% sodium hypochlorite solution for one minute. Afterward, seeds were rinsed three times with sterile distilled water to remove any residual disinfectant [39].

In the mycopriming assay, the disinfected seeds were immersed in the fungal extract (3 mg of extract per mL of sterilized distilled water) for 6 h to ensure optimum uptake of metabolites. Seeds immersed for the same period in sterile distilled water, the solvent used for the extract, served as controls. Sixty seeds per treatment were then individually sown in separate plastic pots (7 × 7 × 6 cm) containing a mixture of substrate and perlite, with a pH of 7.00 ± 0.50, EC of 1.50 ± 0.10 dS m^{-1}, organic matter = 60.0 ± 2.0%, N = 1.29 ± 0.08%, P_2O_5 = 0.58 ± 0.05%, and K_2O = 1.25 ± 0.10%. Before sowing, half of the pots were inoculated with a solution containing the pathogen (100 mL solution per L soil, adjusted to 2×10^4 CFU mL^{-1} of *P. syringae* NCPPB 1464), while the other half received the same volume of a pathogen-free solution. The pots were then placed in a greenhouse for 15 days and watered every 2–3 days until reaching field capacity. Germination was monitored daily.

At the end of the 15-day period, five plants from each treatment combination were randomly harvested. Shoot and root elongation, as well as the number of roots, were measured for each plant. The vigor index of each treatment was calculated by multiplying the respective germination rate (in %) at the end of the test by the shoot length of the corresponding samples. To ensure the validity of the test, this assay was performed simultaneously in triplicate, with each set of 60 seeds considered as a repetition. The experiment took place from 18 February to 4 March 2020.

In the post-emergence test, an additional 60 disinfected seeds were individually sown in pots containing the same substrate mixture, placed in the greenhouse, and watered every 2–3 days until reaching field capacity. After one month, half of the pots received an application of 5 mL of the pathogen solution (2×10^4 CFU mL^{-1} of *P. syringae* NCPPB 1464) per plant. Twelve hours later, each plant was treated with the endophyte by spraying a dose of 1 mL of the extract (3 mg mL^{-1}). A control treatment with 1 mL of sterilized distilled water was included. To assess treatment efficacy, disease severity was measured weekly for one month after application, based on the percentage of damaged leaves, with three leaves per plant randomly selected [40]. To quantify disease progression in each pot, the area under the disease progression curve (AUDPC) was calculated by summing the areas of the corresponding trapezoids. Each period between two consecutive measurements was considered as a unit. After the final measurement, five plants from each treatment were randomly selected for laboratory analysis, where the number and lengths of roots, as well as shoot length, were measured. Herbage and root dry weights were also determined after oven drying the samples at 60 °C until constant weight. To ensure the validity of the results, the assay was conducted simultaneously in duplicate. This experiment was carried out from 18 February to April 30, 2020. The temperature and humidity conditions recorded during both experiments mycopriming and post-emergence are shown in Figure 1.

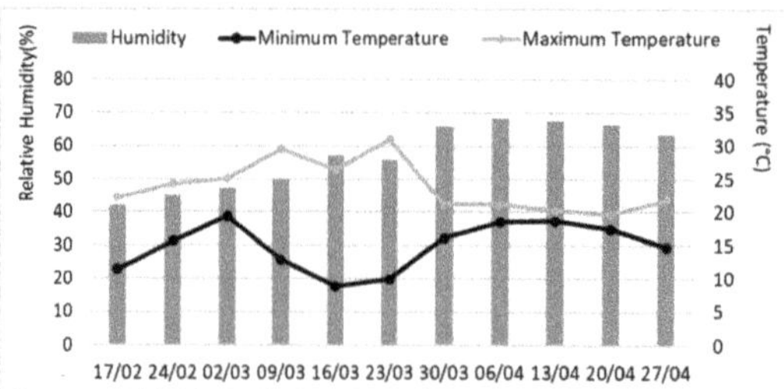

Figure 1. Temperature and relative humidity in the greenhouse during the greenhouse assays. For each date, the average value is shown for each week starting from the specified day.

2.7. Evaluation of Other Protective Traits in *Alternaria leptinellae* E138

The potential of *A. leptinellae* E138 to produce substances directly or indirectly involved in plant protection was assessed using tree key traits: (i) production of phytohormone-like substances, (ii) antioxidant activity and synthesis of phenolic compounds, and (iii) nutrient mobilization, including phosphate solubilization capacity, siderophore production, and ammonia production.

2.7.1. Estimation of Phytohormone-like Substances Production

To determine the production of phytohormone-like substances, we assessed both auxin-like (indoleacetic acid, IAA) and gibberellin-like (gibberellic acid equivalents, GAE) substances in the fungal filtrate. All samples were analyzed in triplicate, and the results were expressed as milligrams of compound (i.e., IAA or GAE, respectively) per milliliter of fungal filtrate. For IAA quantification in the fungal filtrate, a colorimetric test was employed [41]. Salkowski reagent was mixed with the fungal filtrate, and after incubation in the dark for 30 min, the absorbance was measured at 530 nm using a spectrophotometer (JP Selecta UV 3100). The concentration of IAA was determined using a regression equation based on a standard curve of pure IAA (Sigma Aldrich). Additionally, the endophyte was cultured in YMB medium supplemented with L-tryptophan (5 mM), to enhance IAA production [42]. For GAE quantification, a colorimetric method was also employed. Firstly, 15 mL of the fungal filtrate was mixed with 2 mL of zinc acetate (21.9%) and 2 mL of potassium ferrocyanide (10.6%) sequentially at two-minute intervals. The mixture was then centrifuged at 2000 rpm for 15 min. The supernatant was mixed with 5 mL of HCl (30%) and incubated at 20 °C for 75 min. Absorbance was measured at 254 nm, and the results were compared with a calibration curve for gibberellic acid (GA3, Sigma Aldrich) [43]. All samples were analyzed in triplicate.

2.7.2. Determination of the Antioxidant Activity

To assess the antioxidant activity, the fungal extract of *A. leptinellae* E138 was subjected to a DPPH (1,1-diphenyl-2-picrylhydrazyl) assay [44] with slight modifications. A volume of 0.1 mL of the fungal extract at a concentration of 3 mg mL^{-1} was mixed with 2.9 mL of 0.004% aqueous DPPH, and the mixture was incubated in the dark at 25 °C for 30 min. Absorbance was then measured at 517 nm using a spectrophotometer (JP Selecta UV 3100). The DPPH scavenging capacity was calculated using the following formula: % Radical scavenging = (Absorbance Control − Absorbance Sample)/Absorbance Control × 100. All samples were analyzed in triplicate.

2.7.3. Determination of Total Polyphenol Content (TPC)

The Folin–Ciocalteau method was employed to determine the TPC in the fungal extract of *A. leptinellae* E138 [45]. For this analysis, 1 mL of the fungal extract (1 mg mL^{-1}) was mixed with 500 μL of 50% aqueous Folin–Ciocalteau reagent, 1.5 mL of 20% aqueous Na_2CO_3, and 2 mL of distilled water. The mixture was then incubated at room temperature in the dark for 30 min, and the absorbance was measured at 765 nm. The polyphenol content was expressed as milligrams of gallic acid equivalents per gram of fungal extract. All samples were analyzed in triplicate.

2.7.4. Determination of Nutrient Mobilization

Three parameters were evaluated in *A. leptinellae* E138 for nutrient mobilization: phosphate solubilization capacity, siderophore production, and ammonia production.

Phosphate solubilization. For the qualitative assessment, an actively growing plug of *A. leptinellae* E138 was placed in a Petri dish containing the National Botanical Research Institute's phosphate growth medium (NBRIP) amended with 1.5% agar [46,47]. Plates were incubated at 27 °C for 7 days and then examined for a clear halo around the colony. The diameter of this clear zone was measured, and the solubilization capacity of the endophyte was estimated using the following formula: Solubilization Index (%) = (Colony diameter + Clear zone diameter)/(Colony diameter). Three replicates were performed. For quantitative assessment, a piece of active mycelium was inoculated into a flask containing 50 mL of agar-free NBRIP medium and incubated on a shaker at 27 °C and 140 rpm for 15 days. During this period, three aliquots of 10 mL each were taken on days 5, 10, and 15 to assess the pH variation and phosphorus content [48]. The results were compared with a calibration curve of $Ca_3(PO_4)_2$. In both assays, a blank was introduced using a piece of PDA medium without any fungal inoculation and the experiments were conducted in triplicate.

Siderophore production. Siderophore production was estimated using a modified Chrome Azurol S (CAS) universal assay [49]. An actively growing plug of *A. leptinellae* E138 was placed in the center of Petri dishes containing 20 mL of Minimal Medium 9 (MM9) and incubated in a growth chamber at 27 °C for 7 days. After incubation, a modified CAS solution (60.5 mg of CAS, 72.9 mg of HDTMA, 30.24 g of PIPES, and $FeCl_3{\cdot}6H_2O$ in 10 mL 10 mM HCl and agarose 0.9%, *w*/*v*) was added to each Petri dish. After 15 min, a color change was observed in the medium surrounding the siderophore-producing microorganisms. The size of the halo was measured, and the siderophore production was calculated using the same formula as used for the phosphate solubilization.

Additionally, the protocol was repeated with a non-deferrated medium to confirm that siderophore production was not induced under normal conditions. For both assays, a blank was introduced using a plug of uninoculated PDA medium, and three replicates were performed.

Ammonia production. Qualitative assessment of ammonia production was performed by growing *A. leptinellae* E138 in peptone water at 28 °C for 72 h. After incubation, Nessler's reagent was added, and the color change indicated the degree of ammonia production. A pale-yellow color indicated minimal ammonia production, while a deep yellow to brown color indicated maximum ammonia production [50]. The test was also performed in triplicate.

2.8. Mass Spectometry Analysis of the Fungal Extract

For a tentative identification of the metabolites present in the fungal extract, mass spectrometry was performed using an Agilent 6520 Accurate Mass Q-TOf LC/MS system (Agilent, Santa Clara, CA, USA) with an electrospray ionization interface in positive ion mode. The operating parameters were as follows: capillary voltage, 3500 V; fragmenter, 100 V; nebulizer pressure, 35 psig; drying gas temperature, 300 °C; acquisition range 150–800 m/z. Nitrogen was employed as the drying gas at a flow rate of 12.0 L min^{-1}. The system also included a diode array detector operating in the 280 to 350 nm range with a 2 nm step. Samples were eluted on an Agilent Zorbax eEclipse Plus C18 Rapid Resolution column (4.6 × 100 mm, 3.5 μm) maintained at 30 °C. The mobile phase consisted of 0.1% formic acid in ultrapure water (obtained from Millipore Integral-5 purification system) (solvent A) and 0.1% formic acid in acetonitrile (solvent B). A gradient elution was applied as follows: 0–10% B (0 min); 10–100% B (30 min); 100% B isocratic mode (10 min); and for column reconditioning, 100–10% B (1 min) and 10% B (7 min). Both formic acid and acetonitrile were of LC/MS grade. The flow rate was set to 0.30 mL min^{-1}, and the injection volume was 1 μL. By examining the information obtained from the HPLC analysis, including observed mass/charge relationships and proposed formulae, compounds detected by the Q-TOf LC/MS system were tentatively assigned through a literature search.

2.9. Statistical Analysis

One-way ANOVAs and Mixed-effects ANOVAs were used to assess the effect of the filtrate or the extract on bacterial growth and response variables in the greenhouse assays, respectively. The factors considered were endophyte treatment, pathogen inoculation, and their interaction. A Fisher's protected least significant difference (LSD) test at $p \leq 0.05$ was applied for post hoc analyses. Prior to the analysis, assumptions of normal distribution and homoscedasticity were ensured by Shapiro–Wilk and Levene's tests, respectively. All analyses were performed with the Statistix v. 8.10 package (Analytical Software, Tallahassee, FL, USA).

3. Results

3.1. Effect of the Filtrate and the Extract of Alternaria leptinellae E138 on the Bacterial Growth and Quorum Sensing Activity In Vitro

The summary of the one-way ANOVA summary, presented in Table 1, indicates a significant decrease ($p < 0.05$) caused by the endophyte *A. leptinellae* E138 filtrate on the in vitro growth of *Bacillus subtilis* NCTC 8236 and *Pseudomonas syringae* NCPPB 1464, as well as on the quorum sensing activity of *Chromobacterium violaceum* NCTC 9757. In the disc diffusion assay, *A. leptinellae* E138 produced inhibition zones of 0.95 mm and 0.39 mm against each bacterial species, respectively. These values represent 50% and 57.4% of the respective inhibitory effect of penicillin, used as a reference control. The blank (sterilized YM broth medium) showed no inhibitory effect, indicating that the observed inhibition was due to the compounds present in the filtrate. In the evaluation of quorum sensing activity (QSA) using *C. violaceum* NCTC 9757, the *A. leptinellae* E138 extract exhibited an inhibitory effect with an inhibition zone value of 1.38 mm, suggesting a potential role in disrupting bacterial communication mechanisms. In this case, it accounted for 27.6% of the inhibitory effect of the compound (Z-)-4-bromo-5-(bromomethylene)-2(5H)-furanone, used as a control. Positive values for the minimum inhibitory concentration (MIC) were obtained. Thus, the inhibition of *B. subtilis* NCTC 8236 or *P. syringae* NCPPB 1464 was achieved with a concentration of the extract of 125 μg mL^{-1} and 300 μg mL^{-1}, respectively.

Table 1. Effect of *Alternatia leptinellae* E138 filtrates and extracts on the bacterial growth and the quorum sensing activity (QSA).

	Disc Diffusion Assay (mm)		*QSA* Inhibition (mm)	MIC (μg Extract mL^{-1})	
	Bacillus subtilis	*Pseudomonas syringae*		*Bacillus subtilis*	*Pseudomonas syringae*
E138	0.95 ± 0.03 b	0.39 ± 0.04 b	1.38 ± 0.13 b	125.00 ± 25.00	300.00 ± 0.00
Blank [1]	0.00 ± 0.00 c	0.00 ± 0.00 c	0.00 ± 0.00 c	0.00 ± 0.00	0.00 ± 0.00
Control [2]	1.90 ± 0.06 a	0.68 ± 0.04 a	5.00 ± 0.38 a	1.24 ± 0.00	1.24 ± 0.00
df	2	2	2	-	-
Endophyte	689.18 ***	117.72 ***	127.25 ***	-	-

QSA on *Chromobacterium violaceum* NCTC 9757. Values (estimated by the radius of the inhibition zone, in the disc diffusion assay) are expressed as mean ± error standard (n = 4). Different letters for each column indicate significant differences according to LSD (least significant difference) test at $\alpha = 0.05$. [1] Blank: sterilized YM broth medium for the disc diffusion assay, and methanol for the MIC assay. [2] Control: penicillin (1.5 mg mL^{-1}) and (Z-)-4-bromo-5-(bromomethylene)-2(5H)-furanone (20 μg), in the disc diffusion assay and QSA, respectively. A summary of the one-way ANOVAs in the disc diffusion assays is shown at the bottom, indicating the degree of freedom (df), F-value and the level of significance (*** $p \leq 0.001$).

3.2. Effect of Mycopriming with A. leptinellae E138 Extract on the Control of Pseudomonas syringae NCPPB 1464 in Tomato Plants under Greenhouse Conditions

Regarding the germination rate of *S. lycopersicum*, it was significantly affected by the treatment with the endophytic extract and its interaction with the presence or absence of the pathogen *P. syringae* NCPPB 1464 for each day of measurement (from the fourth day until the fifteenth day after sowing), except on the sixth and seventh days (Table 2). This variability in the nature of the relationship between the endophyte extract and the pathogen, manifested in the different F- and p-values among days, tended to stabilize from day 9 onwards.

Table 2. ANOVAs showing the effect of *Alternaria leptinellae* E138 extract, *Pseudomonas syringae* NCPPB 1464, and their interaction on the germination rate of *Solanum lycopersicum* seeds.

	Day 4	Day 5	Day 6	Day 7	Day 8	Day 9
df	1	1	1	1	1	1
Endophyte (E)	8.00 *	25.48 **	180.44 ***	112.45 ***	98.01 ***	484.99 ***
Pathogen (P)	11.99	21.07 *	3.57	12	2.28	4.92
E × P	16.00 *	48.02 *	3.00	3.00	15.93*	99.91 **
	Day 10	Day 11	Day 12	Day 13	Day 14	Day 15
df	1	1	1	1	1	1
Endophyte (E)	264.66 ***	220.67 ***	242.20 ***	72.09 **	72.09 **	72.09 **
Pathogen (P)	7	6.26	12.01	12.01	12.01	12.01
E × P	121.45 **	360.00 **	64.07 *	100.36 **	100.36 **	100.36 **

The degrees of freedom (df), the F-values, and the levels of significance (* $p \leq 0.05$; ** $p \leq 0.01$; *** $p \leq 0.001$) are shown for each day of measurement.

Analysis of germination rates for tomato seeds, as presented in Figure 2, revealed different patterns depending on the presence or absence of *P. syringae* NCPPB 1464 and the application of *A. leptinellae* extract E138 compared to the control group. In the absence of *P. syringae* NCPPB 1464, the control group without extract (Control−) exhibited a steady increase in germination rates from day 4 to day 9, and then stabilized until day 15, reaching a germination rate of around 71% at the end. In contrast, in presence of the pathogen (Control+), there was a significant decrease in germination rates, reaching a maximum (only 53.33%) on day 11, demonstrating the pathological nature of the strain used. The application of the *A. leptinellae* extract (138−) increased the germination rate compared to the controls (Control−) and also reverted the negative impact of the pathogen when it was also inoculated (138+). Thus, regardless of the presence or absence of *P. syringae* NCPPB 1464, the germination rate reached maximum values of 86–91% by day 13 when the endophyte extract was applied. This result demonstrates not only the lack of toxicity of the endophyte extract on tomato seeds, at least in terms of their germination rate, but also a clear biocontrol and plant-growth-promoting activity capable of reverting the negative effects of *P. syringae* NCPPB 1464.

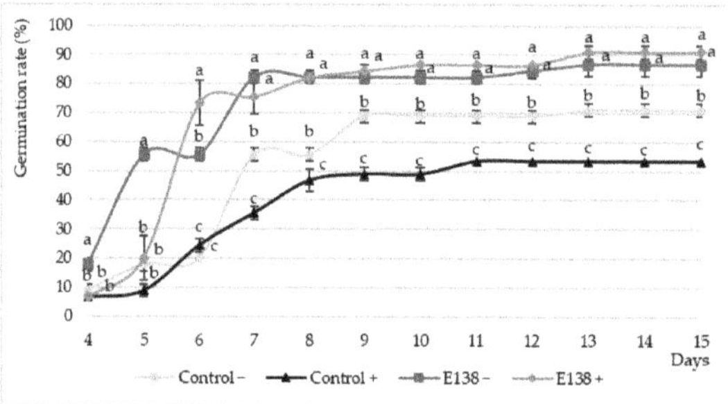

Figure 2. Effect of endophytic extract (E138) on the germination percentage of *Solanum lycopersicum* seeds inoculated (+) and not inoculated (−) with *Pseudomonas syringae* NCPPB 1464 from 4 to 15 days after sowing (12 measurements). Values are expressed as mean (n = 3) ± standard error (error bars). In each day of measurement, different letters indicate significant differences between treatments according to the LSD (least significant difference) test at α = 0.05.

In relation to the growth parameters of tomato plants under greenhouse conditions, both the treatment with *A. leptinellae* E138 extract and inoculation with *P. syringae* NCPPB 1464 significantly affected shoot and root lengths, the number of roots, and the vigor index of seedlings. The interaction between these factors demonstrated significant influence ($p < 0.05$) on shoot and root lengths and vigor index (Table 3). The presence of the pathogen led to a significant decrease in root length and vigor index compared to the control group (root length: 1.43 cm vs. 1.85 cm and vigor index: 466.64 vs. 661.32). Application of the E138 extract not only reverted the negative effect of the pathogen but also resulted in increased parameters compared to the controls (root length: 2.58 cm and vigor index: 924.77). This increase can be attributed to both genuine biocontrol activity and clear plant-growth-promoting effects, as evidenced by superior growth parameters in the absence of the pathogen (root length: 3.48 cm and vigor index: 951.17; Table 3). In this latter case, there was also a significant increase in the number of roots (2.25 vs. 1.00). These increases were 88.11%, 125%, and 43.83%, respectively, compared to seedlings in the control treatment growing in the absence of the pathogen.

Table 3. Growth parameters of tomato plants (shoot length, root length, number of roots, and vigor index) as affected by treatment of seeds with the *Alternaria leptinellae* E138 extract, inoculation with *Pseudomonas syringae* NCPPB 1464, and their interaction under greenhouse conditions.

Endophyte	Pathogen	Shoot Length (cm)	Root Length (cm)	Number of Roots	Vigor Index
Control	No	9.30 ± 0.18	1.85 ± 0.05 c	1.00 ± 0.00 b	661.32 ± 12.65 b
	Yes	8.75 ± 0.25	1.43 ± 0.08 d	1.00 ± 0.00 b	466.64 ± 13.33 c
E138	No	10.98 ± 0.24	3.48 ± 0.05 a	2.25 ± 0.25 a	951.17 ± 20.74 a
	Yes	10.15 ± 0.15	2.58 ± 0.05 b	1.00 ± 0.00 b	924.77 ± 13.67 a
df		1	1	1	1
Endophyte (E)		46.55 ***	443.53 ***	25.00 **	613.66 ***
Pathoten (P)		10.56 *	238.47 ***	25.00 *	42.34 **
E × P		0.68	21.24 *	25.00 *	33.64 *

Values are expressed as mean (n = 5) ± standard error. For each parameter, different letters (if any) indicate significant differences according to LSD (least significant difference) test at α = 0.05. A summary of the ANOVAs is also given at the bottom of the table. The degrees of freedom (df), the F-values, and the levels of significance (* $p \leq 0.05$; ** $p \leq 0.01$; *** $p \leq 0.001$) are given for each parameter.

3.3. *Effect of the Post-Emergence Application of A. leptinellae E138 Extract on the Control of Pseudomonas syringae NCPPB 1464 in Tomato Plants under Greenhouse Conditions*

Regarding post-emergence application, the ANOVAs revealed that both the application of *A. leptinellae* extract E138 and the inoculation with *P. syringae* NCPPB 1464, had a significant influence on all the growth parameters studied (Table 4). The pathogenicity of the *P. syringae* NCPPB 1464 strain used was demonstrated by its inoculation resulting in a significant decrease in all parameters compared to the controls (shoot length: 4.88 cm vs. 26.64 cm, root length: 4.66 cm vs. 16.40 cm, roots per plant: 8.40 vs. 20.20; Table 4). Once again, the application of the endophyte extract not only reverted the negative effects of the pathogen (resulting in a 2.34-fold, 1.42-fold, and 2-fold increase in shoot, root, and total dry matter weights, respectively, compared to the plants inoculated with the pathogen), but also enhanced these growth parameters compared to the non-inoculated controls (with increases of 35.66% and 88.12% for shoot length and root length, respectively). Finally, the AUDPC showed a partial reversion of the symptoms exhibited by plants inoculated with *P. syringae* NCPPB 1464, when treated with the extracts, resulting in a 41.10% reduction in disease severity (Table 4). The non-inoculated plants treated with the endophyte extract did not exhibit any disease symptoms, further demonstrating its non-phytotoxic nature in tomato plants.

Table 4. Effect of the post-emergence application of the extract of *Alternaria leptinellae* E138, the inoculation with the pathogen *Pseudomonas syringae* NCPPB 1464, and their interaction on the different growth traits (shoot and root length; number of roots per plant and number of plants per pot; shoot, root, and total dry weight), and on the disease severity (estimated though the area under the disease progression curve; AUDPC) in tomato plants under greenhouse conditions.

Endophyte	Pathogen Presence	Shoot Length (cm)	Root Length (cm)	Number of Roots	AUDPC
Control	No	26.64 ± 1.71 b	16.40 ± 0.63 b	20.20 ± 0.56 b	0.00 ± 0.00 c
	Yes	4.88 ± 2.99 c	4.64 ± 2.85 c	8.40 ± 5.15 c	66.50 ± 1.29 a
E138	No	36.14 ± 2.10 a	21.54 ± 0.64 a	38.00 ± 0.71 a	0.00 ± 0.00 c
	Yes	38.00 ± 1.24 a	16.16 ± 0.11 b	42.60 ± 0.93 a	39.20 ± 3.06 b
		Shoot dry matter weight (g)	Root dry matter weight (g)	Total dry matter weight (g)	
Control	No	173.05 ± 7.47 b	74.35 ± 3.60 b	247.4 ± 9.16 b	
	Yes	37.72 ± 23.85 c	9.638 ± 5.91 c	47.358 ± 29.66 c	
E138	No	578.35 ± 32.17 a	180.65 ± 14.47 a	759.00 ± 39.37 a	
	Yes	581.05 ± 26.52 a	163.375 ± 7.38 a	744.425 ± 19.92 a	
	df	Shoot Length (cm)	Root Length (cm)	Number of roots	AUDPC
Endophyte (E)	1	229.50 ***	43.47 ***	93.5 ***	67.36 ***
Pathogen (P)	1	11.60 *	21.06 *	1.59	2205.82 ***
E × P	1	25.91 **	4.54 *	11.87 *	43.67 *
Source	df	Shoot dry matter weight (g)	Root dry matter weight (g)	Total dry matter weight (g)	
Endophyte (E)	1	276.01 ***	273.98 ***	412.61 ***	
Pathogen (P)	1	12.44 *	11.92 *	17.98 *	
E × P	1	12.58 *	11.89 *	17.2 *	

Values are expressed as mean (n = 5) ± standard error. For each parameter, different letters indicate significant differences according to LSD (least significant difference) test at $\alpha = 0.05$. A summary of the ANOVAs is also given at the bottom of the table. The degrees of freedom (df), the F-values, and the levels of significance (* $p \leq 0.05$; ** $p \leq 0.01$; *** $p \leq 0.001$) are given for each factor.

3.4. *Traits of A. leptinellae E138 Directly or Indirectly Related to Its Biocontrol Activity*

In terms of phytohormone-like production, the results indicate that the endophyte E138 was capable of producing indole-acetic acid (IAA), particularly when the culture media were supplemented with tryptophan (IAA+, resulting in a threefold increase), and gibberellins (GA3) (Table 5). Regarding antioxidant activity, although *A. leptinellae* E138 produced a total polyphenol content of 68.02 mg GAE (gallic acid equivalent) per gram of extract and 43.35 mg of quercetin equivalent per gram of extract, it exhibited minimal DPPH scavenging activity, with a result of only 1.47% (Table 5). The endophyte also showed a positive potential for ammonia synthesis and phosphate solubilization in qualitative tests.

Table 5. Traits of *Alternaria leptinellae* E138 related to its biocontrol activity (phytohormone-like production, antioxidant activity, and nutrient mobilization capacity).

Trait		
Group	**Activity**	**Mean ± SE**
Phytohormone	IAA (µg mL^{-1})	6.10 ± 0.05
	IAA+ (µg mL^{-1})	19.22 ± 0.20
	GA_3 (µg mL^{-1})	442.88 ± 0.36
Antioxidant activity	DPPH (%)	1.47 ± 0.61
	TPC (mg GAE g^{-1})	68.02 ± 4.46
Nutrient mobilization	Siderophore	−
	P solubilization	+
	Ammonia production	++

IAA, indole-acetic acid in the filtrate; IAA+, indole-acetic acid in the filtrate with culture media supplemented with L-tryptophan; GA3, gibberellic Acid in the filtrate; DPPH, % of scavenging of DPPH in the extract; TPC, total polyphenol content in the extract. For the parameters related to nutrient mobilization, the positive or negative potential to produce each type of compounds is shown (− no production; + production; ++ high production). Quantitative data are expressed as mean ± standard error (n = 3).

Furthermore, in the quantitative assay, the endophyte E138 demonstrated significant phosphorus solubilization capabilities, as evidenced by the increased percentage of solubilized P Olsen after 5, 10, and 15 days. The sustained increase in solubilization over time indicates the efficiency of E138 in rendering phosphorus more available to plants without acidifying the environment (Table 6).

Table 6. Quantitative analysis of the phosphorus solubilization in vitro by *Alternaria leptinellae* E138, showing the percentage of P Olsen solubilized after 5, 10 and 15 days, together with the pH measured at each moment.

	P Solubilization Day 5 (%)	**P Solubilization Day 10 (%)**	**P Solubilization Day 15 (%)**
E138	67.60 ± 0.34 a	68.34 ± 0.64 a	69.67 ± 0.41 a
Control	0.47 ± 0.03 b	0.08 ± 0.10 b	0.10 ± 0.13 b
df	1	1	1
Endophyte	8502.89 ***	16,798.50 ***	37,195.30 ***
	pH day 5	**pH day 10**	**pH day 15**
E138	6.51 ± 0.08	6.70 ± 0.05 a	6.68 ± 0.02 a
Control	6.21 ± 0.05	6.12 ± 0.04 b	5.80 ± 0.00 b
df	1	1	1
Endophyte	5.11	24.14 **	91.58 ***

Values are expressed as mean (n = 3) ± standard error. A summary of ANOVAs is also given at the bottom of the table. Different letters for each column indicate significant differences according to LSD (least significant difference) test at α = 0.05. The degrees of freedom (df), the F-values, and the levels of significance (** $p \leq 0.01$; *** $p \leq 0.001$) are given for each parameter.

3.5. Tentative Identification of Metabolites by Mass Spectometry

As a preliminary attempt to identify the metabolites responsible for the observed plant protection effects, we analyzed the compounds present in the fungal extract obtained through mass spectrometry analysis. Compounds were tentatively assigned by examining mass/charge relationships and proposed formulae through a literature search. Mass spectrometry graphs associated with each compound are displayed in Supplementary Figures (Figures S1–S4). Table 7 provides a comprehensive overview of the compounds identified in the methanolic extracts of *A. leptinellae* E138, detailing their molecular characteristics and activities. As a result, the following compounds were proposed: altechromone A, maculosin, ciclo(L-Phe-L-Pro), and phomopsinone A. All of these compounds were found to have activities consistent with the results obtained in the different experiments conducted in this study, ranging from plant growth promotion to antimicrobial and antifungal activities. However, these preliminary results require further research to unequivocally confirm their identification.

Table 7. Peak assignment for methanolic extracts of *Alternaria leptinellae* E138.

N°	Proposed Formula	Rt (min)	Obs. *m/z*	Proposed Compound	Activity	Reference
Cpd01	$C_{11}H_{10}O_3$	5.344	190.0623	Altechromone A	Plant growth promotion, antimicrobial activity (biofilm inhibition)	[51,52]
Cpd02	$C_{14}H_{16}N_2O_3$	8.836	260.1155	Maculosin	Antibacterial and antioxidant activity	[53]
Cpd03	$C_{14}H_{16}N_2O_2$	12.979	244.1205	Ciclo(L-Phe-L-Pro)	Influence on QSA	[54]
Cpd04	$C_{12}H_{16}O_4$	15.242	224.1042	Phomopsinone A	Antifungal activity	[55,56]

Rt: retention time; Obs *m*/*z*: observed mass/charge relationship; QSA: quorum sensing activity.

4. Discussion

The results of our study explore the potential of *A. leptinellae* E138 as a biocontrol agent against *Pseudomonas syringae* in tomato plants, emphasizing its multifaceted action as an antimicrobial and a plant growth promoter. In this study, we focused on the direct use of the filtrates and extracts produced by the endophyte, as their effectiveness has already been demonstrated in other cases, overcoming the drawbacks associated with using the living organism [57,58]. Results regarding the effect of *A. leptinellae* E138 filtrates on bacterial growth in vitro demonstrate significant biocontrol potential against *B. subtilis* NCTC 8236 and *P. syringae* NCPPB 1464. This was supported by MIC tests and the evaluation of quorum sensing activity (QSA), wherein the endophyte extracts inhibited the growth of both bacteria and disrupted QSA in *C. violaceum* NCTC 9757.

The observed inhibition zones in the disc diffusion assay, although slightly smaller than the reference control (penicillin), underscore the potential of *A. leptinellae* E138 as a source of antimicrobial compounds that may prove valuable in biocontrol strategies against *P. syringae*. Furthermore, the effect of the endophyte on QSA adds another dimension to its biocontrol mechanisms. Disrupting bacterial communication mechanisms may reduce the coordination of pathogenic activities, providing an additional layer of defense against *P. syringae* [59]. This is consistent with the findings of Adonizio et al. [60], who demonstrated the potential of inhibiting the quorum sensing activity of a *P. aeruginosa* bacterial colony to diminish its virulence factor. The MIC values obtained evidenced the efficacy of the *A. leptinellae* E138 extract against *P. syringae* NCPPB 1464, given the low concentration required to inhibit the pathogen (300 $\mu g\ mL^{-1}$). This highlights the practical feasibility of incorporating *A. leptinellae* E138 extract into biocontrol strategies. This agrees with the findings of Singh et al. [61], who demonstrated the anti-biofilm and anti-quorum sensing activity of microbial extracts against *Pseudomonas* sp. Understanding the MIC values is crucial for determining appropriate concentrations in field applications to ensure optimal efficacy against target pathogens.

This antimicrobial activity observed in *A. leptinellae* E138 filtrates and extracts may be attributed to the secondary metabolites produced by the endophyte, which could possess biocontrol properties. Several studies have already demonstrated this ability of endophytes to produce antimicrobial compounds that effectively inhibit the growth of bacterial pathogens [62,63] or disrupt bacterial communication mechanisms [63–65]. This is in clear agreement with the compounds tentatively identified by mass spectrometry in the methanolic fungal extract of *A. leptinellae* E138: altechromone A, maculosin, ciclo(L-Phe-L-Pro), and phomopsinone A. The presence of altechromone A, which has been already shown to promote plant growth and inhibit the formation of bacterial biofilms [51,52], could help explain our results regarding the inhibition of bacterial growth and the overall improvement in plant development, regardless of the presence or absence of *P. syringae*

NCPPB 1464. This effect could have been facilitated by the presence of maculosin, known for its antibacterial and antioxidant properties [53], and ciclo(L-Phe-L-Pro), a compound previously found to disrupt QSA [66]. The identification of phomopsinone A, which has demonstrated antifungal potential [55,56], may suggest that the action of *A. leptinellae* E138 is not solely limited to bacterial control but may also be applicable in strategies against fungal pathogens. This finding aligns with previous studies highlighting the role of endophytic fungi as sources of natural bioactive compounds with diverse biological activities [67–71].

In addition to the effect of the antimicrobial metabolites produced by *A. leptinellae* E138, which may directly explain the biocontrol of *P. syringae* NCPPB 1464, other indirect mechanisms may have contributed to this biocontrol and the observed plant growth promotion. All these direct and indirect traits exhibited by *A. leptinellae* E138, such as the production of antimicrobial compounds, phytohormone-like production, phosphate solubilization, and antioxidant activity, could indicate a multifaceted fungal species that could be considered not only as a biocontrol agent but also as a contributor to plant health and disease resistance, a broader concept proposed by Baron and Rigobelo [72]. Improving plant growth, in addition to increasing fruit yield expectations, may also impact plant fitness and consequently its defensive system [73]. Therefore, plant-growth-promoting traits could also be indirectly considered as biocontrol traits. The production of phytohormones, particularly indoleacetic acid (IAA) and gibberellins (GA3), by *A. leptinellae* E138 is consistent with other studies that have highlighted the role of endophytic fungi in promoting plant growth through their synthesis [74], as these phytohormones play a crucial role in regulating plant growth and development. Additionally, the ability of *A. leptinellae* E138 to enhance phosphate solubilization and ammonia synthesis underscores its potential for nutrient mobilization, consistent with the plant-growth-promoting functions of endophytic microorganisms reported in previous studies [75]. Furthermore, the sustained increase in phosphorus solubilization over time may suggest the efficiency of *A. leptinellae* E138 in making phosphorus more available to plants, which is also crucial for plant growth and development [76]. Finally, the total polyphenol content of the *A. leptinellae* E138 extract may indicate its potential to mitigate oxidative stress in plants, which may also contribute to overall plant health and resilience [77].

Based on the results of the mycopriming assay, the *A. leptinellae* E138 extract significantly increased the germination rates of *S. lycopersicum* seeds, regardless of the presence or absence of the pathogen *P. syringae* NCPPB 1464. This suggests that the *A. leptinellae* E138 extract may be an effective germination stimulant in tomato seeds, facilitating more robust seedling establishment, even in the presence of pathogens. This is supported by the significant increases in root length, number of roots, and vigor index of seedlings observed after its application. Seed treatment with beneficial fungi is gaining popularity due to its potential to contribute to more sustainable agriculture. Various articles have already demonstrated its effectiveness in promoting the development of different plant species, both in terms of growth promotion [78] and protection against biotic [79,80] and abiotic [81,82] stresses. In our case, although the results presented demonstrate a lack of phytotoxicity when used on tomato plants, further studies including a wider range of crops and employing different application conditions could allow for evaluating the actual potential of this product for future commercial use. The prospects are promising based on previous findings that have demonstrated the ability of endophytic extracts and microbial inoculants to enhance plant growth and confer resistance to pathogens [83–85].

The postemergence application of *A. leptinellae* E138 extract exhibited significant effects on various aspects of tomato plant health and disease severity under greenhouse conditions, particularly in presence of the pathogen *P. syringae* NCPPB 1464. The significant improvements in shoot and root length, the number of roots, and the vigor index, even in the presence of *P. syringae* NCPPB 1464, underscore the potential of *A. leptinellae* E138 as a growth-promoting endophyte. These findings align with the concept of utilizing endophytes not only for biocontrol but also to enhance overall plant health and fitness.

Additionally, the area under the disease progression curve (AUDPC) values revealed a significant reduction in disease severity in the presence of the endophyte, emphasizing its potential role in mitigating the impact of *P. syringae*-induced diseases on tomato plants. This is consistent with the findings of Mostafa et al. [86], who demonstrated the impairment of virulence and inhibition of quorum sensing of *P. aeruginosa* through the action of different metabolites such as polyphenols from *Salix tetrasperma*. Similarly, Yoo et al. [87] described the potential of the strain *Aspergillus terreus* JF27 for the biocontrol of *P. syringae* in tomato plants, impacting the same parameters observed in our study. Thus, inoculation with the strain significantly reduced disease severity and enhanced shoot length and weight under greenhouse conditions. This outcome highlights the potential of direct application of the fungal extract, which may have a similar effect to that observed with the fungal inoculation. Therefore, the post-emergence application of *A. leptinellae* E138 extract has demonstrated significant potential to improve tomato plant growth and mitigate the effects of *P. syringae*-induced diseases under greenhouse conditions. The results are consistent with previous research on the beneficial effects of endophytic fungi on plant health and disease resistance and underscore the promising role of *A. leptinellae* E138 in sustainable plant disease management strategies.

5. Conclusions

Our study revealed the multifaceted potential of *Alternaria leptinellae* E138 as a biocontrol agent against *Pseudomonas syringae* in tomato plants, as well as its role as a plant growth promoter. Filtrates and extracts from this endophyte exhibited remarkable antimicrobial activity, disruption of quorum sensing activity, phytohormone-like production, phosphate solubilization, and antioxidant activity. Several secondary metabolites were tentatively identified in the methanolic extract of the endophyte, including altechromone A, maculosin, ciclo(L-Phe-L-Pro), and phomopsinone A, compounds with either plant-growth-promoting or antimicrobial and antifungal activities. Application of the extract via mycopriming increased seed germination and improved growth parameters of tomato seedlings, regardless of the presence or absence of *P. syringae*. Its postemergence application mitigated disease severity and promoted tomato plant growth under greenhouse conditions. This study provides valuable insights into the field of endophytic research and phytopathology, contributing to sustainable plant disease management in agriculture. Further research should delve deeper into the metabolites involved in biocontrol and incorporate field trials encompassing a wider range of environments, crops, and pathogens.

Supplementary Materials: The following supporting information can be downloaded at: https://www.mdpi.com/article/10.3390/horticulturae10040334/s1, Figure S1. ESI-LC-MS mass spectrum extracted from the liquid chromatogram peak (above) and the extracted ion mass spectrum corresponding to the signal of Altechromone A (according to 51,52) at m/z 191.06 (M+H+) (below). Figure S2. ESI-LC-MS mass spectrum extracted from the liquid chromatogram peak (above) and the extracted ion mass spectrum corresponding to the signal of Maculosin (according to 53) at m/z 261.12 (M+H+) (below). Figure S3. ESI-LC-MS mass spectrum extracted from the liquid chromatogram peak (above) and the extracted ion mass spectrum corresponding to the signal of Ciclo(L-Phe-L-Pro; according to 54) at m/z 245.12 (M+H+) (below). Figure S4. ESI-LC-MS mass spectrum extracted from the liquid chromatogram peak (above) and the extracted ion mass spectrum corresponding to the signal of Phomopsinone A (according to 55,56) at m/z 225.11 (M+H+) (below).

Author Contributions: Conceptualization, C.G.-L., O.S. and S.R.; methodology, C.G.-L. and S.R.; formal analysis, C.G.-L.; investigation, C.G.-L.; data curation, O.S.; writing—original draft preparation, C.G.-L. and O.S.; writing—review and editing, O.S. and S.R.; supervision, O.S.; project administration, S.R.; funding acquisition, O.S. and C.G.-L. All authors have read and agreed to the published version of the manuscript.

Funding: Carlos García-Latorre has been financed by a pre-doctoral grant (PD18037) from the Regional Government of Extremadura (Spain) and by the European Social Fund (ESF).

Data Availability Statement: Data are contained within the article and Supplementary Materials.

Conflicts of Interest: The authors declare no conflicts of interest.

References

1. Baker, S.; Al- Assie, A.; Mahmood, R. Determination of Some Morphological Traits and Genetic Factor of Two Species of *Liriomyza* Spp. (*Diptera: Agromyzidae*) on the Climbing Beans and Tomato Crops. *Tikrit J. Agric. Sci.* **2022**, *19*, 57–68. [CrossRef]
2. Thompson, N.M.; Waterton, N.; Armaou, A.; Polston, J.E.; Curtis, W.R. Establishing an Inexpensive, Space Efficient Colony of Bemisia Tabaci MEAM1 Utilizing Modelling and Feedback Control Principles. *J. Appl. Entomol.* **2022**, *146*, 648–658. [CrossRef]
3. Food and Agriculture Organization of the United Nations. FAOSTAT Database [Online]. 2023. Available online: https://www.fao.org/statistics/en/ (accessed on 3 February 2024).
4. Pascual, L.; Blanca, J.M.; Cañizares, J.; Nuez, F. Transcriptomic Analysis of Tomato Carpel Development Reveals Alterations in Ethylene and Gibberellin Synthesis during Pat3/Pat4 Parthenocarpic Fruit Set. *BMC Plant Biol.* **2009**, *9*, 67. [CrossRef]
5. Li, H.; Liu, H.; Gong, X.; Li, S.; Pang, J.; Chen, Z.; Sun, J. Optimizing Irrigation and Nitrogen Management Strategy to Trade off Yield, Crop Water Productivity, Nitrogen Use Efficiency and Fruit Quality of Greenhouse Grown Tomato. *Agric. Water Manag.* **2021**, *245*, 106570. [CrossRef]
6. Panno, S.; Davino, S.; Caruso, A.G.; Bertacca, S.; Crnogorac, A.; Mandić, A.; Noris, E.; Matić, S. A Review of the Most Common and Economically Important Diseases that Undermine the Cultivation of Tomato Crop in the Mediterranean Basin. *Agronomy* **2021**, *11*, 2188. [CrossRef]
7. Silva, G.A.; Picanço, M.C.; Bacci, L.; Crespo, A.L.B.; Rosado, J.F.; Guedes, R.N.C. Control Failure Likelihood and Spatial Dependence of Insecticide Resistance in the Tomato Pinworm, *Tuta Absoluta*. *Pest Manag. Sci.* **2011**, *67*, 913–920. [CrossRef]
8. Schiavi, D.; Balbi, R.; Giovagnoli, S.; Camaioni, E.; Botticella, E.; Sestili, F.; Balestra, G.M. A Green Nanostructured Pesticide to Control Tomato Bacterial Speck Disease. *Nanomaterials* **2021**, *11*, 1852. [CrossRef] [PubMed]
9. Preston, G.M. *Pseudomonas syringae* Pv. Tomato: The Right Pathogen, of the Right Plant, at the Right Time. *Mol. Plant Pathol.* **2000**, *1*, 263–275. [CrossRef]
10. Quaglia, M.; Bocchini, M.; Orfei, B.; D'Amato, R.; Famiani, F.; Moretti, C.; Buonaurio, R. Zinc Phosphate Protects Tomato Plants against *Pseudomonas syringae* Pv. Tomato. *J. Plant Dis. Prot.* **2021**, *128*, 989–998. [CrossRef]
11. Rodriguez, R.J.; White, J.F.; Arnold, A.E.; Redman, R.S. Fungal Endophytes: Diversity and Functional Roles. *New Phytol.* **2009**, *182*, 314–330. [CrossRef]
12. Maciá-Vicente, J.G.; Jansson, H.-B.; Lopez-Llorca, L. V Assessing Fungal Root Colonization for Plant Improvement. *Plant Signal. Behav.* **2009**, *4*, 445–447. [CrossRef] [PubMed]
13. Xia, Y.; Sahib, M.R.; Amna, A.; Opiyo, S.O.; Zhao, Z.; Gao, Y.G. Culturable Endophytic Fungal Communities Associated with Plants in Organic and Conventional Farming Systems and Their Effects on Plant Growth. *Sci. Rep.* **2019**, *9*, 1669. [CrossRef] [PubMed]
14. Morsy, M.; Cleckler, B.; Armuelles-Millican, H. Fungal Endophytes Promote Tomato Growth and Enhance Drought and Salt Tolerance. *Plants* **2020**, *9*, 877. [CrossRef] [PubMed]
15. Amruthesh, K.N.; Murali, M.; Sujatha, H.S. Fungal Endophytes as Growth Promoters and Inducers of Resistance in Tomato (*Lycopersicon esculentum* Mill.) against *Alternaria Solani*: Life Sciences-Botany for Medicinal Science. *Int. J. Life Sci. Pharma Res.* **2021**, *11*, L227–L235. [CrossRef]
16. Constantin, M.E.; de Lamo, F.J.; Vlieger, B.V.; Rep, M.; Takken, F.L.W. Endophyte-Mediated Resistance in Tomato to *Fusarium Oxysporum* Is Independent of ET, JA, and SA. *Front. Plant Sci.* **2019**, *10*, 979. [CrossRef]
17. Mohammad Golam Dastogeer, K.; Oshita, Y.; Yasuda, M.; Kanasugi, M.; Matsuura, E.; Xu, Q.; Okazaki, S. Host Specificity of Endophytic Fungi from Stem Tissue of Nature Farming Tomato (*Solanum lycopersicum* Mill.) in Japan. *Agronomy* **2020**, *10*, 1019. [CrossRef]
18. Abo Nouh, F.; Abu-Elsaoud, A.; Abdel-Azeem, A. The Role of Endophytic Fungi in Combating Abiotic Stress on Tomato. *Microb. Biosyst.* **2021**, *6*, 35–48. [CrossRef]
19. Halo, B.A.; Al-Yahyai, R.; Al-Sadi, A.; Al-Sibani, A. Desert Endophytic Fungi Improve Reproductive, Morphological, Biochemical, Yield and Fruit Quality Characteristics of Tomato under Drought Stress. *Arab Gulf J. Sci. Res.* **2023**, *41*, 638–655. [CrossRef]
20. Coninck, E.; Scauflaire, J.; Gollier, M.; Liénard, C.; Foucart, G.; Manssens, G.; Munaut, F.; Legrève, A. *Trichoderma Atroviride* as a Promising Biocontrol Agent in Seed Coating for Reducing *Fusarium* Damping-off on Maize. *J. Appl. Microbiol.* **2020**, *129*, 637–651. [CrossRef]
21. Rodrigo, S.; Santamaria, O.; Halecker, S.; Lledó, S.; Stadler, M. Antagonism between *Byssochlamys Spectabilis* (Anamorph *Paecilomyces variotii*) and Plant Pathogens: Involvement of the Bioactive Compounds Produced by the Endophyte. *Ann. Appl. Biol.* **2017**, *171*, 464–476. [CrossRef]
22. James, D.; Gleena Mary, C.F.; Mathew, S.K. Evaluation of Culture Filtrates of Endophytic Microorganisms from Tomato against *Ralstonia solanacearum*. *Int. J. Environ. Clim. Chang.* **2022**, *12*, 1476–1481. [CrossRef]
23. García-Latorre, C.; Rodrigo, S.; Santamaría, O. Potential of Fungal Endophytes Isolated from Pasture Species in Spanish Dehesas to Produce Enzymes under Salt Conditions. *Microorganisms* **2023**, *11*, 908. [CrossRef] [PubMed]
24. Imran, M.; Abo-Elyousr, K.A.M.; Mousa, M.A.A.; Saad, M.M. Use of *Trichoderma* Culture Filtrates as a Sustainable Approach to Mitigate Early Blight Disease of Tomato and Their Influence on Plant Biomarkers and Antioxidants Production. *Front. Plant Sci.* **2023**, *14*, 1192818. [CrossRef] [PubMed]

25. García-Latorre, C.; Rodrigo, S.; Santamaria, O. Protective Effects of Filtrates and Extracts from Fungal Endophytes on *Phytophthora cinnamomi* in *Lupinus luteus*. *Plants* **2022**, *11*, 1455. [CrossRef]
26. Zachow, C.; Jahanshah, G.; de Bruijn, I.; Song, C.; Ianni, F.; Pataj, Z.; Gerhardt, H.; Pianet, I.; Lämmerhofer, M.; Berg, G.; et al. The Novel Lipopeptide Poaeamide of the Endophyte *Pseudomonas poae* RE*1-1-14 Is Involved in Pathogen Suppression and Root Colonization. *Mol. Plant-Microbe Interact.* **2015**, *28*, 800–810. [CrossRef] [PubMed]
27. Pastoshchuk, A.; Yumyna, Y.; Zelena, P.; Nudha, V.; Yanovska, V.; Kovalenko, M.; Taran, N.; Patyka, V.; Skivka, L. Beneficial Traits of Grain-Residing Endophytic Communities in Wheat with Different Sensitivity to *Pseudomonas syringae*. *Regul. Mech. Biosyst.* **2021**, *12*, 498–505. [CrossRef]
28. Elsharkawy, M.M.; Khedr, A.A.; Mehiar, F.; El-Kady, E.M.; Baazeem, A.; Shimizu, M. Suppression of *Pseudomonas syringae* Pv. Tomato Infection by Rhizosphere Fungi. *Pest Manag. Sci.* **2021**, *77*, 4350–4356. [CrossRef]
29. Nilsson, R.H.; Larsson, K.H.; Taylor, A.F.S.; Bengtsson-Palme, J.; Jeppesen, T.S.; Schigel, D.; Kennedy, P.; Picard, K.; Glöckner, F.O.; Tedersoo, L.; et al. The UNITE Database for Molecular Identification of Fungi: Handling Dark Taxa and Parallel Taxonomic Classifications. *Nucleic Acids Res.* **2019**, *47*, D259–D264. [CrossRef]
30. Santamaría, O.; Rodrigo, S.; Lledó, S.; Poblaciones, M.J. Fungal Endophytes Associated with *Ornithopus compressus* Growing under Semiarid Conditions. *Plant Ecol. Divers.* **2018**, *11*, 581–595. [CrossRef]
31. Lledó, S.; Rodrigo, S.; Poblaciones, M.J.; Santamaria, O. Biomass Yield, Mineral Content, and Nutritive Value of *Poa Pratensis* as Affected by Non-Clavicipitaceous Fungal Endophytes. *Mycol. Prog.* **2015**, *14*, 67. [CrossRef]
32. García-Latorre, C.; Rodrigo, S.; Marin-Felix, Y.; Stadler, M.; Santamaria, O. Plant-Growth Promoting Activity of Three Fungal Endophytes Isolated from Plants Living in Dehesas and Their Effect on *Lolium multiflorum*. *Sci. Rep.* **2023**, *13*, 7354. [CrossRef] [PubMed]
33. Stadler, M.; Wollweber, H.; Mühlbauer, A.; Henkel, T.; Asakawa, Y.; Hashimoto, T.; Ju, Y.M.; Rogers, J.D.; Wetzstein, H.G.; Tichy, H.V. Secondary Metabolite Profiles, Genetic Fingerprints and Taxonomy of *Daldinia* and Allies. *Mycotaxon* **2001**, *77*, 379–429.
34. Halecker, S.; Surup, F.; Kuhnert, E.; Mohr, K.I.; Brock, N.L.; Dickschat, J.S.; Junker, C.; Schulz, B.; Stadler, M. Hymenosetin, a 3-Decalinoyltetramic Acid Antibiotic from Cultures of the Ash Dieback Pathogen, *Hymenoscyphus pseudoalbidus*. *Phytochemistry* **2014**, *100*, 86–91. [CrossRef] [PubMed]
35. Clinical and Laboratory Standards Institute Clinical and Laboratory Standards Institute. *M100-S24 Performance Standards for Antimicrobial Susceptibipity Testing*; 24th Informational Supplement; Clinical and Laboratory Standards Institute: Wayne, NJ, USA, 2014; ISBN 1562388657.
36. de Sousa Oliveira, K.; de Lima, L.A.; Cobacho, N.B.; Dias, S.C.; Franco, O.L. Mechanisms of Antibacterial Resistance: Shedding Some Light on These Obscure Processes? Shedding Some Light on These Obscure Processes? In *Antibiotic Resistance*; Kon, K., Rai, M., Eds.; Elsevier Inc.: Amsterdam, The Netherlands, 2016; ISBN 9780128036686.
37. Bodede, O.; Shaik, S.; Chenia, H.; Singh, P.; Moodley, R. Quorum Sensing Inhibitory Potential and in Silico Molecular Docking of Flavonoids and Novel Terpenoids from *Senegalia nigrescens*. *J. Ethnopharmacol.* **2018**, *216*, 134–146. [CrossRef] [PubMed]
38. Kalia, V.C.; Patel, S.K.S.; Kang, Y.C.; Lee, J.-K. Quorum Sensing Inhibitors as Antipathogens: Biotechnological Applications. *Biotechnol. Adv.* **2019**, *37*, 68–90. [CrossRef] [PubMed]
39. Zabalgogeazcoa, Í.; Ciudad, A.G.; Vázquez de Aldana, B.R.; Criado, B.G. Effects of the Infection by the Fungal Endophyte *Epichloë festucae* in the Growth and Nutrient Content of *Festuca Rubra*. *Eur. J. Agron.* **2006**, *24*, 374–384. [CrossRef]
40. Wilson, M.; Campbell, H.L.; Ji, P.; Cuppels, D.A. Biological Control of Bacterial Spot of Tomato under Field Conditions at Several Locations in North America. *Phytopathology* **2002**, *92*, 1284–1292. [CrossRef]
41. Devi, K.A.; Pandey, P.; Sharma, G.D. Plant Growth-Promoting Endophyte *Serratia marcescens* AL2-16 Enhances the Growth of *Achyranthes aspera* L., a Medicinal Plant. *HAYATI J. Biosci.* **2016**, *23*, 173–180. [CrossRef]
42. Gordon, S.A.; Weber, R.P. Colorimetric Estimation of Indoleacetic Acid. *Plant Physiol.* **1951**, *26*, 192–195. [CrossRef]
43. Holbrook, A.A.; Edge, W.J.W.; Bailey, F. Spectrophotometric Method for Determination of Gibberellic Acid. *Adv. Chem.* **1961**, *28*, 159–167. [CrossRef]
44. Liu, X.; Dong, M.; Chen, X.; Jiang, M.; Lv, X.; Yan, G. Antioxidant Activity and Phenolics of an Endophytic *Xylaria* Sp. from *Ginkgo Biloba*. *Food Chem.* **2007**, *105*, 548–554. [CrossRef]
45. Yadav, M.; Yadav, A.; Yadav, J.P. In Vitro Antioxidant Activity and Total Phenolic Content of Endophytic Fungi Isolated from *Eugenia jambolana* Lam. *Asian Pac. J. Trop. Med.* **2014**, *7*, S256–S261. [CrossRef] [PubMed]
46. Nautiyal, C.S. An Efficient Microbiological Growth Medium for Screening Phosphate Solubilizing Microorganisms. *FEMS Microbiol. Lett.* **1999**, *170*, 265–270. [CrossRef] [PubMed]
47. Nieva, A.S.; Vilas, J.M.; Gárriz, A.; Maiale, S.J.; Menéndez, A.B.; Erban, A.; Kopka, J.; Ruiz, O.A. The Fungal Endophyte *Fusarium solani* Provokes Differential Effects on the Fitness of Two Lotus Species. *Plant Physiol. Biochem.* **2019**, *144*, 100–109. [CrossRef]
48. Olsen, S.R.; Cole, C.V.; Watanabe, F.S.; Dean, L.A. Estimation of Available Phosphorus in Soil by Extraction with Sodium Bicarbonate. *Open J. Soil Sci.* **1954**, *43*.
49. Pérez-Miranda, S.; Cabirol, N.; George-Téllez, R.; Zamudio-Rivera, L.S.; Fernández, F.J. O-CAS, a Fast and Universal Method for Siderophore Detection. *J. Microbiol. Methods* **2007**, *70*, 127–131. [CrossRef] [PubMed]
50. Singh, P.; Kumar, V.; Agrawal, S. Evaluation of Phytase Producing Bacteria for Their Plant Growth Promoting Activities. *Int. J. Microbiol.* **2014**, *2014*, 426483. [CrossRef] [PubMed]

51. Kimura, Y.; Mizuno, T.; Nakajima, H.; Hamasaki, T. Altechromones A and B, New Plant Growth Regulators Produced by the Fungus, *Alternaria* sp. *Biosci. Biotechnol. Biochem.* **1992**, *56*, 1664–1665. [CrossRef]
52. Qader, M.M.; Hamed, A.A.; Soldatou, S.; Abdelraof, M.; Elawady, M.E.; Hassane, A.S.I.; Belbahri, L.; Ebel, R.; Rateb, M.E. Antimicrobial and Antibiofilm Activities of the Fungal Metabolites Isolated from the Marine Endophytes *Epicoccum nigrum* M13 and *Alternaria alternata* 13A. *Mar. Drugs* **2021**, *19*, 232. [CrossRef]
53. Wang, H.; Guo, Y.; Luo, Z.; Gao, L.; Li, R.; Zhang, Y.; Kalaji, H.M.; Qiang, S.; Chen, S. Recent Advances in *Alternaria* Phytotoxins: A Review of Their Occurrence, Structure, Bioactivity and Biosynthesis. *J. Fungi* **2022**, *8*, 168. [CrossRef]
54. Capon, R.J.; Stewart, M.; Ratnayake, R.; Lacey, E.; Gill, J.H. Citromycetins and Bilains A-C: New Aromatic Polyketides and Diketopiperazines from Australian Marine-Derived and Terrestrial *Penicillium* spp. *J. Nat. Prod.* **2007**, *70*, 1746–1752. [CrossRef] [PubMed]
55. Elbandy, M.; Shinde, P.B.; Hong, J.; Bae, K.S.; Kim, M.A.; Lee, S.M.; Jung, J.H. A-Pyrones and Y Low Pigments from the Sponge-Derived Fungus *Paecilomyces lilacinus*. *Bull. Korean Chem. Soc.* **2009**, *30*, 188–192.
56. Hussain, H.; Krohn, K.; Ahmed, I.; Draeger, S.; Schulz, B.; Di Pietro, S.; Pescitelli, G. Phomopsinones A-D: Four New Pyrenocines from Endophytic Fungus *Phomopsis* sp. *European J. Org. Chem.* **2012**, *2012*, 1783–1789. [CrossRef]
57. El-Sayed, A.S.A.; Ali, G.S. *Aspergillus flavipes* Is a Novel Efficient Biocontrol Agent of *Phytophthora parasitica*. *Biol. Control* **2020**, *140*, 104072. [CrossRef]
58. Pinheiro, E.A.A.; Pina, J.R.S.; Feitosa, A.O.; Carvalho, J.M.; Borges, F.C.; Marinho, P.S.B.; Marinho, A.M.R. Bioprospecting of Antimicrobial Activity of Extracts of Endophytic Fungi from *Bauhinia guianensis*. *Rev. Argent. Microbiol.* **2017**, *49*, 3–6. [CrossRef] [PubMed]
59. Zhu, H.; He, C.C.; Chu, Q.H. Inhibition of Quorum Sensing in *Chromobacterium violaceum* by Pigments Extracted from *Auricularia auricular*. *Lett. Appl. Microbiol.* **2011**, *52*, 269–274. [CrossRef] [PubMed]
60. Adonizio, A.; Kong, K.-F.; Mathee, K. Inhibition of Quorum Sensing-Controlled Virulence Factor Production in *Pseudomonas aeruginosa* by South Florida Plant Extracts. *Antimicrob. Agents Chemother.* **2008**, *52*, 198–203. [CrossRef]
61. Singh, V.K.; Mishra, A.; Jha, B. 3-Benzyl-Hexahydro-Pyrrolo [1,2-a]Pyrazine-1,4-Dione Extracted From *Exiguobacterium indicum* Showed Anti-Biofilm Activity Against *Pseudomonas Aeruginosa* By Attenuating Quorum Sensing. *Front. Microbiol.* **2019**, *10*, 435900. [CrossRef]
62. Harba, M.; Jawhar, M.; Arabi, M.I.E. In Vitro Antagonistic Activity of Diverse *Bacillus* Species Against *Fusarium culmorum* and *F. solani* Pathogens. *Open Agric. J.* **2020**, *14*, 157–163. [CrossRef]
63. Kadam, H.K.; Salkar, K.; Naik, A.P.; Naik, M.M.; Salgaonkar, L.N.; Charya, L.; Pinto, K.C.; Mandrekar, V.K.; Vaz, T. Silica Supported Synthesis and Quorum Quenching Ability of Isoxazolones Against Both Gram Positive and Gram Negative Bacterial Pathogens. *ChemistrySelect* **2021**, *6*, 11718–11728. [CrossRef]
64. Rekha, P.D.; Vasavi, H.S.; Vipin, C.; Saptami, K.; Arun, A.B. A Medicinal Herb Cassia Alata Attenuates Quorum Sensing in *Chromobacterium violaceum* and *Pseudomonas aeruginosa*. *Lett. Appl. Microbiol.* **2017**, *64*, 231–238. [CrossRef] [PubMed]
65. Abbas, H.A.M.; Soliman, W.E.E.; Shaldam, M.A. Perturbation of Quorum Sensing in *Pseudomonas aeruginosa* by Febuxostat. *Adv. Microbiol.* **2018**, *08*, 650–664. [CrossRef]
66. Abdessemed, N.; Staropoli, A.; Zermane, N.; Vinale, F. Metabolic Profile and Mycoherbicidal Activity of Three *Alternaria alternata* Isolates for the Control of *Convolvulus arvensis*, *Sonchus oleraceus*, and *Xanthium strumarium*. *Pathogens* **2021**, *10*, 1448. [CrossRef] [PubMed]
67. Tang, Z.; Wang, Y.; Yang, J.; Xiao, Y.; Cai, Y.; Wan, Y.; Chen, H.; Yao, H.; Shan, Z.; Li, C.; et al. Isolation and Identification of Flavonoid-Producing Endophytic Fungi from Medicinal Plant *Conyza blinii* H.Lév That Exhibit Higher Antioxidant and Antibacterial Activities. *PeerJ* **2020**, *8*, e8978. [CrossRef]
68. Daley, S.K.; Cordell, G.A. Biologically Significant and Recently Isolated Alkaloids from Endophytic Fungi. *J. Nat. Prod.* **2021**, *84*, 871–897. [CrossRef] [PubMed]
69. Singh, A.; Singh, D.K.; Kharwar, R.N.; White, J.F.; Gond, S.K. Fungal Endophytes as Efficient Sources of Plant-Derived Bioactive Compounds and Their Prospective Applications in Natural Product Drug Discovery: Insights, Avenues, and Challenges. *Microorganisms* **2021**, *9*, 197. [CrossRef]
70. Hashem, A.H.; Attia, M.S.; Kandil, E.K.; Fawzi, M.M.; Abdelrahman, A.S.; Khader, M.S.; Khodaira, M.A.; Emam, A.E.; Goma, M.A.; Abdelaziz, A.M. Bioactive Compounds and Biomedical Applications of Endophytic Fungi: A Recent Review. *Microb. Cell Fact.* **2023**, *22*, 107. [CrossRef]
71. Omomowo, I.O.; Amao, J.A.; Abubakar, A.; Ogundola, A.F.; Ezediuno, L.O.; Bamigboye, C.O. A Review on the Trends of Endophytic Fungi Bioactivities. *Sci. African* **2023**, *20*, e01594. [CrossRef]
72. Baron, N.C.; Rigobelo, E.C. Endophytic Fungi: A Tool for Plant Growth Promotion and Sustainable Agriculture. *Mycology* **2022**, *13*, 39–55. [CrossRef]
73. Zhang, Y.; Li, Y.; Liang, S.; Zheng, W.; Chen, X.; Liu, J.; Wang, A. Study on the Preparation and Effect of Tomato Seedling Disease Biocontrol Compound Seed-Coating Agent. *Life* **2022**, *12*, 849. [CrossRef]
74. Ismail, M.A.; Amin, M.A.; Eid, A.M.; Hassan, S.E.D.; Mahgoub, H.A.M.; Lashin, I.; Abdelwahab, A.T.; Azab, E.; Gobouri, A.A.; Elkelish, A.; et al. Comparative Study between Exogenously Applied Plant Growth Hormones versus Metabolites of Microbial Endophytes as Plant Growth-Promoting for *Phaseolus vulgaris* L. *Cells* **2021**, *10*, 1059. [CrossRef] [PubMed]

75. Khan, A.L.; Hamayun, M.; Kang, S.-M.; Kim, Y.-H.; Jung, H.-Y.; Lee, J.-H.; Lee, I.-J. Endophytic Fungal Association via Gibberellins and Indole Acetic Acid Can Improve Plant Growth under Abiotic Stress: An Example of *Paecilomyces formosus* LHL10. *BMC Microbiol.* **2012**, *12*, 3. [CrossRef] [PubMed]
76. Khalmuratova, I.; Choi, D.; Yoon, H.; Yoon, T.; Kim, J. Diversity and Plant Growth Promotion of Fungal Endophytes in Five Halophytes from the Buan Salt Marsh. *J. Microbiol. Biotechnol.* **2021**, *31*, 408–418. [CrossRef] [PubMed]
77. Jaiswal, S.; Ojha, A.; Thakur, P.; Mishra, S.K. Functional Importance of Endophytic Microorganisms in Plant Growth Promotion Bioactive Compound Production for Sustainable Agriculture. *Def. Life Sci. J.* **2023**, *8*, 93–108. [CrossRef]
78. Liu, B.; Liu, X.; Liu, F.; Ma, H.; Ma, B.; Zhang, W.; Peng, L. Growth Improvement of *Lolium multiflorum* Lam. *Induced by Seed Inoculation with Fungus Suspension of Xerocomus badius and Serendipita indica. AMB Express* **2019**, *9*, 145. [CrossRef]
79. Batool, R.; Umer, M.J.; Shabbir, M.Z.; Wang, Y.; Ahmed, M.A.; Guo, J.; He, K.; Zhang, T.; Bai, S.; Chen, J.; et al. Seed Myco-Priming Improves Crop Yield and Herbivory Induced Defenses in Maize by Coordinating Antioxidants and Jasmonic Acid Pathway. *BMC Plant Biol.* **2022**, *22*, 554. [CrossRef] [PubMed]
80. Ma, Y.; Li, Y.; Yang, S.; Li, Y.; Zhu, Z. Biocontrol Potential of *Trichoderma asperellum* Strain 576 against *Exserohilum turcicum* in *Zea mays*. *J. Fungi* **2023**, *9*, 936. [CrossRef] [PubMed]
81. Kalleli, F.; Aissa, E.; M'Hamdi, M. Seed Biopriming with Endophytic Fungi Enhances Germination, Growth, Yield and Fruit Quality of Fennel under Salinity Stress. *Quest J. J. Res. Environ. Earth Sci.* **2022**, *8*, 1–10.
82. Liu, B.; Liu, X.; Liu, F.; Ma, H.; Ma, B.; Peng, L. Stress Tolerance of *Xerocomus badius* and Its Promotion Effect on Seed Germination and Seedling Growth of Annual Ryegrass under Salt and Drought Stresses. *AMB Express* **2021**, *11*, 15. [CrossRef]
83. Puigvert, M.; Solé, M.; López-Garcia, B.; Coll, N.S.; Beattie, K.D.; Davis, R.A.; Elofsson, M.; Valls, M. Type III Secretion Inhibitors for the Management of Bacterial Plant Diseases. *Mol. Plant Pathol.* **2019**, *20*, 20–32. [CrossRef]
84. Vishwakarma, K.; Kumar, N.; Shandilya, C.; Mohapatra, S.; Bhayana, S.; Varma, A. Revisiting Plant–Microbe Interactions and Microbial Consortia Application for Enhancing Sustainable Agriculture: A Review. *Front. Microbiol.* **2020**, *11*, 560406. [CrossRef] [PubMed]
85. Rashad, Y.M.; Abdalla, S.A.; Sleem, M.M. Endophytic *Bacillus subtilis* SR22 Triggers Defense Responses in Tomato against *Rhizoctonia* Root Rot. *Plants* **2022**, *11*, 2051. [CrossRef] [PubMed]
86. Mostafa, I.; Abbas, H.A.; Ashour, M.L.; Yasri, A.; El-Shazly, A.M.; Wink, M.; Sobeh, M. Polyphenols from *Salix Tetrasperma* Impair Virulence and Inhibit Quorum Sensing of *Pseudomonas aeruginosa*. *Molecules* **2020**, *25*, 1341. [CrossRef]
87. Yoo, S.J.; Shin, D.J.; Won, H.Y.; Song, J.; Sang, M.K. Aspergillus Terreus JF27 Promotes the Growth of Tomato Plants and Induces Resistance against Pseudomonas Syringae Pv. Tomato. *Mycobiology* **2018**, *46*, 147–153. [CrossRef]

horticulturae

Review

Crop Nutrition and Soil Fertility Management in Organic Potato Production Systems

Anita Ierna * and Miriam Distefano

Institute of BioEconomy, National Research Council (IBE-CNR), 95126 Catania, Italy; miriam.distefano@cnr.it
* Correspondence: anita.ierna@cnr.it; Tel.: +39-095-7338395

Abstract: Organic farming is a comprehensive production management system that fosters and improves the health of agroecosystems, encompassing biodiversity, biological cycles, and soil biological activity. The potato (*Solanum tuberosum* L.) is a crucial crop in organic farming systems, standing out as one of the most highly demanded organic products on the market. Among all crops, with potatoes, there is a very large yield gap between organic and conventional systems, attributable mainly to its intensive nutrient demands. The present review, considering the most relevant scientific literature worldwide, discusses the contemporary state of knowledge on crop nutrition and soil fertility management in organic potato crop production, analyzing the effects of animal manures, green manures, organic amendments, and biostimulants on organic potato tuber yield and quality. Overall, the main findings show a particular combination is needed to effectively maintain good soil fertility, satisfy the nutritional needs of the crop, and overcome the difference in potato yield between organic and conventional farming methods while meeting consumer demand. This combination entails using an animal manure or leguminous green manure with an organic soil amendment, and even better with a biofertilizer, such as a mycorrhizae-fungus-based one. It also emerged that more targeted studies are needed to select appropriate cultivars for organic potato farming systems to optimize this environmentally friendly production method.

Keywords: potato; organic; crop nutrition; soil fertility; yield; quality

1. Introduction

Interest in organic agriculture is growing worldwide, mainly supported by a strong consumer interest. This can be measured by the value of retail sales of organic products, which grew to an impressive 44.2 billion euros in the European Union (EU) [1]. The market trend is still growing and increasing faster than the organic farmland area. In 2020, in most of the EU-member states the organic farmland area increased; the biggest shares of the organic farmland area in the total utilized area are in Austria (26.5%), Estonia (22.4%), Sweden (20.4%) and Italy (16%) [1]. The potato (*Solanum tuberosum* L.), globally ranked as the fourth most significant food crop, with an annual production of approximately 400 million tons [2], is a crucial crop in organic farming systems, standing out as one of the most highly demanded organic products on the market [3]. Organic potato in EU-28 is cultivated in an area of 31,771 ha (average 2017–2019) [4], of which approximately half are in five countries, namely Germany (9345 ha), France (3370 ha), the Netherlands (1685 ha), United Kingdom (1234 ha), and Belgium (845 ha). Previous research comparing potato yields in conventional and organic production systems has often found significantly lower yields in organic systems [5]. Comparing various European countries across 21 aspects, it was found that organic potato yields accounted for only 70% of those from conventional crops [6]. When considering potatoes as a vegetable, the yield gap was almost 30%, as noted by Ponisio et al. [7]. Results of field experiment trials in potatoes for a long-term period demonstrated that the yield gap between conventional and organic farming systems was due to both less efficient crop protection and fertilization in organic systems [8–10]. It was determined that 48% of the gap in organic potato yields can be attributed to reduced fertilization, and 25% to diseases, primarily late blight caused by *Phytophthora infestans* and the Colorado potato beetle (*Leptinotarsa decemlineata*) [11]. Moreover, the yield gap between organic and conventional fertilization regimes appears to be attributable mainly to the lower and less predictable nitrogen supply in organically fertilized crops [12].

A potato crop has high nutritional needs; indeed, to produce 1 ton of tubers requires approximately 4–5 kg of nitrogen, 0.7–0.8 kg of phosphorus, 5–6 kg of potassium, and significant quantities of microelements [13]. Furthermore, potato crops are characterized by a relatively low ability to uptake available soil mineral nitrogen [14]. Therefore, to better align N requirements with supply, it is common practice in conventional agriculture to split the application of N fertilizers [15,16]. Given this premise, it is understandable that organic potato fertilization is a real challenge. The efficacy of organic fertilization in organic systems depends on various factors, including the type of fertilizer used (quantity, timing), soil conditions, and weather patterns during the plant's growth stages. Unfavorable combinations of soil and climatic conditions can hinder the decomposition of applied fertilizers and the uptake of nutrients, leading to stress and developmental disorders in potato plants. As is known, potato tuber quality is influenced by climate, environment, genetic background, and cultivation management and therefore by the organic production system [17]. In particular, the limited and intermittent availability of nutrients, particularly nitrogen, could notably lead to influencing the tuber quality profile [18].

This review, based on the most relevant scientific literature, discusses the contemporary state of knowledge on crop nutrition and soil fertility management in organic potato crop production and how these can affect tuber yield and quality. Side-by-side comparative studies were considered, including both organic and conventional farming practices or studies in organic potato systems; however, on occasion, research conducted in integrated or sustainable potato crop systems with the aim of including information that could be utilized in organic farming was also taken into consideration. In conducting the literature search for this review, the authors adopted various strategies, focusing on reputable journals indexed in databases such as Scopus, Web of Science, PubMed, and Google Scholar, with publications dated after 2000. Articles were selected based on their relevance to the topic. Overall, it is noted that there is a preponderance of papers on the effects of fertilization in potato organic farming on tuber yield, rather than on tuber quality (Figure 1); however, it is interesting to note that papers on both tuber yield and quality have shown an almost continuous growth trend from 2000 to the present, demonstrating the interest of researchers worldwide in this topic.

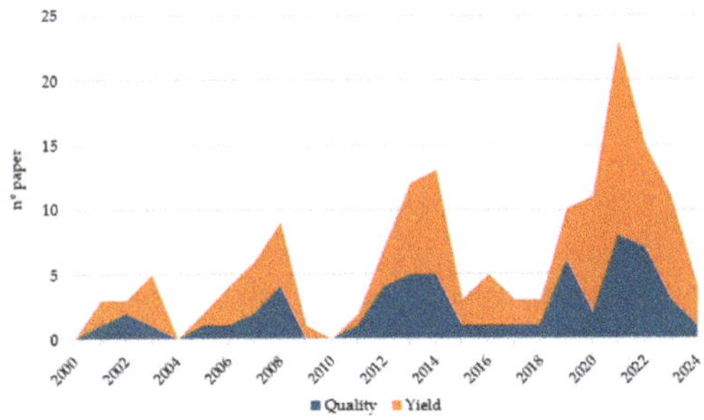

Figure 1. The number of scientific publications per year from 2000 to 2024 indexed in Scopus® database using the keywords "organic farming potato tuber quality" in blue and "organic farming potato tuber yield" in orange.

2. Crop Nutrition and Soil Fertility Management in Organic Potato

In organic potato production, maintaining proper soil fertility is paramount due to the crop's high nutritional demands. Although there is no worldwide regulation governing the use of organic fertilizers, the European Union (EU) has implemented a legal framework that outlines the permitted organic fertilizers, soil conditioners, and nutrients for use in EU organic production. These are detailed in Annex 1 of Regulation (EU) 2018/848 [19]. Following the regulation, organic potato production practices in Europe are based on the principle that no synthetic chemical compounds are applied. This means that organic producers need to adopt different approaches to maintaining soil fertility and plant health, including the following:

- Crop rotation;
- Cultivation of nitrogen-fixing plants and other green manure crops to restore the soil fertility;
- Choosing resistant varieties and breeds and techniques that encourage natural pest control.

Fertility management inputs, such as fertilizers and soil conditioners, should be used only if they adhere to the principles and objectives of organic production, with a strict limitation on the use of chemically synthesized inputs. Within this conceptual-regulatory framework, the recycling of wastes and by-products from plant and animal sources is essential for replenishing nutrients and organic carbon (C) in soils. Organic amendments such as manure and compost are essential for managing soil fertility and promoting the long-term sustainability of crop production. Numerous studies have demonstrated that organic inputs improve soil structure, enhance water retention, support root growth, and boost microbial activity [20]. To mitigate the risk of nitrate contamination, EU Regulation 2018/848 [19] imposes a restriction on the total amount of organic nitrogen that can be applied in organic farming, capping it at 170 kg ha^{-1} year^{-1}. This limit includes different types of organic materials, such as farmyard manure, dried manure, chicken manure, composted manure, solid animal waste, and liquid animal waste.

The chemical composition of the different organic products that are used for crop nutrition and to maintain soil fertility in organic farming can vary greatly. Table 1 reports a summary of the average nutrient content of the most common organic products applied for fertilization in organic potato crops.

Table 1. Summary of the average nutrient content of some organic products (adapted from Gelaye) [21].

Organic Products	Nutrient Content (%)		
	N	P_2O_5	K_2O
Farmyard manure	0.5–1.5	0.2–0.4	0.5–1.0
Crop residues	0.34–0.63	-	1–6
Organic waste	3.03	2.63	1.4
Green manure	40–60	2.5–6.55	0.0327
Vermicompost	2–3	1.55–2.25	1.85–2.25
Compost	2	0.5–1	2
Biogas slurry	1.4–1.8	1.1–1.7	0.8–1.3

In addition to macronutrients such as nitrogen (N), phosphorus (P), and potassium (K), organic products also have micronutrients such as copper (Cu), manganese (Mn), and zinc (Zn). In organic farming, managing fertilization is challenging primarily because nitrogen is supplied indirectly through organic fertilizers, which must undergo mineralization to release mineral nitrogen into the soil [8,11]. Synchronizing the nitrogen supply from these fertilizers with the nitrogen demands of potato crops is often difficult. This synchronization depends on the chemical composition of the organic fertilizers, particularly their nitrogen content and C/N ratio, as well as on the pedo-climatic conditions that influence the decomposition and mineralization of organic matter. Effective nitrogen management, therefore, requires a thorough understanding of the mineralization processes associated with the organic fertilizers being used, ensuring that nitrogen release aligns with the potato crops' peak nitrogen demand periods [22]. In this context, Harraq et al. [23] evaluated nitrogen mineralization and release from various organic fertilizers used in organic farming. Their study involved incubating four commonly applied organic fertilizers—worm compost (5 t ha^{-1}), compost (5 t ha^{-1}), sheep manure (30 t ha^{-1}), and fishmeal (3.5 t ha^{-1}). The highest nitrogen mineralization apparent rates (NMAR) were found to be 62.3% for fishmeal, 38% for compost, 30% for worm compost, and 18% for sheep manure. By the end of the incubation, nitrogen immobilization was observed in both worm compost and compost, attributed to an increase in the C/N ratio.

The present review summarizes the results of the research carried out in organic potatoes to evaluate the effects on the yield and tuber quality attributes deriving from the use of organic products for crop nutrition and soil fertility distinguished in animal manures, green manures, organic amendments, and biostimulants. Figure 2 summarizes the most studied organic products for nutrition and fertility management in organic potatoes.

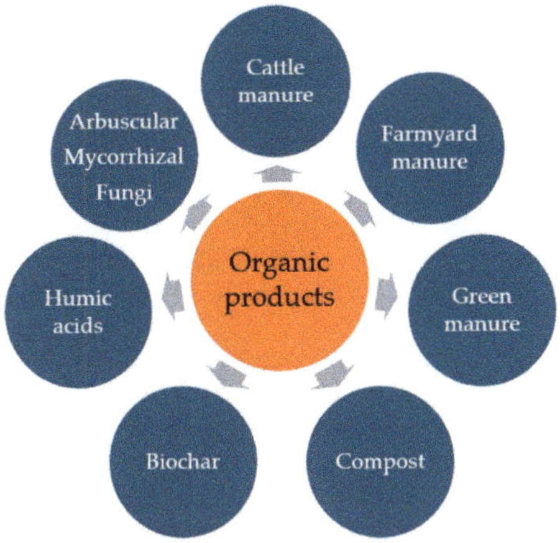

Figure 2. Most common fertilizers, amendments, and biostimulants studied in organic potato systems.

3. Animal Manure in Organic Potato Cultivation

The main animal-derived fertilizers used in potato organic production include poultry, pig, sheep, and cattle manure, along with urine. The utilization of these fertilizers depends on factors such as regional availability, cost, transportation, and handling. Although manures are not sufficient nitrogen sources for achieving maximum potato yields, they do contribute to increasing soil organic carbon levels. Additionally, applying organic manure enhances soil fertility and structure, fostering a more stable ecosystem by supporting soil fauna and flora [24]. Farmyard manure (FYM) consists of a decomposed mixture of animal waste, including dung, urine, litter, and leftover roughage and feed provided to the animals [25]. When fully decomposed, FYM serves as an excellent source of organic carbon, which supports the biotic activity of soil flora and fauna [26]. FYM also has all the macro-and micronutrients needed for plant development including potatoes, playing an important role in organic farming systems for crop nutrition and soil fertility maintenance [27]. However, some researchers have highlighted limitations associated with the use of manure in organic agriculture, particularly in optimizing multiple nutrients for potato crop nutrition [28]. The yield gap between organic and conventional potato systems associated with fertilization seems mainly due to the lower supply/availability of N from FYM compared with NPK mineral fertilizers [29]. Rempelos et al. [30] supported these results by the finding of lower leaf chlorophyll and tuber N content in FYM-fertilized potato crops, although results suggest that K was the main macronutrient limiting yield. In any case, it is very difficult to envisage approaches/strategies that may close the yield gap associated with fertilization, because the maximum manure inputs for organic crops allowed in a given year under EU environmental legislation for nitrate-sensitive zones are equivalent to 170 kg N ha^{-1}. Hajšlová et al. [31] investigated the effects of organic versus conventional farming on potato quality. The field trials were conducted from 1996 to 1999 in two different locations in the Czech Republic, involving eight potato cultivars: Christa, Koruna, Krystala, Rosara (very early maturing), Krasa and Monalisa (both early maturing), and Karin and Rosella (both semi-early maturing). Crop rotation and organic farmyard manure at a rate of 40–45 t ha^{-1} were employed in the organic system fields. The authors reported significant differences between the farming systems in terms of tuber yield and tuber size distribution, but tuber dry matter (−5% in conventional tubers) and starch content were not significantly influenced, in accordance with the findings of Herencia et al. [32]. Additionally, no significant effects were observed on tuber micronutrient contents (B, Fe, Mn, and Zn), consistent with the results of Warman and Havard [33]. Among other parameters characterizing the nutritional profile of tubers, the ascorbic acid content varied from year to year and was only significantly influenced by the farming system in individual years. Conversely, chlorogenic acid, the most abundant phenolic acid in potato tubers, increased by 45% in organically grown tubers, despite significant year-to-year variation. Makaraviciute [34] studied the effect of organic (manure) and mineral fertilizers on the yield and quality of different potato varieties. It was found that the highest potato tuber yields of all the tested varieties were harvested when one-component and complex mineral fertilizers with microelements had been applied, while the lowest when manure (60 t ha^{-1}) had been used in spring. Makaraviciute also reported non-significant differences in potato content in essential amino acids between organically and conventionally grown potato tubers [34].

Wszelaki et al. [35] evaluated mineral and glycoalkaloid concentrations in organically (fertilized with composted dairy manure, 6.5 Mg ha^{-1}) and conventionally (fertilized with mineral 10-20-20 N-P-K, 672 kg ha^{-1}) grown redskin potatoes. They found that in tuber skin and flesh, potassium, magnesium, phosphorus, sulfur, and copper concentrations were significantly higher in the organically grown potatoes, while iron and manganese concentrations were higher in the skin of conventionally grown potatoes. In addition, glycoalkaloid levels tended to be higher in organic potatoes.

Bártová et al. [36] on five cultivars in two localities during three consecutive years, compared organic (fertilized with farmyard composted dairy manure, 40 t ha^{-1}), to conventional (fertilized with common manure, 40 t ha^{-1} + mineral fertilizers: 100 kg ha^{-1} of N, 35 kg ha^{-1} of P and 60 kg ha^{-1} of K) cropping systems. They found that organically produced potato tubers contained a significantly lower content of total nitrogen, crude protein, nitrates, and glycoalkaloids (α-solanine and α-chaconine). Paffrath and Milz [37] conducted field trials, during 2005–2007, to test the effects of organic fertilization, pregermination, and copper treatment on tuber quality and yield performance of the early maturing cultivar Princess. Fertilization with organic nitrogen reduced the starch content of the tubers in the first year of trials; however, this effect was not observed in the subsequent two years, with starch levels remaining comparable to those of non-treated control. Conversely, copper treatment was reported to significantly increase starch levels in the tubers, with an average increase of 7–15% compared to the control group over the three-year period. In a field trial conducted in Italy [38] with two potato cultivars (Agria and Merit), the organic farming system (using an organic mix of manure, feathers, and torrefied bone-meat, 6% N) was compared to conventional farming systems (using ammonium nitrate, 33% N). It was found that organic farming caused a 25% marketable yield reduction with a higher percentage of large tubers under conventional farming; organic potatoes had higher dry matter content, reducing sugars and total protein than conventional ones. A study conducted in Italy [39] compared the performance of three potato cultivars over two seasons under organic and conventional cultivation systems. The organic system used a combination of feathers, torrefied bone and meat meal, dried manure, and hydrolyzed pelt, while the conventional system relied on ammonium nitrate. The results showed that the organic system consistently produced lower total yields compared to the conventional system in both years. However, during the second season, when late blight infection was less severe, the yield gap between the two systems decreased, varying between 7% and 20% depending on the cultivar. In addition, the efficacy of animal-based bio-fertilizers was highlighted in enhancing the total phenolic content of potatoes. In addition, they observed an improved sensory performance after frying, with increased crispness and a lower degree of browning; these organically fertilized potatoes had a lower nitrate content compared to conventionally grown ones, a characteristic that is beneficial for human health, as previously confirmed [40]. Lombardo et al. [41] evaluated the mineral profile in organically (mixture of feathers and torrefied bone and meat meal plus a mixture of dried manure and hydrolyzed pelt) and conventionally (100 kg ha^{-1} N, 70 kg ha^{-1} P_2O_5 and 150 kg ha^{-1} K_2O + 83 kg ha^{-1} N in the form of ammonium nitrate) grown potatoes. Specifically, the organically grown potatoes exhibited higher phosphorus content, even though lower levels of potassium, calcium, sodium, iron, and manganese [41]. In a three-season field trial, organic vs. conventional was compared on five potato genotypes [42], utilizing for fertilization (100 kg ha^{-1} N, 70 kg ha^{-1} P_2O_5, and 150 kg ha^{-1} K_2O), respectively, a mixture of feathers and torrefied bone and meat meal (organic) and NPK synthetic fertilizer (conventional). The organic cultivation system was less productive than the conventional one across the 3 seasons (−5% season I, −50% season II, and −25% season III), due to lower availability of N and to the appearance time and severity level of late blight infection. A different behavior of genotypes was also found. Grudinska et al. [43] on 4 potato cultivars in experimental fields, studied organic (Manure–28 t ha^{-1} + mustard as a catch crop) and conventional (4–5 t plowed rye straw + 1 kg mineral nitrogen per 100 kg straw + mustard as a catch crop; N: 100 kg ha^{-1}, P: 53 k ha^{-1}, K: 150 kg ha^{-1}) fertilization management over two years. They found higher total phenolic content in potato tubers under organic production than in potatoes grown conventionally. Kazimierczak et al. [44], in a controlled field experiment in Poland, evaluated the concentrations of polyphenols, lutein, vitamin C, and nitrates in eight potato cultivars grown organically (fertilized with Manure–28 t ha^{-1} + mustard as a catch crop) and conventionally (fertilized with 4–5 t plowed rye straw + 1 kg mineral nitrogen per 100 kg straw, N: 100 kg ha^{-1}, P: 53 kg ha^{-1}, K: 150 kg ha^{-1}). Higher concentrations of nitrates and lutein were found in conventional compared to the organic tubers, while organic potatoes were, on average, richer in phenolic compounds. Keutgen et al. [45], in three consecutive years and on 10 genotypes, evaluated organic (fertilized with 25 t ha^{-1} composted manure) compared to integrated production system (fertilized with 120 N, 60 P_2O_5, 180 K_2O + 15 t composted manure + foliar fertilization Adob Cu, Adob Mn, Adob S, Basfoliar 36 Extra, or green manures + 90 N, 110 P_2O_5, 180 K_2O). They found higher contents of phenolic compounds, flavonoids, and ascorbic acid in organically grown potato tubers than the conventionally grown potato tubers. Shafeeva et al. [46] studied the impact of cattle manure on potato yields and the quality of the tubers grown under both irrigated and non-irrigated conditions, in the southern forest-steppe region of the Republic of Bashkortostan. Farmyard manure 40, 60, 80, 100, and 120 t ha^{-1} was introduced in spring, during soil treatment. However, it was found that slightly lower application rates, specifically 100 t ha^{-1} in non-irrigated conditions and 40 t ha^{-1} under irrigation, yielded a higher percentage of marketable tubers. The use of cattle manure, also, influenced the quality of potato tubers: the highest starch content was observed with the application of 120 t ha^{-1} of manure. Additionally, the dry matter content of the tubers was affected by manure, with the highest levels observed at 40 t ha^{-1} in non-irrigated conditions and 100 t ha^{-1} under irrigation, while nitrate levels in the study were under the established safe limits. Haase et al. [47] investigated the impact of N and K fertilization, preceding crop, pre-sprouting, and cultivar on the quality attributes of organically grown potatoes destined for processing into French fries or crisps during two years at two sites. The fertilization treatments consisted of the following: cattle manure (considered as the reference for potassium and nitrogen content), potassium sulfate (40% K), potassium sulfate combined with horn grits, horn grits alone (14% N), and a control treatment with no fertilizer. The cultivars used in the study were Agria and Marlen, both tested as French fries and crisps. Regardless of the other factors under study, it was found that in the first year, the fry color values for fries from crops treated with cattle manure were significantly lower than those from the control group with no fertilizer. In the second year, however, cattle manure application resulted in significantly higher fry color values compared to those from crops treated with pure horn grits. Ierna et al. [48] investigated the effects of organic versus conventional cultivation on the quality of early potato tubers over two seasons. Their findings, consistent with a previous report [49] indicated that the use of a mixture of meat and bone meal and dried manure in organic farming produced tubers with more attractive skin color, greater skin thickness and firmness, and higher dry matter and total phenolics content. However, these organically grown tubers exhibited lower ascorbic acid content and antioxidant activity compared to conventionally grown tubers. Considering the effects of the cultivation system on the visual appearance and sensory profile of cooked tubers, the same authors reported that organically grown early potato tubers, cultivated with dried manure and animal-based fertilizer, exhibited improved skin appearance and enhanced sensory traits when boiled, fried, and baked [50].

As a rule, the availability of nitrogen present in organic manures is very low (about 3.0%) compared to inorganic fertilizers, and this is the reason why animal manures generally fail to satisfy the nitrogen needs of the potato. Many experimental works have demonstrated that the addition of minerals to organic fertilizers, especially of animal origin, improves production performances considerably. In an experimental plot at the University of La Laguna, Tenerife, Spain over two years [51] treatments of fertilizing were evaluated on cv. Cara with a single supply or binary combinations of commercial, farmyard manure, sheep manure compost, ammonium nitrate, and "multicote" (17-17-10 NPK commercial coated fertilizer). Fertilizer rates were determined to supply 168 kg of N ha^{-1}, in accordance with normal rates of N used by farmers; each fertilizer of binary combinations supplied 50% of N. Potato yields were similar in all the treatments during the first year of cultivation, although in the second year potatoes that received ammonium nitrate, "multicote", sheep manure compost plus "multicote" or farmyard manure plus "multicote" gave significantly higher yields than those subjected to the other treatments. Peñaloza Monroy et al. [52] carried out a field trial to evaluate the response of three levels of chicken manure (2, 3, 4 t ha^{-1}) added to an inorganic fertilization treatment (154 N, 356 P, 60 K) in four potato cultivars. The results showed that the cv. Rosita and Ágata in 4 t ha^{-1} of chicken manure, had more stems and greater fresh weights of foliage and produced the highest tuber yields. In the long-term experiment in a four-course crop rotation, Baniuniene & Zekaite [53] studied the effects on potatoes of cultivation with and without farmyard manure (40 t ha^{-1}) combined with various mineral fertilization ($N_0P_0K_0$; $N_{90}P_{90}$; $P_{90}K_{120}$; $N_{90}K_{120}$; N_{90} $P_{90}K_{120}$). The long-term experimental data suggest that farmyard manure (FYM) increased potato tuber yield by 35–82%, depending on fertilizer combinations. Mineral fertilizer efficacy on the background without FYM was up to 28% higher. The highest tuber yield increases were obtained by using fertilizer combinations with nitrogen. Potato crops applied with only mineral fertilizers contained higher starch and dry matter contents in tubers compared with those applied with FYM and mineral fertilizers. Fertilizer combinations with potassium tended to reduce starch and the dry matter content in tubers. Caliskan et al. [54] in a field experiment conducted at Hatay, Turkey, studied the effects of farmyard manure (0, 10, 20, 30, 40, and 50 t ha^{-1}) and mineral fertilization (no or 200-90-90 kg ha^{-1} N-P-K, respectively) on the growth and yield of early potato crops. The application of farmyard manure had positive effects on growth and yield with or without mineral fertilization. Tuber yield increased as the farmyard manure levels increased under non-mineral fertilized conditions while no significant increase was obtained from 40 and 50 t manure ha^{-1} rates in mineral-fertilized plots. The application of farmyard manure significantly increased the dry-matter content of tubers, and this was more evident in mineral fertilized plots. Other researchers have also explored the impact of nutrient supply and management on potato production. For instance, researchers found that a combined application of farmyard manure (FYM) at 18.75 t ha^{-1} and NPK (25:25:25 kg ha^{-1}) significantly improved total tuber yield [55]. Similarly, the joint use of 40 t of FYM and 300 kg of nitrogen fertilizer resulted in the highest total dry weight, plant height, and dry matter content. Additionally, the maximum average tuber weight and yield were obtained from 40 t of FYM and 200 kg of nitrogen fertilizer [56]. In India, investigations into the effects of nitrogen and FYM on potato productivity, nutritional content, and quality revealed positive responses to increasing rates of both FYM and nitrogen fertilizers [57]. This response may be attributed to enhanced metabolite synthesis. Furthermore, the combined application of 24 t of FYM per hectare and 180 kg of nitrogen per hectare yielded the highest tuber production. Interaction effects between FYM and nitrogen fertilizers consistently outperformed separate applications. A study conducted in Jimma District, Ethiopia, in 2016 assessed different rates of organic and inorganic fertilizers on potato growth and yield components. The combined use of 20 t ha^{-1} of FYM and 70 kg ha^{-1} of nitrogen fertilizer resulted in the largest plant and leaf diameter [58]. In another Ethiopian study, the potato cultivar 'Jalene' was subjected to a combined experiment involving farmyard manure (ranging from 0 to 20 t ha^{-1}) and nitrogen (at rates of 0, 35, and 70 kg ha^{-1}). The results demonstrated that the interaction between different levels of farmyard manure and nitrogen significantly influenced various parameters, including leaf area, plant height, leaf length, leaf number, leaf diameter, stem diameter, and stem number. Notably, the control treatment exhibited the minimum values for all these parameters [59]. Furthermore, a 2020 study conducted in the North Shewa Zone of the Amhara Regional State in Ethiopia aimed to assess the impact of farmyard manure (FYM) at rates of 0, 4.5, 9, and 13.5 t ha^{-1}, along with inorganic fertilizers at rates of 0, 81.7, 163.4, and 245.1 kg ha^{-1}, on the growth and yield of irrigated potatoes. The results indicated that the combined application of 245.1 kg ha^{-1} of inorganic fertilizers with 13.5 t ha^{-1} of FYM resulted in the highest marketable and total tuber yields [60]. Atanaw and Zewide [61], in a paper on the effect of the combined application of inorganic nitrogen, phosphorus, and potassium (NPK) and cattle manure fertilizers on vegetative growth, yield components, and improved productivity of the soil, claimed that animal manure such as cattle manure has positively beneficial effects on vegetative growth, yield, and tuber quality. Therefore, a combined application of mineral NPK and cattle manure is essential to sustain high yields, better tuber quality, and more profit, as well as to improve soil fertility.

4. Green Manure in Organic Potato Cultivation

Green manuring consists of incorporating green plants into the soil, either by cultivating them in the same field (cover crops) or by using plants that were grown elsewhere at the green stage just before flowering. Researchers suggest intercropping green manure with potatoes as a complementary crop, as it has the potential to enhance crop productivity and improve soil fertility [62,63]. Utilizing cover crops plays a multifaceted role in organic agriculture by maintaining soil organic matter, enhancing soil health, mitigating erosion, and optimizing nutrient management, thereby increasing nutrient availability [64]. In organic potato agro-ecosystems, a variety of cover crop families and species have been introduced to provide different services based on the characteristics of the agro-ecosystem. Generally, non-leguminous cover crops such as sunflower, crucifers, and cereals such as rye and barley, offer benefits such as organic matter generation, weed suppression, and erosion control. Non-leguminous cover crops can also function as nitrogen (N) catch crops, optimizing N utilization throughout the rotation system (e.g., rye) [65,66], and capturing N that might otherwise leach from the soil [67]. Furthermore, certain non-leguminous cover crops such as winter rye, ryegrass, brassicas, and buckwheat, have demonstrated the ability to reduce soil-borne diseases when incorporated into rotations with potatoes [68]. On the contrary, leguminous crops can serve as full-season cover crops either with a cereal nurse crop (e.g., small red clover undersown in oats, barley, or wheat) or as the sole cover crop in the year preceding potatoes. These crops are typically terminated as green manures, with estimates suggesting that the impact of a preceding leguminous cover crop is comparable to applying N fertilizers at a rate ranging from approximately 20 to 150 kg N ha^{-1} [69]. It is widely acknowledged that the release of N from decomposing green manure residues can coincide well with plant uptake, potentially enhancing N uptake efficiency and crop yield while mitigating N leaching losses [70]. For these reasons, leguminous cover crops play a crucial role in the fertility management of organic potato farming, replacing chemically synthesized nitrogenous fertilizers and being particularly beneficial when preceding potatoes, which have high N requirements [71]. Uchino et al. [72] evaluated the effect of fertilization and inter-seeding cover crops on the growth of main crops and weeds and the stability of weed suppression under four-year rotational organic farming. Fertilization involved a fully ripened compost (N: 0.725%, P_2O_5: 0.59%, K_2O: 0.86% *w*/*w* in fresh weight) at 45 t ha^{-1} on average over the 4 years of study and 1 t ha^{-1} of fermented organic fertilizer (N 5%, P_2O_5 6%, K_2O 4%). In an organic farming system, researchers interseeded two cover crops—winter rye (*Secale cereale* L.) and hairy vetch (*Vicia villosa* Roth)—in furrows alongside potato (*Solanum tuberosum* L.), soybean (*Glycine max* Merr.), and maize (*Zea mays* L.). The study revealed that effective and stable weed suppression could be achieved without compromising the yield of main crops when cover crops were interseeded with adequate fertilization.

The choice of cover crop species is crucial, as demonstrated by a separate investigation conducted in Central Italy (Viterbo) over a two-year period, which examined yield and weed control effects in potato crops following different cover crops [73]. In a three-year chick-pea/potato/tomato rotation, each crop was preceded by seven different soil managements: five cover crops (rapeseed, Italian ryegrass, hairy vetch, snail medick, and subclover), one unfertilized weedy fallow (cover crop absent), and one control (weedy fallow fertilized with mineral N at a rate of 170 kg ha^{-1} for potato). Two different weed control regimes in potatoes were also applied [weed-free crop (1 inter-row hoeing + 1 hilling up + manual weeding on the row); mechanical control (1 inter-row hoeing + 1 hilling up)]. The potato crops following the cover crops were only fertilized with green manure. Following subclover and hairy vetch the potato crop yield was similar to that obtained by mineral N-P-K fertilization. In a study conducted by Larkin and Halloran [74], researchers evaluated potential disease-suppressive crops using four different production management approaches: cover crop, green manure, harvested crop-residue-incorporated, and harvested crop-residue-not-incorporated. These trials were part of potato rotation experiments, and the focus was on their impact on yield. Mustard blend, sudangrass, and rapeseed rotations led to yield increases (ranging from 6% to 11%) compared to a simple rotation control. Notably, all rotation crops managed as green manures consistently achieved higher yields (6% to 13%) than other management practices. Among them, the combination of mustard blend managed as a green manure was particularly effective, resulting in a 25% yield increase relative to a soybean cover crop.

In a field experiment by Tein et al. [29], potato (*Solanum tuberosum* L.) was part of a five-crop rotation in which red clover (*Trifolium pratense* L.), winter wheat (*Triticum aestivum* L.), peas (*Pisum sativum* L.), potato and barley (*Hordeum vulgare* L.) followed each other simultaneously on the same field. The field experiment was performed with six different farming systems as follows: two organic and four conventional. In both organic farming systems, catch crops were used to provide organic green manure; in the second organic system, fully composted cattle manure at a rate of 40 t ha^{-1} was also added as a fertilizer. The four conventional farming systems differed in the amounts of mineral fertilizers used as follows: $N_0P_0K_0$ (control), $N_{50}P_{25}K_{95}$ (low), $N_{100}P_{25}K_{95}$ (average), or $N_{150}P_{25}K_{95}$ (high). Fresh tuber yields were significantly lower under organic systems than conventional. However, the other organic system, involving green manure and composted cattle manure, significantly increased the average soil organic carbon (C) and phosphorus (P) concentrations after potato cultivation, whereas the conventional system that received mineral nitrogen fertilizers showed fluctuations in soil nitrogen (N) total concentration, average soil organic C and P concentrations, the soil pH and potassium after potato cultivation and magnesium (Mg) concentrations in relation to N dose of fertilizer. Tein et al. [29] reported that integrating potatoes into a three-year crop rotation experiment with red clover (*Trifolium pratense* L.), winter wheat (*Triticum aestivum* L.), peas (*Pisum sativum* L.), and barley (*Hordeum vulgare* L.) was beneficial also for potato quality. Indeed, the average

P concentrations were higher in organic potato tubers compared to other systems, while the Mg content in organic tubers was surpassed only by the treatment involving organic farming systems supplemented with fully composted cattle manure. In a study by Nyamdavaa and Friedel [75], the impact of different preceding crops, cover crops, and manure application on the agronomic performance of potatoes was investigated over two consecutive years in an organic farming system. The experiment evaluated lucerne, field pea, and spring barley as pre-crops, non-leguminous species, and a mix was applied as a cover crop (a bare fallow treatment was adopted as a control in comparison to cover crop) along with farmyard manure application. Notably, the subsequent crop response to preceding crops was minimal, with no significant difference in tuber yields (fresh tuber, marketable, and dry matter) observed between legume pre-crops and barley. However, both cover crops and manure slightly increased tuber dry matter yield in the first year, while the second year showed no effect on dry matter yield. Interestingly, the percentage of small-sized tubers (<35 mm in diameter) increased notably in the second year compared to the first. Additionally, cover crops used as green manures contributed to a higher percentage of large-sized tubers (>65 mm in diameter) over the two-year period. In a study conducted in the highlands of Awi Agro-Ecological Zone [76], two rotation systems and four levels of organic treatments: V1 = 0 t ha^{-1} farmyards manure (FYM), V2 = 2.5 t ha^{-1} fresh sesbania green manure (FSB), V3 = 5 t ha^{-1} FYM, and V4 = 5 t ha^{-1} FYM +2.5 t ha^{-1} FSB were factorially arranged in fixed plots for three years. Among all, the highest total potato tuber yield was obtained at the combined application of 5 t ha^{-1} FYM +2.5 t ha^{-1} FSB. The treatment combination increased total potato tuber yield by 140% and 41% over that of the first and the second years. The effect of cover crops plus added composted manure was also studied on the soil microbial hydrolytic activity [77]. A study investigated the impact of five-field crop rotations within two distinct farming systems: organic (Org II, which included winter cover crops and composted manure, compared to Org 0 as the control) and conventional (Conv II, which used mineral nitrogen at a rate of 150 kg ha^{-1}, compared to Conv 0 as the control) on crop productivity and soil health. After potato cultivation, the lowest and highest decrease in the soil microbial hydrolytic activity was seen in Org II and Conv II systems, respectively. Soil organic carbon (SOC) and total nitrogen (N) levels were found to be higher in organic farming systems, with no significant changes observed after potato cultivation. In a field study conducted on Prince Edward Island [78], researchers examined the impact of applying pen-pack cow (*Bos taurus*) manure at a rate of 20 Mg ha^{-1} and the use of cover crops—comprising grasses, legumes, or a mixture of both—with red clover (*Trifolium pratense* L.) serving as the control, on nitrate dynamics, soil nitrogen supply capacity, and subsequent potato yields. On average, red clover accumulated 88% more total nitrogen than the grass-legume mixtures; however, this did not lead to higher potato yields. In contrast, pearl millet and sorghum sudangrass (*Sorghum bicolor* and *Sorghum bicolor* var. Sudanese) were associated with lower soil nitrate levels compared to red clover but resulted in higher total potato yields. The incorporation of manure increased both total and marketable yields by 28% and 26%, respectively, and enhanced soil nitrogen supply capacity by an average of 44%. Rittl et al. [79], in a field with regular potato growing, examined the effects of different organic amendments with or without winter rye as a cover crop. They found that the cover crop did not affect soil organic matter, but increased tuber yields in the second year, and reduced the severity of potato diseases by 10% in post-harvest potatoes in both years. The use of cover crops and manure in organic farming can produce comparable outcomes to those of conventional farming with low nitrogen input, as also demonstrated by the research results of Margus et al. [80]. In their study conducted in Estonia from 2018 to 2020, potato crop performances under three organic (Org) and four conventional (N_0, N_{50}, N_{100}, N_{150}) system treatments were assessed. The crop rotation sequence included spring barley (*Hordeum vulgare* L.) with under-sown red clover, red clover (*Trifolium pratense* L.), winter wheat (*Triticum aestivum* L.), pea (*Pisum sativum* L.), and potato (*Solanum tuberosum* L.). The organic farming system included three treatments: Org I (organic control), Org II (organic crop rotation with winter cover crops), and Org III (organic crop rotation with winter cover crops and supplemented with composted cattle manure). Over the three-year trial period, conventional systems showed an average yield that was 25% higher than that of organic systems. However, yields within the organic systems demonstrated greater stability. The Org I treatment achieved the same dry matter yield as the N_0 system, indicating that the application of chemical plant protection without fertilization did not yield any positive effect on growth. Annually, the yield in the Org III system was equivalent to that of the N_{50} system. There were no significant differences in the number of tubers per plant across the different farming systems, though notable differences were observed between the trial years. Tubers from conventional systems had a lower starch content compared to those from organic systems. Mamiev and colleagues [81] conducted a comprehensive study to improve crop productivity and quality using green manure crops, manure, and mineral fertilizers while enhancing soil fertility. In the foothill zone of North Ossetia-Alania, green manure improved the bulk density, porosity, and air capacity in the arable soil layer and consequently increased moisture content in the 0–30 cm soil layer. Over a three-year period, the combined use of straw, nitrogen fertilizers, and green manure resulted in the highest potato yield, with an increase of 34.0%, compared to the control and also improved tubers quality, with increases in dry matter content, starch, and vitamin C. A three-year organic crop rotation was set up in a field with sandy loam soil, with a cover crop of rye and vetch grown over the three autumn/winter seasons for green manure, followed by potato and lettuce (first year), Swiss chard and turnip (second year), and Portuguese cabbage and carrot (third year) [82]. Nitrogen (N) mineralization was determined by field incubation in response to green manure (GM), with 20 and 40 t ha^{-1}, farmyard manure (FYM) (C20 and C40), and GM with 1 and 2 t ha^{-1} of commercial organic fertilizer (CF1 and CF2). The second season crops (lettuce, turnip, and carrot) yields were higher for the treatment C40 compared to all other treatments because most of the commercial fertilizer was mineralized during the previous crop. Swiss chard, grown in a short season (54 days), produced a higher yield for CF2 compared with C40. However, this was not true for potatoes (first year), probably because of increased manure mineralized N recovery during the longer growing season for the potatoes (124 days), nor for the cabbage (third year), which had a short growing season (56 days), because of increased N availability with continuous composted and green manure application. This study highlighted that field incubation can be used to assess mineralization rates and that the fast N release of commercial fertilizers increased the yield of the first crop of the year, whereas the slowly released N of FYM increased the yield of both crops of the year, with lower risk of N loss. In a study by Drakopoulos and colleagues [83,84], the effects of different organic amendments (solid cattle manure, lucerne pellets, and grass/clover silage) on crop performance and nitrogen utilization in organic potato cultivation were examined. Notably, plant-based fertilizers demonstrated improved nitrogen utilization compared to animal-based manures, and lucerne pellets consistently yielded the highest crop output, irrespective of tillage practices. A study conducted at the Intermountain Research and Extension Center in Tulelake examined the impact of cover crops and organic amendments (including composts, manures, bloodmeal, and soymeal) on soil nitrogen availability for potato production. The research evaluated various cover crop species, three planting dates, and multiple cover crop combinations [85]. Notably, vetches and field peas, when used as green manure, effectively met the in-season nitrogen requirements of potatoes. These cover crops, whether grown individually or in mixtures with non-legume species, yielded potato crops comparable in quality and yield to those cultivated using conventional fertilizers.

5. Organic Amendments in Organic Potato Cultivation

The incorporation of organic amendments is a common practice in organic potato production systems [86]. Organic amendments such as compost, vermicompost, and biochar are considered to be soil conditioners rather than a source of nutrients, and their release of mineral elements, mainly nitrogen, is thus limited. They genuinely could improve soil fertility, physically, chemically, and biologically. Physically, soil that has been amended with compost or vermicompost has improved aeration, porosity, bulk density, and water retention; in addition, chemical characteristics including pH, electrical conductivity, and organic matter content are also enhanced [21]. Table 2 reports some examples of applied amendments in organic potato cultivation.

Table 2. Impact of different amendments in organic potato cultivation.

Amendments	Impact on Tuber Yield	Impact on Tuber Quality	Location	References
Swine manure-sawdust compost	decrease		Canada	Lynch et al. [20]
Worm-compost	increase		Morocco	Harraq et al. [23]
Compost + green manure	increase		Italy	Canali et al. [68]
Biochar + liquid digestate	-		Norway	Rittl et al. [79]
Food residues + MSWC	increase		Italy	Passoni & Borin [87]
Compost	increase	-	Canada	Carter et al. [88]
Compost	increase		Egypt	El-Sayed et al. [89]
Pulp-fiber compost	increase		Canada	Fahmy et al. [90]
Different types of compost	no effect		Canada	Wilson et al. [91]
Compost		increase phenolics, flavonoids, anthocyanins	Lithuania	Vaitkevičienė et al. [92]
On-farm compost (fish waste, seaweed, pine bark)	increase	increase reducing sugars	Spain	Illera-Vives et al. [93]
MSWC [1] and paper mill biosolid compost	decrease		Canada	Alam et al. [94]
MSWC	increase	increase micronutrients	Spain	Escobedo-Monge et al. [95]
On-farm prepared compost	increase		Belgium	Willekens et al. [96]
Farmyard compost, compost tea	increase	increase dry matter, specific gravity	Egypt	El-Tantawy et al. [97]
Conifer-based compost + green manure+ biofertilizer	increase		United States	Bernard et al. [98]
Biochar	increase		Egypt	Youssef et al. [99]
Biochar	increase		Nepal	Upadaya et al. [100]
Biochar, biochar + compost	increase		Canada	Mawof et al. [101]
Biochar + recommended NPK	increase		Canada	Farooque et al. [102]
Biochar + organic fertilizer	increase		China	Hou et al. [103]
Biochar + recommended NPK	increase	increase dry matter, specific gravity	Bangladesh	Mollick et al. [104]
Biochar + vermicompost + bonemeal, others	increase		India	Singh & Siddique [105]

[1] MSWC = Municipal Solid Waste Compost.

5.1. Compost

Composting stands out as the premier strategy for managing organic solid waste, relying on the bio-oxidative decomposition of original organic materials. This process transforms organic waste into humified, stable, odorless, and pathogen-free material, suitable for enhancing degraded soils. Compost is highly regarded in organic agriculture due to its profitability, environmental friendliness, and ease of handling [106]. The ideal compost for potato cultivation is open-textured, rich in organic matter, high in nutrients, and neutral in pH. Compost provides a slow-release source of organic nutrients, which can contribute to a nutrient-driven yield response. However, compost degradation occurs slowly, and the amount of nitrogen (N) mineralized from compost in the first year after application is minimal, as evidenced by the crop's very low apparent nitrogen recovery (ANR) [107]. Specifically, compost contains a significant portion of total nitrogen, with over 90% generally not being immediately available for plant uptake [108]. Most of this nitrogen is bound within the organic N-pool, while the mineral nitrogen, which is readily accessible to plants, constitutes less than 2% of the total nitrogen content. Studies have shown that in the year following compost application, the available nitrogen accounts for less than one-fifth of the total nitrogen applied [109]. Consequently, compost alone is insufficient to meet the nitrogen needs of crops in the short term, making it unsuitable as the sole nitrogen source. In their evaluation of nitrogen application through organic amendments in an organic potato field, Lynch et al. [20] observed a decline in tuber yield after applying swine manure-sawdust compost with a C/N ratio of 22, attributing this reduction to nitrogen immobilization. The C/N ratio is a commonly used, though approximate, indicator of net nitrogen mineralization potential. Organic materials with a C/N ratio below 20, which decompose more readily, tend to release nitrogen during decomposition, while those with a C/N ratio above 20 are more likely to temporarily immobilize nitrogen [110]. In a study assessing various composts as nitrogen sources in crop successions, including potatoes and catch crops, Passoni and Borin [87] found that crop response and nitrogen uptake were minimally affected by compost fertilization. On the other hand, Carter et al. [88] reported an increase in tuber yield that exceeded the maximum yield achieved with nitrogen application alone in an experiment exploring the effects of compost on a potato rotation. This "non-nitrogen" compost yield effect was attributed to a slight yet significant enhancement in soil water-holding capacity. In addition, El-Sayed et al. [89] indicated that the organic production of potatoes using 23.8 t ha^{-1} of compost could be an alternative to conventional production with commercial organic fertilizers without significant reduction in yield and quality. In a field experiment conducted at two sites in Atlantic Canada over two years, the impacts of two organic amendments (commercial hog manure–sawdust compost applied at 300 kg total N ha^{-1}, pelletized poultry manure applied at 600 kg total N ha^{-1}, compared to an unamended control) on Shepody potato yield, quality and soil mineral nitrogen dynamics under organic management were studied [20]. Relatively high tuber yields and crop N uptake were achieved for unamended soil in those site-years when soil moisture was non-limiting. Compost resulted in higher total yields than un-amended control in one of three site-years, although the apparent recovery of N from compost was negligible. Fahmy et al. [90] observed that the addition of pulp-fiber residue compost increased the availability of phosphorus and potassium for potato plants. Specifically, tuber yield improved under supplementary irrigation conditions, while no significant change was noted in rainfed settings compared to unamended plots. The adoption of green manure practices and the recycling of organic materials offer a promising alternative to conventional synthetic fertilizer-based management systems, supporting sustainable potato production without exacerbating potential environmental risks associated with nitrogen leaching [68]. A research study conducted in rain-fed potato production in New Brunswick, Canada, investigated the impact of various compost products on tuber yield and plant nutrient availability. The study included a no-compost control and five compost types: marine with shells compost, poultry manure compost, forestry residues compost, municipal source-separated organic waste compost, and forestry waste and poultry manure compost [91]. In a small plot trial, mature compost products slightly increased plant nitrogen availability, while immature compost led to net nitrogen immobilization. Despite these differences in nutrient availability, there was no significant effect on tuber yield. In on-farm trials spanning 19 site-years, compost did not significantly enhance yield, suggesting that any short-term nutrient benefits from wood waste and manure-based composts are unlikely to impact crop yield. However, there may be potential to reduce nutrient application rates, particularly for potassium. Vaitkevičienė et al. [92] compared several colored potato cultivars grown organically with 30 t ha^{-1} of compost to those grown conventionally using complex fertilizers (112 kg ha^{-1} nitrogen, 56 kg ha^{-1} phosphorus, and 136 kg ha^{-1} potassium). Their findings indicated that potato tubers cultivated under organic conditions contained higher levels of polyphenols, phenolic acids, chlorogenic acid, p-coumaric acid, caffeic acid, flavonoids, and anthocyanins compared to those grown conventionally.

Illera-Vives et al. [93] investigated the impact of three compost application rates (32 t ha^{-1}, 43 t ha^{-1}, and 65 t ha^{-1}) derived from composted fish waste, seaweed, and pine bark, comparing them to a mineral fertilizer, a certified organic fertilizer made from dehydrated broiler litter, and a control with no fertilizer on potato yield and the chemical composition of potato tubers, petioles, and folioles. The compost showed a strong fertilizing effect, even outperforming the mineral fertilizer. The 65 t ha^{-1} compost treatment increased total production by 53% compared to the control and by 30% relative to the mineral fertilizer. It also performed similarly to dehydrated broiler litter and significantly reduced the proportion of non-commercial potatoes (caliber <35 mm) compared to the control. Notably, only the 65 t ha^{-1} compost treatment altered the tubers' chemical composition by increasing their reducing sugar content.

In Eastern Canada, a five-year study compared four treatments—control, inorganic fertilizer (FERT), municipal solid food waste compost (MSW), and paper mill biosolid compost (PMB)—in organic potato rotations [94]. The study found that soil amendments significantly affected potato yields. Total nitrogen uptake by potatoes was 89, 115, 107, and 147 kg N ha^{-1} for the control, MSW, PMB, and FERT treatments, respectively, with higher uptake levels observed when potatoes followed red clover (119–124 kg N ha^{-1}) compared to an oat/pea/vetch mixture (107–108 kg N ha^{-1}). Plant nitrogen use efficiency was measured at 299, 263, 263, and 235 for the control, MSW, PMB, and FERT treatments, respectively.

In a study conducted by Escobedo-Monge and colleagues [95], the impact of municipal solid waste compost was compared to other organic and inorganic amendments in terms of potato production, quality, and yield across four potato genotypes. The addition of all amendments, particularly urban waste compost, resulted in raised potato production and elevated levels of macronutrients, micronutrients, and heavy metals in the soil, with a progressive accumulation observed in the tubers.

A long-term fertilization study involving a four-year crop rotation of maize, potatoes, spring barley, and red clover was conducted across two fields, with a one-year offset between them, to assess the impact of various fertilization methods. These included on-farm prepared compost primarily made from vegetable residues (applied in both single and double doses), farmyard manure, slurry, and a combination of slurry with composted municipal waste [96]. In one growing season, potato yields were notably higher for both farm compost treatments, while soil nitrate levels were significantly lower. However, during another season characterized by early, severe, and consistent leaf blight disease pressure, yields were reduced, and the differences between treatments were less pronounced. This study highlighted that using mature compost can promote quicker early-season potato crop development, which is crucial for achieving adequate yields in organic potato farming. A field experiment was conducted (North Nile delta region), to investigate the potential of applying farmyard compost and its compost tea on potato crops (*Solanum tuberosum* L.) [97]. Compared to the control, four application methods (ore compost, compost tea, sediment, and sediment + extract combination) were studied. The application of compost tea enhanced dry shoot yield, fresh tuber yield, tuber dry matter content, crude protein, and tuber-specific gravity in both seasons. Additionally, the combination of sediment and extract significantly increased tuber dry matter and crude protein in both seasons. Incorporating farmyard manure into the soil boosted both total and available nitrogen, phosphorus, and potassium levels in the soil. In organic cropping systems, the synergistic effect of green manure and various organic amendments can significantly enhance soil fertility and yield while also mitigating the risk of nutrient leaching, particularly nitrogen

(N) [111,112]. Canali et al. [68] conducted a field experiment in Tuscany, Central Italy, aimed at evaluating the impact of cattle farmyard manure and compost, applied at three different rates (0, 50, and 100 kg N ha^{-1}), in conjunction with a legume green manure cover crop (*Trifolium subterraneum* L.), on potato nutrition, tuber yield, N uptake, and use efficiency. Their findings suggest that eco-functional intensification in organic potato cropping systems can be achieved by leveraging the combined effects of legume green manure and organic amendments. Bernard and colleagues [98] investigated three distinct disease-suppressive management practices. These included a rotation crop of *Brassica napus* (rapeseed) as green manure, compost amendment derived from conifers, and the evaluation of three biological control organisms (*Trichoderma virens, Bacillus subtilis*, and *Rhizoctonia solani* hypovirulent isolate Rhs1A1). The study assessed these practices individually and in combination at sites managed under both organic and conventional systems. They found that compost amendment had variable effects on tuber disease. However, the combination of rapeseed rotation and compost amendment consistently improved yield at both sites. Moreover, this integrated approach not only mitigated disease but also enhanced overall crop productivity.

5.2. Vermicompost

Vermicompost is produced through the decomposition process involving various worm species, such as red wigglers, white worms, and other earthworms, resulting in a blend of bedding materials, decaying organic matter, and vermicast. Studies show that vermicompost generally contains higher concentrations of macronutrients and micronutrients than traditional compost, with nutrients such as nitrogen, phosphorus, potassium, calcium, magnesium, and trace elements present in more soluble forms. Moreover, vermicompost offers a beneficial load of enzymatic–bacterial microbes, which provide suppressive and antibiotic effects against pathogens, along with growth-regulating substances and humic acids that enhance plant growth [113]. These characteristics improve soil fertility in physical, chemical, and biological terms, leading to higher crop yields. In a pot experiment, Harraq et al. [23] investigated the impact of applying varying amounts of worm compost (0, 5, 10, and 15 t ha^{-1}) and fishmeal (0, 1.75, and 3.5 t ha^{-1}) on potato growth and yield. Fishmeal significantly improved leaf number, plant height, dry matter, aerial nitrogen uptake, tuber count, and yield, while worm compost primarily affected leaf number and tuber yield. The optimal yield (554.4 g per plant; 23.1 t ha^{-1}) was achieved with a combination of 10 t ha^{-1} worm compost and 1.75 t ha^{-1} fishmeal, resulting in maximum underground and total nitrogen uptake at harvest, supported by nitrogen mineralization apparent rates of 71% for fishmeal and 64.3% for worm compost.

5.3. Biochar

Biochar is a carbon-rich solid produced through the pyrolysis of plant or animal biomass under conditions of limited or partial oxygen. Various feedstocks can be used to produce biochar, including agricultural residues such as rice straw, corn cobs, and wheat residues, as well as animal waste, sewage sludge, municipal solid waste, and biogas residues. The chemical properties of biochar are influenced by the type of feedstock and the processing parameters, such as pyrolysis temperature, heating rate, residence time, particle size, reactor type, and pressure conditions [114]. Due to its unique characteristics, biochar is utilized in various applications, including waste management, adsorption functions, soil amendment in agriculture, catalysis, carbon sequestration, and energy storage. Biochar can enhance fertilizer effectiveness for plant uptake and reduce soil nutrient loss. Nutrient loading onto biochar is crucial for improving fertilizer efficiency, as pristine biochar contains few nutrients [115]. Youssef et al. [99], examined the effect of biochar addition on the production of potatoes grown under sandy soil conditions in integrated farming conditions in Egypt. The study included 12 treatments, which were combinations of three potato cultivars (Accent, Cara, and Spunta) and four biochar application rates (0, 1.25, 2.50, and 5.00 m^3/feddan; 1 feddan = 0.42 ha). The results indicated that the potato cultivars Cara and Spunta recorded the highest values for morphological characteristics, leaf content of Ca, Mn, and Cu, as well as tuber yield and quality. The Cara cultivar achieved the maximum dry weight of various plant parts, the Spunta cultivar exhibited the highest concentrations of photosynthetic pigments in leaf tissues and the greatest starch content in tubers. Conversely, the Accent cultivar displayed the lowest values for these parameters and had the lowest nitrate content in tubers. Upadhyay et al. [100] investigated the impact of various biochar types (including *Lantana camara*, *Ipomoea carnea*, rice husk, and sawdust) on potato growth and yield attributes. Overall, biochar positively influenced total yield. However, even when applied at 0.3% and 1% levels, biochar did not fully prevent the survival or breeding of cyst nematode species in potatoes [116].

In a two-year field study, Mawof et al. [101] evaluated the impact of biochar, compost, and a biochar-compost mixture on soil properties and potato yield when irrigated with wastewater. The treatments included a factorial combination of three levels of barley straw biochar amendment (0%, 1%, and 3%) and two levels of mixed green and kitchen waste compost amendment (0% and 7.5%). Compared to the unamended control, all amendment treatments significantly improved soil physico-chemical properties and crop yield, though they did not affect plant growth or physiological parameters. Soil cation exchange capacity, soil organic matter, and pH were significantly enhanced by compost amendment, with the combination of 3% biochar + 7.5% compost producing some of the best outcomes. In an Atlantic Canada representative soil suitable for potato cultivation, a trial was conducted using 8000 cm^2 lysimeters. The study evaluated the impact of four soil amendment treatments: (1) control (no added nutrients), (2) biochar, (3) synthetic fertilizer with recommended NPK, and (4) biochar combined with recommended NPK [102]. The biochar amendment improved soil micro- and macro-nutrients, soil organic matter, pH, and cation exchange capacity (ECE), whereas the maximum potato yield was achieved by the combined application of biochar and synthetic fertilizer. In a combined application experiment [103] involving biochar (B) and organic fertilizer (O) with four concentration gradients using the equal carbon ratio method, it was found that soil fertility was improved by the combined application of biochar and organic fertilizer, with the best effect being achieved at a ratio of B:O = 1:2. When compared to the control, the relative abundance and diversity index of soil bacteria were significantly improved by the treatment at B:O = 1:2. Bacterial diversity directly affected the potato yield, while soil fertility indirectly affected potato yield by influencing the soil bacterial diversity. Mollick et al. [104] conducted a field experiment in Bangladesh to assess the effect of biochar on the yield and quality of potato tubers. The study included nine treatments: a control (no chemical fertilizer or biochar), a recommended fertilizer dose (RFD), a biochar-only treatment, and six combinations of biochar with the recommended fertilizer dose on the Damian cultivar. In general, the combination biochar-RFD significantly increased plant height, tuber weight, tuber yield, tuber dry matter content, tuber-specific gravity, and soil organic carbon when compared to control, RDF, or biochar alone. In a research [105] the impact of biochar-based organic amendment was analyzed on the growth and yield of the potato crop. Different combinations of biochar base organic amendment with recommended doses of fertilizer (RDF), bone meal (BM), vermi-compost (VC), and poultry manure (PM) were studied. Among all the thirteen treatment combinations of biochar base organic amendment, a treatment combination of biochar base organic amendment consisting of 25% RDF + 75% (BM + VC + PM) + 40% Biochar resulted in maximum emergence percentage, plant height, LAI, days to maturity, number of haulm plant-1, average weight of tuber plant^{-1} and tuber yield. Singh et al. [117] investigated the effects of different combinations of biochar, vermicompost, poultry manure, and bone meal on the soil health and quality of the cultivar Kufri Pukhraj, which is widely cultivated in India. Among the various treatments, the combination of a 25% recommended dose of fertilizer with 75% bone meal, vermicompost, and poultry manure plus 40% biochar exhibited the best performance: highest values for pH, electrical conductivity, organic carbon, soil microbial biomass, nitrogen content, labile carbon, and particulate organic carbon. Additionally, this combination achieved the highest starch content and the greatest quantity of A-grade potato tubers. Rittl et al. [79] examined the effects of organic materials: one application of biochar mixed with liquid digestate (BLD), solid digestate (SD), or farmyard manure (FYM). Organic amendments increased soil organic matter, especially for FYM and BLD, and though they did not affect potato yield or quality, the proportion of marketable potatoes tended to be higher in the amended soil. Worth mentioning is a study carried out in Karangreja District, Indonesia, in which the effects on the quality of potato cultivar Atlantic of three types of charcoal derived from wood, husk, and coconut shell, applied at varying concentrations in addition to commercial organic fertilizer "Petroganik", compared to chemical fertilizers were studied [118]. Regardless of the types of charcoal and concentrations, they found that tubers of organically fertilized plants showed lower moisture content and higher ash content compared to those of chemically fertilized plants.

6. Biostimulants in Organic Potato Cultivation

Plant biostimulants are defined as follows: "EU fertilising product the function of which is to stimulate plant nutrition processes independently of the product's nutrient content with the sole aim of improving one or more of the following characteristics of the plant and/or the plant rhizosphere: (1) nutrient use efficiency, (2) tolerance resistance to abiotic stress, (3) quality characteristics, or (4) availability of confined nutrients in the soil or rhizosphere" [119].

Many diverse natural substances and chemical derivatives of natural or synthetic compounds, as well as beneficial microorganisms, are cataloged as plant biostimulants including (i) humic substances; (ii) plant- or animal-based protein hydrolysates; (iii) macro- and micro-algal extracts; (iv) silicon; (v) arbuscular mycorrhizal fungi (AMF); and (vi) plant-growth-promoting rhizobacteria (PGPR) belonging to the genus *Azotobacter*, *Azospirillum*, and *Rhizobium* spp. [120]. Among several plant biostimulants, the most studied in organic cultivation potato systems are the humic substances, Arbuscular mycorrhizal fungi, and biofertilizers.

6.1. Humic Substances

Humic preparations containing humic acids, such as the organic product Rosahumus from Rosier, S.A., offer a promising alternative for soil enhancement. Rosahumus, sourced from leonardites, comprises roughly 85% humic acids and contains notable amounts of potassium (12% KK_2O) and iron (0.6% Fe). In studies carried out from 2018 to 2020, Rosahumus was applied to the soil before tuber planting and again just before ridging, using a spray solution of 4.5 kg ha^{-1} dissolved in 300 L ha^{-1} of water [121]. This application resulted in a notable increase in tuber yield compared to the control group that did not use humic preparations. Similar results were observed in a study by Rizk et al. [122], where the addition of humic acid to irrigation water enhanced plant growth parameters and potato tuber yield. Furthermore, research has shown that treatments involving humic acid and algae extracts lead to significant improvements in potato plant morphology and yield. The application of humic acid in combination with Alga 600 and Sea Force 2 resulted in substantially higher values for vegetative features and yield characteristics [123]. These findings underscore the potential of humic preparations and their synergistic effects with algae extracts in enhancing potato productivity and quality. Studies by Man-Hong et al. [124], demonstrated that the application of humic acid led to significant increases in tuber yield, particularly under varying levels of water retention. Specifically, yield improvements ranged from 34.5% to 63.5% under 45% water retention, 35.9% to 37.3% under 60% water retention, and 23.4% to 27.1% under 75% water retention. By contrast, research by Suh et al. [125] did not observe significant differences in tuber number or total yield after spraying potato plants with fulvic acid at different intervals post-tuber planting compared to the control group. However, when humic acid was applied to the soil before planting potato tubers at rates of 40 and 80 g m^{-2}, an increase in the weight of large tubers was noted. Furthermore, studies by Ekin et al. [126] demonstrated that treating potato tubers with a combination of humic acid and plant growth-promoting rhizobacteria (PGPR) strains, specifically *Bacillus megaterium* and *Bacillus subtilis*, led to significant improvements in plant growth, tuber yield, and overall tuber quality. These findings underscore the potential benefits of combining humic acid with microbial inoculants for enhancing potato productivity and quality. Wadas and Dziugieł [127] found that the use of humic and fulvic acids in HumiPlant (leonardite extract), as well as the seaweed extracts Bio algeen S90 (*Ascophyllum nodosum*) and Kelpak SL (*Ecklonia maxima*), had a significant effect on tuber starch content in some potato cultivars but did not affect the content of total sugars (glucose, fructose, and sucrose), monosaccharides (glucose and fructose), or sucrose.

6.2. Arbuscular Mycorrhizal Fungi

The association of arbuscular mycorrhizal fungi (AMF) with most terrestrial plants is geographically ubiquitous and occurs over a broad ecological range. AMFs are particularly important in organic and/or sustainable farming systems that rely on biological processes rather than agrochemicals to control plant diseases [128]. Of particular importance is the bioprotection conferred to plants against many soil-borne pathogens such as species of *Aphanomyces, Cylindrocladium, Fusarium, Macrophomina, Phytophthora, Pythium, Rhizoctonia, Sclerotinia, Verticillium* and *Thielaviopsis* and various nematodes by AM fungal colonization of the plant root [129]. AMF plays a crucial role as fertility-promoting microorganisms in soils, enhancing soil health and bolstering crop productivity. Specifically, incorporating AMF can yield numerous benefits: (i) expanding the root system's reach by over 100 times; (ii) enhancing the uptake of immobile soil mineral nutrients; (iii) mitigating abiotic stresses such as drought and temperature fluctuations; and (iv) promoting soil aggregation, which is crucial for improving soil structure and preventing erosion [130]. Thus, AMF application emerges as pivotal in organic farming, as it enables plants to thrive through enhanced assimilation of low-concentration mineralized soil nutrients [131]. On potato crops, several scientific studies to assess the impact of AMF on production showed that trial inoculation resulted in higher yields and larger tubers than treatments using conventional chemical fertilizers. Mycorrhizal potato plants were reported to show improved growth and development, pathogen resistance, and productivity compared to non-inoculated potato plants [132]. Furthermore, it is acknowledged that AMF isolates may exhibit host genotype specificity [133]. Douds et al. [134] investigated the impact of inoculating potato plants (*Solanum tuberosum* L. cv. Superior) with arbuscular mycorrhizal (AM) fungi in a field characterized by very high phosphorus availability (375 $\mu g\ g^{-1}$ soil) over two growing seasons. The inoculation treatments included a commercially available inoculum containing *Glomus intraradices*, mixed species inocula produced on-farm using compost and vermiculite, and a control treatment of freshly prepared compost and vermiculite mixture. The study compared two farming systems: one using conventional chemical fertilizers and the other using dairy manure composted with leaves to meet recommended nutrient levels. In the first year, AM fungi inoculation significantly boosted tuber yields by 33% under conventional fertilizer application and 45% with compost addition compared to the controls in each system. The second year showed a smaller response, with yields of inoculated plants being 10 to 20% higher than controls. The effectiveness of AM fungi in large-scale potato production was further supported by an analysis of 231 field trials using the same AMF inoculant (*Rhizophagus irregularis* DAOM 197198) over a four-year period in North America and Europe under real-field conditions [135]. In a study by Khosravifar et al. [136], potato plants were inoculated with *Claroidoglomus etunicatum* and *Rhizophagus intraradices*. The presence of *R. intraradices* led to a significant increase in tuber yield (32.5–36.0%) compared to non-inoculated control plants, with a maximum root colonization percentage of 54.2%. Arbuscular mycorrhizal fungi (AMF) have been shown to enhance the drought, salinity, and disease tolerance of potato plants by improving water and nutrient uptake and enhancing overall soil structure [137]. Lombardo et al. [138] explored the impact of mycorrhizal inoculants in combination with either full or reduced fertilizer doses on the yield and physiological characteristics of three early potato cultivars grown organically in highly calcareous and alkaline soils. The study revealed that AMF symbiosis improved potato tolerance to limestone stress compared to non-inoculated plants, as evidenced by enhanced potential quantum efficiency of photosystem II, better plant gas-exchange parameters, and increased marketable yield, attributed to a higher number of tubers per plant and greater average tuber weight. The effectiveness of AMF was particularly notable when half the usual fertilizer dose was applied in areas where soil conditions were less conducive to potato growth. Furthermore, Lombardo et al. [139] demonstrated that soil mycorrhization can significantly enhance the quality of organically grown early potato tubers, especially in low-fertility soils such as calcareous soils, by improving the plant's uptake of essential minerals. Their research highlighted the role of soil mycorrhization in increasing the accumulation of Na, Cu, Mn, and P in tubers, as well as lowering the Na/K ratio, though these effects varied depending on the cultivar and location. In a farmer's experiment in Iran, Abdulkhadum et al. [140] studied the addition of mycorrhiza (control, and 7.0 g plant^{-1}) and three kinds of organic fertilizer: chicken waste (10 t ha^{-1}), humic fertilizer (3 t ha^{-1}), and vermicompost (4 t ha^{-1}). They found that mycorrhiza increased plant height and chlorophyll content; the results of adding organic fertilizer showed that poultry waste significantly improved plant height, leaf area, chlorophyll, number of marketable tubers total yield, dry matter, starch, and N, K in tubers. The mutual interaction therapy (mycorrhiza + organic manure and chicken waste) improved vegetative, quantity, and quality features, plant height, leaf area, chlorophyll percentage of dry matter, the number of marketable tubers, and the total yield.

6.3. Biofertilizers

Apart from Arbuscular mycorrhizal fungi in potatoes, other biofertilizers containing microorganisms that augment the supply of primary nutrients to the main crop have also been applied in organic potato systems [141]. The application of microbiological preparations has been associated with the elimination of putrefactive processes, dissolution of mineral compounds inaccessible to plants, and enhancement of soil fertility and structure, all of which may contribute to increased potato yield. Abou-Hussein et al. [142] demonstrated that a biofertilizer containing *Candida tropicalis*, *Pseudomonas aeruginosa*, and *Bacillus megatherium*, along with the commercial biofertilizer Microbin, increased plant height, leaf number, and both fresh and dry weights of leaves and stems. Additionally, the combination of compost with chicken manure and biofertilizers enhanced dry matter content, specific gravity, and total carbohydrate levels.

A study investigated the effects of combining bio-fertilizers, which included nitrogen fixers (*Azospirillum brasilense* and *Azotobacter chroococcum*), phosphorus-solubilizing bacteria (*Bacillus megatherium* and vesicular-arbuscular mycorrhiza), and potassium-solubilizing bacteria (*Bacillus cereus*), with varying compost rates, compared to the application of mineral fertilizers at 285.6 kg N, 178.5 kg P_2O_5, and 357 kg K_2O per hectare, on potato crops [143]. The results showed significant increases in both total and marketable potato yields in plots treated with 50% of the recommended mineral fertilizers combined with 23.8 t ha^{-1} compost, whether or not bio-fertilizers were included. Additionally, plots receiving 35.7 t ha^{-1} of compost alone also demonstrated improved yields compared to the control plots that received full mineral fertilizer doses along with 11.9 t ha^{-1} of compost.

Moreover, El-Sayed et al. [143] suggested the adoption of compost combined with bio-fertilizer as an alternative to conventional production, obtaining a lower content of K and better storability in organic potatoes. Minin et al. [144], studied the effects of the level of mineral nutrition provided by compost (from 0 to 160 kg N ha^{-1}) and bio preparations (*Flavobacterin* which has an N fixer attribute, and Vitaplan and Kartofen, which have bio fungicide features). The use of both microbiological preparations and compost gave approximately the same effect and increased yields by 35–37% compared to the control. In their research, Bernard and colleagues [98] investigated three distinct disease-suppressive management practices: (1) a *Brassica napus* (rapeseed) green manure rotation crop, (2) conifer-based compost amendment, and (3) the application of three biological control organisms (*Trichoderma virens*, *Bacillus subtilis*, and *Rhizoctonia solani* hypovirulent isolate Rhs1A1). The study was conducted at sites with both organic and conventional management histories, focusing on their impact on soilborne diseases and tuber yield. Notably, rapeseed rotation led to a reduction in observed soilborne diseases (including stem canker, black scurf, common scab, and silver scurf) by 10% to 52% in at least one year at both sites. Compost amendment exhibited variable effects on tuber diseases but consistently increased yield at both locations. The biocontrol effects on disease varied; for instance, Rhs1A1 decreased black scurf at the conventional site, while *T. virens* mitigated multiple diseases at the organic site in at least one year. Interestingly, the combination of rapeseed rotation with compost amendment not only reduced disease incidence but also enhanced yield. However, biocontrol additions resulted in only marginal additive effects. AlHadidi et al. [145], in a two-year experiment on potato cultivar "Desiree", investigated the possible role of seven different treatments of microbial inoculates including AMF, PGPR, and *Trichoderma* spp. to improve the potato tubers production and quality under irrigated and non-irrigated conditions. The results indicated that the mycorrhizal colonization intensity and mycorrhizal frequency were increased in non-irrigated treated and the control plants over the two years. Most of the treatments did not show arbuscular abundance in both years, except the mixture of *Rhizophagus irregularis* MucL41833 + *Pseudomonas brassicacearum* (41%), *R. irregularis* MucL41833 + *Paraburkholderia phytofirmans* (24%), and *Trichoderma asperelloides* A (33%). For the tuber yield, the treated plants under irrigated conditions had the highest mean. In the first year, *P. phytofirmans* PSJN had the highest yield under irrigation conditions, while in the second year, *R. irregularis* MucL41833 showed the highest yield in the irrigated conditions as well, followed by the control treatment in both years. Starch content was similar between different treatments in both years, without differences with control plants. In terms of total phosphorus content, the control treatment in the second year had the highest total phosphorus content in the irrigated conditions followed by the mixture treatment of the microbial inoculates treated plants *R. irregularis* MucL41833 and *P. phytofirmans*. In a comprehensive study, various biological amendments were assessed for their impact on potato cultivation in greenhouse and field trials conducted in Maine [146]. The evaluated amendments included commercial biocontrol agents, microbial inoculants, mycorrhizae, and an aerobic compost tea. These amendments were tested both individually and in conjunction with different crop rotations. Notably, the amendments effectively introduced microorganisms into the soil, influencing microbial populations and activity based on the specific organisms added. Over a period of 2 to 24 weeks, significant changes in soil microbial community characteristics were observed. While the effects of the amendments were most pronounced early on (2 weeks post-amendment), the impact of crop treatment became more significant in later assessments (10- and 24-weeks post-amendment). In field trials, the outcomes varied. Some microbial inoculants and biostimulants had no significant impact, whereas the application of arbuscular mycorrhizae led to a reduction in stem canker and black scurf by 17–28%. Additionally, in three different two-year crop rotations (barley/ryegrass, barley/clover, and potato, all followed by potato), biological amendments demonstrated a reduction in soilborne diseases and an improvement in yield in certain rotations, while other rotations showed no significant effects. Soil-applied aerobic compost tea, and its combination with a mixture of beneficial microorganisms, reduced stem canker, black scurf, and common scab on tubers by 18–33% and increased yield by 20–23% in the barley/ryegrass rotation but had no significant impact in the other rotations. The mixture of beneficial microorganisms also reduced disease by 20–32% in the barley/clover rotation only. None of the amendments significantly reduced disease in continuous potato plots. Both crop rotation and amendment treatments significantly affected soil microbial community characteristics, with rotation effects being more dominant. Castillo et al. [147] carried out a field trial to assess the effect of the biofertilizer Twin N (containing bacteria of *Azospirillum* genera in a concentration of 1×10^{11} UFC g^{-1}) on mycorrhizal fungal parameters. Utilizing the potato variety Desireé, three treatments were evaluated: T0 (control with chemical fertilization at planting, CFS), T1 (CFS + 50 kg N ha^{-1} at ridging), and T2 (CFS + 1 L Twin-N ha^{-1} at flowering). Notably, the Twin-N treatment exhibited the lowest root colonization and a reduced number of spores in the rhizosphere, yet it resulted in a significant increase in tuber yield. Twin-N appeared to enhance the effectiveness of native mycorrhizal fungi. Although differences were observed in disease incidence (specifically dry rot and craters), incorporating the biofertilizer Twin-N into potato cultivation can be a partial substitute for chemical nitrogen fertilization.

7. Conclusions

Nutrition and fertility management are challenging for organic crops with high nutritional demands such as potatoes. In particular, the poor availability of nitrogen for the plants in the products used in organic farming is the main cause of the yield gap between organic and conventional. In organic potato production, maintaining proper soil fertility is paramount, given the efficacy of organic fertilization in organic systems depends on various factors, including the type of fertilizer used (quantity, timing), soil conditions, and weather patterns during the plant's growth stages. Current scientific studies on the use of practices and techniques in organic potato crop production are presented in this review, highlighting the high level of research performed on crop nutrition and soil fertility management. Among organic products, farmyard manure is what has been traditionally used in potato organic systems, but compared to the conventional one, in general, it leads to a marked reduction in yield. Nonetheless, the use of farmyard manure in organic potato cultivation strongly improves soil fertility, making its application a means of supporting long-term environmental preservation. From the numerous studies conducted on potatoes, it has emerged that cover cropping or green manure application can have various beneficial effects on the cropping system such as the increase of soil nutrient content and weed suppression, even if the species used for covering is of great importance; in fact, the best results were obtained with leguminous cover crops, which thanks to biological N fixation ensure that potato yields are not dissimilar to those obtained using mineral fertilizers. Substantial organic potato yields can also be achieved by combining green manure with farmyard manure. Organic amendments such as compost, provide crops with a limited amount of nutrients, primarily nitrogen, necessitating supplementation with organic fertilizers such as leguminous green manure or farmyard manure; in any case, compost does enhance soil fertility. The potential of using biochar to improve macro and micronutrients available for the plant, soil structure, water retention, root growth, and microbial activity also emerged; however, achieving acceptable potato yields is only possible if biochar is combined with other organic products such as bone meal or poultry manure. More recently, biostimulants and biofertilizers (especially arbuscular mycorrhizal fungi) have received increasing attention from the scientific community for their complementary role in potato organic farming. They are useful to enhance soil fertility and, by improving the absorption of water and essential minerals from the soil (above all P), also promote plant growth; in addition, they play a role in the suppression of crop pests and diseases, particularly soil-borne fungal diseases, all of which contribute to increasing organic potato yield. In general, from the literature studied in the review, it also emerged that potato tubers coming from plants fertilized with organic products contain higher sugar and polyphenols content compared to tubers from plants fertilized with mineral fertilizers; some contradictory results are also reported on mineral tuber content, vitamin C, sensory properties.

8. Future Research Directions

The application of farmyard manure in organic potato farming greatly improves soil fertility, making it a key practice for long-term environmental sustainability. However, the main challenge in maintaining this practice is the substantial amount of organic manure needed per hectare, which potato growers find difficult to produce. Therefore, emphasizing the development of efficient manure production technologies is advised.

Numerous studies on potatoes have revealed that cover cropping or the use of green manure can significantly benefit the cropping system. Nonetheless, the success of cover cropping largely depends on context-specific management practices, such as the selection of species, timing of seeding, and termination methods. Unfortunately, many of these factors are not well understood, along with the ecophysiological impacts of cover crops on soil and plant nutrition. It is therefore essential for the scientific community to intensify efforts to address these knowledge gaps. In particular, large-scale, long-term studies employing a multidisciplinary approach are crucial for advancing future research.

A crucial factor in successful organic potato production is the selection of the right genotype. Currently, organic potato farming predominantly relies on varieties developed for conventional agriculture, which are often ill-suited for organic systems, leading to various negative outcomes. It is hoped then that useful studies and research will be carried out to reach the breeding or selection of new varieties characterized by high nutrient uptake efficiency, particularly of nitrogen, while also meeting consumer preferences for quality characteristics.

Author Contributions: Conceptualization, A.I.; resources, data curation, writing—original draft preparation, illustrations and tables preparation, A.I. and M.D.; review and editing, A.I.; supervision, A.I. All authors have read and agreed to the published version of the manuscript.

Funding: This research received no external funding.

Data Availability Statement: No new data were created or analyzed in this study. Data sharing is not applicable to this article.

Conflicts of Interest: The authors declare no conflicts of interest.

References

1. Willer, H.; Meier, C.; Schlatter, B.; Dietemann, L.; Kemper, L.; Trávníček, J. (Eds.) *The World of Organic Agriculture. Statistics and Emerging Trends 2022*; Research Institute of Organic Agriculture FiBL, Frick, and IFOAM–Organics International: Bonn, Germany, 2022; pp. 20–31.
2. FAO 2021. FAOSTAT Online Database. Available online: https://www.fao.org/faostat/#home (accessed on 1 July 2024).
3. Tamm, L.; Smit, A.B.; Hospers, M.; Janssens, S.R.M.; Buurma, J.S.; Molgaard, J.-P.; Lærke, P.E.; Hansen, H.H.; Hermans, A.; Bødker, L.; et al. *Assessment of the Socio-Economic Impact of Late Blight and State-of-the-Art Management in European Organic Potato Production Systems*; FiBL: Frick, Switzerland, 2004; Available online: http://orgprints.org/2936 (accessed on 15 May 2024).
4. Eurostat. *Agriculture, Forestry and Fishery Statistics*, 2020 ed.; Eurostat: Luxembourg, 2020; Available online: https://ec.europa.eu/eurostat/documents/3217494/12069644/KS-FK-20-001-EN-N.pdf (accessed on 1 June 2024).
5. Djaman, K.; Sanogo, S.; Koudahe, K.; Allen, S.; Saibou, A.; Essah, S. Characteristics of organically grown compared to conventionally grown potato and the processed products: A review. *Sustainability* **2021**, *13*, 6289. [CrossRef]
6. Zarzyńska, K.; Trawczyński, C.; Pietraszko, M. Environmental and agronomical factors limiting differences in potato yielding between organic and conventional production system. *Agriculture* **2023**, *13*, 901. [CrossRef]
7. Ponisio, L.C.; M'Gonigle, L.K.; Mace, K.C.; Palomino, J.; De Valpine, P.; Kremen, C. Diversification practices reduce organic to conventional yield gap. *Proc. R. Soc. B Biol. Sci.* **2015**, *282*, 20141396. [CrossRef]
8. Finckh, M.R.; Schulte-Geldermann, E.; Bruns, C. Challenges to organic potato farming: Disease and nutrient management. *Potato Res.* **2006**, *49*, 27–42. [CrossRef]
9. Rempelos, L.; Barański, M.; Sufar, E.K.; Gilroy, J.; Shotton, P.; Leifert, H.; Średnicka-Tober, D.; Hasanaliyeva, G.; Rosa, E.A.S.; Hajslova, J.; et al. Effect of climatic conditions, and agronomic practices used in organic and conventional crop production on yield and nutritional composition parameters in potato, cabbage, lettuce and onion; results from the long-term NFSC-trials. *Agronomy* **2023**, *13*, 1225. [CrossRef]
10. Lehesranta, S.J.; Koistinen, K.M.; Massat, N.; Davies, H.V.; Shepherd, L.V.; McNicol, J.W.; Cakmak, I.; Cooper, J.; Lück, L.; Kärenlampi, S.O.; et al. Effects of agricultural production systems and their components on protein profiles of potato tubers. *Proteomics* **2007**, *7*, 597–604. [CrossRef]
11. Möller, K.; Habermeyer, J.; Zinkernagel, V.; Reents, H.J. Impact and interaction of nitrogen and Phytophthora infestans as yield-limiting and yield-reducing factors in organic potato (*Solanum tuberosum* L.) crops. *Potato Res.* **2006**, *49*, 281–301. [CrossRef]
12. Palmer, M.W.; Cooper, J.; Tétard-Jones, C.; Średnicka-Tober, D.; Barański, M.; Eyre, M.; Shotton, P.N.; Volakakis, N.; Cakmak, I.; Ozturk, L.; et al. The influence of organic and conventional fertilisation and crop protection practices, preceding crop, harvest year and weather conditions on yield and quality of potato (*Solanum tuberosum*) in a long-term management trial. *Eur. J. Agron.* **2013**, *49*, 83–92. [CrossRef]
13. Ierna, A.; Mauromicale, G. Potato growth, yield and water productivity response to different irrigation and fertilization regimes. *Agric. Water Manag.* **2018**, *201*, 21–26. [CrossRef]

14. Ierna, A.; Mauromicale, G. Sustainable and profitable nitrogen fertilization management of potato. *Agronomy* **2019**, *9*, 582. [CrossRef]
15. Foti, S.; Mauromicale, G.; Ierna, A. Response of seed-grown globe artichoke to different levels of nitrogen fertilization and water supplies. *Acta Hortic.* **2005**, *681*, 237–242. [CrossRef]
16. Ierna, A.; Lombardo, G.M.; Mauromicale, G. Yield, nitrogen use efficiency and grain quality in durum wheat as affected by nitrogen fertilization under a Mediterranean environment. *Exp. Agric.* **2016**, *52*, 314–329. [CrossRef]
17. Rembiałkowska, E. Quality of plant products from organic agriculture. *J. Sci. Food Agric.* **2007**, *87*, 2757–2762. [CrossRef]
18. Bártová, V.; Bárta, J.; Švajner, J.; Diviš, J. Soil nitrogen variability in relation to seasonal nitrogen cycling and accumulation of nitrogenous components in starch processing potatoes. *Acta Agric. Scand. Sect. B-Soil Plant Sci.* **2012**, *62*, 70–78. [CrossRef]
19. Regulation (EU) 2018/848 of the European Parliament and of the Council of 30 May 2018 on organic production and labelling of organic products and repealing Council Regulation (EC) No 834/2007. Available online: https://eur-lex.europa.eu/legal-content/EN/TXT/?uri=OJ:L:2018:150:TOC (accessed on 30 June 2024).
20. Lynch, D.H.; Zheng, Z.; Zebarth, B.J.; Martin, R.C. Organic amendment effects on tuber yield, plant N uptake and soil mineral N under organic potato production. *Renew. Agric. Food Syst.* **2008**, *23*, 250–259. [CrossRef]
21. Gelaye, Y. Effect of combined application of organic manure and nitrogen fertilizer rates on yield and yield components of potato: A review. *Cogent Food Agric.* **2023**, *9*, 2217603. [CrossRef]
22. Cerrato, M.E.; Leblanc, H.A.; Kameko, C. Potencial de mineralización de nitrógeno de Bokashi, compost y lombricompost producidos en la Universidad Earth. *Tierra Trop.* **2007**, *3*, 183–197.
23. Harraq, A.; Sadiki, K.; Bourioug, M.; Bouabid, R. Organic fertilizers mineralization and their effect on the potato "*Solanum tuberosum*" performance in organic farming. *J. Saudi Soc. Agric. Sci.* **2022**, *21*, 255–266. [CrossRef]
24. Sofo, A.; Zanella, A.; Ponge, J.F. Soil quality and fertility in sustainable agriculture, with a contribution to the biological classification of agricultural soils. *Soil Use Manag.* **2022**, *38*, 1085–1112. [CrossRef]
25. Suryawanshi, P.; Pagar, V. Organic Nutrient, Resources and Its Fortification. In *Advances in Agricultural and Horticultural Sciences*; Department of Agricultural Entomology, Shri Vaishnav Institute of Agriculture: Indore, India, 2022; pp. 171–183.
26. Sarker, T.C.; Zotti, M.; Fang, Y.; Giannino, F.; Mazzoleni, S.; Bonanomi, G.; Cai, Y.; Chang, S.X. Soil aggregation in relation to organic amendment: A synthesis. *J. Soil Sci. Plant Nutr.* **2022**, *22*, 2481–2502. [CrossRef]
27. Mäder, P.; Fliessbach, A.; Dubois, D.; Gunst, L.; Fried, P.; Niggli, U. Soil fertility and biodiversity in organic farming. *Science* **2002**, *296*, 1694–1697. [CrossRef]
28. Hlisnikovský, L.; Menšík, L.; Kunzová, E. The effect of soil-climate conditions, farmyard manure and mineral fertilizers on potato yield and soil chemical parameters. *Plants* **2021**, *10*, 2473. [CrossRef] [PubMed]
29. Tein, B.; Kauer, K.; Eremeev, V.; Luik, A.; Selge, A.; Loit, E. Farming systems affect potato (*Solanum tuberosum* L.) tuber and soil quality. *Field Crops Res.* **2014**, *156*, 1–11. [CrossRef]
30. Rempelos, L.; Cooper, J.; Wilcockson, S.; Eyre, M.; Shotton, P.; Volakakis, N.; Orr, C.H.; Leifert, C.; Gatehouse, A.M.R.; Tétard-Jones, C. Quantitative proteomics to study the response of potato to contrasting fertilisation regimes. *Mol. Breed.* **2013**, *31*, 363–378. [CrossRef]
31. Hajšlová, J.; Schulzova, V.; Slanina, P.; Janne, K.; Hellenäs, K.E.; Andersson, C.H. Quality of organically and conventionally grown potatoes: Four-year study of micronutrients, metals, secondary metabolites, enzymic browning and organoleptic properties. *Food Addit. Contam.* **2005**, *22*, 514–534. [CrossRef] [PubMed]
32. Herencia, J.F.; García-Galavís, P.A.; Dorado, J.A.R.; Maqueda, C. Comparison of nutritional quality of the crops grown in an organic and conventional fertilized soil. *Sci. Hortic.* **2011**, *129*, 882–888. [CrossRef]
33. Warman, P.R.; Havard, K.A. Yield, vitamin and mineral contents of organically and conventionally grown potatoes and sweet corn. *Agric. Ecosyst. Environ.* **1998**, *68*, 207–216. [CrossRef]
34. Makaraviciute, A. Effect of organic and mineral fertilizers on the yield and quality of different potato varieties. *Agric. Food Sci.* **2003**, *1*, 197–209.
35. Wszelaki, A.L.; Delwiche, J.F.; Walker, S.D.; Liggett, R.E.; Scheerens, J.C.; Kleinhenz, M.D. Sensory quality and mineral and glycoalkaloid concentrations in organically and conventionally grown redskin potatoes (*Solanum tuberosum*). *J. Sci. Food Agric.* **2005**, *85*, 720–726. [CrossRef]
36. Bártová, V.; Diviš, J.; Bárta, J.; Brabcová, A.; Švajnerová, M. Variation of nitrogenous components in potato (*Solanum tuberosum* L.) tubers produced under organic and conventional crop management. *Eur. J. Agron.* **2013**, *49*, 20–31. [CrossRef]
37. Paffrath, A.; Milz, E. Effect of presprouting, organic nitrogen-fertilization and copper-treatment on yield and quality of potato in organic farming. *Gesunde Pflanz.* **2008**, *60*, 1–4. [CrossRef]
38. Maggio, A.; Carillo, P.; Bulmetti, G.S.; Fuggi, A.; Barbieri, G.; De Pascale, S. Potato yield and metabolic profiling under conventional and organic farming. *Eur. J. Agron.* **2008**, *28*, 343–350. [CrossRef]
39. Lombardo, S.; Monaco, A.L.; Pandino, G.; Parisi, B.; Mauromicale, G. The phenology, yield and tuber composition of 'early' crop potatoes: A comparison between organic and conventional cultivation systems. *Renew. Agric. Food Syst.* **2013**, *28*, 50–58. [CrossRef]
40. Lombardo, S.; Pandino, G.; Mauromicale, G. Nutritional and sensory characteristics of "early" potato cultivars under organic and conventional cultivation systems. *Food Chem.* **2012**, *133*, 1249–1254. [CrossRef]

41. Lombardo, S.; Pandino, G.; Mauromicale, G. The mineral profile in organically and conventionally grown "early" crop potato tubers. *Sci. Hortic.* **2014**, *167*, 169–173. [CrossRef]
42. Ierna, A.; Parisi, B. Crop growth and tuber yield of "early" potato crop under organic and conventional farming. *Sci. Hortic.* **2014**, *165*, 260–265. [CrossRef]
43. Grudzińska, M.; Czerko, Z.; Zarzyńska, K.; Borowska-Komenda, M. Bioactive compounds in potato tubers: Effects of farming system, cooking method, and flesh color. *PLoS ONE* **2016**, *11*, e0153980. [CrossRef] [PubMed]
44. Kazimierczak, R.; Średnicka-Tober, D.; Hallmann, E.; Kopczyńska, K.; Zarzyńska, K. The impact of organic vs. conventional agricultural practices on selected quality features of eight potato cultivars. *Agronomy* **2019**, *9*, 799. [CrossRef]
45. Keutgen, A.J.; Wszelaczyńska, E.; Pobereżny, J.; Przewodowska, A.; Przewodowski, W.; Milczarek, D.; Tatarowska, B.; Flis, B.; Keutgen, N. Antioxidant properties of potato tubers (*Solanum tuberosum* L.) as a consequence of genetic potential and growing conditions. *PLoS ONE* **2019**, *14*, e0222976. [CrossRef]
46. Shafeeva, E.; Komissarov, A.; Ishbulatov, M.; Safin, K.; Davletshin, F. Cattle Manure Disposal on Potato Growing Fields in Conditions of the Southern Forest-Steppe of the Bashkortostan Republic. *Pol. J. Environ. Stud.* **2021**, *30*, 4195–4203. [CrossRef]
47. Haase, T.; Schüler, C.; Haase, N.U.; Heß, J. Suitability of organic potatoes for industrial processing: Effect of agronomical measures on selected quality parameters at harvest and after storage. *Potato Res.* **2007**, *50*, 115–141. [CrossRef]
48. Ierna, A.; Parisi, B.; Melilli, M.G. Overall quality of "early" potato tubers as affected by organic cultivation. *Agronomy* **2022**, *12*, 296. [CrossRef]
49. Ierna, A.; Melilli, M.G.; Parisi, B. Effects of organic versus conventional cultivation system on nutritional quality of early potato tubers. In Proceedings of the III International Organic Fruit Symposium and I International Organic Vegetable Symposium, Catania, Italy, 14–17 December 2021; Volume 1354, pp. 63–70.
50. Ierna, A.; Melilli, M.G.; Parisi, B. Sensory characteristics of raw and cooked early potato tubers as affected by organic cultivation. In Proceedings of the III International Organic Fruit Symposium and I International Organic Vegetable Symposium, Catania, Italy, 14–17 December 2021; Volume 1354, pp. 375–382.
51. Alvarez, C.E.; Amin, M.; Hernández, E.; González, C.J. Effect of compost, farmyard manure and/or chemical fertilizers on potato yield and tuber nutrient content. *Biol. Agric. Hortic.* **2006**, *23*, 273–286. [CrossRef]
52. Peñaloza Monroy, J.; Reyes Ramírez, A.K.; González Huerta, A.; Pérez López, D.D.J.; Sangerman-Jarquín, D.M. Fertilización orgánica con tres niveles de gallinaza en cuatro cultivares de papa. *Rev. Mex. Cienc. Agrícolas* **2019**, *10*, 1139–1149.
53. Baniuniene, A.; Zekaite, V. The effect of mineral and organic fertilizers on potato tuber yield and quality. *Agron. Vēstis Latv. J. Agron.* **2008**, *11*, 202–206.
54. Caliskan, M.E.; Kilic, S.; Gunel, E.; Mert, M. Effect of farmyard manure and mineral fertilization on growth and yield of early potato (*Solanum tuberosum*) under the Mediterranean conditions in Turkey. *Indian J. Agron.* **2004**, *49*, 198–200. [CrossRef]
55. Islam, M.M.; Akhter, S.; Majid, N.M.; Ferdous, J.; Alam, M.S. Integrated nutrient management for potato (*'Solanum tuberosum'*) in grey terrace soil (Aric Albaquipt). *Aust. J. Crop Sci.* **2013**, *7*, 1235–1241.
56. Asghari, M.T.; Mir, R.; Fard, A. The effect of farmyard manure and nitrogen fertilizer on some characteristics of potato (*Solanum tuberosum* var. Agria). *Biharean Biol.* **2015**, *9*, 81–84.
57. Bashir, U.; Qureshi, F. Effect of nitrogen and farmyard manure on yield, nutrient content and quality of potato (*Solanum tuberosum* L.). *Biolife* **2014**, *2*, 786–791.
58. Asfaw, F. Effect of integrated soil amendment practices on growth and seed tuber yield of potato (*Solanum tuberosum* L.) at Jimma Arjo, Western Ethiopia. *J. Nat. Sci. Res.* **2016**, *6*, 38–63.
59. Mama, A.; Jeylan, J.; Aseffa, A.W. Effects of different rates of organic and inorganic fertilizer on growth and yield components of potato (*Solanum tuberosum* L.) in Jimma Are, southwest Ethiopia. *Int. J. Res.* **2016**, *4*, 115–121.
60. Alemayehu, M.; Jemberie, M.; Yeshiwas, T.; Aklile, M. Integrated application of compound NPS fertilizer and farmyard manure for economical production of irrigated potato (*Solanum tuberosum* L.) in highlands of Ethiopia. *Cogent Food Agric.* **2020**, *6*, 1724385. [CrossRef]
61. Atanaw, T. Israel Zewide. Fertility Management on Potato (*Solanum tuberosum* L.). *Crop. Res. Rev. J. Crop Sci. Technol.* **2021**, *10*, 33–46.
62. Gitari, H.I.; Gachene, C.K.; Karanja, N.N.; Kamau, S.; Nyawade, S.; Schulte-Geldermann, E. Potato-legume intercropping on a sloping terrain and its effects on soil physico-chemical properties. *Plant Soil* **2019**, *438*, 447–460. [CrossRef]
63. Nedunchezhiyan, M.; Suja, G.; Ravi, V. tropical root-and tuber crops-based cropping systems—A review. *Curr. Hortic.* **2022**, *10*, 14–22. [CrossRef]
64. Lu, Y.C.; Watkins, K.B.; Teasdale, J.R.; Abdul-Baki, A.A. Cover crops in sustainable food production. *Food Rev. Int.* **2000**, *16*, 121–157. [CrossRef]
65. Reents, H.J.; Möller, K. Effect of different green manure catch crops grown after peas on nitrate dynamics in soils and on yield and quality of subsequent potatoes and wheat. In Proceedings of the 13th International IFOAM Scientific Conference, Basel, Switzerland, 28–31 August 2000; Alföldi, T., Lockeretz, W., Niggli, U., Eds.; IFOAM: Zürich, Switzerland, 2000; pp. 73–75.
66. Larkin, R.P.; Griffin, T.S. Control of soilborne potato diseases using Brassica green manures. *Crop Prot.* **2007**, *26*, 1067–1077. [CrossRef]

67. Jégo, G.; Martinez, M.; Antigüedad, I.; Launay, M.; Sanchez-Pérez, J.M.; Justes, E. Evaluation of the impact of various agricultural practices on nitrate leaching under the root zone of potato and sugar beet using the STICS soil–crop model. *Sci. Total Environ.* **2008**, *394*, 207–221. [CrossRef]
68. Canali, S.; Ciaccia, C.; Tittarelli, F. Soil fertility management in organic potato: The role of green manure and amendment applications. In *Sustainable Potato Production: Global Case Studies*; Springer: Dordrecht, The Netherlands, 2012; pp. 453–469.
69. Ledgard, S.F. Nitrogen cycling in low input legume-based agriculture, with emphasis on legume/grass pastures. *Plant Soil* **2001**, *228*, 43–59. [CrossRef]
70. Båth, B.; Ekbladh, G.; Ascard, J.; Olsson, K.; Andersson, B. Yield and nitrogen uptake in organic potato production with green manures as pre-crop and the effect of supplementary fertilization with fermented slurry. *Biol. Agric. Hortic.* **2006**, *24*, 135–148. [CrossRef]
71. Sincik, M.; Turan, Z.M.; Göksoy, A.T. Responses of potato (*Solanum tuberosum* L.) to green manure cover crops and nitrogen fertilization rates. *Am. J. Potato Res.* **2008**, *85*, 150–158. [CrossRef]
72. Uchino, H.; Iwama, K.; Jitsuyama, Y.; Ichiyama, K.; Sugiura, E.R.I.; Yudate, T.; Nakamura, S.; Gopal, J. Effect of interseeding cover crops and fertilization on weed suppression under an organic and rotational cropping system: 1. Stability of weed suppression over years and main crops of potato, maize and soybean. *Field Crops Res.* **2012**, *127*, 9–16. [CrossRef]
73. Campiglia, E.; Paolini, R.; Colla, G.; Mancinelli, R. The effects of cover cropping on yield and weed control of potato in a transitional system. *Field Crops Res.* **2009**, *112*, 16–23. [CrossRef]
74. Larkin, R.P.; Halloran, J.M. Management effects of disease-suppressive rotation crops on potato yield and soilborne disease and their economic implications in potato production. *Am. J. Potato Res.* **2014**, *91*, 429–439. [CrossRef]
75. Nyamdavaa, T.; Friedel, J.K. Effect of pre-crops, cover crops and manure application on organic potato production in a field experiment in eastern Austria. *Acta Fytotech. Zootech.* **2016**, *18*, 16–18. [CrossRef]
76. Shibabaw, A.; Alemayehu, G.; Adgo, E.; Asch, F.; Freyer, B. Effects of organic manure and crop rotation system on potato (*Solanum tuberosum* L.) tuber yield in the highlands of Awi Zone. *Ethiop. J. Sci. Technol.* **2018**, *11*, 1–18. [CrossRef]
77. Eremeev, V.; Talgre, L.; Kuht, J.; Mäeorg, E.; Esmaeilzadeh-Salestani, K.; Alaru, M.; Loit, E.; Runno-Paurson, E.; Luik, A. The soil microbial hydrolytic activity, content of nitrogen and organic carbon were enhanced by organic farming management using cover crops and composts in potato cultivation. *Acta Agric. Scand. Sect. B—Soil Plant Sci.* **2020**, *70*, 87–94. [CrossRef]
78. Nyiraneza, J.; Chen, D.; Fraser, T.; Comeau, L.P. Improving soil quality and potato productivity with manure and high-residue cover crops in Eastern Canada. *Plants* **2021**, *10*, 1436. [CrossRef] [PubMed]
79. Rittl, T.F.; Grønmyr, F.; Bakken, I.; Løes, A.K. Effects of organic amendments and cover crops on soil characteristics and potato yields. *Acta Agric. Scand. Sect. B—Soil Plant Sci.* **2023**, *73*, 13–26. [CrossRef]
80. Margus, K.; Eremeev, V.; Loit, E.; Runno-Paurson, E.; Mäeorg, E.; Luik, A.; Talgre, L. Impact of farming system on potato yield and tuber quality in Northern Baltic Sea climate conditions. *Agriculture* **2022**, *12*, 568. [CrossRef]
81. Mamiev, D.; Abaev, A.; Tedeeva, A.; Khokhoeva, N.; Tedeeva, V. Use of green manure in organic farming. In *IOP Conference Series: Earth and Environmental Science*; IOP Publishing: Bristol, UK, 2019; Volume 403, p. 012137.
82. Pinto, R.; Brito, L.M.; Coutinho, J. Organic production of horticultural crops with green manure, composted farmyard manure and organic fertiliser. *Biol. Agric. Hortic.* **2017**, *33*, 269–284. [CrossRef]
83. Drakopoulos, D.; Scholberg, J.M.; Lantinga, E.A.; Tittonell, P.A. Influence of reduced tillage and fertilization regime on crop performance and nitrogen utilization of organic potato. *Org. Agric.* **2016**, *6*, 75–87.
84. Drakopoulos, D.; Scholberg, J.M.; Lantinga, E.A. Influence of reduced tillage and fertilisation regime on soil quality indicators in an organic potato production system. *Biol. Agric. Hortic.* **2018**, *34*, 132–140. [CrossRef]
85. Wilson, R.; Culp, D.A.; Peterson, S.; Nicholson, K.; Geisseler, D. Cover crops prove effective at increasing soil nitrogen for organic potato production. *Calif. Agric.* **2019**, *73*, 79–89. [CrossRef]
86. Grandy, A.S.; Porter, G.A.; Erich, M.S. Organic amendment and rotation crop effects on the recovery of soil organic matter and aggregation in potato cropping systems. *Soil Sci. Soc. Am. J.* **2002**, *66*, 1311–1319. [CrossRef]
87. Passoni, M.; Borin, M. Effects of different composts on soil nitrogen balance and dynamics in a biennial crop succession. *Compos. Sci. Util.* **2009**, *17*, 108–116. [CrossRef]
88. Carter, M.R.; Sanderson, J.B.; MacLeod, J.A. Influence of compost on the physical properties and organic matter fractions of a fine sandy loam throughout the cycle of a potato rotation. *Can. J. Soil Sci.* **2004**, *84*, 211–218.
89. El-Sayed, S.F.; Hassan, H.A.; El-Mogy, M.M. Impact of bio-and organic fertilizers on potato yield, quality and tuber weight loss after harvest. *Potato Res.* **2015**, *58*, 67–81.
90. Fahmy, S.H.; Sharifi, M.; Hann, S.W.; Chow, T.L. Crop productivity and nutrient bioavailability in a potato-based three-year rotation as affected by composted pulp fiber residue application and supplemental irrigation. *Commun. Soil Sci. Plant Anal.* **2010**, *41*, 744–756.
91. Wilson, C.; Zebarth, B.J.; Burton, D.L.; Goyer, C.; Moreau, G.; Dixon, T. Effect of diverse compost products on potato yield and nutrient availability. *Am. J. Potato Res.* **2019**, *96*, 272–284. [CrossRef]
92. Vaitkevičienė, N.; Kulaitienė, J.; Jarienė, E.; Levickienė, D.; Danillčenko, H.; Średnicka-Tober, D.; Rembiałkowska, E.; Hallmann, E. Characterization of bioactive compounds in colored potato (*Solanum tuberosum* L.) cultivars grown with conventional, organic, and biodynamic methods. *Sustainability* **2020**, *12*, 2701. [CrossRef]

93. Illera-Vives, M.; Seoane Labandeira, S.; Iglesias Loureiro, L.; López-Mosquera, M.E. Agronomic assessment of a compost consisting of seaweed and fish waste as an organic fertilizer for organic potato crops. *J. Appl. Phycol.* **2017**, *29*, 1663–1671. [CrossRef]
94. Alam, M.Z.; Lynch, D.H.; Sharifi, M.; Burton, D.L.; Hammermeister, A.M. The effect of green manure and organic amendments on potato yield, nitrogen uptake and soil mineral nitrogen. *Biol. Agric. Hortic.* **2016**, *32*, 221–236. [CrossRef]
95. Escobedo-Monge, M.A.; Aparicio, S.; Escobedo-Monge, M.F.; Marugán-Miguelsanz, J.M. Long-term effects of the application of urban waste compost and other organic amendments on *Solanum tuberosum* L. *Agronomy* **2020**, *10*, 1575. [CrossRef]
96. Willekens, K.; De Vliegher, A.; Vandecasteele, B.; Carlier, L. Effect of compost versus animal manure fertilization on crop development, yield and nitrogen residue in the organic cultivation of potatoes. In Proceedings of the Second Scientific Conference of the International Society of Organic Agriculture Research (ISOFAR), Modena, Italy, 18–20 June 2008; pp. 576–579.
97. El-Tantawy, I.M.; El-Ghamry, A.M.; Habib, A.H. Impact of farmyard manure and manure compost tea on potato yield and soil fertility. *J. Soil Sci. Agric. Eng.* **2009**, *34*, 669–678. [CrossRef]
98. Bernard, E.; Larkin, R.P.; Tavantzis, S.; Erich, M.S.; Alyokhin, A.; Gross, S.D. Rapeseed rotation, compost and biocontrol amendments reduce soilborne diseases and increase tuber yield in organic and conventional potato production systems. *Plant Soil* **2014**, *374*, 611–627. [CrossRef]
99. Youssef, M.E.S.; Al-Easily, I.A.S.; AS Nawar, D. Impact of biochar addition on productivity and tubers quality of some potato cultivars under sandy soil conditions. *Egypt. J. Hortic.* **2017**, *44*, 199–217. [CrossRef]
100. Upadhyay, K.P.; Dhami, N.B.; Sharma, P.N.; Neupane, J.D.; Shrestha, J. Growth and yield responses of potato (*Solanum tuberosum* L.) to biochar. *J. Agric. Sci.* **2020**, *2*, 244–253.
101. Mawof, A.; Prasher, S.; Bayen, S.; Nzediegwu, C. Effects of biochar and biochar-compost mix as soil amendments on soil quality and yield of potatoes irrigated with wastewater. *J. Soil Sci. Plant Nutr.* **2021**, *21*, 2600–2612. [CrossRef]
102. Farooque, A.A.; Zaman, Q.; Abbas, F.; Hammad, H.M.; Acharya, B.; Easu, T. How can potatoes be smartly cultivated with biochar as a soil nutrient amendment technique in Atlantic Canada? *Arab. J. Geosci.* **2020**, *13*, 336. [CrossRef]
103. Hou, J.; Xing, C.; Zhang, J.; Wang, Z.; Liu, M.; Duan, Y.; Zhao, H. Increase in potato yield by the combined application of biochar and organic fertilizer: Key role of rhizosphere microbial diversity. *Front. Plant Sci.* **2024**, *15*, 1389864. [CrossRef]
104. Mollick, M.O.A.; Paul, A.K.; Alam, I.; Sumon, M.M. Effect of biochar on yield and quality of potato (*Solanum tuberosum*) tuber. *Int. J. Bio-Resour. Stress Manag.* **2020**, *11*, 445–450. [CrossRef]
105. Singh, J.; Siddique, A. Impact of biochar base organic amendment on morphological changes and yield of potato (*Solanum tuberosum* L.). *Spec. Ugdym.* **2022**, *2*, 865–871.
106. Bian, B.; Hu, X.; Zhang, S.; Lv, C.; Yang, Z.; Yang, W.; Zhang, L. Pilot-scale composting of typical multiple agricultural wastes: Parameter optimization and mechanisms. *Bioresour. Technol.* **2019**, *287*, 121482. [CrossRef]
107. Sharifi, M.; Lynch, D.H.; Zebarth, B.J.; Zheng, Z.; Martin, R.C. Evaluation of nitrogen supply rate measured by in situ placement of Plant Root Simulator™ probes as a predictor of nitrogen supply from soil and organic amendments in potato crop. *Am. J. Potato Res.* **2009**, *86*, 356–366. [CrossRef]
108. Amlinger, F.; Götz, B.; Dreher, P.; Geszti, J.; Weissteiner, C. Nitrogen in biowaste and yard waste compost: Dynamics of mobilisation and availability—A review. *Eur. J. Soil Biol.* **2003**, *39*, 107–116. [CrossRef]
109. Zhang, M.; Heaney, D.; Henriquez, B.; Solberg, E.; Bittner, E. A four-year study on influence of biosolids/MSW cocompost application in less productive soils in Alberta: Nutrient dynamics. *Compos. Sci. Util.* **2006**, *14*, 68–80. [CrossRef]
110. Whitmore, A.P. Determination of the mineralization of nitrogen from composted chicken manure as affected by temperature. *Nutr. Cycl. Agroecosystems* **2007**, *77*, 225–232. [CrossRef]
111. Nyiraneza, J.; Snapp, S. Integrated management of inorganic and organic nitrogen and efficiency in potato systems. *Soil Sci. Soc. Am. J.* **2007**, *71*, 1508–1515. [CrossRef]
112. Cambardella, C.A.; Moorman, T.B.; Singer, J.W. Soil nitrogen response to coupling cover crops with manure injection. *Nutr. Cycl. Agroecosystems* **2010**, *87*, 383–393. [CrossRef]
113. Lim, S.L.; Wu, T.Y.; Lim, P.N.; Shak, K.P.Y. The use of vermicompost in organic farming: Overview, effects on soil and economics. *J. Sci. Food Agric.* **2015**, *95*, 1143–1156. [CrossRef]
114. Edussuriya, R.; Rajapaksha, A.U.; Jayasinghe, C.; Pathirana, C.; Vithanage, M. Influence of biochar on growth performances, yield of root and tuber crops and controlling plant-parasitic nematodes. *Biochar* **2023**, *5*, 68. [CrossRef]
115. Ding, Y.; Liu, Y.; Liu, S.; Li, Z.; Tan, X.; Huang, X.; Zeng, G.; Zhou, L.; Zheng, B. Biochar to improve soil fertility. A review. *Agron. Sustain. Dev.* **2016**, *36*, 36. [CrossRef]
116. Ebrahimi, N.; Viaene, N.; Vandecasteele, B.; D'Hose, T.; Debode, J.; Cremelie, P.; De Tender, C.; Moens, M. Traditional and new soil amendments reduce survival and reproduction of potato cyst nematodes, except for biochar. *Appl. Soil Ecol.* **2016**, *107*, 191–204. [CrossRef]
117. Singh, J.; Kumar, P.; Siddique, A. Biochar-based organic amendments on soil health, nutrient status and quality of potato (*Solanum tuberosum*). *Plant Sci. Today* **2024**, *11*, 102–108. [CrossRef]
118. Wibowo, C.; Wijaya, K.; Biyantara, A.L. Effect of organic fertilizer and application of charcoal on quality of potato tuber variety Atlantic. In *IOP Conference Series: Earth and Environmental Science*; IOP Publishing: Bristol, UK, 2021; Volume 653, p. 012125.
119. Regulation (EU) 2019/1009 of the European Parliament and of the Council of 5 June 2019 Laying down Rules on the Making Available on the Market of EU Fertilising Products and Amending Regulations (EC) No 1069/2009 and (EC) No 1107/2009 and

Repealing Regulation (EC) No 2003/2003. 2019. Available online: https://eur-lex.europa.eu/legal-content/EN/TXT/?uri=celex:32019R1009 (accessed on 30 May 2024).

120. Rouphael, Y.; Colla, G. Toward a sustainable agriculture through plant biostimulants: From experimental data to practical applications. *Agronomy* **2020**, *10*, 1461. [CrossRef]
121. Mijwel, A.K.; Al-redha, A.A.; Bresim, T.H. Response of some traits of potato (*Solanum tuberosum* L.) to the addition of the ground palm remnants compost and foliar with humic acid. *Plant Arch.* **2019**, *19*, 94–97.
122. Rizk, F.A.; Shaheen, A.M.; Singer, S.M.; Sawan, O.A. The productivity of potato plants affected by urea fertilizer as foliar spraying and humic acid added with irrigation water. *Middle East J. Agric. Res.* **2013**, *2*, 76–83.
123. Sarhan, T.Z. Effect of humic acid and seaweed extracts on growth and yield of potato plant (*Solanum tuberosum* L) Desiree cv. *Mesop. J. Agric.* **2011**, *39*, 19–25.
124. Man-Hong, Y.; Lei, Z.; Sheng-Tao, X.; McLaughlin, N.B.; Jing-Hui, L. Effect of water soluble humic acid applied to potato foliage on plant growth, photosynthesis characteristics and fresh tuber yield under different water deficits. *Sci. Rep.* **2020**, *10*, 7854. [CrossRef]
125. Suh, H.Y.; Yoo, K.S.; Suh, S.G. Tuber growth and quality of potato (*Solanum tuberosum* L.) as affected by foliar or soil application of fulvic and humic acids. *Hortic. Environ. Biotechnol.* **2014**, *55*, 183–189. [CrossRef]
126. Ekin, Z. Integrated use of humic acid and plant growth promoting rhizobacteria to ensure higher potato productivity in sustainable agriculture. *Sustainability* **2019**, *11*, 3417. [CrossRef]
127. Wadas, W.; Dziugieł, T. Quality of new potatoes (*Solanum tuberosum* L.) in response to plant biostimulants application. *Agriculture* **2020**, *10*, 265. [CrossRef]
128. Gosling, P.; Hodge, A.; Goodlass, G.; Bending, G.D. Arbuscular mycorrhizal fungi and organic farming. *Agric. Ecosyst. Environ.* **2006**, *113*, 17–35. [CrossRef]
129. Harrier, L.A.; Watson, C.A. The potential role of arbuscular mycorrhizal (AM) fungi in the bioprotection of plants against soil-borne pathogens in organic and/or other sustainable farming systems. *Pest Manag. Sci. Former. Pestic. Sci.* **2004**, *60*, 149–157. [CrossRef]
130. Scavo, A.; Abbate, C.; Mauromicale, G. Plant allelochemicals: Agronomic, nutritional and ecological relevance in the soil system. *Plant Soil* **2019**, *442*, 23–48. [CrossRef]
131. Mäder, P.; Edenhofer, S.; Boller, T.; Wiemken, A.; Niggli, U. Arbuscular mycorrhizae in a long-term field trial comparing low-input (organic, biological) and high-input (conventional) farming systems in a crop rotation. *Biol. Fertil. Soils* **2000**, *31*, 150–156. [CrossRef]
132. Wu, F.; Wang, W.; Ma, Y.; Liu, Y.; Ma, X.; An, L.; Feng, H. Prospect of beneficial microorganisms applied in potato cultivation for sustainable agriculture. *Afr. J. Microbiol. Res.* **2013**, *7*, 2150–2158.
133. Duffy, E.M.; Cassells, A.C. The effect of inoculation of potato (*Solanum tuberosum* L.) microplants with arbuscular mycorrhizal fungi on tuber yield and tuber distribution. *Appl. Soil Ecol.* **2000**, *15*, 137–144. [CrossRef]
134. Douds, D.D., Jr.; Nagahashi, G.; Reider, C.; Hepperly, P.R. Inoculation with arbuscular mycorrhizal fungi increases the yield of potatoes in a high P soil. *Biol. Agric. Hortic.* **2007**, *25*, 67–78. [CrossRef]
135. Hijri, M. Analysis of a large dataset of mycorrhiza inoculation field trials on potato shows highly significant increases in yield. *Mycorrhiza* **2016**, *26*, 209–214. [CrossRef] [PubMed]
136. Khosravifar, S.; Farahvash, F.; Aliasgharzad, N.; Yarnia, M.; Khoei, F.R. Effects of different irrigation regimes and two arbuscular mycorrhizal fungi on some physiological characteristics and yield of potato under field conditions. *J. Plant Nutr.* **2020**, *43*, 2067–2079. [CrossRef]
137. Chifetete, V.W.; Dames, J.F. Mycorrhizal interventions for sustainable potato production in Africa. *Front. Sustain. Food Syst.* **2020**, *4*, 593053. [CrossRef]
138. Lombardo, S.; Abbate, C.; Pandino, G.; Parisi, B.; Scavo, A.; Mauromicale, G. Productive and physiological response of organic potato grown under highly calcareous soils to fertilization and mycorrhization management. *Agronomy* **2020**, *10*, 1200. [CrossRef]
139. Lombardo, S.; Scavo, A.; Abbate, C.; Pandino, G.; Parisi, B.; Mauromicale, G. Mycorrhizal inoculation improves mineral content of organic potatoes grown under calcareous soil. *Agriculture* **2021**, *11*, 333. [CrossRef]
140. Abdulkadhum, M.H.; Manea, A.I.; Mahmoud, S.S. Effect of Mycorrhizal Addition and Organic Fertilization on the Growth and Yield of Potato in Autumn Season. In *IOP Conference Series: Earth and Environmental Science*; IOP Publishing: Bristol, UK, 2023; Volume 1262, p. 042004. [CrossRef]
141. Aloo, B.; Makumb, B.; Mbega, E. Plant growth promoting rhizobacterial biofertilizers for sustainable crop production: The past, present, and future. *Front. Plant Sci.* **2022**, *13*, 1002448. [CrossRef]
142. Abou-Hussein, S.D.; El-Oksh, I.; El-Shorbagy, T.; El-Bahiry, U.A. Effect of chicken manure, compost and biofertilizers on vegetative growth, tuber characteristics and yield of potato crop. *Egypt. J. Hortic.* **2002**, *29*, 135–149.
143. Sayed, F.; Hassien, A.; Muhamed, M.; Ahimed, A. Growth, yield and nutrient concentration of potato plants. *Am.-Eurasian J. Agric. Environ. Sci.* **2014**, *7*, 36–43.
144. Minin, V.B.; Popov, V.D.; Maksimov, D.A.; Ustroev, A.A.; Melnikov, S.P.; Papushin, E. Developing of modern cultivation technology of organic potatoes. *Agron. Res.* **2020**, *18*, 1359–1367.
145. Al Hadidi, N.; Hasan, H.; Pap, Z.; Tóth, F.; Papp, O.; Drexler, D.; Ganszky, D.; Kappel, N. Beneficial Effects of Microbial Inoculation to Improve Organic Potato Production under Irrigated and Non-Irrigated Conditions. *Int. J. Agric. Biol.* **2024**, *31*, 57–64.

146. Larkin, R.P. Relative effects of biological amendments and crop rotations on soil microbial communities and soilborne diseases of potato. *Soil Biol. Biochem.* **2008**, *40*, 1341–1351. [CrossRef]
147. Castillo, C.; Huenchuleo, M.J.; Michaud, A.; Solano, J. Micorrización en un cultivo de papa adicionado del biofertilizante Twin-N establecido en un Andisol de la Región de La Araucanía. *Idesia* **2016**, *34*, 39–45. [CrossRef]

horticulturae

MDPI

Article

Effects of Ripening Phase and Cultivar under Sustainable Management on Fruit Quality and Antioxidants of Sweet Cherry

Iulia Mineață [1], Otilia Cristina Murariu [2,*], Sorina Sîrbu [1], Alessio Vincenzo Tallarita [3], Gianluca Caruso [3] and Carmen Doina Jităreanu [4,*]

[1] Research Station for Fruit Growing, 3 Ion Vodă cel Viteaz Street, Miroslava, 707305 Iasi, Romania; iulia_mineata@yahoo.com (I.M.); sorinas66@yahoo.com (S.S.)

[2] Department of Food Technology, "Ion Ionescu de la Brad" Iasi University of Life Sciences, M. Sadoveanu Alley, 700490 Iasi, Romania

[3] Department of Agricultural Sciences, University of Naples Federico II, Via Università 100, 80055 Naples, Italy; alessiovincenzo.tallarita@unina.it (A.V.T.); gcaruso@unina.it (G.C.)

[4] Department of Plant Science, "Ion Ionescu de la Brad" Iasi University of Life Sciences, M. Sadoveanu Alley, 700490 Iasi, Romania

* Correspondence: otilia.murariu@iuls.ro (O.C.M.); doina.jitareanu@iuls.ro (C.D.J.)

Abstract: Sweet cherry grown under sustainable management produces highly valuable fruits, whose quality shows important biochemical and morphological changes during ripening. Research was carried out in Iasi (Romania), with the aim to assess the quality characteristics of the sweet cherry fruits of three cultivars (Van, Andreiaș, Margonia), grown in an inner or outer position inside the tree crown, at the pre-ripening or full ripeness phase. In 2022, the colour component a* showed higher values in cv. Van and Andreias red fruits and in an inner position, whereas the components L* and b* at the full ripeness phase were highest in cv. Margonia. The dry matter and total soluble solids contents increased from the pre-ripening to the full ripeness phase and were highest in cv. Van sweet cherry fruits; the DM of fruit from the outer part of crown was higher than that of fruit from the inner part at the pre-ripening phase. The content of phenolics was the highest in cv. Margonia fruits at the pre-ripening stage and in cv. Van at the full ripeness phase and higher in the inner tree crown zones. The cultivar Margonia generally showed the highest vitamin C content in both years and development phases. The yellow fruit cv. Margonia mostly showed the highest values of chlorophyll a and b. The fruit's content of carotene, lycopene, and anthocyanins was generally the highest in the red fruits of cv. Andreias. The examined sweet cherry cultivars showed a high variability in fruit nutritional quality and proved to be a rich source of bioactive compounds with antioxidant potential.

Keywords: *Prunus avium* L.; organic farming; vitamin C; polyphenols; carotenoids; chlorophylls

1. Introduction

Sweet cherry (*Prunus avium* L.) is a fruit tree species belonging to the family Rosaceae, particularly appreciated for its fruits which are the first fresh fruits of the year [1] and show premium quality characteristics in terms of colour, sweetness, sourness, firmness, and nutrients that are beneficial to health [2].

The consumer preference for sweet cherries is mainly influenced by taste and colour, but in recent years also by their nutritional value and the content of bioactive compounds such as anthocyanins, polyphenols, carotenoids, vitamins C and E, and antioxidants [3,4]. Moreover, an increasing percentage of consumers have been paying attention to aspects related to the whole production chain's sustainability in terms of the balance between resource input and yield output, impact on the environment, nutritional value of the products, and respect of workers' rights.

There are important factors, i.e., the cultivar, climate conditions, ripening stage, and crop management, that influence the content of bioactive compounds, starting from the growth stage of the fruit until the moment of harvesting [3].

Physiologically, sweet cherry is a non-climacteric fruit and goes through three stages of development before ripening: stage I, which begins after flowering and is followed by a rapid growth and division of the mesocarp cells; stage II, with the volume of the fruit augmenting slowly and the endocarp becoming lignified; and stage III, when the fruit's volume increases quickly again, and ripening occurs with biochemical changes [5,6].

In the last phenological stage of development, i.e., the full ripening, sweet cherry fruits undergo important biochemical and morphological changes, including the enhancement of colour and sugar content [7], which are the main indicators of maturity for commercial harvesting. However, fruit maturity varies within and between single trees and depends on the cultivar, meteorological conditions [8], agronomic practices [9], canopy management, fruit load [10], and fertilization [11]. The mentioned factors need to be examined from the perspective of crop system sustainability, mainly from the technical, nutritional, and environmental points of view.

Physicochemical studies are crucial to appropriately manage pre-harvest and harvest technology for sweet cherry production worldwide [12].

According to previous works related to some *Prunus* species, variations in quality parameters and fruit size reportedly depend on the position of the fruit within the crown of the tree [13,14]. Particularly, the location should be taken into account of both the fruit inside the tree, i.e., the sunny or shady side, and the tree within the orchard [15].

Previous studies [9,16–18] highlighted that the ripening stage influences the quality of sweet cherries by visibly changing the fruit colour and skin intensification, measurable by the parameters L* (luminosity), a* (red–green), and b* (yellow–blue), which turn from green to red or yellow depending on the cultivar, as chlorophyll pigments are degraded and anthocyanins accumulate.

Dry matter and total soluble solids contents increase during the development of fruits [18,19], which become much sweeter during ripening due to the sugar concentration increase, while acids (malic acid) remain relatively constant or decrease [1].

The purpose of this study was to evaluate the effects of the cultivar and position of the fruits inside the tree crown, at the pre-ripening or full ripeness phases, on the quality parameters of organically grown sweet cherry fruits.

2. Materials and Methods

2.1. Experimental Protocol and Growing Conditions

Research was carried out in 2022 and 2023 at the experimental field of the Research Station for Fruit Growing (RSFG) in Iași, North-East Romania (47°20′ N; 27°60′ E and 165 m altitude), in a sweet cherry orchard planted in a meso-relief with a gently sloping plateau and predominantly cambic chernozem soil with good natural drainage [20], pH 7.3, 4.0% humus content, and 0.19% N.

The meteorological conditions at the experimental field during the research years (2022–2023) were monitored by an Agroexpert system and are presented in Table 1.

Table 1. Average monthly values of temperature, rainfall, and relative humidity during the experimental period (RSFG Iași, Romania, 2022–2023).

Month	Air Temperature (°C) a*		Precipitation (mm) L*		Relative Humidity (%)	
	2022	2023	2022	2023	2022	2023
March	3.2	6.6	56.9	5.8	60.0	58.0
April	10.0	9.7	58.0	2.0	59.1	66.0
May	16.6	16.2	17.4	5.0	50.5	54.0
June	21.9	20.7	26.6	10.0	53.6	53.0
July	23.2	23.3	27.8	91.5	50.0	51.0

The experimental protocol was based on the factorial combination between three cultivars (Van, Andreiaș, Margonia) and two fruit positions inside the tree crown (inner, outer). A split plot design was used for the treatment distribution in the field, with three replicates, with a density of 500 plants ha^{-1}.

Sweet cherry samples were harvested from the sampled trees at two different fruit development phases: pre-ripening, with fruit at about 90% of their final size and beginning colouring; and full ripeness.

The three cultivars of sweet cherry (Van, Andreiaș, and Margonia) were provided by the Experimental Orchard Fields of RSFG, Iași. The trees of the mentioned cultivars were reared on high-vigour rootstocks of *Prunus mahaleb* L., spaced 5 m along the rows, which were 4 m apart, and trained as a free palmette. The soil was ploughed using a rotary tiller along the rows and covered by grass between the rows.

The cultivar Van originated from Canada and is an early and very productive cultivar, with medium to large fruits (7.5–8.5 g), a globular shaped, bright red colour, stony pink-red flesh, and a sweet acidic taste [21].

Andreiaș and Margonia are two new cultivars of sweet cherry, created and certified at RSFG Iași. Andreiaș has large (8.5–9.5 g), cordiform, red-brown fruit, almost black when fully ripe, and the cultivar Margonia has yellow, cordiform, slightly flattened fruit, elongated at the tip, with an average weight of 8.0–9.0 g [22].

The crops were organically managed [23], using the formulations permitted under the European regulation both for fertilization and plant protection.

2.2. Sampling Procedures

Fruit samples were harvested in all the plots, put in biodegradable bags, and immediately transported under appropriate conditions to the laboratory for analysis (Figure 1).

Figure 1. Sweet cherry cultivars at the two fruit development phases examined.

2.3. Qualitative Analyses

2.3.1. Measurement of Epidermis Colour

The fruit's epidermis colour was measured using a Minolta CR 400/410C portable colorimeter (Konica Minolta Measuring Instruments, München, Germany), referring to the CIE L^* a^* b^* uniform colour space. The value of L^* reflects the colour lightness, a^* the proportions of red (positive values) and green (negative values), and b^* the proportion of yellow (positive values) and blue (negative values). The parameter L^* defines the colour intensity, while a^* and b^* define the colour chromaticity [24,25].

2.3.2. Total Soluble Solids (TSS)

TSS content was determined by the refractive index of the fruit juice, measured using a Zeiss precision portable refractometer (Jena, Germany), and reported in °Brix.

2.3.3. Dry Matter

The total dry matter content was measured by drying 5 g of fruit samples (without stone) in an oven at 103 ± 2 °C until constant weight [26].

2.3.4. pH and Titratable Acidity

The pH value of the liquid samples was determined using the potentiometric method using a Hanna Instruments (Laval, QC, Canada) pH meter at 20 °C. Due to the similarity of the liquid fruit sample preparation, pH and titratable acidity (TA) were measured simultaneously [27].

To determine the titratable acidity (TA), the samples were homogenized with distilled water and titrated with 0.1 N NaOH until reaching pH 8.1. The results were expressed as percentage of malic acid content and calculated using Formula (1) below:

$$\text{TA (\% malic acid)} = (\text{V} \times \text{N} \times 100 \times 0.0067)/\text{m} \quad (1)$$

where N is the normality of NaOH; V is the volume of NaOH (mL) used in the titration; m is the weight of the sweet cherry sample used (g); and 0.0067 is the conversion factor for malic acid [12]. TA has been calculated in terms of the most common acid in the sample, which for different cultivars of sweet cherry is usually malic acid [15].

2.3.5. Carotene and Lycopene

The identification of carotene and lycopene in sweet cherry fruits (both peel and pulp) was carried out spectrophotometrically with petroleum ether. The ether extract was read by a UV/Vis Specord 210 Plus spectrophotometer (Analytik Jena, Jena, Germany) at 472 nm for lycopene and 452 nm for carotene. The obtained values were integrated into the following formulas [28].

$$\begin{aligned} \text{Carotene} &= \text{A452} \times 19.96, \text{ in mg } 100 \text{ g}^{-1} \\ \text{Lycopene} &= \text{A472} \times 14.495, \text{ in mg } 100 \text{ g}^{-1} \end{aligned} \quad (2)$$

2.3.6. Anthocyanin Content

The anthocyanin content was determined using a spectrophotometric version method based on the differential pH of 1.0 to 4.5 [29,30] with an acidulated alcoholic solution and measurement at a 515 nm wavelength by a Specord 210 PLUS UV-VIS spectrophotometer (Analytik Jena, Germany) [31].

2.3.7. Chlorophyll Pigments

The content of chlorophyll pigments (chlorophyll *a* and *b*) in the sweet cherry fruits was determined by the spectrophotometric method described by Jităreanu et al. [32], based on the light absorption capacity of the acetonic extract (1%) and analysed by a computer-aided spectrophotometer at 647 and 663 nm absorption based on Lichtenthaler's equations [33,34]:

$$\begin{aligned} Ca &= 12.25 \times \text{A663} - 2.79 \times \text{A647} \\ Cb &= 21.50 \times \text{A647} - 5.10 \times \text{A663} \\ Ca + b &= 7.15 \times \text{A663} - 18.71 \times \text{A647} \end{aligned} \quad (3)$$

2.3.8. Total Phenolic Content

The determination of the content of total phenolic compounds was carried out according to the Folin–Ciocalteu [35] by reading values on a spectrophotometer at a 750 nm wavelength. The results reveal the antioxidant activity and are expressed as mg of gallic acid equivalents (GAEs) per 100 g of sample [36].

2.3.9. Vitamin C

In order to determine the vitamin C content of the sweet cherry samples, the 2.6-dichloroindophenol titrimetric method of juice analysis was used (AOAC Method 967.21) [37]. Samples were fused with 40 mL of 2% oxalic acid, and after titration (2.6-dichloroindophenol), the results were expressed in mg ascorbic acid 100 g^{-1} fruit [38].

2.4. Statistical Analysis

The data were statistically processed by two-way analysis of variance, and Duncan's multiple range test was performed for mean separation at $p \leq 0.05$ using the SPSS software version 29. The standard deviation (STDEV) and the Pearson coefficient of variation (COVAR) were also determined [39].

3. Results and Discussion

As no significant interactions arose between the two experimental factors applied, i.e., cultivar and fruit position inside the tree crown, only their simple effects are presented.

The colour of fresh sweet cherry fruits is among the primary factors influencing consumer choice, because it is an important quality attribute, also serving to estimate the maturity and harvesting stage [40].

With regard to the fruit's chromatic parameters (L^*, a^*, b^*), in 2022, the component a^* showed significantly higher values in the red fruits of the cultivars Van and Andreias and in the inner position inside the tree crown, both at the pre-ripening and full ripeness phase (Table 2). In the same year, the components L^* and b^* were significantly affected by cultivar only at the full ripeness phase, with the highest levels recorded in the yellow fruits of the cultivar Margonia. In 2023 (Table 3), the cultivars Van and Margonia attained higher values of the three colour components at the pre-ripening phase, whereas similar trends to the previous year were recorded at the full ripeness phase. No colour components were significantly influenced by the fruit's position inside the tree crown.

Table 2. Fruit skin colour of three sweet cherry cultivars at two development phases, examined in 2022.

Treatment	Pre-Ripening Phase			Full Ripeness Phase		
	L*	a*	b*	L*	a*	b*
Cultivar						
Van	56.7	0.57 a	38.1	20.4 b	17.1 a	1.7 b
Andreiaș	56.7	0.43 b	39.9	19.8 b	15.7 a	1.5 b
Margonia	57.0	0.32 c	40.7	60.8 a	0.9 b	35.3 a
	n.s.		n.s.			
Fruit position						
Inner	58.7	0.59	41.2	33.6	12.0	13.3
Outer	54.9	0.28	37.9	33.8	10.4	12.4
	n.s.	*	n.s.	n.s.	*	n.s.
Av.	56.79	0.44	39.55	33.67	11.22	12.83
STDEV	2.15	0.25	2.25	21.01	8.20	17.43
COVAR s%	3.79	56.66	5.68	62.39	73.06	135.82
Min.	54.32	0.11	37.18	19.75	0.14	1.00
Max.	59.09	0.74	42.55	60.79	19.30	36.26

n.s.: not significant; * significant at $p \leq 0.05$. Within each column, values followed by different letters are statistically different according to Duncan test at $p \leq 0.05$, $n = 3$.

Compared to the year 2022, in the meteorological conditions recorded in 2023, there was generally a decrease in the values of colour parameters of Andreias fruits at the pre-ripening phase; moreover, the component a^* was not affected by the fruit's position inside the crown.

As for the comparison between the two development phases, the yellow fruits of the cultivar Margonia showed a stability of values, whereas the red fruits of cv. Van and Andreias exhibited a decrease in L^* and b^* components and an increase in the a^* component during the ripening process.

The differences between the two phenological phases or between the inner and outer crown area can be useful in evaluating the fruit ripening process to compare the quality of cultivars and the influence of climatic conditions and to determine the time required to reach full maturity. In this respect, compared to the year 2022, in 2023, the time interval from the pre-ripening stage to full ripening was longer due to the lower average temperature, which caused a slower but more efficient accumulation of anthocyanin pigments in the red fruits, associated with higher values of the a^* parameter.

Table 3. Fruit skin colour of three sweet cherry cultivars at two development phases, examined in 2023.

Treatment	Pre-Ripening Phase			Full Ripeness Phase		
	L*	a*	b*	L*	a*	b*
Cultivar						
Van	57.5 a	0.67 a	38.5 a	20.5 b	17.5 a	1.6 b
Andreiaș	51.8 b	0.35 b	30.4 b	20.7 b	17.8 a	2.0 b
Margonia	58.2 a	0.72 a	38.3 a	59.2 a	1.9 b	35.4 a
Fruit position						
Inner	56.3	0.59	36.0	33.6	12.4	13.4
Outer	55.3	0.56	35.4	33.3	12.3	12.6
	n.s.	n.s.	n.s.	n.s.	n.s.	n.s.
Av.	55.8	0.6	35.7	33.4	12.3	13.0
STDEV	3.3	0.3	4.2	19.5	8.1	17.3
COVAR s%	6.0	43.4	11.8	58.2	65.6	133.6
Min.	51.3	0.1	30.2	20.0	1.9	0.9
Max.	59.3	0.8	39.8	60.3	18.1	36.9

n.s.: not significant. Within each column, values followed by different letters are statistically different according to Duncan test at $p \leq 0.05$, $n = 3$.

The chromatic values recorded in our investigation are consistent with those detected by Goncalves et al. [41] in sweet cherry fruits of the cultivar Van at the partial and full ripeness phases. The results of the present research are also similar to those obtained by Girad and Kopp [42], who stated that darker sweet cherries (lower L^* values) tended to be less red (lower a^* value) and less yellow (lower b^* value). The maximum values of the colour component L^* corresponded to the minimum values of the hue parameters a^* and b^*. Similar colour variations during fruit ripening also occurred in sour cherries, with the skin colour parameters $L^*$and b^* decreasing uniformly [40,43].

Both in 2022 (Table 4) and in 2023 (Table 5), at the two development phases examined, the dry matter (DM) and total soluble solids (TSS) contents of cv. Van sweet cherry fruits were highest (on average, 17.7% and 12.9 °Brix, respectively), except in 2022, when at the full ripeness phase, the DM of Van did not significantly differ from that of Andreias. Moreover, the DM of the fruit in the outer brighter part of crown was significantly higher than in the inner part in both years but only at the pre-ripening phase.

Table 4. Dry matter and total soluble solids content of fruits from three sweet cherry cultivars at two development phases, examined in 2022.

Treatment	Pre-Ripening Phase		Full Ripeness Phase	
	DM (%)	TSS (°Brix)	DM (%)	TSS (°Brix)
Cultivar				
Van	13.1 a	10.5 a	21.4 a	14.8 a
Andreiaș	11.5 b	9.5 b	20.9 a	12.4 b
Margonia	11.9 b	9.4 b	18.0 b	12.5 b
Fruit position inside the crown				
Inner	11.6	9.5	19.9	13.5
Outer	12.8	10.1	20.2	13.0
	*	n.s.	n.s.	n.s.
Av.	12.17	9.80	20.06	13.23
STDEV	1.07	0.66	1.66	1.29
COVAR s%	8.81	6.77	8.29	9.71
Min.	10.78	9.03	17.93	11.90
Max.	13.30	10.81	21.80	14.80

TSS: total soluble solids; DM: dry matter content. n.s.: not significant; * significant at $p \leq 0.05$. Within each column, values followed by different letters are statistically different according to Duncan test at $p \leq 0.05$, $n = 3$.

Table 5. Dry matter and total soluble solids content of fruits from three sweet cherry cultivars at two development phases, examined in 2023.

Treatment	Pre-Ripening Phase		Full Ripeness Phase	
	DM (%)	TSS (°Brix)	DM (%)	TSS (°Brix)
Cultivar				
Van	13.5 a	10.8 a	22.7 a	15.3 a
Andreiaș	11.5 b	9.2 b	20.5 b	12.5 b
Margonia	11.7 b	9.5 b	19.7 b	12.7 b
Fruit position inside the crown				
Inner	11.5	9.6	20.4	13.2
Outer	12.9	10.1	21.6	13.8
	*	n.s.	n.s.	n.s.
Av.	12.20	9.85	20.97	13.50
STDEV	1.27	1.12	1.53	1.58
COVAR s%	10.37	11.39	7.29	11.73
Min.	10.69	8.00	19.20	12.08
Max.	13.90	10.93	23.53	16.33

TSS: total soluble solids; DM: dry matter content. n.s.: not significant; * significant at $p \leq 0.05$. Within each column, values followed by different letters are statistically different according to Duncan test at $p \leq 0.05$, $n = 3$.

Either the dry matter or the soluble solids content of sweet cherry fruits of the three cultivars examined always increased from the pre-ripening to the full ripeness phase, independently of the year, with the top values being those of cv. Van in 2023 (22.7% and 15.3 °Brix).

Total soluble solids (TSS), predominantly estimating the level of dissolved sugars (sucrose, glucose, and fructose) in fruit juice but also including a small number of other compounds such as organic acids, soluble amino acids, and minerals [15], are an important quality parameter that has a remarkable influence on the fruit taste.

Consistently with our results, in previous research carried out on sweet cherry cultivars, both the DM and TSS contents increased during the fruit ripening, in the same agricultural area and with the same cultivars as in the present experiment [44] or in other locations [17,45–47].

Many scientific investigations have shown that the TSS vary between 11 °Brix and 25 °Brix in sweet cherries in several areas of the world. In the Spanish cultivars Sweetheart and Picota, a TSS range of 14.0–23.2 °Brix was reported by Serradilla et al. [48] and Gonzalez-Gomez et al. [49]. In the cultivars Bing and Summit, average TSS values of 13.3 and 19.6 °Brix were recorded, respectively [12]. Moreover, Wen et al. [50] detected TSS values between 17.8 and 20.0 °Brix for the Chinese sweet cherry cultivars Rainier, Hongyan, and Hongdeng.

In previous research, the meteorological trend in the last month of fruit formation exerted the greatest influence on TSS accumulation in sweet cherry fruit, regardless of the ripening phase [51,52], which has an important impact on the flavour evaluation [53].

Important physicochemical parameters, such as pH, titratable acidity (TA), and nutraceuticals (total polyphenols, ascorbic acid, and pigments) influence the consumer's perception of fruit quality [18,44].

Neither the pH or titratable acidity values in the sweet cherry fruits (on average, 3.87 and 1.04 g 100 g^{-1} malic acid, respectively) were significantly influenced by the cultivar and ripening process, both in 2022 and 2023 (Tables 6 and 7). In previous research, sweet cherry juice had a higher pH than the sour one [54,55], and the levels of both the juice's pH and total acidity were proven to also be genotype-dependant [56].

Acidity is essential for the consumer's perception of a good sweet cherry flavour [11].

According to other studies carried out on sweet cherries [45–47], TA widely ranged between 0.39% and 0.87%, but based on Crisosto et al.'s reports [16], the framing limits of TA in ripe sweet cherries vary from 0.5 to 1.0 mg malic acid 100 g^{-1} f.w.

Table 6. Biochemical parameters of fruits of three sweet cherry cultivars at two development phases, examined in 2022.

Treatment	Pre-Ripening Phase				Full Ripeness Phase			
	pH	TA	TP	Vit C	pH	TA	TP	Vit C
Cultivar								
Van	3.83	1.12	124.1 c	5.3 c	3.76	0.95	77.2 a	17.5
Andreiaș	4.02	1.05	156.2 b	9.0 b	3.85	0.92	65.6 b	17.5
Margonia	4.03	1.15	186.9 a	9.8 a	3.85	0.87	63.8 b	18.8
	n.s.	n.s.			n.s.	n.s.		n.s.
Fruit position								
Inner	3.96	1.14	156.8	7.7	3.80	0.92	75.4	17.0
Outer	3.95	1.07	154.8	8.3	3.84	0.91	62.3	18.8
	n.s.	n.s.	n.s.	n.s.	n.s.	n.s.	*	*
Av.	3.96	1.11	155.75	8.00	3.82	0.91	68.85	17.92
STDEV	0.10	0.08	28.82	2.28	0.06	0.04	11.70	1.28
COVAR s%	2.60	7.39	18.50	28.50	1.54	4.68	17.00	7.15
Min.	3.82	1.00	122.43	4.50	3.74	0.85	62.11	16.00
Max.	4.04	1.21	193.33	10.50	4.00	0.98	92.13	19.50

TA: titratable acidity, as mg malic acid 100 g^{-1} f.w.; TP: total phenolic content, as mg GAE 100 g^{-1} f.w.; vit C: vitamin C, as mg 100 g^{-1} f.w. n.s.: not significant; * significant at $p \leq 0.05$. Within each column, values followed by different letters are statistically different according to Duncan test at $p \leq 0.05$, $n = 3$.

Table 7. Biochemical parameters of fruits of three sweet cherry cultivars at two development phases, examined in 2023.

Treatment	Pre-Ripening Phase				Full Ripeness Phase			
	pH	TA	TP	Vit C	pH	TA	TP	Vit C
Cultivar								
Van	3.85	1.15	145.8 b	5.5 c	3.77	0.97	75.8 a	17.3 b
Andreiaș	3.71	1.12	152.8 b	8.0 b	3.90	1.01	60.0 b	16.5 b
Margonia	4.04	1.21	178.4 a	10.0 a	4.00	0.95	60.1 b	20.3 a
	n.s.	n.s.			n.s.	n.s.		
Fruit position								
Inner	3.88	1.20	164.4	8.0	3.85	0.94	74.5	17.5
Outer	3.85	1.12	153.6	7.7	3.92	1.01	56.2	18.5
	n.s.	n.s.	n.s.	n.s.	n.s.	n.s.	*	n.s.
Av.	3.87	1.16	159.00	7.83	3.89	0.97	65.31	18.00
STDEV	0.16	0.08	17.02	2.75	0.12	0.06	14.33	2.51
COVAR s%	4.08	7.27	10.70	35.12	3.08	5.87	21.94	13.94
Min.	3.66	1.03	142.05	5.00	3.75	0.87	54.68	14.00
Max.	4.08	1.26	182.67	11.50	4.00	1.02	92.99	20.50

TA: titratable acidity, as mg malic acid 100 g^{-1} f.w.; TP: total phenolic content, as mg GAE 100 g^{-1} f.w.; vit C: vitamin C, as mg 100 g^{-1} f.w. n.s.: not significant; * significant at $p \leq 0.05$. Within each column, values followed by different letters are statistically different according to Duncan test at $p \leq 0.05$, $n = 3$.

In both studied years, the contents of phenolic compounds (Tables 6 and 7) were significantly the highest in cv. Margonia fruits at the pre-ripening phase (on average, 182.7 mg GAE 100 g^{-1} f.w.) and cv. Van at the full ripeness phase (on average, 76.5 mg GAE 100 g^{-1} f.w.). Moreover, on average, phenolics showed an over 50% decrease from the pre-ripening to the full ripeness phase in both years.

The polyphenol content was higher in the fruits located in the inner part of the crown but only at the full ripeness phase.

Sweet cherry fruits are considered a very good source of phenolic compounds that are beneficial in a human diet [51]. In the present experiment, the variation in polyphenols contents in fruits mainly depended on the examined cultivars and development phase.

Similar values of TP as in the present research were detected in the ripe fruits of some sweet cherry cultivars by other authors [51,57,58], but in the premature stage, the fruits had a much higher content.

Sweet cherry fruits are appreciated by consumers for their good taste but also for their many healthy nutritional properties, such as the high vitamin C content [51].

As can be observed in Tables 6 and 7, in our study, the cultivar Margonia showed the highest vitamin C content in both years and development phases (14.7 mg·100 g^{-1} f.w., on average), though no significant differences between the cultivars were found at the full ripeness phase in 2022. Only in the latter case, a significant difference was recorded between the positions inside the crown, because the outer sweet cherry fruits had higher vitamin C contents than the inner ones. The fully ripe fruits showed a more than doubled vitamin C content compared to the unripe ones.

Contrary to the reports by Średnicka-Tober [59], indicating that the fruits of sweet cherry were characterized by a significantly higher content of vitamin C, up to 42.9 mg 100 g^{-1} f.w., in other studies [51,60], depending on the cultivar and the meteorological conditions of the research year, the average content of vitamin C in sweet cherries varied between 5.3 and 20.3 mg 100 g^{-1} f.w.

The phenological phase of full ripeness in sweet cherry fruits influences the content of secondary compounds, such as pigments, which enhance the nutritional and health value of a fruit and, in addition, their evolution visually appeals consumers in terms of fruit colouration [61].

As for the fruit's chlorophyll pigments, in both years (Tables 8 and 9), the yellow fruit cv. Margonia showed the highest values of chlorophyll a and b, either at the pre-ripening (22.0 mg 100 g^{-1} f.w. chlorophyll a and 36.9 mg 100 g^{-1} f.w. chlorophyll b, on average) or at the full ripeness phase (3.9 mg 100 g^{-1} f.w. chlorophyll a and 6.8 mg 100 g^{-1} f.w. chlorophyll b, on average), but in 2023 (Table 9), it did not significantly differ from Andreias at the full ripeness phase. Moreover, only in the latter context was a significant difference recorded between the two fruit positions inside the tree crown, with the outer fruits attaining higher chlorophyll a contents than the inner ones.

Differently from chlorophyll pigments, in both years (Tables 8 and 9), the fruit's contents of carotene and lycopene were highest in the red fruits of cv. Andreias (0.39 mg 100 g^{-1} f.w. carotene and 0.31 mg 100 g^{-1} f.w. lycopene, as an average of the year and development phase), but in 2022 at the full ripeness phase, the carotene level did not significantly differ from that of cv. Margonia. The anthocyanin trend was exactly the same as that of lycopene, but its values were much higher at the full ripeness phase in both years (38.2 mg 100 g^{-1} f.w., on average) compared to the pre-ripening phase (1.5 mg 100 g^{-1} f.w., on average).

Table 8. Pigment content of fruits of three sweet cherry cultivars at two development phases, examined in 2022.

Treatment	Pre-Ripening Phase					Full Ripeness Phase				
	Chl a	**Chl b**	**Car**	**Lyc**	**Anth**	**Chl a**	**Chl b**	**Car**	**Lyc**	**Anth**
	mg 100 g^{-1} f.w.									
Cultivar										
Van	4.9 c	9.3 c	0.23 b	0.06 b	0.73 b	1.45 c	2.55 c	0.11 b	0.25 b	21.6 b
Andreiaș	17.7 b	27.1 b	0.45 a	0.27 a	1.24 a	2.31 b	5.20 b	0.28 a	0.35 a	36.9 a
Margonia	22.2 a	36.5 a	0.09 c	0.09 b	0.34 c	3.11 a	6.41 a	0.28 a	0.07 c	4.7 c
Fruit position										
Inner	14.6	25.2	0.26	0.14	0.74	2.24	4.58	0.22	0.21	22.2
Outer	15.2	23.4	0.25	0.14	0.79	2.33	4.86	0.22	0.22	19.9
	n.s.	n.s.	n.s.	n.s.	n.s.	n.s.	n.s.	n.s.	n.s.	*
Av.	14.91	24.30	0.26	0.28	0.77	2,29	4.72	0.22	0.22	21.05
STDEV	8.06	12.63	0.17	0.33	0.44	0.75	1.78	0.09	0.13	14.58
COVAR s%	54.09	51.97	66.68	118.41	56.78	32.58	37.72	40.49	58.07	69.26
Min.	4.84	8.71	0.07	0.05	0.21	1.41	2.24	0.10	0.05	4.50
Max.	23.44	37.35	0.48	0.92	1.47	3.21	6.44	0.29	0.35	40.52

Chl a: chlorophyll a, as; Chl b: chlorophyll b; Car: carotene; Lyc: lycopene; Anth: anthocyanin. n.s.: not significant; * significant at $p \leq 0.05$. Within each column, values followed by different letters are statistically different according to Duncan test at $p \leq 0.05$, $n = 3$.

Table 9. Pigment content of fruits of three sweet cherry cultivars at two development phases, examined in 2023.

Treatment	Pre-Ripening Phase					Full Ripeness Phase				
	Chl a	Chl b	Car	Lyc	Anth	Chl a	Chl b	Car	Lyc	Anth
	mg 100 g^{-1} f.w.									
Cultivar										
Van	5.9 c	9.1 c	0.29 b	0.11 b	0.87 b	1.74 b	3.65 b	0.11 b	0.08 c	21.6 b
Andreiaș	15.6 b	25.8 b	0.48 a	0.35 a	1.75 a	4.65 a	7.72 a	0.36 a	0.25 a	39.4 a
Margonia	21.8 a	37.3 a	0.10 c	0.07 c	0.46 c	4.70 a	7.12 a	0.11 b	0.16 b	2.3 c
Fruit position										
Inner	15.0	25.1	0.29	0.16	0.90	3.54	5.65	0.19	0.15	22.8
Outer	13.8	23.0	0.29	0.19	1.15	3.85	6.68	0.20	0.17	19.4
	n.s.	n.s.	n.s.	*	*	*	n.s.	n.s.	*	*
Av.	14.42	24.04	0.29	0.17	1.03	3.69	6.16	0.19	0.16	21.09
STDEV	7.52	13.70	0.18	0.14	0.69	1.59	2.11	0.13	0.08	16.85
COVAR s%	52.17	56.98	63.59	78.23	67.18	43.13	34.22	66.94	46.69	79.90
Min.	5.86	8.77	0.08	0.10	0.40	1.01	2.52	0.10	0.10	2.14
Max.	23.27	41.03	0.49	0.35	2.30	4.90	7.73	0.37	0.26	44.19

Chl a: chlorophyll a; Chl b: chlorophyll b; Car: carotene; Lyc: lycopene; Anth: anthocyanin. n.s.: not significant; * significant at $p \leq 0.05$. Within each column, values followed by different letters are statistically different according to Duncan test at $p \leq 0.05$, $n = 3$.

As for the comparison between the two positions inside the tree crown, the content of lycopene was higher in the outer fruits only in 2023 at both the development phases; the anthocyanin level was higher in the outer fruits at the pre-ripening phase only in 2022, and in the inner ones at the full ripeness phase in both years.

The total contents of carotene and lycopene during the ripening period did not univocally change and, according to Abers and Wrolstad's study [62], they can decrease due to the augmentation of the total anthocyanins, which tend to mask the colour of carotenoids and degrade the chlorophyll due to the development of dark pigmented compounds.

Anthocyanins, belonging to the subclass of water-soluble flavonoids, are one of the main phenolic compounds in sweet cherries and is responsible for the red colouring of the fruit epidermis and pulp [15] and associated with the high antioxidant potential of sweet cherry fruits [63].

During the two years of the present study, a remarkable increase in the content of anthocyanins from the pre-ripening to full ripeness phase was recorded in the cultivars with red fruits (Van and Andreiaș), whereas these compounds showed a low accumulation in the yellow fruits of cv. Margonia at the full ripeness phase.

In previous studies regarding conventionally grown sweet cherry fruits, a variability of anthocyanins arose between cultivars, from 12.1 (cultivar Elton Heart) to 91.6 mg 100 g^{-1} f.w. (cv. Kordia) [59]. Ferretti et al. [60] showed that cv. Regina fruits contained 41.0 mg 100 g^{-1} f.w. anthocyanins, and in another study [51], 54.0 mg 100 g^{-1} f.w. anthocyanins were detected in red sweet cherry fruits on average.

4. Conclusions

In the context of sustainable production management carried out in the present research, significant quality differences arose between the organically grown *Prunus avium* L. fruits of the examined cultivars, which proved to be a very good source of biologically active compounds with beneficial nutritional value to human health.

When the fruits ripened and the colour changed, increases were recorded in the values of the quality indices that are highly appreciated by consumers, such as the content of soluble sugars, vitamin C, carotenoids, and anthocyanins, although a higher content of polyphenolic compounds was found at the pre-ripening phase.

The biochemical processes involved in the fruit ripening process led to several changes in quality parameters, mainly influenced by the variation in seasonal meteorological conditions and cultivar, but also by the fruit's position inside the tree crown.

The fruits of the locally created cultivar Andreiaș were the richest in antioxidants, and the latter outcome suggests an opportunity to both valorise and include the mentioned cultivar in breeding strategies under an organic management programme.

Author Contributions: Conceptualization, I.M., O.C.M. and C.D.J.; methodology, I.M. and O.C.M.; software, I.M. and O.C.M.; validation, O.C.M., G.C. and C.D.J.; formal analysis, I.M., A.V.T. and S.S; investigation, I.M., S.S., A.V.T. and O.C.M.; resources, I.M., O.C.M. and C.D.J.; data curation, I.M., O.C.M. and G.C.; writing—original draft preparation, I.M.; writing—review and editing, O.C.M., G.C. and C.D.J.; visualization, O.C.M., G.C. and C.D.J.; supervision, C.D.J. All authors have read and agreed to the published version of the manuscript.

Funding: This research received no external funding.

Data Availability Statement: The original contributions presented in the study are included in the article, further inquiries can be directed to the corresponding authors.

Conflicts of Interest: The authors declare no conflicts of interest.

References

1. Habib, M.; Bhat, M.; Dar, B.N.; Wani, A.A. Sweet cherries from farm to table: A review. *Crit. Rev. Food Sci. Nutr.* **2017**, *57*, 1638–1649. [CrossRef] [PubMed]
2. Esti, M.; Cinquanta, L.; Sinesio, F.; Moneta, E.; Di Matteo, M. Physicochemical and sensory fruit characteristics of two sweet *cherry* cultivars after cool storage. *Food Chem.* **2002**, *76*, 399–405. [CrossRef]
3. Quero-García, J.; Lezzoni, A.; Puławska, J.; Lang, G. *Cherries: Botany, Production and Uses*; CABI: Wallingford, UK, 2017. Available online: https://www.cabi.org/bookshop/book/9781780648378 (accessed on 22 August 2023).
4. Ballistreri, G.; Continella, A.; Gentile, A.; Amenta, M.; Fabroni, S.; Rapisarda, P. Fruit quality and bioactive compounds relevant to human health of sweet *cherry* (*Prunus avium* L.) cultivars grown in Italy. *Food Chem.* **2013**, *140*, 630–638. [CrossRef] [PubMed]
5. Andrews, P.K.; Li, S. Partial Purification and Characterization of.beta.-D-Galactosidase from Sweet *Cherry*, a Nonclimacteric Fruit. *J. Agric. Food Chem.* **1994**, *42*, 2177–2182. [CrossRef]
6. Knoche, M.; Beyer, M.; Peschel, S.; Oparlakov, B.; Bukovac, M.J. Changes in strain and deposition of cuticle in developing sweet *cherry* fruit. *Physiol. Plant.* **2004**, *120*, 667–677. [CrossRef]
7. Tudela, J.; Luchsinger, L.; Artés-Hdez, F.; Artés, F. 'Ambrunes' sweet *cherry* quality factors change during ripening. *Acta Hortic.* **2005**, *667*, 529–534. [CrossRef]
8. Gonçalves, B.; Landbo, A.-K.; Knudsen, D.; Silva, A.P.; Moutinho-Pereira, J.; Rosa, E.; Meyer, A.S. Effect of Ripeness and Postharvest Storage on the Phenolic Profiles of Cherries (*Prunus avium* L.). *J. Agric. Food Chem.* **2004**, *52*, 523–530. [CrossRef] [PubMed]
9. Mozetič, B.; Simčič, M.; Trebše, P. Anthocyanins and hydroxycinnamic acids of Lambert Compact cherries (*Prunus avium* L.) after cold storage and 1-methylcyclopropene treatment. *Food Chem.* **2006**, *97*, 302–309. [CrossRef]
10. Whiting, M.D.; Lang, G.; Ophardt, D. Rootstock and Training System Affect Sweet *Cherry* Growth, Yield, and Fruit Quality. *HortScience* **2005**, *40*, 582–586. [CrossRef]
11. Crisosto, C.H.; Johnson, R.S.; Day, K.; DeJong, T. Preharvest Factors Affecting Postharvest Stone Fruit Quality. *HortScience* **1995**, *30*, 751A. [CrossRef]
12. Hayaloglu, A.A.; Demir, N. Phenolic compounds, volatiles, and sensory characteristics of twelve sweet *cherry* (*Prunus avium* L.) cultivars grown in Turkey. *J. Food Sci.* **2016**, *81*, 7–18. [CrossRef]
13. Lopresti, J.; Goodwin, I.; Holford, P.; McGlasson, B.; Golding, J. Study on the relationship between sugar concentration and cell number and size distribution in nectarine. *Acta Hortic.* **2015**, *1084*, 667–674. [CrossRef]
14. Lopresti, J.; Goodwin, I.; Stefanelli, D.; Holford, P.; McGlasson, B.; Golding, J. Understanding the factors affecting within-tree variation in soluble solids concentration in peaches and nectarines. *Acta Hortic.* **2016**, *1130*, 249–256. [CrossRef]
15. Ricardo-Rodrigues, S.; Laranjo, M.; Agulheiro-Santos, A.C. Methods for quality evaluation of sweet *cherry*. *J. Sci. Food Agric.* **2022**, *103*, 463–478. [CrossRef] [PubMed]
16. Crisosto, C.H.; Crisosto, G.M.; Metheney, P. Consumer acceptance of 'Brooks' and 'Bing' cherries is mainly dependent on fruit SSC and visual skin color. *Postharvest Biol. Technol.* **2003**, *28*, 159–167. [CrossRef]
17. Muskovics, G.; Felföldi, J.; Kovács, E.; Perlaki, R.; Kállay, T. Changes in physical properties during fruit ripening of Hungarian sweet *cherry* (*Prunus avium* L.) cultivars. *Postharvest Biol. Technol.* **2006**, *40*, 56–63. [CrossRef]
18. Magri, A.; Malorni, L.; Cozzolino, R.; Adiletta, G.; Siano, F.; Picariello, G.; Cice, D.; Capriolo, G.; Nunziata, A.; Di Matteo, M.; et al. Agronomic, Physicochemical, Aromatic and Sensory Characterization of Four Sweet *Cherry* Accessions of the Campania Region. *Plants* **2023**, *12*, 610. [CrossRef] [PubMed]
19. Kovács, E.; Muskovics, G.; Perlaki, R. Relationship of colour and other quality parameters of sweet *cherry* during development and ripening. *Acta Aliment.* **2009**, *38*, 415–426. [CrossRef]

20. Iurea, E.; Sîrbu, S.; Corneanu, G.; Corneanu, M. Results obtained from sweet *cherry* breeding in Iaşi, Romania. *J. Appl. Life Sci. Environ.* **2022**, *187*, 333–341. [CrossRef]
21. Budan, S.; Grădinariu, G. Cireșul. In *Ion Ionescu de la Brad*; Editura: Iași, Romania, 2000; p. 264.
22. Corneanu, G.; Cârdei, E.; Corneanu, M.; Sîrbu, S.; Iurea, E. *Cercetare, Inovare și Progres în Pomicultura Ieșeană*; Editura Altfel: Iași, România, 2012; p. 99.
23. *Regulation (EU) 2018/848 of the European Parliament and of the Council of 30 May 2018 on Organic Production and Labelling of Organic Products*; Document 32018R0848; European Union: Brussels, Belgium, 2018.
24. Ibraheem, N.A.; Hasan, M.M.; Khan, R.Z.; Mishra, P.K. Understanding Color Models: A Review. *ARPN J. Sci. Technol.* **2012**, *2*, 265–275.
25. Rutkowski, K.; Łysiak, G.P. Weather Conditions, Orchard Age and Nitrogen Fertilization Influences Yield and Quality of *'Łutówka'* Sour *Cherry* Fruit. *Agriculture* **2022**, *12*, 2008. [CrossRef]
26. Escribano, S.; Biasi, W.V.; Lerud, R.; Slaughter, D.C.; Mitcham, E.J. Non-destructive prediction of soluble solids and dry matter content using NIR spectroscopy and its relationship with sensory quality in sweet cherries. *Postharvest Biol. Technol.* **2017**, *128*, 112–120. [CrossRef]
27. Murariu, O.C.; Brezeanu, C.; Jităreanu, C.D.; Robu, T.; Irimia, L.M.; Trofin, A.E.; Popa, L.-D.; Stoleru, V.; Murariu, F.; Brezeanu, P.M. Functional Quality of Improved Tomato Genotypes Grown in Open Field and in Plastic Tunnel under Organic Farming. *Agriculture* **2021**, *11*, 609. [CrossRef]
28. Brezeanu, C.; Brezeanu, P.M.; Stoleru, V.; Irimia, L.M.; Lipsa, F.D.; Teliban, G.-C.; Ciobanu, M.M.; Murariu, F.; Puiu, I.; Branca, F.; et al. Nutritional Value of New Sweet *Pepper* Genotypes Grown in Organic System. *Agriculture* **2022**, *12*, 1863. [CrossRef]
29. Lee, J.; Durst, R.W.; Wrolstad, E.R.; Eisele, T.; Giusti, M.M.; Hofsommer, H.; Koswig, S.; Krueger, A.D.; Kupina, S.; Martin, S.K.; et al. Determination of Total Monomeric Anthocyanin Pigment Content of Fruit Juices, Beverages, Natural Colorants, and Wines by the pH Differential Method: Collaborative Study. *J. AOAC Int.* **2005**, *88*, 1269–1278. [CrossRef] [PubMed]
30. Teng, Z.; Jiang, X.; He, F.; Bai, W. Qualitative and Quantitative Methods to Evaluate Anthocyanins. *eFood* **2020**, *1*, 339–346. [CrossRef]
31. Zdremţan, M. *Conservarea Legumelor şi Fructelor—Îndrumător de Lucrări Practice*; Editura Universităţii: Bucharest, Romania; Aurel Vlaicu University: Arad, Romania, 2008; p. 153, ISBN 978-973-752-290-0.
32. Jitareanu, C.D.; Toma, L.D.; Slabu, C.; Marta, A.E. Effect of weather conditions on photosynthetic and flavonoid pigment contents in leaves of grapevine cultivars during growing season. *J. Food Agric. Environ.* **2011**, *9*, 793–798.
33. Lichtenthaler, H.K. Chlorophylls and carotenoids: Pigments of photosynthetic biomembranes. *Methods Enzymol.* **1987**, *148*, 350–382. [CrossRef]
34. Wellburn, A.R. The Spectral Determination of Chlorophylls a and b, as well as Total Carotenoids, Using Various Solvents with Spectrophotometers of Different Resolution. *J. Plant Physiol.* **1994**, *144*, 307–313. [CrossRef]
35. Dziadek, K.; Kopeć, A.; Piątkowska, E.; Leszczyńska, T.; Pisulewska, E.; Witkowicz, R.; Bystrowska, B.; Francik, R. Identification of polyphenolic compounds and determination of antioxidant activity in extracts and infusions of buckwheat leaves. *Eur. Food Res. Technol.* **2017**, *244*, 333–343. [CrossRef]
36. Caruso, G.; Stoleru, V.V.; Munteanu, N.C.; Sellitto, V.M.; Teliban, G.C.; Burducea, M.; Tenu, I.; Morano, G.; Butnariu, M. Quality Performances of Sweet Pepper under Farming Management. *Not. Bot. Horti Agrobot. Cluj-Napoca* **2019**, *47*, 458–464. [CrossRef]
37. AOAC International. *Official Methods of Analysis*, 20th ed.; AOAC International: Rockville, MD, USA, 2016.
38. Nielsen, S.S. Vitamin C Determination by Indophenol Method. In *Nielsen's Food Analysis Laboratory Manual*; Food Science Text Series; Ismail, B.P., Nielsen, S.S., Eds.; Springer: Cham, Switzerland, 2017; pp. 143–146.
39. Botu, I.; Botu, M. *Tehnica Experimetală în Horticultură și Ecologie (Elemente de Bază)*; Editura Conphys, Rm.: Vâlcea, România, 2010; p. 139.
40. Pedišić, S.; Levaj, B.; Dragović-Uzelac, V.; Škevin, D.; Skendrović Babojelić, M. Color parameters and total anthocyanins of sour cherries (*Prunus cerasus* L.) during ripening. *Agric. Conspec. Sci.* **2009**, *74*, 259–262.
41. Gonçalves, B.; Silva, A.P.; Moutinho-Pereira, J.; Bacelar, E.; Rosa, E.; Meyer, A.S. Effect of ripeness and postharvest storage on the evolution of colour and anthocyanins in cherries (*Prunus avium* L.). *Food Chem.* **2007**, *103*, 976–984. [CrossRef]
42. Girard, B.; Kopp, T.G. Physicochemical Characteristics of Selected Sweet *Cherry* Cultivars. *J. Agric. Food Chem.* **1998**, *46*, 471–476. [CrossRef] [PubMed]
43. Molaeafard, S.; Jamei, R.; Marjani, A.P. Co-pigmentation of anthocyanins extracted from sour *cherry* (*Prunus cerasus* L.) with some organic acids: Color intensity, thermal stability, and thermodynamic parameters. *Food Chem.* **2021**, *339*, 128070. [CrossRef]
44. Petriccione, M.; Mastrobuoni, F.; Pasquariello, M.S.; Zampella, L.; Nobis, E.; Capriolo, G.; Scortichini, M. Effect of Chitosan Coating on the Postharvest Quality and Antioxidant Enzyme System Response of Strawberry Fruit during Cold Storage. *Foods* **2015**, *4*, 501–523. [CrossRef]
45. Corneanu, M.; Iurea, E.; Sîrbu, S. Biological properties and fruit quality of sweet *cherry* (*Prunus avium* L.) cultivars from Rom. *Assortment. Agron. Res.* **2020**, *18*, 2353–2364. [CrossRef]
46. Corneanu, M.; Iurea, E.; Sirbu, S. Comparison of Five New Sweet *Cherry* Cultivars Bred in Romania, with their Parental Forms. *J. Hortic. Res.* **2021**, *29*, 1–8. [CrossRef]

47. Sîrbu, S.; Beceanu, D.; Corneanu, G.; Petre, L.; Anghel, R.M.; Iurea, E. Fruit quality evaluation of some sweet *cherry* cultivars in Iasi, Romania. In *Lucrări Științifice*; Universitatea de Științe Agricole Și Medicină Veterinară Ion Ionescu de la Brad: Iași, Romania, 2009; Volume 52, pp. 419–422.
48. Serradilla, M.J.; Martín, A.; Ruiz-Moyano, S.; Hernández, A.; López-Corrales, M.; De Guía Córdoba, M. Physicochemical and sensorial characterisation of four sweet *cherry* cultivars grown in Jerte Valley (Spain). *Food Chem.* **2012**, *133*, 1551–1559. [CrossRef]
49. González-Gómez, D.; Lozano, M.; Fernández-León, M.F.; Bernalte, M.J.; Ayuso, M.C.; Rodríguez, A.B. Sweet cherry phytochemicals: Identification and characterization by HPLC-DAD/ESI-MS in six sweet-*cherry* cultivars grown in Valle del Jerte (Spain). *J. Food Compos. Anal.* **2010**, *23*, 533–539. [CrossRef]
50. Wen, Y.-Q.; He, F.; Zhu, B.-Q.; Lan, Y.-B.; Pan, Q.-H.; Li, C.-Y.; Reeves, M.J.; Wang, J. Free and glycosidically bound aroma compounds in *cherry* (*Prunus avium* L.). *Food Chem.* **2014**, *152*, 29–36. [CrossRef] [PubMed]
51. Hallmann, E.; Rozpara, E. The estimation of bioactive compounds content in organic and conventional sweet *cherry* (*Prunus avium* L.). *J. Res. Appl. Agric. Eng.* **2017**, *62*, 141–145. Available online: https://bibliotekanauki.pl/articles/336048.pdf (accessed on 28 June 2024).
52. Ivanova, I.; Serdiuk, M.; Malkina, V.; Bandura, I.; Kovalenko, I.; Tymoshchuk, T.; Tonkha, O.; Tsyz, O.; Mushtruk, M.; Omelian, A. The study of soluble solids content accumulation dynamics under the influence of weather factors in the fruits of cherries. *Potravin. Slovak J. Food Sci.* **2021**, *15*, 350–359. [CrossRef]
53. Drake, S.R.; Fellman, J.K. Indicators of Maturity and Storage Quality of 'Rainier' Sweet *Cherry*. *HortScience* **1987**, *22*, 283–285. [CrossRef]
54. Siddiq, M.; Iezzoni, A.; Khan, A.; Breen, P.; Sebolt, A.M.; Dolan, K.D.; Ravi, R. Characterization of New Tart *Cherry* (*Prunus cerasus* L.): Selections Based on Fruit Quality, Total Anthocyanins, and Antioxidant Capacity. *Int. J. Food Prop.* **2011**, *14*, 471–480. [CrossRef]
55. Chaovanalikit, A.; Wrolstad, R.E. Total Anthocyanins and Total Phenolics of Fresh and Processed Cherries and Their Antioxidant Properties. *J. Food Sci.* **2004**, *69*, FCT67–FCT72. [CrossRef]
56. Blando, F.; Oomah, B.D. Sweet and sour cherries: Origin, distribution, nutritional composition and health benefits. *Trends Food Sci. Technol.* **2019**, *86*, 517–529. [CrossRef]
57. Usenik, V.; Fajt, N.; Mikulic-Petkovsek, M.; Slatnar, A.; Stampar, F.; Veberic, R. Sweet *Cherry* Pomological and Biochemical Characteristics Influenced by Rootstock. *J. Agric. Food Chem.* **2010**, *58*, 4928–4933. [CrossRef]
58. Kim, D.-O.; Heo, H.J.; Kim, Y.J.; Yang, H.S.; Lee, C.Y. Sweet and Sour *Cherry* Phenolics and Their Protective Effects on Neuronal Cells. *J. Agric. Food Chem.* **2005**, *53*, 9921–9927. [CrossRef]
59. Średnicka-Tober, D.; Ponder, A.; Hallmann, E.; Głowacka, A.; Rozpara, E. The Profile and Content of Polyphenols and Carotenoids in Local and Commercial Sweet *Cherry* Fruits (*Prunus avium* L.) and Their Antioxid. *Act. Vitro. Antioxid.* **2019**, *8*, 534. [CrossRef]
60. Ferretti, G.; Bacchetti, T.; Belleggia, A.; Neri, D. *Cherry* Antioxidants: From Farm to Table. *Molecules* **2010**, *15*, 6993–7005. [CrossRef] [PubMed]
61. Overbeck, V.; Schmitz, M.; Blanke, M. Non-Destructive Sensor-Based Prediction of Maturity and Optimum Harvest Date of Sweet *Cherry* Fruit. *Sensors* **2017**, *17*, 277. [CrossRef] [PubMed]
62. Abers, J.; Wrolstad, R. Causative factors of color deterioration in strawberry preserves during processing and storage. *J. Food Sci.* **1979**, *44*, 75–81. [CrossRef]
63. Liu, Y.; Liu, X.; Zhong, F.; Tian, R.; Zhang, K.; Zhang, X.; Li, T. Comparative Study of Phenolic Compounds and Antioxidant Activity in Different Species of Cherries. *J. Food Sci.* **2011**, *76*, C633–C638. [CrossRef] [PubMed]

horticulturae

MDPI

Article

Exploring the Benefits of Greek Tomato Landraces for Development of High Yield and Quality Hybrids in Organic Farming Systems

Rafail Tagiakas [1], Efthalia Vardaki [1], Evangelos Katsanoulas [1], Stavroula Kostoula [1], Athanasios Mavromatis [1], Ioannis Xynias [1] and Ilias D. Avdikos [2,3,*]

[1] Labaratory of Genetics and Plant Breeding, School of Agriculture, Aristotle University, 54124 Thessaloniki, Greece; tagiakasr@agro.auth.gr (R.T.); eevardaki@agro.auth.gr (E.V.); ekatsano@agro.auth.gr (E.K.); staur_kostoula@yahoo.gr (S.K.); amavromat@agro.auth.gr (A.M.); ioannis_xynias@hotmail.com (I.X.)

[2] Laboratory of Vegetable Crop Science, Department of Agriculture, International Hellenic University, Sindos, 57400 Thessaloniki, Greece

[3] Laboratory of Agrobiotechnology and Inspection of Agricultural Products, Department of Agriculture, International Hellenic University, Sindos, 57400 Thessaloniki, Greece

* Correspondence: avdikos.elias@ihu.gr

Abstract: Nowadays, there is a lack of high-performing genetic materials specifically developed to thrive under low-input conditions and meet the demands of organic agriculture. This study aims to evaluate seven improved Greek tomato landraces and their (F1) hybrids designed for low-input farming systems as a potential solution to the shortage of superior varieties for organic farming in specific environments. This evaluation focuses on yield components and fruit quality parameters. According to the results, experimental genetic material matches or exceeds the performance of the commercial (F1) hybrid Formula and outperforms the control variety cv. Macedonia (pure line), indicating that selected inbred lines (S.I.L.s) originated from landraces that developed under low-input conditions and are better suited to organic farming systems. For instance, the improved line (S.I.L.) from landrace Pantaroza pink has only a 13% lower total yield compared to (F1) commercial hybrid Formula, whereas the (F1) hybrid between Pantaroza pink × Kardia vodiou surpasses it, by 23%. In conclusion, tomato landraces are an important gene pool for seed development for organic or low-input farming systems, especially under the circumstances of climate change, presenting increasing challenges. Improving these local genetic materials using mild breeding approaches under low-input conditions leads to the production of new selected pure lines (S.I.L.s) along with their (F1) hybrids, which could ensure high-quality products with satisfactory yield and low requirements, offering an innovative pathway toward sustainable agriculture.

Keywords: climate change; hybrid; landraces; low inputs; organic farming; tomato; yield

1. Introduction

Solanum lycopersicum, commonly known as tomato, holds a prominent place in global agriculture, serving as a staple horticultural crop in many regions [1], a key ingredient in traditional Mediterranean culinary [2], and a vital source of nutrition due to its rich content of vitamins, minerals, proteins, and antioxidants [3,4]. Cultivation of tomato plants covers 5.4 million hectares with a global production of 192 million tons [5]. As a result, it is undeniable that tomato is one of the most economically important vegetable crops.

Cultivated tomato has been extensively modified throughout its evolutionary history and domestication process [6], resulting in low genetic diversity. Thus, tomato landraces are central to the success and future potential of tomato breeding programs. Today, landraces are known to constitute "dynamic populations" of cultivated plant species, with historical origin, local adaptation, distinct identity, genetic diversity, and lack of genetic improvement, and are still associated with landraces farming systems [7]. Landraces are known for their ability to thrive in environments with specific conditions and for their great genetic diversity, even within populations. This genetic diversity is invaluable for future improvement efforts, as well as for the development of new agricultural systems and the creation of new products [8].

Consumers genuinely value tomato landraces, which are highly esteemed for their exceptional organoleptic qualities [9]. People appreciate the distinctive traits of aroma and taste that are uniquely present in landraces, setting them apart from modern commercial varieties. Additionally, breeders often utilize landraces because they have been selected over many years under the harsh conditions of specific climates, such as the Mediterranean summer. Consequently, they may offer a valuable genetic pool for improving tomato crop tolerance to various biotic and abiotic stresses [10]. Moreover, landraces are ideal genetic material for environmentally friendly projects [11], as long as they have been evolved within sustainable traditional farming systems [12], and are likely to be well suited for low-input or organic production methods [13,14].

Heterosis, or hybrid vigor, is crucial for tomatoes, as it enhances yield, quality, disease resistance, and growth. It was first observed in 1876, when Charles Darwin, through his experiments, found that hybrids created via cross-pollination demonstrated superior traits compared to self-pollinated plants [15]. More precisely, heterosis is the phenomenon in which the progeny resulting from crosses between different cultivars or species exhibit heightened biomass, accelerated growth rate, and enhanced fertility surpassing either parent [16].

In contemporary times, tomato hybrids dominate the market in both fresh consumption and processed products. According to Bai and Lindhout [17], the first (F1) hybrid tomato cultivar was released in 1946. Since then, tomato hybrids began to spread and have come to command the trade market. Hybrid varieties are a solution for increasing tomato production [18], and many are resistant to major diseases [19]. Moreover, the popularity of certain commercial tomato hybrids is also attributed to the nutritional value of their fruit. It is also indisputable that in today's market, some of the fundamental priorities for fruits are a uniform appearance and a long shelf life [20].

A major problem facing breeders is that the intensification of agricultural practices in recent decades has relied heavily on high-input systems, including extensive use of fertilizers, pesticides, and irrigation. Over 95% of organic production depends on crop varieties bred for high-input systems, often lacking traits essential for success in organic and low-input farming [21]. However, numerous studies show that genotypes developed for optimal performance in high-input systems often fail to sustain equivalent yield levels under conditions of reduced inputs or environmental stress [22,23]. Thus, producing new S.I.L.s or F1 hybrids through breeding within low-input systems could yield new type of cultivars optimized for organic farming, offering an innovative pathway toward sustainable agriculture.

The aim of this study is to evaluate the performance of improved S.I.L.s under low-input farming systems originating from Greek tomato landraces as well as their F1 hybrid combinations, enhancing the distinct phenotypical traits, in order to find a possible solution to the emerging and escalating issue of the lack of superior varieties that are suitable specifically for organic farming. This research follows the principles of the European Green Deal [24], which emphasizes the transition to sustainable agricultural practices, including the development of high-yielding, high-quality, and distinctive organic varieties adapted to the challenges of low-input and organic systems. This study also aims to improve the resilience and sustainability of agricultural production by addressing this gap in organic tomato production, particularly in the face of increasing environmental and economic challenges. If the evaluation of the genetic varieties developed under low-input conditions in terms of yield, quality, and nutritional value confirms their superiority under similar cultivation conditions, it could point the way toward a more sustainable and resilient organic agriculture.

2. Materials and Methods

2.1. Genetic Material

The experiments took place during two experimental culture periods (2023–2024). Specifically, the Greek tomato S.I.L.s used into these experiments for cultivation and evaluation were selected from previous research, when the "Pure Line Selection" method was applied [25]. These S.I.L.s originated from landraces that initially constituted local populations, within which plant selections were made for plants that exhibited desirable yield and fruit quality characteristics. Applying this mild breeding approach for five consecutive selection cycles resulted in the pure lines (S.I.L.s) used in this experiment. Crosses among these pure lines (S.I.L.s) were also performed, and (F1) hybrid combinations were generated and involved in this study.

The genetic material of the first year of experimentation is presented in Table 1. It consists of 18 genotypes, of which 7 are improved landraces, 9 (F1) are hybrid combinations derived from diallel crosses, and 2 are controls, Macedonia and the commercial hybrid Formula F1. The landrace Macedonia is one of the best-known pure lines grown in organic farming systems in northern Greece and it has been improved and maintained by the Agricultural Research Center of Northern Greece (ARCNG). Regarding the second control of the experiment, Formula F1 is a commercial hybrid sold by the Golden West company and stands out for its high yield and uniformity in the field [26].

Table 1. Genetic material used in the first year of the experiment.

Experiment Code	Genetic Material
1	Formula F1 (Control I)
2	Macedonia (Control II)
3	Kardia vodiou
4	Aspros lotos
5	Pantaroza red
6	Pantaroza pink
7	Lemonati
8	Lotos
9	Karabola
10	Aspros lotos × Kardia vodiou F1
11	Pantaroza pink × Kardia vodiou F1
12	Pantaroza pink × Aspros lotos F1
13	Lemonati × Kardia vodiou F1
14	Lemonati × Aspros lotos F1
15	Lemonati × Pantaroza red F1
16	Macedonia × Lemonati F1
17	Lotos × Kardia vodiou F1
18	Lotos × Aspros lotos F1

After the evaluation carried out in the first year, the two hybrids that excelled in productivity and quality traits were selected and re-evaluated the following year alongside their parents (Table 2) (Figures 1 and 2). The controls used were the same as in the first experimental year. A total of 7 genetic materials were evaluated in the second year.

Table 2. Genetic material used in the second year of the experiment.

Experiment Code	Genetic Material
1	Formula F1 (Control I)
2	Macedonia (Control II)
3	Pantaroza pink
4	Kardia vodiou
5	Lemonati
6	Pantaroza pink × Kardia vodiou F1
7	Lemonati × Kardia vodiou F1

Figure 1. Genetic material used in the second year of the experiment.

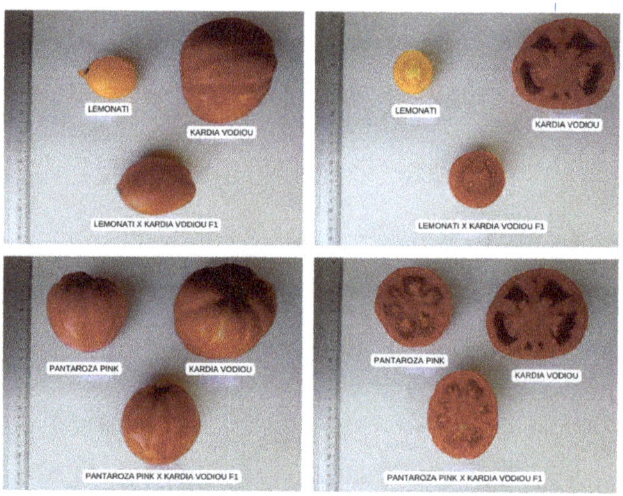

Figure 2. Morphology of tomato fruits of selected parents and their hybrids.

2.2. Experimental Design and Low-Input Cultivation System

The two experiments were carried out at the Faculty of Agriculture of the Aristotle University of Thessaloniki farm, in greenhouses maintained by the Laboratory of Genetics and Plant Breeding. In the first year, the experimental design was a randomized complete block design (RCBD) with three replicates and 15 plants per genotype to minimize the effect of environmental variability. The same design (RCBD with three replicates and 15 plants per genotype) was used in the following year. In both years, the planting distance was 50 cm between plants within rows and 1 m between rows.

During both years of experimentation, the low-input farming system was used, which is an environmentally friendly crop management system with lower costs for the producer and a lower environmental footprint for the environment compared to the conventional intensive farming system. In the present experiment, low-input conditions included the use of reduced amounts of fertilizers, water, and pesticide applications. The fertilizers used were exclusively organic (poultry manure). The pesticides used were those prescribed by organic sustainable crop management (*Bacillus thuringiensis, fatty acid potassium salt, pyrethrins*). More specifically, organic manure fertilization was applied at 3 t/ha (dry weight) and plant protection was based on chemical compounds that are compatible with the principles of organic agriculture, such as the use of copper hydroxide and sulfur. Irrigation was applied once a week during the spring months and twice a week during the summer months, using a drip irrigation system, providing a total of 7 L of water per plant per week. In conventional farming, irrigation frequency is at least four times per week, with a total water supply of 14 L per plant per week.

2.3. Observations, Measurements, and Traits Evaluation

During the two years of experimentation, measurements and observations were taken concerning the genetic material's yield potential and traits related to fruit quality.

More specifically, to evaluate the yield potential of the plants, the total fruit yield and the earliness of the fruits were studied. All measurements taken for this purpose were made at the individual plant level. Total yield was defined as the total weight of fruit produced by each individual plant in the experiment. Furthermore, the components of the yield were measured and recorded, such as the number of fruits produced by each plant and the average individual fruit weight.

During the first year of experimentation, four harvests were carried out. The yield was estimated from each and the total number of harvests, while the earliness was calculated, including the yield from the two first harvests. The four harvests were carried out at 90, 105, 115, and 150 days after transplanting the plants into the greenhouse. Similarly, the same harvesting schedule was followed in the second year of experimentation as in the first year.

To study the category of fruit quality, a sample of ripe fruits (the first and second fruit of the third inflorescence, which were at the red ripening stage) was harvested from each plant in the experiment. The measurements were related to the physicochemical characteristics of the fruit. The physicochemical characteristics analyzed in this study included the determination of pH and soluble solids (Brix). The pH determination methodology followed was homogenization of the fruit in a blender and then filtration of the solution using paper. About 10 mL of the filtrate was used to measure pH with a digital pH meter. Regarding the determination of soluble solids, a small amount of sauce (about 10 g) was placed on the glass plate of a refractometer (Atago PR-100, Minato-ku, Tokyo, Japan) to measure the soluble solid components.

2.4. Statistical Analysis

For statistical analysis of the results, variability was calculated by applying a fully randomized group design (ANOVA SPSS28) and the means were compared by Duncan's multiple range test ($\alpha = 0.05$). Regarding the measured traits, the vigor or depression of the genetic materials of the experiment compared to the "FORMULA F1" control was calculated. Furthermore, the standardized mean (x/s) was determined for behavioral stability, where (x) is defined as the mean and (s) is the standard deviation [27].

In addition, for the estimation of the differentiation, the heterosis was calculated for all the traits under study in the following ways:

1. Relative heterosis (%) = [(F1 − MP)/MP] × 100.

Heterosis in terms of the intergenic value, i.e., the behavior of the hybrid, is in relation to the average behavior of the two parents [28].

2. Heterobeltiosis (%) = [(F1 − BP)/BP] × 100.

Heterosis in terms of the best parent, i.e., the behavior of the hybrid, is in relation to the behavior of the best parent involved in the cross [28,29].

Heterosis towards the best (F1) hybrid/or commercial variety, i.e., the behavior of the hybrid, is in relation to the behavior of the best commercial variety, which is adapted to the crop area.

Here, F1 = hybrid behavior, MP = interparental value (P1 + P2)/2, and BP = best parent behavior.

3. Results

3.1. Evaluation of Yield Components—1st Year of Experimentation

Table 3 shows the averages of early production for the 18 genotypes evaluated in the first year of the experiment in terms of g/plant, number of fruits/plant, and g/fruit. Concerning the first characteristic (g/plant), it is observed that eight out of nine hybrids and four out of five landraces evaluated in this study showed superiority over "Formula F1". In particular, the Pantaroza pink × Kardia vodiou F1 hybrid stood out and had the best earliness, as it yielded an average of 3400.33 g/plant, which is statistically significantly different from all other genotypes and superior to the "Formula F1" control by 154%. The lowest early yield was obtained by the variety Pantaroza red, with 757.00 g, a value not statistically significantly different from the control Macedonia. The genetic materials with the highest levels of stability of performance for this trait were Aspros lotos × Kardia vodiou F1 and Pantaroza pink × Kardia vodiou F1 (8.04 and 7.86).

Table 3. Early fruit yield of the 1st year of experimentation (g/plant, number of fruits/plant, and g/fruit), where x: mean of measured traits, V/D: vigor/depression (% of commercial hybrid "Formula F1"), and x/s: stability of performance of genetic materials.

	Genetic Material	Early Fruit Yield								
		g/Plant			Number of Fruits/Plant			g/Fruit		
		x	V/D	x/s	x	V/D	x/s	x	V/D	x/s
1.	Formula F1 (Control I)	1339.00 defg *	-	1.70	9.50 def	-	2.97	137.60 bc	-	2.74
2.	Macedonia (Control II)	833.85 fg	62%	2.38	6.35 f	67%	2.67	134.00 bc	97%	6.39
3.	Kardia vodiou	1513.90 de	113%	3.09	7.33 ef	77%	3.35	206.13 a	150%	9.07
4.	Aspros lotos	1357.00 defg	101%	3.44	11.33 cdef	119%	3.43	122.37 bcd	89%	4.93
5.	Pantaroza red	757.00 g	57%	1.70	9.77 def	103%	1.67	79.83 ef	58%	3.22

Table 3. *Cont.*

Genetic Material		Early Fruit Yield								
		g/Plant			Number of Fruits/Plant			g/Fruit		
		x	V/D	x/s	x	V/D	x/s	x	V/D	x/s
6.	Pantaroza pink	2281.57 bc	170%	5.60	15.47 bcd	163%	3.17	155.03 b	113%	5.53
7.	Lemonati	1520.73 de	114%	2.94	27.77 a	292%	4.08	52.60 f	38%	9.85
8.	Lotos	1571.23 de	117%	5.01	11.20 cdef	118%	6.04	140.67 bc	102%	5.97
9.	Karabola	1465.33 def	109%	2.62	9.60 def	101%	2.77	152.07 b	111%	8.25
10.	Aspros lotos × Kardia vodiou F1	2571.90 b	192%	8.04	18.67 bc	196%	5.60	139.17 bc	101%	14.59
11.	Pantaroza pink × Kardia vodiou F1	3400.33 a	254%	7.86	18.03 bc	190%	4.12	202.03 a	147%	3.38
12.	Pantaroza pink × Aspros lotos F1	1902.10 cde	142%	2.17	16.23 bcd	171%	2.89	114.10 bcd	83%	4.72
13.	Lemonati × Kardia vodiou F1	1996.90 bcd	149%	5.36	23.13 ab	244%	2.48	101.57 cde	74%	2.42
14.	Lemonati × Aspros lotos F1	1305.30 efg	97%	1.91	17.33 bcd	182%	1.80	79.43 ef	58%	5.51
15.	Lemonati × Pantaroza red F1	1763.13 cde	132%	3.78	21.83 ab	230%	3.11	83.27 def	61%	7.40
16.	Macedonia × Lemonati F1	1421.67 def	106%	2.79	15.23 bcde	160%	2.21	102.30 cde	74%	3.24
17.	Lotos × Kardia vodiou F1	1749.43 cde	131%	2.87	13.67 cdef	144%	2.73	132.67 bc	96%	3.92
18.	Lotos × Aspros lotos F1	1847.77 cde	138%	4.02	11.77 cdef	124%	5.23	155.90 b	113%	8.22
	Average	1723.49	128%	-	14.94	158%	-	126.93	92%	-

* Varieties with the same letter within a column indicate no significant differences according to Duncan test (a = 0.05).

For the trait number of fruits per plant in early production, the variety Lemonati stood out, with an average of 27.77 fruits per plant, followed by its hybrids Lemonati × Kardia vodiou F1 and Lemonati × Pantaroza red F1, which showed no statistically significant differences. On the other hand, the lowest values were obtained by the variety Macedonia (6.35 fruits/plant) and landrace Kardia Vodiou (7.33 fruits/plant). The stability of performance (x/s) was also calculated, with values ranging from 1.67 (Pantaroza red) to 6.04 (Lotos).

The results for the average fruit weight (g) in early production are inversely proportional to those of the number of fruits per plant. Specifically, the varieties that stood out are Kardia vodiou, with an average fruit weight of 206.13 g, and the hybrid Pantaroza pink × Kardia vodiou F1, with 202.03 g. In the last positions came Lemonati, with 52.60 g, lagging behind the control "Formula F1" by 62%. Regarding performance stability, the highest value was shown by Aspros lotos × Kardia vodiou F1 (14.59), while the lowest value was demonstrated by Lemonati × Kardia vodiou F1 (2.42).

The evaluation of the total yield (g/plant) revealed that the most significant production was that of Pantaroza pink × Kardia vodiou (F1), with a total production of 4322.57 g and a 132% vigor as compared to Formula (F1), followed by the hybrids Lemonati × Kardia vodiou (F1) and Aspros lotos × Kardia vodiou (F1) (Table 4). Among the landraces, Pantaroza pink had a better yield of 2803.57 g. The stability of performance ranged from 2.21 to 7.53, with the lowest value corresponding to the Pantaroza pink × Aspros lotos F1 and the highest value to Pantaroza pink × Kardia vodiou F1.

Table 4. Total fruit yield of the 1st year of experimentation (g/plant, number of fruits/plant, and g/fruit), where x: mean of measured traits, V/D: vigor/depression (% of commercial hybrid "Formula F1"), and x/s: stability of performance of genetic materials.

	Genetic Material	Total Fruit Yield								
		g/Plant			Number of Fruits/Plant			g/Fruit		
		x	V/D	x/s	x	V/D	x/s	x	V/D	x/s
1.	Formula F1 (Control I)	1864.25 def *	-	3.15	18.75 def	-	3.94	97.60 defgh	-	6.53
2.	Macedonia (Control II)	1314.65 def	71%	3.01	12.50 f	67%	3.86	103.15 defgh	106%	6.59
3.	Kardia vodiou	1944.67 def	104%	4.05	11.50 f	61%	5.34	168.17 ab	172%	6.33
4.	Aspros lotos	1596.00 def	86%	3.77	14.90 ef	79%	3.64	109.27 cdefg	112%	5.31
5.	Pantaroza red	2522.40 bcde	135%	3.60	28.50 cde	152%	4.51	90.00 efgh	92%	4.15
6.	Pantaroza pink	2803.57 bcd	150%	4.18	20.23 def	108%	3.11	143.93 abc	147%	5.37
7.	Lemonati	2244.33 cdef	120%	2.65	55.67 a	297%	2.70	38.90 i	40%	5.74
8.	Lotos	2186.13 cdef	117%	5.29	19.00 def	101%	5.78	117.50 cdef	120%	5.25
9.	Karabola	2076.93 cdef	111%	3.74	17.00 def	91%	4.26	121.47 cdef	124%	14.36
10.	Aspros lotos × Kardia vodiou F1	3115.10 bc	167%	5.80	24.67 cdef	132%	4.55	128.37 cde	132%	8.34
11.	Pantaroza pink × Kardia vodiou F1	4322.57 a	232%	7.53	26.13 cdef	139%	4.53	173.37 a	178%	3.97
12.	Pantaroza pink × Aspros lotos F1	2834.67 bcd	152%	2.21	28.20 cde	150%	2.77	99.50 defgh	102%	5.36
13.	Lemonati × Kardia vodiou F1	3336.67 b	179%	4.74	43.43 ab	232%	2.49	89.87 efgh	92%	2.36
14.	Lemonati × Aspros lotos F1	2582.00 bcde	139%	4.42	38.97 bc	208%	2.91	70.43 hi	72%	4.28
15.	Lemonati × Pantaroza red F1	2845.90 bcd	153%	6.86	38.00 bc	203%	6.39	75.80 gh	78%	7.96
16.	Macedonia × Lemonati F1	2519.43 bcde	135%	3.09	30.87 bcd	165%	2.87	84.07 fgh	86%	3.58
17.	Lotos × Kardia vodiou F1	2504.13 bcde	134%	4.19	21.67 def	116%	3.93	118.73 cdef	122%	4.10
18.	Lotos × Aspros lotos F1	2389.77 bcde	128%	3.77	17.53 def	94%	5.21	135.60 bcd	139%	6.95
	Average	2535.20	137%	-	26.37	144%	-	109.55	111%	-

* Varieties with the same letter within a column indicate no significant differences according to Duncan test (a = 0.05).

In the number of fruits per plant for the total production trait, the top positions were occupied by the L.L. Lemonati, with 55.67, and its F1 hybrid Lemonati × Kardia vodiou, with 43.43. On the other hand, Kardia vodiou occupied last place, with an average number of fruits per plant of 11.50. In terms of stability of performance, values ranged from 2.49 (Lemonati × Aspros lotos (F1) to 6.39 (Lemonati × Pantaroza red (F1)).

Finally, the average fruit weight for total production was calculated, and the F1 hybrid Pantaroza pink × Kardia vodiou (173.37 g and 78% superiority over "Formula F1"), as well as its parental lines, Kardia vodiou (168.17 g) and Pantaroza pink (143.93 g), stood out. Last, again, was Lemonati, with 38.90 g, which was expected due to the small fruit size provided by this variety.

3.2. Evaluation of Quality Traits—1st Year of Experimentation

Regarding fruit quality (Table 5), total soluble solids (°Brix) were calculated, and the values ranged from 3.47 (Lemonati × Pantaroza red F1) to 5.03 (Macedonia). Eight out of eighteen genetic materials had equal levels of total soluble solids, ranging from 4.40 to 5.03 °Brix, surpassing the levels of the control Formula F1. Furthermore, the pH was measured; the values ranged from 4.11 (Pantaroza pink) to 4.37 (Lotos × Aspros lotos F1), with the highest measurement not being statistically significantly different to that of Formula F1.

Table 5. Average fruit values (x) for pH and total soluble solids (°Brix) of 1st year of experimentation, and vigor/depression (V/D—% of commercial hybrid "Formula F1") for genetic materials.

	Genetic Material	Total Soluble Solids (°Brix)		pH	
		x	V/D	x	V/D
1.	Formula F1 (Control I)	4.10 cdef *	-	4.31 abcd	-
2.	Macedonia (Control II)	5.03 a	123%	4.15 de	96%
3.	Kardia vodiou	4.23 bcde	103%	4.25 abcde	99%
4.	Aspros lotos	4.57 abc	111%	4.27 abcde	99%
5.	Pantaroza red	3.57 ef	87%	4.15 de	96%
6.	Pantaroza pink	3.77 def	92%	4.11 e	95%
7.	Lemonati	4.13 cdef	101%	4.18 cde	97%
8.	Lotos	4.53 abc	111%	4.23 abcde	98%
9.	Karabola	4.57 abc	111%	4.20 bcde	98%
10.	Aspros lotos × Kardia vodiou F1	4.63 abc	113%	4.33 abc	101%
11.	Pantaroza pink × Kardia vodiou F1	4.17 cdef	102%	4.36 ab	101%
12.	Pantaroza pink × Aspros lotos F1	4.27 bcde	104%	4.23 abcde	98%
13.	Lemonati × Kardia vodiou F1	4.40 abcd	107%	4.29 abcd	100%
14.	Lemonati × Aspros lotos F1	4.77 abc	116%	4.18 cde	97%
15.	Lemonati × Pantaroza red F1	3.47 f	85%	4.22 abcde	98%
16.	Macedonia × Lemonati F1	4.07 cdef	99%	4.32 abc	100%
17.	Lotos × Kardia vodiou F1	4.17 cdef	102%	4.27 abcde	99%
18.	Lotos × Aspros lotos F1	4.93 ab	120%	4.37 a	101%
	Average	4.30	105%	4.25	99%

* Varieties with the same letter within a column indicate no significant differences according to Duncan test (a = 0.05).

3.3. Evaluation of Yield Components—2nd Year of Experimentation

Table 6 shows the average early production of the seven genetic materials. Formula F1, yielding an average of 2772.00 g, and the Pantaroza pink × Kardia vodiou hybrid, yielding 2080.00 g, stood out among the materials. Therefore, the hybrids yielded higher early production than their respective parents. In terms of stability of performance (x/s), Formula F1 was the most stable material.

As for the trait of the number of fruits per plant for early production, Lemonati (18.27 fruits/plant and 39% higher than Formula F1), Lemonati × Kardia vodiou F1 (14.13 fruits/plant and 7% higher than Formula F1), and Formula F1 (13.18 fruits/plant) excelled compared to all the other genetic materials. In contrast, Kardia vodiou showed the lowest value (4.75 fruits/plant). The stability of performance (x/s) was calculated, revealing values between 1.24 (Pantaroza pink × Kardia vodiou F1) and 4.75 (Formula F1).

Table 6. Early fruit yield of 2nd year of experimentation (g/plant, number of fruits/plant, and g/fruit), where x: mean of measured traits, V/D: vigor/depression (% of commercial hybrid "Formula F1"), and x/s: stability of performance of genetic materials.

Genetic Material		Early Fruit Yield								
		g/Plant			Number of Fruits/Plant			g/Fruit		
		x	V/D	x/s	x	V/D	x/s	x	V/D	x/s
1.	Formula F1 (Control I)	2772.00 a *	-	5.07	13.17 ab	-	4.75	214.40 bc	-	7.14
2.	Macedonia (Control II)	1096.58 c	40%	3.50	4.62 c	35%	3.40	264.11 ab	123%	8.40
3.	Pantaroza pink	1521.40 bc	55%	1.76	6.67 c	51%	1.70	235.33 ab	110%	3.54
4.	Kardia vodiou	1455.33 bc	53%	2.30	4.75 c	36%	2.52	305.34 a	142%	5.38
5.	Lemonati	1017.93 c	37%	2.22	18.27 a	139%	2.45	54.50 d	25%	9.20
6.	Pantaroza pink × Kardia vodiou F1	2080.00 ab	75%	1.77	9.02 bc	68%	1.24	278.38 ab	130%	3.45
7.	Lemonati × Kardia vodiou F1	1830.20 bc	66%	2.85	14.13 ab	107%	2.31	140.96 c	66%	2.86
	Average	1711.19	71%	-	10.36	88%	-	210.75	98%	-

* Varieties with the same letter within a column indicate no significant differences according to Duncan test (a = 0.05).

Regarding the average fruit weight (g) of early production, Kardia vodiou stood out, with an average fruit weight of 305.34 g, which was expected as it is the largest fruiting variety. The hybrid Pantaroza pink × Kardia vodiou F1, along with Macedonia (control) and Pantaroza pink, showed no significant differences compared to Kardia vodiou in the rankings. In last place was Lemonati (54.50 g), which is the smallest of the varieties studied. With regard to stability of performance, the highest value in this measurement was obtained by Lemonati (9.20).

The assessment of the total yield, as presented in Table 7, indicated that following Formula F1 (4186.16 g), the highest yield was achieved from the hybrid Pantaroza pink × Kardia vodiou F1, which recorded a total yield of 3143.06 g, demonstrating no statistically significant difference. The subsequent rankings are as follows: Lemonati, Lemonati × Kardia vodiou F1, and Pantaroza pink, with Macedonia placed last. Furthermore, in terms of performance stability, the values ranged from 1.97 to 3.60. The lowest value corresponded to the genetic material Pantaroza pink × Kardia vodiou F1, whereas the highest value was associated with the control variety Macedonia.

Upon analyzing the total fruit yield per plant, the Lemonati variety, once again, demonstrated remarkable superiority, achieving an average of 61.53 fruits per plant, thereby exceeding the control Formula F1 by 195%. Conversely, the Kardia vodiou recorded the lowest performance, with an average of 8.47 fruits per plant. The stability of performance (x/s) ranged from 1.44 for Pantaroza pink × Kardia vodiou F1 to 3.98 for Formula F1.

In summary, with respect to the average fruit weight of total production, Kardia Vodiou (267.67 g), Pantaroza pink × Kardia vodiou F1 (251.67 g), Pantaroza pink (218.49 g), and the control variety Macedonia (201.82 g) surpassed all others. Furthermore, these genotypes demonstrated superior performance compared to Formula F1 by 34% and 26%, respectively. As expected, Lemonati presented the lowest average weight at 43.70 g per fruit. Regarding stability, Macedonia and Formula F1, along with Kardia vodiou, exhibited the highest values, ranging from 5.73 to 7.60, whereas the Lemonati × Kardia vodiou F1 genotype recorded the lowest value at 3.00.

Table 7. Total fruit yield of 2nd year of experimentation (g/plant, number of fruits/plant, and g/fruit), where x: mean of measured traits, V/D: vigor/depression (% of commercial hybrid "Formula F1"), and x/s: stability of performance of genetic materials.

	Genetic Material	Total Fruit Yield								
		g/Plant			Number of Fruits/Plant			g/Fruit		
		x	V/D	x/s	x	V/D	x/s	x	V/D	x/s
1.	Formula F1 (Control I)	4186.16 a *	-	3.26	20.83 bc	-	3.98	199.41 b	-	7.54
2.	Macedonia (Control II)	1730.50 c	41%	3.60	9.48 d	46%	3.50	201.82 ab	101%	7.60
3.	Pantaroza pink	2441.41 bc	58%	2.11	11.93 d	57%	1.67	218.49 ab	110%	3.22
4.	Kardia vodiou	2309.45 bc	55%	2.19	8.47 d	41%	2.96	267.67 a	134%	5.73
5.	Lemonati	2679.86 bc	64%	2.61	61.53 a	295%	2.90	43.70 d	22%	5.26
6.	Pantaroza pink × Kardia vodiou F1	3143.06 ab	75%	1.97	13.67 cd	66%	1.44	251.67 ab	126%	4.55
7.	Lemonati × Kardia vodiou F1	2634.14 bc	63%	3.09	22.00 b	106%	2.36	130.42 c	65%	3.00
	Average	2782.16 c	69%	-	21.71	86%	-	186.89	96%	-

* Varieties with the same letter within a column indicate no significant differences according to Duncan test (a = 0.05).

3.4. Evaluation of Quality Traits—2nd Year of Experimentation

Regarding the quality attributes, both Brix and pH levels were assessed (refer to Table 8). The Brix values varied from 3.97 for Lemonati to 5.34 for Macedonia, with Kardia vodiou recording a notable average that was not statistically significantly different to that of Macedonia. Concerning the average pH levels, all genotypes exhibited no statistically significant differences (refer to Table 8).

Table 8. Average fruit values (x) for pH and total soluble solids (°Brix) of 2nd year of experimentation, and vigor/depression (V/D—% of commercial hybrid "Formula F1") for genetic materials.

	Genetic Material	Total Soluble Solids (°Brix)		pH	
		x	V/D	x	V/D
1.	Formula F1 (Control I)	4.27 de *	-	4.30 a	-
2.	Macedonia (Control II)	5.34 a	125%	4.48 a	104%
3.	Pantaroza pink	4.43 cd	104%	4.27 a	99%
4.	Kardia vodiou	5.03 ab	118%	4.47 a	104%
5.	Lemonati	5.97 e	93%	4.37 a	102%
6.	Pantaroza pink × Kardia vodiou F1	4.73 bc	111%	4.40 a	102%
7.	Lemonati × Kardia vodiou F1	4.73 bc	111%	4.50 a	105%
	Average	4.64	110%	4.40 a	103%

* Varieties with the same letter within a column indicate no significant differences according to Duncan test (a = 0.05).

3.5. Estimation of Heterosis—1st Year of Experimentation

Heterosis is a phenomenon that significantly enhances crop yields, particularly in high-input agricultural systems. It is essential to evaluate the degrees of heterosis when utilizing landraces as parental lines, particularly within low-input organic cultivation systems. This assessment is crucial to ascertain the potential benefits of employing pure-line varieties (S.I.L.s) (when heterosis levels are minimal) or hybrids (when heterosis levels are pronounced) in such context systems. A potentially more beneficial form of heterosis for assessing the comparative advantages of utilizing landraces versus their hybrids in the production of organic products within analogous cultivation systems is heterobeltiosis. This concept quantifies the extent to which the (F1) hybrid surpasses its most superior parental line in performance. According to Table 9, regarding the early yield of genetic materials, the heterosis values exhibited both positive and negative outcomes. Notably, the hybrids Aspros lotos × Kardia vodiou F1 and Pantaroza pink × Kardia vodiou F1 showcased the highest heterosis values across all categories, with heterobeltiosis rates of 70% and 49%, respectively. In contrast, negative values were recorded for the hybrids Lemonati × Aspros lotos F1 and Pantaroza pink × Aspros lotos F1, which displayed heterobeltiosis rates of −14% and −17%, respectively.

Table 9. Analysis of heterosis for early fruit yield of 1st year of experimentation (g/plant, number of fruits/plant, and g/fruit).

	Genetic Material	Heterosis—Early Fruit Yield					
		g/Plant		Number of Fruits/Plant		g/Fruit	
		Relative Heterosis	Heterobeltiosis	Relative Heterosis	Heterobeltiosis	Relative Heterosis	Heterobeltiosis
1.	Aspros lotos × Kardia vodiou F1	79%	70%	100%	65%	−15%	−32%
2.	Pantaroza pink × Kardia vodiou F1	79%	49%	58%	17%	12%	−2%
3.	Pantaroza pink × Aspros lotos F1	5%	−17%	21%	5%	−18%	−26%
4.	Lemonati × Kardia vodiou F1	32%	31%	32%	−17%	−21%	−51%
5.	Lemonati × Aspros lotos F1	−9%	−14%	−11%	−38%	−9%	−35%
6.	Lemonati × Pantaroza red F1	55%	16%	16%	−21%	26%	4%
7.	Macedonia × Lemonati F1	21%	−7%	−11%	−45%	10%	−24%
8.	Lotos × Kardia vodiou F1	13%	11%	48%	22%	−23%	−36%
9.	Lotos × Aspros lotos F1	26%	18%	4%	4%	19%	11%

In the fruit number per plant of early yield, the hybrid Aspros lotos × Kardia vodiou F1 exhibited heterosis rates greater than 50% in all types of heterosis, with 100% relative heterosis and 65% heterobeltiosis. The hybrids Pantaroza pink × Kardia vodiou, Pantaroza pink × Aspros lotos, and Lotos × Aspros lotos also showed positive but low values in all types of heterosis.

Finally, in the fruit weight per plant trait, the values of relative heterosis ranged from −23% to 26%, and heterobeltiosis from −51% to 11%, with most of the hybrids showing negative values.

Concerning the heterosis of the total yield from the first year of experimentation, the materials generally showed positive values (Table 10). More specifically, the relative heterosis ranged from 19% to 82%, and the heterobeltiosis ranged from 1% to 60%. The hybrids Aspros lotos × Kardia vodiou F1, Pantaroza pink × Kardia vodiou F1, and Lemonati × Kardia vodiou F1 exhibited high values in heterobeltiosis, reaching 60%, 54%, and 49%, respectively.

Table 10. Analysis of heterosis for total fruit yield of 1st year of experimentation (g/plant, number of fruits/plant, and g/fruit).

Genetic Material	Heterosis—Total Fruit Yield					
	g/Plant		Number of Fruits/Plant		g/Fruit	
	Relative Heterosis	Heterobeltiosis	Relative Heterosis	Heterobeltiosis	Relative Heterosis	Heterobeltiosis
1. Aspros lotos × Kardia vodiou F1	76%	60%	87%	66%	−7%	−24%
2. Pantaroza pink × Kardia vodiou F1	82%	54%	65%	29%	11%	3%
3. Pantaroza pink × Aspros lotos F1	29%	1%	61%	39%	−21%	−31%
4. Lemonati × Kardia vodiou F1	59%	49%	29%	−22%	−13%	−47%
5. Lemonati × Aspros lotos F1	34%	15%	10%	−30%	−5%	−36%
6. Lemonati × Pantaroza red F1	19%	13%	−10%	−32%	18%	−16%
7. Macedonia × Lemonati F1	42%	12%	−9%	−45%	18%	−18%
8. Lotos × Kardia vodiou F1	21%	15%	42%	88%	−17%	−29%
9. Lotos × Aspros lotos F1	26%	9%	3%	−8%	20%	15%

Regarding the number of fruits per plant, which corresponds to the total production trait, heterobeltiosis exhibited positive values in the hybrids Pantaroza pink × Kardia vodiou F1 (29%), Pantaroza pink × Aspros lotos F1 (39%), Aspros lotos × Kardia vodiou F1 (66%), and Lotos x Kardia vodiou F1 (88%).

Concerning the trait of fruit weight per plant, the relative heterosis exhibited variations ranging from −21% to 20%, while the heterobeltiosis displayed a range from −47% to 15%. The hybrids Pantaroza pink × Kardia vodiou F1 and Lotos × Aspros lotos F1 demonstrated positive but low values in both categories of heterosis.

According to Table 11, the values of heterosis varied in minimal percentages, both positive and negative, with regard to fruit quality characteristics. The highest heterobeltiosis values were observed in the Lotos × Aspros lotos F1 hybrid (8%) and the Pantaroza pink × Kardia vodiou F1 hybrid (3%), pertaining to total soluble solids and pH, respectively.

Table 11. Analysis of heterosis for pH and total soluble solids (°Brix) of 1st year of experimentation.

Genetic Material	Heterosis—Quality Traits			
	Total Soluble Solids (°Brix)		pH	
	Relative Heterosis	Heterobeltiosis	Relative Heterosis	Heterobeltiosis
1. Aspros lotos × Kardia vodiou F1	5%	1%	2%	1%
2. Pantaroza pink × Kardia vodiou F1	4%	−1%	4%	3%
3. Pantaroza pink × Aspros lotos F1	2%	−7%	1%	−1%
4. Lemonati × Kardia vodiou F1	5%	4%	2%	1%
5. Lemonati × Aspros lotos F1	10%	4%	−1%	−2%

Table 11. Cont.

Genetic Material	Heterosis—Quality Traits			
	Total Soluble Solids (°Brix)		pH	
	Relative Heterosis	Heterobeltiosis	Relative Heterosis	Heterobeltiosis
6. Lemonati × Pantaroza red F1	−10%	−16%	1%	1%
7. Macedonia × Lemonati F1	−11%	−1%	4%	3%
8. Lotos × Kardia vodiou F1	−5%	−8%	1%	0%
9. Lotos × Aspros lotos F1	8%	8%	3%	2%

3.6. Estimation of Heterosis—2nd Year of Experimentation

During the second year of experimentation, the two hybrids selected on the basis of their performance and the heterosis levels from the initial year to continue the evaluation process demonstrated positive values in early yield for relative heterosis (40% and 48%) and heterobeltiosis (37% and 26%), indicating their superiority in comparison to their parental lines (Tables 12 and 13). However, regarding overall yield (Table 13), a positive value for heterobeltiosis was recorded only in the Pantaroza pink × Kardia vodiou F1 hybrid.

Table 12. Analysis of heterosis for early fruit yield of 2nd year of experimentation (g/plant, number of fruits/plant, and g/fruit).

Genetic Material	Heterosis—Early Fruit Yield					
	g/Plant		Number of Fruits/Plant		g/Fruit	
	Relative Heterosis	Heterobeltiosis	Relative Heterosis	Heterobeltiosis	Relative Heterosis	Heterobeltiosis
1. Pantaroza pink × Kardia vodiou F1	40%	37%	58%	35%	3%	−9%
2. Lemonati × Kardia vodiou F1	48%	26%	23%	−23%	−22%	−54%

Table 13. Analysis of heterosis for total fruit yield of 2nd year of experimentation (g/plant, number of fruits/plant, and g/fruit).

Genetic Material	Heterosis—Total Fruit Yield					
	g/Plant		Number of Fruits/Plant		g/Fruit	
	Relative heterosis	Heterobeltiosis	Relative heterosis	Heterobeltiosis	Relative heterosis	Heterobeltiosis
1. Pantaroza pink × Kardia vodiou F1	32%	29%	34%	15%	4%	−6%
2. Lemonati × Kardia vodiou F1	6%	−2%	−37%	−64%	−16%	−51%

Regarding the number of fruits per plant, the hybrid Pantaroza pink × Kardia vodiou F1 was the only one with positive values in early and total yield for heterobeltiosis, reaching 35% and 15%, respectively. Regarding the fruit weight trait, the values for the two types of heterosis for both early and total yield were similar and predominantly negative.

Concerning the values of the heterosis of the fruit quality traits (Table 14), it appears that the hybrid Lemonati × Kardia vodiou F1 recorded a high relative heterosis (35%) but exhibited negative heterobeltiosis. The values of fruit pH heterosis showed low values, ranging between −2% and 2%.

Table 14. Analysis of heterosis for pH and total soluble solids (°Brix) of 2nd year of experimentation.

	Genetic Material	Heterosis—Quality Traits			
		Total Soluble Solids (°Brix)		pH	
		Relative Heterosis	Heterobeltiosis	Relative Heterosis	Heterobeltiosis
1.	Pantaroza pink × Kardia vodiou F1	0%	7%	1%	−2%
2.	Lemonati × Kardia vodiou F1	35%	−6%	2%	1%

4. Discussion

The crisis within contemporary conventional agriculture constitutes a global concern, affecting both developed and developing nations. The existing agricultural system predominantly depends on financial capital and technology to attain elevated productivity and competitiveness, resulting in numerous economic, social, and environmental challenges [30]. Recently, conventional agriculture has been linked to a multitude of environmental challenges that pose significant threats to ecosystems, including soil degradation, water pollution, and a decline in biodiversity. In recent decades, there has been a global proliferation of (F1) hybrid crop varieties aimed at overcoming the productivity limitations inherent in traditional agricultural cultivars. However, these hybrids frequently lack the desirable organoleptic properties that consumers seek. The extensive adoption of (F1) hybrid varieties has resulted in a significant loss of genetic diversity as landraces have become increasingly obsolete [31].

In several European Union member states, the cultivation of fruits and vegetables represents a crucial segment of agricultural output. The value of tomato production within the European Union was recorded at EUR 7.8 billion in 2022, constituting nearly one-fifth (19.1%) of the overall value of fresh vegetable production. In Greece, almost one-third (31.7%) of the output value of the agricultural industry in 2022 came from fruit and fresh vegetables [32]. These issues can be seen in Greek agriculture, which is greatly impacted by horticultural crops. The tomato stands out among them, as it is the horticultural crop with the most extensive surface area, as well as the one that contributes the most to the production value of Greek horticultural crops. Specifically, in Greece, 6,298,000 tons of tomatoes were produced in 2022, making it the most significant horticultural crop in the country. Considering that Greeks are among the nations with the highest per capita tomato consumption and that Greece possesses remarkable biodiversity with numerous landraces, the demand for high-yielding genetic materials becomes evident. These materials are expected to deliver superior flavor, variability in characteristics such as color and shape, and, ultimately, high nutritional value.

In accordance with the provisions of the EU Biodiversity Strategy for 2030, it is mandated that at least 25% of the European Union's agricultural land be designated for organic farming by the year 2030, alongside a significant increase in organic aquaculture. To date, over 95% of organic production relies on crop varieties that have been bred for high-input systems, such as hybrids, which frequently exhibit a deficit of essential traits required for success in organic and low-input farming practices [21]. Nonetheless, many studies indicate that genotypes designed for peak performance in high-input systems frequently do not maintain similar yield levels when faced with lower inputs or environmental stress [33,34]. Thus, creating landraces or hybrids developed by them using organic practices may provide farmers with high-yielding genetic materials that offer excellent nutritional value for sustainable agriculture.

This study aims to assess the performance concerning yield components and quality traits of improved Greek tomato landraces and their hybrids in low-input farming systems. This assessment seeks to address the growing challenge of a shortage of superior varieties specifically designed for organic farming and suited to particular cultivation environments. In the first year of the experiment, nearly all genetic materials demonstrated higher early plant yield and early total fruit number per plant compared to the commercial F1 hybrid Formula (which was produced under and for conventional cultivation systems) and the Greek commercial variety-control Macedonia. The hybrid Pantaroza pink × Kardia vodiou F1 significantly outperformed in early yield, with 3400.33 g per plant, followed by the hybrid Aspros lotos × Kardia vodiou F1 and the improved landrace Pantaroza pink, which was also the parent of the best-performing hybrid in terms of earliness.

Concerning total production, it was again observed that the genetic materials pure lines–S.I.L.-improved landraces, or hybrids derived from them, outperformed both the commercial F1 hybrid and the local variety Macedonia. Similar results were presented by Terzopoulos et al. [14], where tomato landraces under low-input cultivation exhibited similar behavior to the commercial hybrid control. Notably, the commercial F1 hybrid has been one of the most popular tomato hybrids in Greece for several years [35]. The highest yields (up to 4322.57 g per plant) were achieved by the F1 hybrids Pantaroza pink × Kardia vodiou, Lemonati × Aspros lotos, and Aspros lotos × Kardia vodiou, exceeding the commercial hybrid's yield by up to 132%. At the same time, the improved landraces outperformed the commercial F1 hybrid by up to 50%, with yields 10% to 35% lower than those of the three aforementioned hybrids. The highest values for the fruit number per plant trait were recorded for Lemonati and the hybrids derived from it, while the highest fruit weight was associated with Kardia vodiou or its F1 hybrid derivatives. The strong performance of landraces under low-input systems is also noted in previous research on Greek traditional tomato cultivars [25], as well as in different plant species such as bread wheat [36]. Similarly, work on the Maroon rice landraces showed that they can outperform commercial varieties in low-input environments, achieving notably higher yield (as 2600 kg/ha) than the commercial rice cultivars (as 1205 kg/ha) in the same farmer-managed fields [37].

In the second year of the experiment, the hybrid Pantaroza pink × Kardia vodiou F1 was the only genetic material that matched the earliness and yield of the commercial F1 hybrid, significantly surpassing the yield of the control variety Macedonia. Regarding fruit quality traits, specifically fruit's soluble solids content, both hybrids and improved landraces exhibited levels comparable to those of Macedonia, and significantly higher than those of the commercial F1 hybrid Formula, in both experimental years. There are many previous studies showing that farming practices can affect quality traits of tomatoes [38], with organic methods leading to higher phenolic content but not total soluble solids (°Brix) [39]. Similar results concerning the quality of the fruit come from Southern Italy, where two tomato landraces were studied for their shelf-life, high-quality fruit, and productivity characteristics under water-deficit conditions [40].

Given that some genetic materials equaled or exceeded the performance characteristics of the commercial hybrid Formula F1 and significantly outperformed the variety Macedonia, it can be concluded that the development of genetic materials under organic cultivation systems favors their suitability for such environments. For instance, the improved landrace Pantaroza pink showed just a 13% lower total yield as average estimated for both years, compared to the commercial hybrid Formula F1, while the hybrid Pantaroza pink × Kardia vodiou F1 outperformed it by 23%.

Estimating heterosis, and especially heterobeltiosis, would help determine whether cultivating improved pure-line local varieties or their hybrids is more advantageous for such systems, given the cost and difficulty of hybrid development. Regarding fruit quality characteristics, heterobeltiosis levels show that hybrids do not justify their use in organic environments. In terms of earliness and total yield, the hybrids Aspros Lotos × Kardia vodiou, Pantaroza pink × Kardia vodiou, and Lemonati × Kardia vodiou exhibited heterobeltiosis ranging from 31% to 70%. In such cases, cultivating these materials as hybrids rather than pure-line landraces (parental lines) is preferable. However, other combinations do not justify F1 hybrid use. The results of our study are consistent with those of other researchers in the field, as recent studies have found significant heterobeltiosis for fruit yield per plant in tomato experiments [41–43].

In the second experimental year, the hybrid Pantaroza pink × Kardia vodiou F1 exhibited hybrid vigor of 32% and 29% in early and total production, respectively, while Lemonati × Kardia vodiou showed no heterobeltiosis in total yield. Based on the results of the two years, the hybrid Pantaroza pink × Kardia vodiou F1 is the only one recommended for cultivation in organic systems.

Despite these promising findings, certain limitations must be acknowledged. This study was conducted over two years and in a specific geographical and environmental context, which may limit the generalizability of the results to other regions with different climatic or soil conditions. Additionally, while this study evaluated key agronomic and quality traits, further research is needed to assess the long-term stability of these genotypes across multiple seasons and diverse organic farming environments. Furthermore, the economic feasibility of adopting such improved landraces or hybrids in large-scale organic production should be explored, including aspects related to seed availability, cost-effectiveness, and farmer acceptance.

Future research should also investigate the genetic mechanisms underlying the superior performance of certain landrace-derived hybrids, as well as their resilience to biotic and abiotic stresses commonly encountered in organic farming. A more comprehensive approach integrating breeding, agronomic practices, and economic assessments would further strengthen the case for utilizing these genetic materials in sustainable agricultural systems.

5. Conclusions

In conclusion, tomato landraces are perhaps the most crucial gene pool for developing new types of varieties intended for cultivation in organic systems, especially nowadays, as climate change exacerbates challenges. This study demonstrated that the two years of experimentation identified both selected inbred lines derived from landraces and hybrids developed from them, which excelled in productive potential and fruit quality characteristics under low-input cultivation conditions. Developing such improved, adapted varieties for sustainable organic systems, even hybrids originated from crosses among landraces when they demonstrate significant superiority, could offer new approaches for producing high-quality, distinct products with unique characteristics, such as diverse fruit shape, color, sensory, and diet properties. This can provide added value and adequately compensate farmers for cultivation in organic farming systems, offering an innovative pathway toward sustainable agriculture.

Author Contributions: Conceptualization, R.T., E.V., A.M. and I.D.A.; methodology, R.T., S.K., I.X., A.M. and I.D.A.; software, R.T., E.V., E.K. and I.D.A.; validation, R.T., I.X., A.M. and I.D.A.; formal analysis, R.T. and I.D.A.; investigation, R.T., E.V., E.K., S.K., I.X., A.M. and I.D.A.; resources, R.T., I.X., A.M. and I.D.A.; data curation, R.T., E.V., E.K. and I.D.A.; writing—original draft preparation, R.T., A.M. and I.D.A.; writing—review and editing, R.T., E.V., E.K., A.M. and I.D.A.; visualization, R.T., E.V. and I.D.A.; supervision, I.X., A.M. and I.D.A.; project administration, A.M. and I.D.A.; funding acquisition, A.M. and I.D.A. All authors have read and agreed to the published version of the manuscript.

Funding: EIP-AGRI: "Ideas, put into practice, with success"—EUROPEAN COMMISION GR06RDNP001: Rural development for Greece evaluation of reconstructed greek tomato hybrids in two farming systems for yield potential, fruit quality and nutritional value in perfecture of Hepirus.

Data Availability Statement: We do not want to share our research data.

Conflicts of Interest: The authors declare no conflicts of interest.

References

1. Thakur, R.; Mohanty, S.; Sethy, P.K.; Patro, N.; Sethy, P.; Acharya, A.A. Detection of Tomato Leaf Ailment Using Convolutional Neural Network Technique. In *Lecture Notes in Networks and Systems*; Marriwala, N., Tripathi, C., Jain, S., Kumar, D., Eds.; Springer: Singapore, 2023; Volume 588, pp. 193–202. [CrossRef]
2. Trombino, S.; Cassano, R.; Procopio, D.; Di Gioia, M.L.; Barone, E. Valorization of Tomato Waste as a Source of Carotenoids. *Molecules* **2021**, *26*, 5062. [CrossRef]
3. Ravi, M. History of Tomato (Poor Man's Apple). *IOSR J. Humanit. Soc. Sci. (IOSR-JHSS)* **2017**, *22*, 31–34.
4. Kumar, A.; Kumar, V.; Gull, A.; Nayik, G. Tomato (*Solanum lycopersicon*). In *Antioxidants in Vegetables and Nuts—Properties and Health Benefits*, 1st ed.; Nayik, G.A., Gull, A., Eds.; Springer Nature Singapore Pte Ltd.: Singapore, 2020; pp. 191–207. [CrossRef]
5. Food and Agriculture Organization of the United Nations (FAO). Plant Variety Database. 2023. Available online: https://food.ec.europa.eu/plants_en (accessed on 20 February 2025).
6. Blanca, J.; Montero-Pau, J.; Sauvage, C.; Bauchet, G.; Illa, E.; Diez, M.J.; Francis, D.; Causse, M.; van der Knapp, E.; Canizares, J. Genomic variation in tomato, from wild ancestors to contemporary breeding accessions. *BMC Genom.* **2015**, *16*, 257. [CrossRef] [PubMed]
7. Villa, T.C.C.; Maxted, N.; Scholten, M.; Ford-Lloyd, B. Defining and identifying crop landraces. *Plant Genet. Resour.* **2005**, *3*, 373–384. [CrossRef]
8. Marone, D.; Russo, M.A.; Mores, A.; Ficco, D.B.M.; Laidò, G.; Mastrangelo, A.M.; Borrelli, G.M. Importance of landraces in cereal breeding for stress tolerance. *Plants* **2021**, *10*, 1267. [CrossRef] [PubMed]
9. Felföldi, Z.; Ranga, F.; Roman, I.A.; Sestras, A.F.; Vodnar, D.C.; Prohens, J.; Sestras, R.E. Analysis of Physico-Chemical and Organoleptic Fruit Parameters Relevant for Tomato Quality. *Agronomy* **2022**, *12*, 1232. [CrossRef]
10. Fullana-Pericàs, M.; Conesa, M.; Douthe, C.; El Aou-ouad, H.; Ribas-Carbó, M.; Galmés, J. Tomato landraces as a source to minimize yield losses and improve fruit quality under water deficit conditions. *Agric. Water Manag.* **2019**, *223*, 1267. [CrossRef]
11. Puneeth, G.M.; Gowthami, R.; Katral, A.; Laxmisha, K.M.; Vasudeva, R.; Singh, G.P.; Archak, S. On-farm crop diversity, conservation, importance and value: A case study of landraces from Western Ghats of Karnataka, India. *Sci. Rep.* **2024**, *14*, 10712. [CrossRef]
12. Thanopoulos, R.; Negri, V.; Pinheiro de Carvalho, M.A.A.; Petrova, S.; Chatzigeorgiou, T.; Terzopoulos, P.; Ralli, P.; Suso, M.J.; Bebeli, P.J. Landrace legislation in the world: Status and perspectives with emphasis in EU system. *Genet. Resour. Crop Evol.* **2024**, *71*, 957–997. [CrossRef]
13. Mavromatis, A.; Arvanitoyannis, I.; Chatzitheodorou, V.; Khah, E.; Korkovelos, A.; Goulas, C. Landraces versus Commercial Common Bean Cultivars under Organic Growing Conditions: A Comparative Study Based on Agronomic Performance and Physicochemical Traits. *Eur. J. Hortic. Sci.* **2006**, *72*, 214–219.
14. Terzopoulos, P.J.; Walters, S.A.; Bebeli, P.J. Evaluation of Greek Tomato Landrace Populations for Heterogeneity of Horticultural Traits. *Eur. J. Hortic. Sci.* **2009**, *74*, 24–29.
15. Hochholdinger, F.; Baldauf, J.A. Heterosis in plants. *Curr. Biol.* **2018**, *28*, 1089–1092. [CrossRef]
16. Birchler, J.A.; Yao, H.; Chudalayandi, S.; Vaiman, D.; Veitia, R.A. Heterosis. *Plant Cell* **2010**, *22*, 2105–2112. [CrossRef]
17. Bai, Y.; Lindhout, P. Domestication and breeding of tomatoes: What have we gained and what can we gain in the future? *Ann. Bot.* **2007**, *100*, 1085–1094. [CrossRef] [PubMed]
18. Devkota, S.; Shrestha, S.L.; Dhakal, D.D.; Shakya, S.M.; Pandey, A. Evaluation of Tomato Hybrids for Yield Attributes under Khumaltar Condition. *J. Inst. Agric. Anim. Sci.* **2018**, *35*, 191–196. [CrossRef]
19. Saleem, M.Y.; Akhtar, P.; Iqbal, Q.; Asghar, M.; Shoaib, M. Development of tomato hybrids with multiple disease tolerance. *Pak. J. Bot.* **2016**, *48*, 771–778.
20. Sánchez, A.S.; Flores, P.; Hernández, V.; Sánchez, E.; Molina, E.; López, N.; Rodríguez-Burruezo, A.; Fenoll, J.; Hellín, P. Fruit Agronomic and Quality Traits of Tomato F1 Hybrids Derived from Landraces Varieties. *Horticulturae* **2024**, *10*, 440. [CrossRef]
21. van Bueren, E.L.; Jones, S.; Tamm, L.; Murphy, K.; Myers, J.; Leifert, C.; Messmer, M. The need to breed crop varieties suitable for organic farming, using wheat, tomato and broccoli as examples: A review. *NJAS Wagening. J. Life Sci.* **2010**, *58*, 193–205. [CrossRef]
22. Ceccarelli, S. Specific adaptation and breeding for marginal conditions. *Euphytica* **1994**, *77*, 205–219. [CrossRef]

23. Foulkes, M.J.; Sylvester-Bradley, R.; Scott, R.K. Evidence for differences between winter wheat cultivars in acquisition of soil mineral nitrogen and uptake and utilization of applied fertilizer nitrogen. *J. Agric. Sci.* **1998**, *130*, 29–44. [CrossRef]
24. Paull, J. Organic Agriculture in Europe: EU Sets Goal of Growing Organic Farmland from 10% to 25% by 2030. *Eur. J. Agric. Food Sci.* **2024**, *6*, 26–31. [CrossRef]
25. Tagiakas, R.; Avdikos, I.; Goula, A.; Koutis, K.; Nianiou-Obeidat, I.; Mavromatis, A. Characterization and evaluation of Greek tomato landraces for productivity and fruit quality traits related to sustainable low-input farming systems. *Front. Plant Sci.* **2022**, *13*, 994530. [CrossRef]
26. Avdikos, I.D.; Nteve, G.-M.; Apostolopoulou, A.; Tagiakas, R.; Mylonas, I.; Xynias, I.N.; Papathanasiou, F.; Kalaitzis, P.; Mavromatis, A.G. Analysis of Re-Heterosis for Yield and Fruit Quality in Restructured Hybrids, Generated from Crossings among Tomato Recombinant Lines. *Agronomy* **2021**, *11*, 822. [CrossRef]
27. Fasoula, V.; Fasoula, D. Principles underlying genetic improvement for high and stable crop yield potential. *Field Crops Res.* **2002**, *75*, 191–209. [CrossRef]
28. Walton, P.D. Heterosis in Spring Wheat 1. *Crop Sci.* **1971**, *11*, 422–424. [CrossRef]
29. Meredith, W.R., Jr.; Bridge, R.R. Heterosis and gene action in cotton, *Gossypium hirsutum* L. 1. *Crop Sci.* **1972**, *12*, 304–310. [CrossRef]
30. Rosset, P.M.; Altieri, M.A. Agroecology versus input substitution: A fundamental contradiction of sustainable agriculture. *Soc. Nat. Resour.* **1997**, *10*, 283–295. [CrossRef]
31. Brugarolas, M.; Martínez-Carrasco, L.; Martínez-Poveda, A.; Ruíz, J.J. A competitive strategy for vegetable products: Landraces varieties of tomato in the local market. *Span. J. Agric. Res.* **2009**, *7*, 294–304. [CrossRef]
32. Eurostat. The Fruit and Vegetable Sector in the EU—A Statistical Overview. Available online: https://ec.europa.eu/eurostat/statistics-explained/index.php?title=The_fruit_and_vegetable_sector_in_the_EU_-_a_statistical_overview (accessed on 20 January 2025).
33. Murphy, K.; Lammer, D.; Lyon, S.; Carter, B.; Jones, S.S. Breeding for organic and low-input farming systems: An evolutionary–participatory breeding method for inbred cereal grains. *Renew. Agric. Food Syst.* **2005**, *20*, 48–55. [CrossRef]
34. Carbonell, P.; Salinas, J.F.; Alonso, A.; Grau, A.; Cabrera, J.A.; García-Martínez, S.; Ruiz, J.J. Effect of low inputs and salinity on yield and quality—A 3 year study in virus-resistant tomato (*Solanum lycopersicum* L.) breeding lines and hybrids. *Sci. Hortic.* **2020**, *260*, 108889. [CrossRef]
35. Avdikos, I.; Tagiakas, R.; Tsouvaltzis, P.; Mylonas, I.; Xynias, I.; Mavromatis, A. Comparative Evaluation of Tomato Hybrids and Inbred Lines for Fruit Quality Traits. *Agronomy* **2021**, *11*, 609. [CrossRef]
36. Korpetis, E.; Ninou, E.; Mylonas, I.; Ouzounidou, G.; Xynias, I.N.; Mavromatis, A.G. Bread Wheat Landraces Adaptability to Low-Input Agriculture. *Plants* **2023**, *12*, 2561. [CrossRef] [PubMed]
37. Pinas, N.M.; Tjoe Awie, J.R.; Dongstra, R.E.; Maat, H.; Schranz, M.E.; van de Loosdrecht, M.S.; van Andel, T. Yield and growth duration of Maroon rice landraces measured in landraces settings. *Genet. Resour. Crop Evol.* **2024**, *72*, 2211–2226. [CrossRef]
38. Santangelo, E.; Bergonzoli, S.; Papetti, P.; Mazzucato, A.; Del Giudice, A.; Scarfone, A.; Beni, C. The Profile of Bioactive Compounds in Two Tomato Lines (*hp-2* and *Atv_aft*) When Grown under Low-Input Farming Conditions with or without Mulching and Intercropping. *Agriculture* **2023**, *13*, 280. [CrossRef]
39. Györe-Kis, G.; Deák, K.; Lugasi, A.; Csúr-Vargaa, A.; Helyes, L. Comparison of conventional and organic tomato yield from a three-year-term experiment. *Acta Aliment.* **2012**, *41*, 486–493. [CrossRef]
40. Tranchida-Lombardo, V.; Cigliano, R.A.; Anzar, I.; Landi, S.; Palombieri, S.; Colantuono, C.; Bostan, H.; Termolino, P.; Aversano, R.; Batelli, G.; et al. Whole-genome re-sequencing of two Italian tomato landraces reveals sequence variations in genes associated with stress tolerance, fruit quality and long shelf-life traits. *DNA Res.* **2018**, *25*, 149–160. [CrossRef] [PubMed]
41. Camposeco-Montejo, N.; Robledo-Torres, V.; Flores-Naveda, A. Estimation of Heterosis and Heterobeltiosis in Interpoblational Hybrids of Husk Tomato (Physalis ixocarpa Brot.). *Rev. Tecnol. Marcha* **2020**, *33*, 91–101. [CrossRef]
42. Sonagara, M.K.; Patel, J.N.; Acharya, R.R.; Sharma, D.D. Exploitation of heterobeltiosis in tomato (*Solanum lycopersicum* L.). *Int. J. Curr. Microbiol. Appl. Sci.* **2020**, *9*, 674–685. [CrossRef]
43. Mishra, A.; Nandi, A.; Thriveni, V.; Das, S.; Mohanty, I.C.; Pattanayak, S.K. Heterosis in tomato (*Solanum lycopersicum*) hybrids for growth, yield and quality traits. *Pharma Innov.* **2021**, *10*, 220–229.

horticulturae

MDPI

Article

The Application of Conventional and Organic Fertilizers During Wild Edible Species Cultivation: A Case Study of Purslane and Common Sowthistle

Efraimia Hajisolomou [1], Giannis Neofytou [1], Spyridon A. Petropoulos [2] and Nikolaos Tzortzakis [1,*]

[1] Department of Agricultural Sciences, Biotechnology and Food Science, Cyprus University of Technology, Limassol 3603, Cyprus

[2] Department of Agriculture, Crop Production and Rural Environment, University of Thessaly, 38446 Volos, Greece

* Correspondence: nikolaos.tzortzakis@cut.ac.cy

Abstract: The introduction of alternative crops, including wild edible and medicinal plants, in organic cultivation systems presents an attractive approach to producing healthy and high-quality products due to their content in beneficial compounds and increased nutritional value. The current study evaluated the impact of organic and conventional fertilization on the growth, quality, nutrient status and stress response of the two wild edible species, e.g., purslane (*Portulaca oleracea* L.) and common sowthistle (*Sonchus oleraceus* L.), under field conditions. The fertilization treatments included the following: a control (NoFert) treatment with no fertilizers added, base dressing with conventional fertilization (CoFert), base dressing with organic fertilization (OrFert), base dressing and side dressing with conventional fertilization (OrFert + S_{CoFert}) and base dressing and side dressing with organic fertilization (CoFert + S_{CoFert}). Organic fertilization was carried out using a commercial vinasse-based organic fertilizer. In both purslane and common sowthistle, the application of organic fertilizers provided comparable or even enhanced plant growth traits, macronutrient content (i.e., P and K for purslane, and N for sowthistle) and quality (i.e., total soluble solids) compared to the application of conventional fertilizers. On the other hand, conventional fertilization with supplementary fertilization positively influenced the plant growth of purslane (i.e., plant height and stems biomass), as well as its physiological parameters (i.e., chlorophylls content), total phenolics content and antioxidant capacity (i.e., DPPH and FRAP). Similarly, conventional fertilization led to increased total phenolics and antioxidants in common sowthistle, while variable effects were observed regarding plant physiology, stress response and antioxidant capacity indices. In conclusion, the use of organic fertilization in both purslane and common sowthistle exhibited a performance similar to that of conventional fertilization, although further optimization of fertilization regimes is needed to improve the quality of the edible products.

Keywords: organic farming; sustainable fertilization; *Portulaca oleracea*; *Sonchus oleraceus*; total phenolic compounds; antioxidant capacity; wild edible species

1. Introduction

Addressing the global demand for safe and healthy food in the context of a continuously growing population presents a critical challenge for modern agriculture [1,2]. This issue demands not only an increase in global agricultural production to protect food security, but also the adoption of practices that ensure food safety and nutritional quality [3,4]. The current strategy involves highly intensive crop production systems [5], which rely heavily on the application of excessive fertilization methods, predominantly utilizing inorganic fertilizers to enhance crop yields [6]. However, this approach is accompanied by a range of environmental and health implications, including the degradation of soils, loss of nutrients via leaching, increase in greenhouse gasses emission and eventually the pollution of water, air and soil [3,7].

Recently, there has been a significant increase in consumers' demand for products derived from organic cultivation practices [8,9]. Organic practices prohibit the use of inorganic fertilizers, and utilize eco-friendly fertilizers such as organic manure which are within the scope of sustainability and biodiversity and promote resources conservation [3,10,11]. In this context, wild plant species hold significant ecological value, particularly in the context of soil degradation, as they possess inherent natural adaptations to challenging soil and climate conditions [12]. Additionally, organic farming has the potential to produce healthier and higher-quality plant products, while also maintaining a reduced environmental footprint [5,6]. This is particularly relevant for wild edible (WEP) and medicinal (MP) plants, as their true value derives from their quality and beneficial compounds [3,13]. In this regard, the introduction of alternative crops such as WEPs and MPs to organic crop production aligns with the scope of this strategy, as these species are characterized by low requirements in terms of inputs (e.g., fertilizers, agrochemicals, irrigation) [14].

Portulaca oleracea (L.) commonly known as purslane, is an annual herbaceous plant, exhibiting prostate or erect growth. It is classified within the Portulacaceae family and demonstrates a worldwide distribution [15,16]. This wild edible plant serves as a common component of the Mediterranean diet, often used as a salad garnish, and is widely appreciated for its high nutritional value [17]. Purslane has also been utilized as a traditional medicine in various countries, due to its antiseptic, antioxidant, antibacterial, antidiabetic, anticancer and wound-healing properties [18,19]. It is characterized by a high content of omega-3 fatty acids, minerals, proteins, carbohydrates, tocopherols, ascorbic acid and carotenoids [17]. Recognized for its diverse traits, purslane is increasingly acknowledged as a "wonder crop", although it remains highly underutilized [19].

Sonchus oleraceus (L.), commonly known as common sowthistle, is a member of the genus *Sonchus*, which includes various annual, biennial and perennial herbaceous plants, typically found across Asia, Africa and Europe [20]. Although reported to be a damaging, invasive weed, common sowthistle has been a part of the human diet in numerous geographical areas, due to its therapeutic and nutritional properties [21,22]. Common sowthistle is also regarded as an important medicinal plant due to its high content of various phytochemical compounds such as sesquiterpene lactones, flavonoids, flavonols and phenols, as well as fatty acids, carotenoids and minerals [23,24]. Sowthistle plant extract has also been shown to have anti-diabetic, antioxidant and antibacterial activity [25,26], and has been utilized as a post-harvest preservative in fresh-cut vegetables [27], and as a bio-herbicide [28].

Given the potential of common sowthistle and purslane as potential candidates for commercial cropping, it is essential to ensure optimal yields and enhanced quality parameters, in the context of implementing sustainable agricultural practices. These practices will help to mitigate the environmental impact of the current agricultural sector, driven by a booming interest in certified organic products [29]. Therefore, the current study was conducted to evaluate the efficacy of a commercially available organic fertilizer, in comparison to conventional inorganic fertilization on the growth and quality features of purslane and common sowthistle plants grown under field conditions. Additionally, supplementary fertilization was administered in both scenarios (i.e., conventional and organic supplementary fertilization), to investigate the effect of the fertilization regime and determine the nutrients requirements of the tested wild edible species.

2. Materials and Methods

2.1. Plant Material, Soil Analysis and Experimental Conditions

Portulaca oleracea and *Sonchus oleraceus* seedlings were produced in the nursery of the Cyprus University of Technology. Young seedlings were transplanted to the field at the growth stage of 2–3 leaves and 4–5 cm in height (approximately 3 weeks after sowing) during the spring of 2024 in a commercial organic farm at Anogyra, Limassol,

Cyprus (34°44′16.53′ N; 32°44′40.86′ E, 427 m above sea level). The climate of this region is dry with an average midday temperature and air humidity during the summer months of approximately 30 °C and 54%, respectively. The soil had a clay–loam texture, with the majority of the sieved particles (~80%) being sized at <4.75 mm. Furthermore, the soil had 4.61% of organic matter; 60.48% of available $CaCO_3$; a pH of 7.66; and an EC of 1.10 mS cm^{-1}. Moreover, the minerals content was as follows; N of 2.459 g kg^{-1}, P of 0.057 g kg^{-1}, K of 0.364 g kg^{-1}, Ca of 6.304 g kg^{-1}, Mg of 0.604 g kg^{-1}, and Na of 0.119 g kg^{-1}.

2.2. Cultivation Procedures

The experiment field used was 190 m^2 (10 m × 19 m). Young seedlings were transplanted into soil plots and arranged in triple rows at distances of 0.33 m between rows and 0.33 m within the same row. The experiment was carried out in a completely randomized design, where the experimental farm was divided into five treatments: (i) no fertilizers (NoFert), (ii) conventional fertilization–base dressing (CoFert), (iii) organic fertilization–base dressing (OrFert), (iv) base and side dressing with conventional fertilization (CoFert + S_{CoFert}) and (v) base and side dressing with organic fertilization (OrFert + S_{OrFert}). Conventional fertilization was carried out using Fertiflow 20–20−20 (ICL Fertilizers, Tel Aviv-Yafo, Israel), which contains N (20%) in the form of ammonium nitrogen (2.7%) and urea (17.3%), P_2O_5 (20%) and K_2O (20%). Organic fertilization was carried out using Phenix (Hello Nature, Italy), a vinasse-based fertilizer which contains N (6.0%), P_2O_5 (8.0%), K_2O (15.0%), MgO (3.0%) and is characterized by a high organic matter content (50%). A total of 30 plants were cultivated per plot (3.3 m^2) and species, while two replicated plots per species were also used, resulting in 60 individual plants per treatment and per plant species. No pesticides were used in the present study.

The plants were irrigated every day with approx. 0.2 L $plant^{-1}$ for 10 min or according to the plant's needs (irrigated water electrical conductivity of 0.93 dS m^{-1}; pH of 8.1), depending on the environmental conditions. The base dressing of the conventional (20–20−20) and the organic fertilizer was conducted on 24 April 2024, and both fertilizers were broadcast before transplanting, and were lightly incorporated in the soil. For the CoFert + S_{CoFert} and OrFert + S_{OrFert} treatments, additional fertigation was applied on 30 April 2024, 15 May 2024 and 28 May 2024 (Table S1).

2.3. Plant Growth and Physiology

At the end of the experiment (21 June 2024), several measurements were conducted. Plants (six replicates) were harvested at 1 cm above soil; their upper fresh weight was weighed (g), dried at 42 °C to constant weight and the total dry matter content (%) was then calculated.

Leaf photochemistry features were also assessed; relative chlorophyll content values were obtained with an optical chlorophyll meter (SPAD-502, Minolta, Osaka, Japan). Additionally, fresh plant tissue (six replications/treatment; each replication was a pool of two plants tissue; 0.1 g) was used for chlorophyll extraction, with the addition of 7 mL methanol 100% as described by Richardson et al. [30]. The absorbance was then measured at 470, 653 and 666 nm (Multiskan GO, Thermo Fischer Scientific Oy, Vantaa, Finland). Photosynthetic leaf pigments, chlorophyll a (Chl a), chlorophyll b (Chl b), total chlorophyll (total Chl) and carotenoids content were then calculated (mg g^{-1} fresh weight).

2.4. Total Phenolics Content, Antioxidant Capacity, Flavonoids and Ascorbic Acid

To determine total phenolics and antioxidant capacity, plant tissue samples (six replications per treatment) were collected, weighed (approximately 0.9 g) and stored at −20 °C. An extraction procedure was followed by the addition of 10 mL of methanol 50% to each sample, and the mixture was homogenized (ULTRA TURRAK, IKA-Werke, Staufen, Germany). The samples were then incubated in an ultrasound water bath (WB-11, Witeg, Wertheim, Germany), and centrifuged at 4045× *g* and 4 °C for 15 min (Primo R Centrifuge,

Thermo Fisher Scientific Oy, Finland). Finally, the supernatant was collected and stored at −20 °C and measured within 12 h.

The antioxidant capacity of the methanol plant extracts was determined by using the assays of 2,2-diphenyl-1-picrylhydrazyl (DPPH) and ferric-reducing antioxidant power (FRAP), as previously described by Chrysargyris et al. [31], as well as the 2,2′-azino-bis (3-ethylbenzothiazoline-6-sulphonic acid) (ABTS•+) assay as described previously [32]. Multiple radical scavenging assays were used, to provide a more comprehensive and reliable assessment of the plant's antioxidant properties, particularly when different fertilization methods were applied. It helps to account for the complexity of plant chemistry and the influence of different agricultural practices; in fact, it is recommended to employ at least two distinct methods [33]. The three spectrophotometric methods selected assayed the free radical-scavenging activity (DPPH, ABTS•+), and the ability to reduce ferric ions (FRAP) [34].

For DPPH determination, the scavenging activity was measured by bleaching with a 0.3 M DPPH solution. Afterwards, 1 mL of the DPPH solution was mixed with 15 μL of the extract and 1985 μL methanol (50%), and incubated in the dark for 30 min. A control was prepared using 2 mL of buffer solution with 1 mL DPPH solution, a curve blank using 1 mL methanol (100%) and ethanol 100%. Blanks for each sample were prepared using 15 μL of plant extract, 1985 μL of 50% methanol and 1 mL of 100% methanol. Absorbance measurements were conducted at 517 nm using Multiskan GO (Thermo Fisher Scientific Oy, Vantaa, Finland). The FRAP assay was also conducted, wherein Fe^{3+} was reduced to Fe^{2+} in the presence of antioxidants in a redox-linked colorimetric reaction. A FRAP solution was prepared using 2.5 mL $CH_3COONa{\cdot}3H_2O$ (0.3 mol L^{-1}, pH 3.6), 0.25 mL TPTZ (Tripyridil-s-triazine, 10 mmol L^{-1}), 0.25 mL $FeCl_3$ (40 mmol L^{-1}) and 15 μL of the plant tissue extract. A blank was also prepared. The absorbance was then estimated at 734 nm. Finally, the ABTS•+ assay, which measures the capacity of antioxidants to scavenge the ABTS•+ radical cation, was also conducted through the addition of 3 mL of ABTS•+ solution which was added to 50 μL of the extracts. Absorbance was then measured at 734 nm. The results were expressed as the Trolox (6-Hydroxy-2,5,7,8-tetramethylchromane-2-carboxylic acid) equivalent (mg Trolox g^{-1} of fresh weight).

Total phenolic content was measured using the Folin–Ciocalteu method. The Folin–Ciocalteu reagent (125 μL) was mixed with 1375 μL water and 250 μL of the sample. The samples were stirred and were left for 5 min before adding 1250 μL Na_2CO_3 7%. Following incubation in the dark for 1 h, the samples were optometrically measured at 755 nm and the results were expressed as gallic acid equivalents (μmol GAE g^{-1} of fresh weight), as described previously by Tzortzakis et al. [35].

Total flavonoids were measured using the aluminum chloride colorimetric method. Firstly, 5% $NaNO_2$ (Merck, Darmstadt, Germany) and plant tissue extracts were incubated for 6 min. Afterwards, following a wait time of 5 min, 0.15 mL $AlCl_3$ (10%) solution was added. After an additional 5 min wait time, 0.5 mL NaOH (1 M) was added, and distilled water was used to achieve a final volume of 2.5 mL. After thorough mixing, the absorbance of the solution was measured at 510 nm (Multiskan GO, Thermo Fischer Scientific Oy, Finland). The results were expressed as rutin equivalents per gram of fresh weight (mg rutin g^{-1} FW).

Finally, the ascorbic acid (AA) determination required approximately 1 g of plant tissue, which was weighted and homogenized (ULTRA TURRAX, IKA-Werke, Staufen, Germany) with 10 mL 4% oxalic acid. Afterwards, the samples were centrifuged at 4500× *g* and 25 °C for 15 min (Primo R Centrifuge, Thermo Fisher Scientific Oy, Finland). Finally, 10 mL of the supernatant was titrated against a DCIP dye solution (2,6-dichloro-indophenol). The titrate of the dye solution was calculated using a standard ascorbic acid solution (1 mg mL^{-1}). The results were expressed as mg of AA per 100 g of fresh weight (mg AA 100 g^{-1} FW) [36].

2.5. Lipid Peroxidation and Hydrogen Peroxide Content

The lipid peroxidation in terms of malondialdehyde (MDA) was determined by using the thiobarbituric (TBA) acid reaction. Briefly, MDA content was determined by homogenizing the plant tissue (0.2 g) with 0.1% trichloroacetic acid (TCA) and the extract was centrifuged at 15,000× *g* for 10 min. The extract (0.5 mL) was then mixed with 1.5 mL of 0.5% TBA in 20% TCA, and was incubated at 95 °C for 25 min. Following cooling in an ice bath, the absorbance was determined at 532 nm. The results were expressed as micromole of H_2O_2 per gram of fresh weight (μmol g^{-1}). H_2O_2 was determined as previously described [37]. Plant tissue (0.2 g) was ground in ice cold 0.1% TCA, and centrifuged at 15,000× *g* for 15 min. The supernatant (0.5 mL) was collected and mixed with 0.5 mL potassium-phosphate buffer with a pH of 7, and 1 mL of 1 M potassium iodide. Afterwards, the absorbance was measured at 390 nm. The results were expressed as nmol of MDA g^{-1} of fresh weight (nmol g^{-1}).

2.6. Plant Tissue and Nutrient Analysis

At the end of the experiment, samples from the upper parts of the plant (leaves and stems) were used to determine the mineral content, in six replications per treatment (three pooled plants per replication). Plant tissue was dried to constant weight (at 42 °C), ash burned (Carbolite AAF 1100, GERO, Mayen, Germany) at 450 °C for 6 h and then acid digested (2 M HCl). The nutrient content was determined as follows: potassium (K) and sodium (Na) were determined photometrically (Flame photometer, Lasany Model 1832, Lasany International, Haryana, India); phosphorus (P) was determined spectrophotometrically (Multiskan GO, Thermo Fischer Scientific Oy, Finland); and nitrogen (N) was determined by the Kjeldahl method (BUCHI, Digest automat K-439 and Distillation Kjeldahl K-360, Flawil, Switzerland). Data were expressed in g kg^{-1} of dry weight.

2.7. Statistical Methods

To test the effect of fertilization regimes on plant growth, physiological parameters and chemical composition of purslane and common sowthistle plants, a one-way analysis of variance (ANOVA) was used, followed by a comparison of the means with the Duncan's multiple range test at $p < 0.05$ with the fertilization regime as a fixed factor and all the tested parameters as responses. All data were checked for normality and homoscedasticity using Shapiro–Wilk and Bartlett's test, respectively. Statistical analysis was performed using IBM SPSS version 22. Measurements were applied in four to six biological replications/treatment (each replication consisted of a pool of three individual measures/samples), depending on the parameters, as indicated in the respective Tables and Figures.

3. Results

3.1. Purslane and Sowthistle Plant Growth

The effect of the studied fertilization regimes on purslane and common sowthistle plant growth and physiology are presented in Table 1. In the case of purslane, plant height was significantly ($p < 0.05$) increased (15.6% and 17.3%, respectively) with base and side dressing in both OrFert and CoFert treatments (CoFert + S_{CoFert} and OrFert + S_{OrFert}) compared to the NoFert application (control). Purslane plants grown with the NoFert application produced the lowest leaf and stem fresh weights. Furthermore, the addition of a side dressing in the CoFert treatment (CoFert + S_{CoFert}) increased the fresh weight of stems, whereas the opposite result was observed in the respective treatment with organic fertilizer (OrFert + S_{OrFert}). In accordance with the abovementioned results, the NoFert application produced purslane plants with the lowest total fresh biomass weight, while the highest values were recorded for the OrFert and CoFert + S_{CoFert} treatments. Interestingly, the application of a side dressing decreased the total fresh biomass production of plants grown under the OrFert + S_{OrFert} regime. The lowest leaf dry matter content was observed for the OrFert application, while OrFert + S_{OrFert} increased the values of this parameter by

35.2%. Finally, no significant differences were observed in the case of the stems and total dry matter content.

Table 1. The effect of fertilization regimes on purslane height (cm), leaf, stems and total plant fresh weight (FW; g plant^{-1}), and dry matter content (DM; %), and common sowthistle plant height (cm), leaf number, total plant fresh weight (FW; g plant^{-1}), and dry matter content (DM; %). Values are means (±SE) of six replicates for each treatment.

Plant	Parameters	NoFert *	CoFert	OrFert	CoFert + S_{CoFert}	OrFert + S_{OrFert}
Purslane	**Plant height**	46.00 ± 2.08 c	49.17 ± 1.30 bc	46.17 ± 1.49 c	56.83 ± 2.06 a	54.17 ± 2.95 ab
	Leaves FW	72.85 ± 17.45 b	172.65 ± 14.39 a	168.47 ± 24.61 a	145.00 ± 18.92 a	123.98 ± 12.31 ab
	Stems FW	243.17 ± 49.00 c	422.70 ± 24.40 b	599.43 ± 58.36 a	606.47 ± 66.29 a	403.52 ± 45.95 b
	Plant FW	316.02 ± 47.30 c	595.35 ± 35.28 ab	767.90 ± 74.95 a	751.47 ± 82.91 a	527.50 ± 56.79 b
	Leaves DM	9.20 ± 0.91 ab	8.62 ± 0.13 abc	7.14 ± 0.65 c	7.60 ± 0.22 bc	9.65 ± 0.45 a
	Stems DM	7.11 ± 0.87	7.92 ± 0.57	6.86 ± 0.67	6.66 ± 0.16	8.03 ± 0.41
	Total DM	7.48 ± 0.71	8.09 ± 0.43	6.93 ± 0.65	6.85 ± 0.17	8.43 ± 0.39
Sowthistle	**Plant height**	23.67 ± 1.28 ab	20.75 ± 0.73 b	24.58 ± 1.57 a	21.00 ± 0.58 b	22.00 ± 0.76 ab
	Leaf number	17.83 ± 2.04 ab	16.83 ± 1.56 b	18.67 ± 1.41 ab	23.83 ± 2.77 a	19.17 ± 1.58 ab
	Plant FW	36.37 ± 2.58	35.08 ± 2.72	33.38 ± 4.51	46.82 ± 7.49	49.17 ± 12.91
	Plant DM	15.61 ± 0.75 b	15.43 ± 0.72 b	18.30 ± 0.60 a	16.43 ± 0.81 b	15.14 ± 0.21 b

* No fertilization (NoFert), conventional fertilization–base dressing (CoFert), organic fertilization–base dressing (OrFert), conventional fertilization with side dressing of CoFert (CoFert + S_{CoFert}) and organic fertilization with side dressing of OrFert (OrFert + S_{OrFert}); significant differences ($p < 0.05$) among the treatments are indicated by different letters in the same row.

In common sowthistle, the highest plant height was observed for plants treated with OrFert, while both CoFert and CoFert + S_{CoFert} decreased plant height compared to the abovementioned treatment. Furthermore, the highest and lowest numbers of leaves were observed for CoFert + S_{CoFert} and CoFert, respectively, showing an increase of 41.6% in the case of CoFert + S_{CoFert} treatment. Regarding the plant dry matter content, it was beneficially affected under the OrFert treatment, where an increase of up to 20.9% was recorded compared to the rest of the treatments, Finally, the fresh weight of plants remained unaffected by the examined fertilization regimes.

3.2. Purslane and Sowthistle Plant Physiology

Table 2 presents the effects of the examined fertilization regimes on selected purslane and common sowthistle physiological parameters. In the case of purslane, the application of CoFert + S_{CoFert} resulted in decreased chlorophyll fluorescence compared to the NoFert application, while, remarkably, the same treatment produced plants with the highest content of Chl a, Chl b, total chlorophylls and carotenoids. Furthermore, the content of chlorophylls (Chl a, Chl b, total chlorophylls) was the lowest with the application of CoFert, while the introduction of a side dressing to this treatment (CoFert + S_{CoFert}) increased these values. Finally, the carotenoids-to-total chlorophylls ratio was significantly increased with the application of the CoFert treatment, compared to the other applications. Both SPAD and the Chl a-to-Chl b ratio remained unaffected by the examined parameters.

In common sowthistle, the application of a side dressing in both types of fertilizer (e.g., CoFert + S_{CoFert} and OrFert + S_{OrFert}) resulted in a decreased chlorophyll content compared to the NoFert application (except for CoFert + S_{CoFert} where no significant differences were recorded), whereas the highest values were recorded for the NoFert and OrFert treatments. Moreover, CoFert + S_{CoFert} and OrFert + S_{OrFert} resulted in an increased ratio of Chl a to Chl b, while CoFert increased significantly SPAD values, especially when compared to the NoFert treatment. Finally, chlorophyll fluorescence (Fv Fm^{-1}), total carotenoids content and the ratio of total carotenoids to chlorophylls content were not affected by the studied fertilization regimes.

Table 2. The effect of different fertilization regimes on purslane and common sowthistle SPAD value, leaf chlorophyll fluorescence (Fv Fm^{-1}), chlorophyll content (Chl a, Chl b, total Chl; mg g^{-1} fresh weight), carotenoids content (Total Car, mg g^{-1} fresh weight), Chl a:Chl b and Car:Chl ratios. Values are means (±SE) of six replicates for each treatment.

Plant	Parameters	NoFert *	CoFert	OrFert	CoFert + S_{CoFert}	OrFert + S_{OrFert}
Purslane	SPAD	33.38 ± 0.47	34.85 ± 1.07	33.82 ±1.84	35.75 ± 1.46	35.02 ± 2.40
	Fv Fm^{-1}	0.71 ± 0.00 a	0.69 ± 0.02 ab	0.71 ± 0.01 ab	0.67 ± 0.01 b	0.69 ± 0.01 ab
	Chl a	0.34 ± 0.05 b	0.21 ± 0.02 c	0.38 ± 0.04 b	0.57 ± 0.05 a	0.28 ± 0.03 bc
	Chl b	0.12 ± 0.01 b	0.07 ± 0.00 c	0.14 ± 0.01 b	0.19 ± 0.02 a	0.10 ± 0.01 bc
	Total Chl	0.45 ± 0.06 b	0.29 ± 0.02 c	0.51 ± 0.05 b	0.76 ± 0.06 a	0.39 ± 0.04 bc
	Total Car	0.06 ± 0.01 b	0.05 ± 0.00 b	0.07 ± 0.01 b	0.11 ± 0.01 a	0.06 ± 0.01 b
	Chl a:Chl b	2.81 ± 0.11	2.85 ± 0.06	2.69 ± 0.13	2.93 ± 0.04	2.72 ± 0.05
	Car:Chl	0.14 ± 0.01 b	0.18 ± 0.01 a	0.14 ± 0.01 b	0.15 ± 0.00 b	0.16 ± 0.01 b
Sowthistle	SPAD	47.97 ± 2.82 b	54.72 ± 1.91 a	50.78 ± 0.47 ab	49.70 ± 2.83 ab	53.07 ± 0.88 ab
	Fv Fm^{-1}	0.76 ± 0.01	0.74 ± 0.01	0.75 ± 0.01	0.74 ± 0.01	0.76 ± 0.01
	Chl a	1.05 ± 0.02 a	1.00 ± 0.05 a	1.06 ± 0.01 a	0.96 ± 0.04 ab	0.88 ± 0.05 b
	Chl b	0.65 ± 0.05 a	0.56 ± 0.05 ab	0.63 ± 0.02 a	0.48 ± 0.03 b	0.46 ± 0.05 b
	Total Chl	1.70 ± 0.06 a	1.56 ± 0.09 ab	1.69 ± 0.03 a	1.45 ± 0.06 b	1.34 ± 0.10 b
	Total Car	0.12 ± 0.01	0.11 ± 0.01	0.11 ± 0.00	0.12 ± 0.01	0.11 ± 0.01
	Chl a:Chl b	1.65 ± 0.09 b	1.83 ± 0.08 ab	1.69 ± 0.04 b	2.02 ± 0.07 a	1.95 ± 0.08 a
	Car:Chl	0.072 ± 0.007	0.075 ± 0.007	0.065 ± 0.003	0.080 ± 0.006	0.082 ± 0.003

* No fertilization (NoFert), conventional fertilization–base dressing (CoFert), organic fertilization–base dressing (OrFert), conventional fertilization with side dressing of CoFert (CoFert + S_{CoFert}) and organic fertilization with side dressing of OrFert (OrFert + S_{OrFert}); significant differences ($p < 0.05$) among the treatments are indicated by different letters in the same row.

3.3. Purslane and Sowthistle Plant Macronutrient Content

Table 3 presents the macronutrients content of purslane and sowthistle plants grown under different fertilization regimes. In purslane, the N content in leaves and stems was the highest for the treatment of OrFert + S_{OrFert}, while the lowest one was recorded for the NoFert treatment. In the case of leaf P, K and Na content, the highest values were reported for the OrFert treatment, being significantly different from the rest of the treatments. Interestingly, the additional application of a side dressing with organic fertilizer (OrFert + S_{OrFert}) decreased the P, K and Na contents, compared to the base dressing of the same fertilizer (OrFert). The P content in leaves was the lowest with the application of the CoFert fertilization, while an increase in P content was observed with the combined application of base and side dressing (CoFert + S_{CoFert}). Both NoFert and CoFert treatments produced plants with the same content of P and K in stems, while the highest P and K contents were recorded for the CoFert + S_{CoFert} and OrFert + S_{OrFert} treatments, respectively. Finally, the application of OrFert significantly increased Na content in stems, compared to the rest of fertilization regimes.

In common sowthistle, the NoFert treatment significantly decreased (up to 22.0%) leaf N content, compared to the other treatments. Furthermore, P content was significantly influenced by the examined treatments, as the OrFert treatment produced plants with the lowest P content, although an increase of 5.2% was recorded for the OrFert + S_{OrFert} treatment. Moreover, the highest P content was recorded for the CoFert + S_{CoFert} treatment, presenting an increase of 24.6% compared to the treatment of base dressing with the same fertilizer, and an overall increase of up to 34.4% compared to the rest of the treatments. The highest and lowest K content in leaves were recorded for the CoFert + S_{CoFert} and OrFert treatments, respectively. Furthermore, the side dressing with both fertilizers increased the K content. Finally, CoFert produced plants with the lowest Na content, while the addition of side dressing increased the Na content, resulting in similar values to the respective OrFert treatment (OrFert + S_{OrFert}).

Table 3. The effect of different fertilization regimes on purslane and sowthistle macronutrients content (g kg^{-1}). Values are means (±SE) of six replicates for each treatment.

Plant	Parameters	NoFert *	CoFert	OrFert	CoFert + S_{CoFert}	OrFert + S_{OrFert}
Purslane	Leaves N	18.57 ± 0.64 c	23.94 ± 0.21 b	23.36 ± 0.19 b	24.17 ± 0.43 b	28.23 ± 0.72 a
	Leaves P	1.90 ± 0.06 d	1.48 ± 0.05 e	4.23 ± 0.06 a	3.09 ± 0.05 b	2.85 ± 0.02 c
	Leaves K	88.58 ± 0.84 c	94.44 ± 1.36 b	99.39 ± 0.90 a	93.57 ± 0.21 b	90.58 ± 0.90 c
	Leaves Na	5.01 ± 0.04 b	5.01 ± 0.07 b	5.30 ± 0.07 a	4.51 ± 0.02 c	4.53 ± 0.07 c
	Stems N	8.93 ± 0.44 e	15.47 ± 0.30 b	13.68 ± 0.70 c	10.97 ± 0.19 d	16.84 ± 0.32 a
	Stems P	0.59 ± 0.04 c	0.71 ± 0.02 c	1.15 ± 0.12 b	1.06 ± 0.01 b	1.39 ± 0.07 a
	Stems K	116.40 ± 4.68 c	116.04 ± 2.25 c	123.71 ± 3.34 bc	137.05 ± 1.13 a	127.77 ± 2.06 b
	Stems Na	1.92 ± 0.05 b	1.94 ± 0.05 b	2.53 ± 0.18 a	2.06 ± 0.02 b	2.18 ± 0.08 b
Sowthistle	Leaves N	24.31 ± 0.27 b	29.03 ± 0.19 a	29.66 ± 0.90 a	29.63 ± 0.47 a	29.20 ± 0.18 a
	Leaves P	2.09 ± 0.02 b	2.07 ± 0.04 b	1.92 ± 0.04 c	2.58 ± 0.02 a	2.02 ± 0.02 b
	Leaves K	63.01 ± 0.51 b	59.93 ± 0.91 c	52.26 ± 0.39 e	65.20 ± 0.97 a	56.15 ± 0.54 d
	Leaves Na	2.78 ± 0.01 b	2.32 ± 0.04 d	2.68 ± 0.03 c	3.58 ± 0.03 a	3.60 ± 0.04 a

* No fertilization (NoFert), conventional fertilization–base dressing (CoFert), organic fertilization–base dressing (OrFert), conventional fertilization with side dressing of CoFert (CoFert + S_{CoFert}) and organic fertilization with side dressing of OrFert (OrFert + S_{OrFert}); significant differences ($p < 0.05$) among the treatments are indicated by different letters in the same row.

3.4. Total Soluble Solids Content, Total Acidity and Sweetness Index in Purslane and Common Sowthistle Plants

Total soluble solids (TSSs), total acidity (TA), and sweetness index (TSSs/TA) were affected by the different fertilization regimes in both tested plants, e.g., purslane and common sowthistle (Table 4). In purslane, TSSs exhibited decreased values with the NoFert and CoFert treatments, whereas a contrasting trend was recorded for TA where NoFert and CoFert recorded the highest values. Finally, the OrFert + S_{OrFert} treatment recorded the lowest overall values for the sweetness index, compared to the rest of the treatments.

Table 4. The effect of different fertilization regimes on purslane and sowthistle total soluble solids (TSSs; %), total acidity (TA; % malic acid) and sweetness index (TSSs:TA ratio). Values are means (±SE) of six replicates for each treatment.

Plant	Parameters	NoFert *	CoFert	OrFert	CoFert + S_{CoFert}	OrFert + S_{OrFert}
Purslane	TSS	1.23 ± 0.04 b	1.23 ± 0.08 b	1.43 ± 0.11 a	1.53 ± 0.03 a	1.53 ± 0.02 a
	TA	3.17 ± 0.31 a	2.75 ± 0.03 a	2.10 ± 0.27 b	2.03 ± 0.14 b	1.34 ± 0.05 c
	TSS/TA	0.58 ± 0.04 a	0.51 ± 0.03 a	0.45 ± 0.08 a	0.46 ± 0.03 a	0.31 ± 0.01 b
Sowthistle	TSS	0.58 ± 0.05	0.58 ± 0.03	0.55 ± 0.03	0.50 ± 0.04	0.48 ± 0.03
	TA	2.62 ± 0.20 a	2.26 ± 0.19 ab	1.83 ± 0.10 c	1.99 ± 0.04 bc	1.21 ± 0.05 d
	TSS/TA	1.56 ± 0.19 b	1.83 ± 0.25 b	2.09 ± 0.29 ab	1.69 ± 0.14 b	2.71 ± 0.20 a

* No fertilization (NoFert), conventional fertilization–base dressing (CoFert), organic fertilization–base dressing (OrFert), conventional fertilization with side dressing of CoFert (CoFert + S_{CoFert}) and organic fertilization with side dressing of OrFert (OrFert + S_{OrFert}); significant differences ($p < 0.05$) among the treatments are indicated by different letters in the same row.

In common sowthistle, TSSs remained unaffected by the examined fertilization regimes, with values that ranged from 0.48 to 0.58%. Total acidity was the highest for the NoFert treatment, while the lowest value was observed for the OrFert + S_{OrFert} treatment, showing a decrease of 116.5% compared to the NoFert treatment. Finally, the sweetness index was the highest for the OrFert + S_{OrFert} treatment, being significantly different from the rest of the treatments (except for the OrFert treatment where no significant differences were recorded).

3.5. Total Phenolics Content, Antioxidant Capacity, Flavonoids Content and Ascorbic Acid Content in Purslane and Common Sowthistle Plants

The different fertilization regimes significantly affected the antioxidant capacity, as well as the content of total phenolics, flavonoids and ascorbic acid, for purslane and common sowthistle (Figures 1 and 2, respectively).

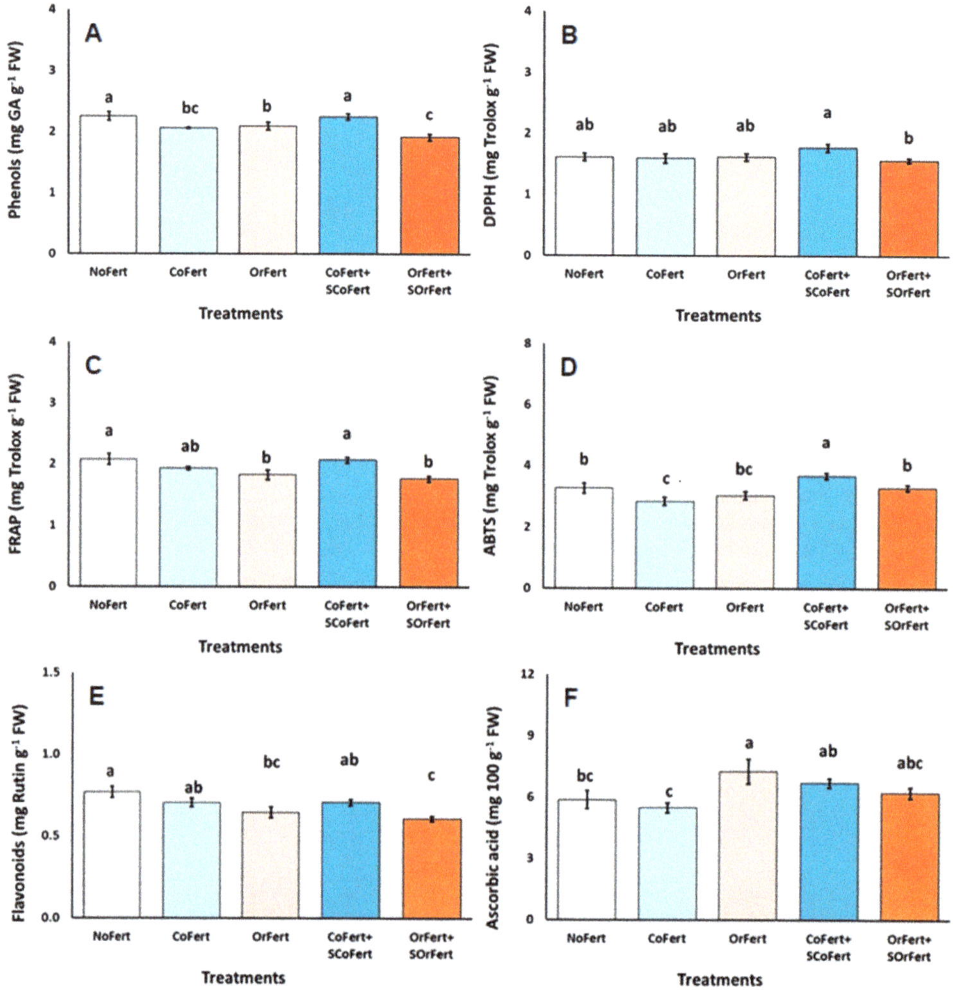

Figure 1. The effect of different fertilization regimes on purslane's (**A**) total phenols content (mg g^{-1} FW) and antioxidant capacity according to (**B**) DPPH, (**C**) FRAP, and (**D**) ABTS•+ (mg Trolox g^{-1} FW); (**E**) flavonoids content (mg Rutin g^{-1} FW) and (**F**) ascorbic acid content (mg 100 g^{-1} FW); no fertilization (NoFert), conventional fertilization–base dressing (CoFert), organic fertilization–base dressing (OrFert), conventional fertilization with side dressing of CoFert (CoFert + S_{CoFert}) and organic fertilization with side dressing of OrFert (OrFert + S_{OrFert}); significant differences ($p < 0.05$) among the treatments are indicated by different letters above the vertical bars. Values are means (±SE) of six replicates for each treatment.

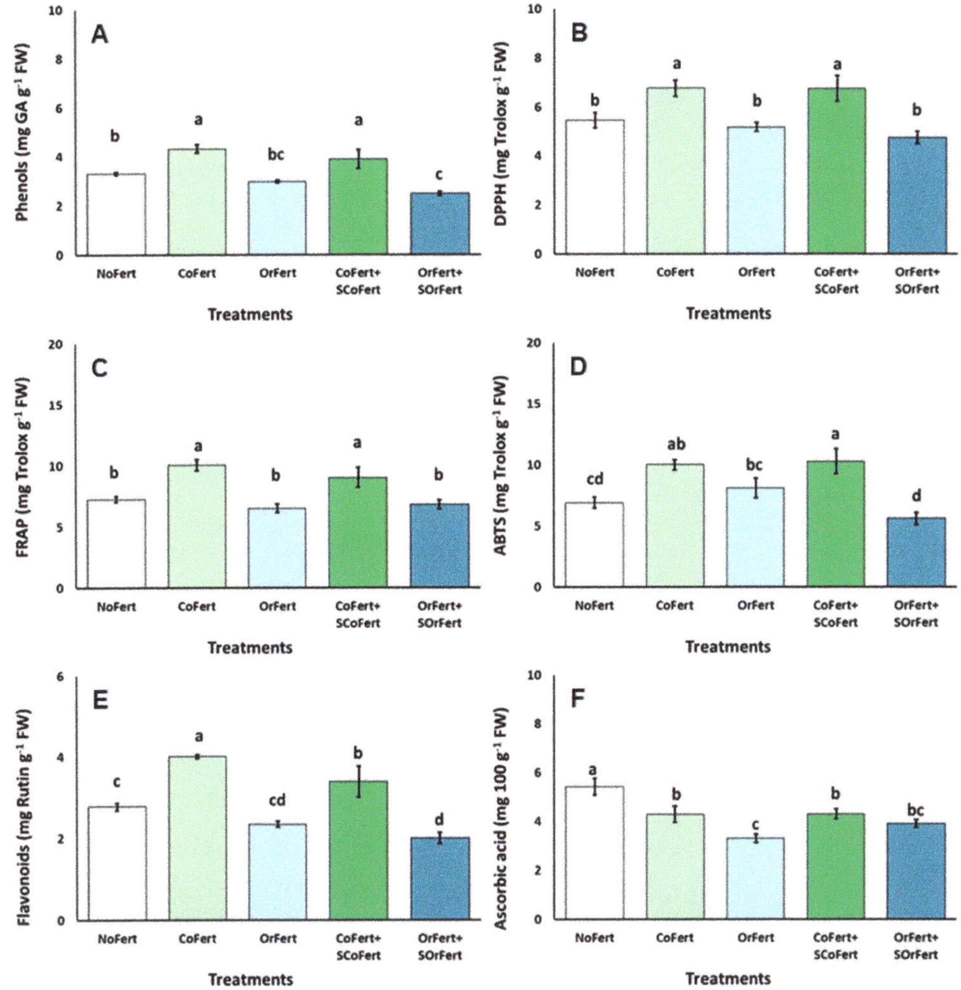

Figure 2. The effect of different fertilization regimes on common sowthistle's (**A**) total phenols (mg GA g^{-1} FW) and antioxidant capacity according to (**B**) DPPH, (**C**) FRAP, (**D**) ABTS•+ (mg Trolox g^{-1} FW); (**E**) flavonoids (mg Rutin g^{-1} FW) and (**F**) ascorbic acid content (mg 100 g^{-1} FW); no fertilization (NoFert), conventional fertilization–base dressing (CoFert), organic fertilization-base dressing (OrFert), conventional fertilization with side dressing of CoFert (CoFert + S_{CoFert}) and organic fertilization with side dressing of OrFert (OrFert + S_{OrFert}); significant differences ($p < 0.05$) among applications are indicated by different letters above the vertical bars. Values are means (±SE) of six replicates for each treatment.

In purslane, the highest total phenolics content was reported for the NoFert and CoFert + S_{CoFert} treatment, and the lowest for the OrFert + S_{OrFert} treatment, while the application of side dressing exhibited an increase compared to the base dressing with the conventional fertilizer (CoFert treatment), and a decrease for the organic fertilizer (OrFert treatment) (Figure 1A). Furthermore, the highest and lowest DPPH values were recorded for the CoFert + S_{CoFert} and OrFert + S_{OrFert} treatments, respectively (Figure 1B). Concerning FRAP, the highest values were found for the application of the NoFert and CoFert + S_{CoFert} treatments, whereas the application of organic fertilizer (both OrFert and OrFert + S_{OrFert})

resulted in decreased FRAP values (Figure 1C). ABTS•+ also exhibited varied response, as the lowest content was observed for the CoFert treatment, whereas the application of CoFert + S_{CoFert} presented the highest overall values for ABTS•+ (Figure 1D). The NoFert treatment presented the highest content of flavonoids, without being significantly different from the CoFert and CoFert + S_{CoFert} treatments (Figure 1E). In contrast, the OrFert + S_{OrFert} treatment presented the lowest flavonoids, without differing significantly from the OrFert treatment. Finally, the OrFert treatment resulted in the highest ascorbic acid content, which did not differ significantly from both CoFert + S_{CoFert} and OrFert + S_{OrFert} treatments, whereas CoFert presented the lowest content, without being different from the NoFert treatment (Figure 1F).

In common sowthistle, the total phenolics content was the highest for the CoFert and CoFert + S_{CoFert} treatments, whereas the lowest ones were recorded for the OrFert + S_{OrFert} treatment, without being significantly different from the OrFert treatment (Figure 2A). Similarly, the antioxidant capacity of common sowthistle plants, as assayed by DPPH (Figure 2B), FRAP (Figure 2C) and ABTS•+ (Figure 2D), increased with the application of CoFert and CoFert + S_{CoFert}, compared to the rest of the treatments, without being significantly different from the CoFert treatment, whereas the lowest antioxidant capacity values were recorded for the OrFert + S_{OrFert} treatment, being not significantly different from the NoFert treatment. Flavonoids recorded a notable increase with the application of the CoFert treatment, whereas the OrFert + S_{OrFert} treatment produced plants with the lowest flavonoids content (Figure 2E). Finally, the ascorbic acid content of plants was influenced by the examined treatments, as it significantly decreased when the fertilizer treatments were applied, especially in the case of OrFert where the lowest values were recorded (reduction up to 64.7%) (Figure 2F).

3.6. Lipid Peroxidation and Hydrogen Peroxide Content in Purslane and Common Sowthistle Plants

The effects of different fertilization regimes on the examined damage indicators (H_2O_2 and MDA) are presented in Figures 3 and 4, for purslane and common sowthistle plants, respectively. In purslane, the production of H_2O_2 in the leaf tissues was the highest for the NoFert treatment, whereas no significant differences were observed with the application of the OrFert treatment (Figure 3A). In contrast, plants treated with CoFert + S_{CoFert} and OrFert + S_{OrFert}, recorded the lowest H_2O_2 production. Finally, MDA was the lowest for the CoFert, whereas the highest values were observed for the OrFert + S_{OrFert} treatment, without being significantly different from either the NoFert or CoFert + S_{CoFert} treatments (Figure 3B).

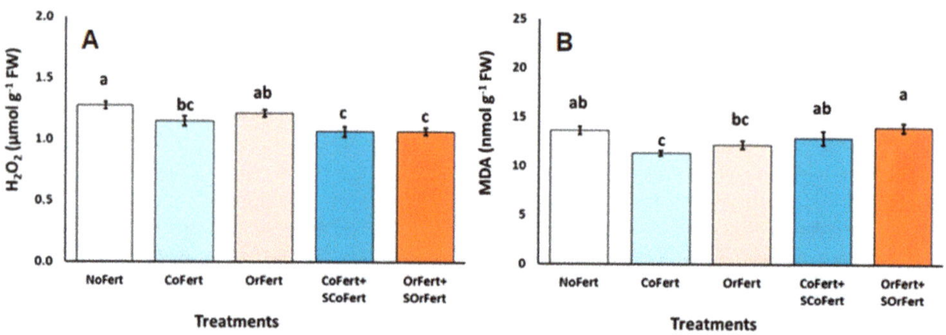

Figure 3. The effect of different fertilization regimes on (**A**) hydrogen peroxide–H_2O_2 (μmol g^{-1}) and (**B**) lipid peroxidation–MDA (nmol g^{-1}) of purslane plants. No fertilization (NoFert), conventional fertilization–base dressing (CoFert), organic fertilization–base dressing (OrFert), conventional fertilization with side dressing of CoFert (CoFert + S_{CoFert}) and organic fertilization with side dressing of OrFert (OrFert + S_{OrFert}); significant differences ($p < 0.05$) among applications are indicated by different letters above the vertical bars. Values are means (±SE) of six replicates for each treatment.

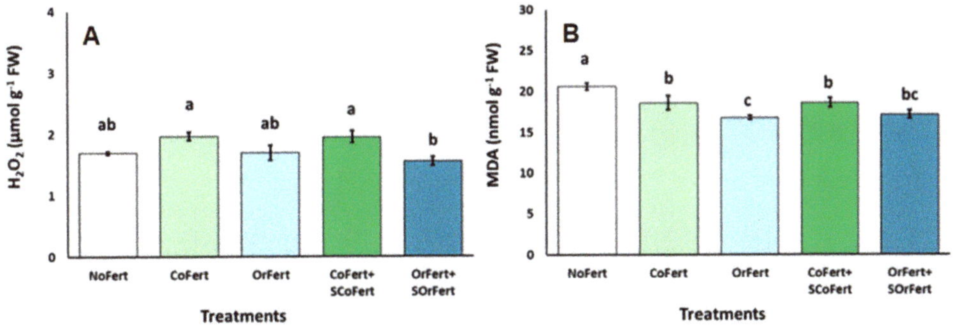

Figure 4. The effect of different fertilization regimes on (**A**) hydrogen peroxide–H_2O_2 (µmol g^{-1}) and (**B**) lipid peroxidation–MDA (nmol g^{-1}) of common sowthistle plants. No fertilization (NoFert), conventional fertilization–base dressing (CoFert), organic fertilization–base dressing (OrFert), conventional fertilization with side dressing of CoFert (CoFert + S_{CoFert}) and organic fertilization with side dressing of OrFert (OrFert + S_{OrFert}); significant differences ($p < 0.05$) among applications are indicated by different letters above the vertical bars. Values are means (±SE) of six replicates for each treatment.

In common sowthistle, the production of H_2O_2 was the highest for the CoFert and CoFert + S_{CoFert} treatments, whereas the OrFert + S_{OrFert} treatment recorded the lowest overall values (Figure 4A). Finally, MDA significantly decreased when plants were treated with fertilizers, regardless of the application regime, compared to the NoFert treatment, with no significant differences being observed between base dressing and the combined base and side dressing for both fertilizers (Figure 4B).

4. Discussion

Organic farming and the application of organic fertilizers falls within the context of sustainability and biodiversity conservation [3,10]. The organic cultivation of alternative crops such as WEPs and MPs is particularly relevant, as such species are adapted to biotic and abiotic stresses, and are characterized by low requirements in terms of inputs, which facilitates the application of organic farming practices [14,15]. In this regard, the current research was conducted to examine and promote the use of sustainable cultivation methods using as case study of two very promising WEPs and MPs, e.g., purslane and common sowthistle. Considering that fertilization has a crucial role in the growth and quality of crops and eventually in the crop yield, it is imperative to maintain an appropriate balance of inputs within the context of sustainable and organic agriculture [12,38].

Previous studies have highlighted the importance of fertilization on the growth and physiology of purslane and sowthistle plants. For example, Carrascosa et al. [12] indicated that the application of organic compost extracts, applied at rates based on N content in the nutrient solution (e.g., 300 mg L^{-1}), supported the growth of purslane in line with the use of inorganic N-P-K fertilization at 100-100-100 mg L^{-1}, in terms of leaf, stem and aerial biomass fresh and dry weights. A similar performance was also reported, in terms of total biomass production, in comparison to several inorganic fertilization regimes with fluctuating N-P-K rates, although varying results were obtained for different purslane genotypes. At the same time, increased N levels using inorganic fertilizers (e.g., 600 mg L^{-1}) resulted in a significantly greater biomass. This is also aligned with the findings of Nastou et al. [39], who demonstrated the positive impact of increased rates of N on the growth of different purslane genotypes. The current study reported similar results for purslane plants, as conventional and organic fertilization produced plants with equivalent growth traits (e.g., aerial biomass production), although in some cases, organic fertilization exceeded the performance of inorganic fertilization, and matched the performance of base and side dressing with conventional fertilizers. In the case of common sowthistle, Carrascosa et al. [12]

reported that the total aerial biomass of plants treated with organic fertilization was similar to various inorganic N-P-K applications (i.e., 300-100-100, 300-200-100, 300-300-100, 300-200-200 mg L^{-1}), although fertilization regimes with increased N (600 mg L^{-1}) and K (300 mg L^{-1}) recorded increased plant biomass. In addition, the application of alternative fertilization (insect frass) at three rates (0.5, 1.0, and 2.0% w/w) produced common sowthistle plants with a similar fresh weight to those that received calcium ammonium nitrate (applied at 100 kg N ha^{-1}) [40]. Moreover, previous reports suggest that vinasse-based fertilizers exhibited a better supplementation of N in the short term, compared to frass fertilizers [41]. In the current study, common sowthistle showed no significant differences in fresh weight production among the examined treatments, while the application of OrFert produced plants with increased height and dry matter content. In the studies of both Karkanis et al. [40] and Cardarelli et al. [41], the use of supplementary organic fertilization in the form of a side dressing did not yield remarkable increase, as its application (i.e., fertilizer side broadcasting) in a pelleted form may not have provided sufficient nutrients within the short duration of the experiments. Moreover, Rodrigues et al. [42] suggested that composted and pelleted amendments have a low N content, and the application of organic fertilization (Phenix) provided a portion of N early in the cultivation period, while the rest of the N remained available for a longer period. This is applicable to various organic amendments, as these materials release nutrients gradually, and therefore plants assimilate them more effectively due to decreased leaching [43].

Fertilization is fundamental in the synthesis of chlorophylls, which directly affects the rate of photosynthesis [44]. In the study by Chrysargyris et al. [5], the application of conventional fertilizers resulted in *Mentha spicata* (L.) plants with higher SPAD values, compared to the application of organic fertilization. Similarly, in the study of Karkanis et al. [40] it was observed that common sowthistle plants that received conventional fertilizers exhibited higher relative chlorophyll content (SPAD) values, compared to ones that received various rates of organic insect frass, which was attributed to the increased rates of N uptake by common sowthistle plants in the former case. However, in the current study both purslane and common sowthistle plants that received organic fertilization exhibited similar SPAD and chlorophylls to conventional fertilization. This could be attributed to the adequate N provided by the vinasse-based organic fertilizer within the short duration of the cultivation period, which appears to not have affected N availability, a finding which is in agreement with previous reports [41,42]. Similar results were obtained for the use of organic fertilization in the cultivation of *Pinellia ternata* (Thunb.), as SPAD values remained at similar levels compared to conventional fertilization [45]. The same researchers also reported that the increase in organic fertilizer rates improved SPAD values. The nitrogen delivered by the organic farming method is primarily organic and not completely accessible to the plants, which may explain poor plant development, a reduction in chlorophyll content and reduced photosynthetic activity [46]. Furthermore, Fontana et al. [47] showed that besides total nitrogen availability, the nitrogen form may alter the chlorophyll levels in purslane plants grown in hydroponics.

Organic farming utilizes organic fertilizers and amendments that enhance the soil's organic matter content, whilst facilitating a gradual release of nutrients, thereby improving their uptake and accumulation [48]. In the current study, the macronutrient content of purslane and common sowthistle plants was significantly influenced by the different fertilizer regimes tested. Previous studies have shown that vinasse-based organic fertilizers can support plant nutrition, especially during the initial growth and in short cultivation cycles, due to the fast mineralization and release of minerals in the post application period [41]. In the case of common sowthistle, various reports suggest that the application of organic fertilization affected the uptake of nutrients, compared to the application of conventional fertilization. Specifically, Karkanis et al. [40] reported that the use of yellow mealworm frass increased the contents of P over the application of calcium ammonium nitrate. However, in the study by Carrascosa et al. [12], decreased P contents were observed in the leaves of common sowthistle plants, compared to the application of inorganic N-P-K fertilization.

Similarly, in the current study, the common sowthistle leaves had decreased levels of P with the application of organic fertilizers, while side dressing slightly increased the P levels. Additionally, both studies of Karkanis et al. [40] and Carrascosa et al. [12] reported that macronutrients such as N, Mg, and Ca remained unaffected by the application of organic fertilizers. Similarly, the organic fertilization applied in the current study produced common sowthistle plants with a high N content in their leaves, similar to the application of conventional fertilization. Purslane was also found to have comparable N levels in both organic and conventional fertilization, while the use of a side dressing improved the uptake of N, P and K, as purslane leaves and stems had increased macronutrient levels compared to the rest of the treatments.

In terms of quality, purslane exhibited increased TSSs and sweetness index values for the application of organic fertilization, compared to conventional fertilization, and matched the results obtained with the use of side dressing with the same fertilizer. Similar findings were reported in previous studies with wild, medicinal plants. For example, Bakhtiari et al. [49] reported that the use of organic (vermicompost) fertilization, with or without additional NPK, produced *Satureja macrantha* C.A.Mey. plants with higher TSSs contents compared to conventional fertilization.

Both purslane and common sowthistle plants are valued for their diverse array of phytochemical constituents, as well as for their high content of polyphenols, which have been shown to exhibit various antioxidant, anti-inflammatory and antimicrobial properties [19,50]. When examining radical scavenging activity, a single assay may offer a limited understanding of the overall antioxidant properties of plant material. This is particularly relevant for medicinal and wild edible plants, as their complex chemical composition may lead to inconsistent results, depending on the test employed. Furthermore, the proper assessment of the antioxidant capacity from these species is crucial to promote the application of functional foods, pharmaceuticals and food additives [34]. Previous studies have reported the significant effect of organic fertilization applications on the antioxidant capacity, flavonoids and total phenolics [5,51]. In the case of purslane, a previous report suggested that the application of organic materials (waste fermentation residue after evaporation) did not affect the plants' total phenols content. Similarly, in the current study, purslane showed no significant differences in terms of total phenols, antioxidant capacity (DPPH, FRAP, ABTS•+) and flavonoids, when organic fertilization was applied compared to conventional fertilization. Similar results were obtained by Chrysargyris et al. [5], who examined the antioxidant capacity and total phenols content of *M. spicata* (L.) plants grown in conventional and organic cultivation systems. Lv et al. [52] also reported similar levels of total phenols in peppermint (*M. piperita* L.) and cinnamon (*Cinnamomum verum* L.), but the individual phenolic molecules were changed when plants grown under organic and conventional cultivation practices. In addition, in coriander (*Coriandrum sativum* L.), Serri et al. [51] reported that total phenolics and flavonoids content depended on the type of fertilizer applied, and the application of vermicompost and biophosphate produced plants with similar total phenolics and flavonoids contents compared to conventional NPK fertilization. This corroborates the notion that the fertilization applied in the current experiment adequately provided plants with balanced nutrition during the short period of cultivation. Moreover, the results indicate that the antioxidant capacity of purslane is an intrinsic property of the species, as reported previously in response to organic fertilization [53]. However, contrasting results were observed in the case of common sowthistle, as total phenolics, antioxidant capacity and flavonoids were decreased with the application of organic fertilization compared to conventional fertilization. In contrast, numerous studies have demonstrated that organic fertilizers may enhance the content of phenolics, and the antioxidant capacity in various plant species [54,55]. Interestingly, a decrease in the total phenolics, antioxidant capacity (DPPH, FRAP, ABTS•+) and total phenolics of both purslane and common sowthistle were observed with the application of OrFert + S_{OrFert}, compared to CoFert + S_{OrFert}. Similarly, in the organic cultivation of barley (*Hordeum vulgare* L.) a decrease in phenolics and flavonoids was observed for well-fertilized plants with organic fertilization [56].

Research investigating the stress response of plants subjected to organic fertilization has yielded variable results, depending on plant species, cultivation methods and fertilizer type. Previously reported data suggest that the use of organic fertilization (e.g., manure) can alleviate oxidative damage and reduce the MDA content in beetroot plants grown under stress conditions [57]. In Chinese cabbage (*Brassica rapa* L. subsp. *pekinensis*), the application of organic fertilization led to an equal or decreased MDA content compared to treatments used conventional fertilization [58]. Similarly, in ginger (*Zingiber officinale* Rosc.), the use of vermicompost and poultry manure decreased the H_2O_2 contents of plants, compared to the application of conventional NPK fertilization [59]. In our study the application of organic fertilization in purslane had a similar effect in MDA and H_2O_2; however, H_2O_2 increased with the application of a side dressing, in both conventional and organic fertilization regimes in levels similar to the control treatment (NoFert), thus indicating the low nutrient requirements for the species which can be covered with base dressing and the negative effects of additional fertilization. In contrast, in common sowthistle, the application of organic fertilization decreased H_2O_2, a finding which is similar to the abovementioned literature reports and indicates that organic practices may trigger the defense system of plants and alleviate stress responses [60].

5. Conclusions

The current study evaluated the application of conventional and organic fertilization in purslane (*P. oleracea* L.) and common sowthistle (*S. oleraceus* L.) plants cultivated in field conditions. Regarding purslane, the application of organic fertilizers provided comparable or even enhanced plant growth traits, compared to the application of conventional fertilizers. Conventional fertilization with supplementary fertilization in the form of a side dressing positively influenced plant growth (i.e., plant height and stems FW) physiological parameters (i.e., chlorophylls content), total phenolics content and antioxidant capacity (i.e., DPPH and FRAP). Interestingly, organic fertilization presented comparable or improved results in terms of macronutrient content (i.e., P and K), quality (i.e., TSS) and ascorbic acid content. In the case of common sowthistle, organic fertilization was presented as a promising alternative to conventional fertilization. Specifically, the growth-related traits of common sowthistle plants treated with organic fertilization were comparable (i.e., leaf number and plant FW) or improved (i.e., plant height and dry matter content), compared to conventional fertilizers. Although decreased K and P contents were observed, organic fertilization provided the plants with adequate N, an element that is typically in shortage in organic cropping systems. On the other hand, total phenolics, antioxidant capacity and flavonoids were decreased with the application of organic fertilization compared to conventional fertilization. Generally, variable effects were observed regarding plant physiology, stress response and antioxidant capacity, based on the fertilization regime and the application of supplementary fertilization (e.g., side dressing). The use of organic fertilization in both purslane and common sowthistle exhibited performance similar to those of conventional fertilization, indicating that organic farming practices are ideal for crops with low input requirements. However, future studies should focus on the long-term optimization of fertilizer doses, as well as the integration of components such as biochar, and beneficial organisms such as mycorrhiza.

Supplementary Materials: The following supporting information can be downloaded at: https://www.mdpi.com/article/10.3390/horticulturae10111222/s1, Table S1: The amounts of fertilizers and nutrients applied for each treatment (kg ha^{-1}).

Author Contributions: Conceptualization, N.T.; methodology, S.A.P. and N.T.; software, G.N.; validation, E.H., G.N. and S.A.P.; formal analysis, E.H. and G.N.; investigation, E.H. and G.N.; resources, N.T.; data curation, G.N. and N.T.; writing—original draft preparation, G.N.; writing—review and editing, S.A.P. and N.T.; visualization, S.A.P.; supervision, N.T.; project administration, N.T.; funding acquisition, N.T. All authors have read and agreed to the published version of the manuscript.

Funding: Financial support was provided by PRIMA (grant Number Prima2019-11, PRIMA/0009/2019, P2P/PRIMA/1218/0006, 01DH20006, Prima2019-12, STDF Valuefarm, 18-3-2021, TUBITAK-119N494, 301/18 October 2020, PCI2020-112091) a program supported by the European Union with co-funding by the Funding Agencies RIF—Cyprus and by the General Secretariat for Research and Technology of Greece.

Data Availability Statement: Data available on request due to privacy.

Conflicts of Interest: The authors declare no conflicts of interest.

References

1. Ventura, Y.; Eshel, A.; Pasternak, D.; Sagi, M. The Development of Halophyte-Based Agriculture: Past and Present. *Ann. Bot.* **2015**, *115*, 529–540. [CrossRef]
2. Łuczaj, Ł.; Pieroni, A.; Tardío, J.; Pardo-De-Santayana, M.; Sõukand, R.; Svanberg, I.; Kalle, R. Wild Food Plant Use in 21st Century Europe: The Disappearance of Old Traditions and the Search for New Cuisines Involving Wild Edibles. *Acta Soc. Bot. Pol.* **2012**, *81*, 359–370. [CrossRef]
3. Rostaei, M.; Fallah, S.; Carrubba, A.; Lorigooini, Z. Organic Manures Enhance Biomass and Improve Content, Chemical Compounds of Essential Oil and Antioxidant Capacity of Medicinal Plants: A Review. *Heliyon* **2024**, *10*, e36693. [CrossRef]
4. Seppelt, R.; Klotz, S.; Peiter, E.; Volk, M. Agriculture and Food Security under a Changing Climate: An Underestimated Challenge. *iScience* **2022**, *25*, 105551. [CrossRef]
5. Chrysargyris, A.; Koutsoumpeli, E.; Xylia, P.; Fytrou, A.; Konstantopoulou, M.; Tzortzakis, N. Organic Cultivation and Deficit Irrigation Practices to Improve Chemical and Biological Activity of *Mentha Spicata* Plants. *Agronomy* **2021**, *11*, 599. [CrossRef]
6. Du, J.; Yu, Y.; Tang, C.; Zong, K.; Zhang, S.; Zhang, Q.; Fang, L.; Li, Y. Organic Fertilizers Increase the Proportion of Saprotrophs Favoring Soil Nitrification under Medicinal Plants *Fritillaria thunbergii*. *Ind. Crops Prod.* **2024**, *219*, 119129. [CrossRef]
7. Kopittke, P.M.; Menzies, N.W.; Wang, P.; McKenna, B.A.; Lombi, E. Soil and the Intensification of Agriculture for Global Food Security. *Environ. Int.* **2019**, *132*, 105078. [CrossRef]
8. Röös, E.; Mie, A.; Wivstad, M.; Salomon, E.; Johansson, B.; Gunnarsson, S.; Wallenbeck, A.; Hoffmann, R.; Nilsson, U.; Sundberg, C.; et al. Risks and Opportunities of Increasing Yields in Organic Farming. A Review. *Agron. Sustain. Dev.* **2018**, *38*, 14. [CrossRef]
9. Shafie, F.A.; Rennie, D. Consumer Perceptions Towards Organic Food. *Procedia Soc. Behav. Sci.* **2012**, *49*, 360–367. [CrossRef]
10. Tscharntke, T.; Grass, I.; Wanger, T.C.; Westphal, C.; Batáry, P. Beyond Organic Farming—Harnessing Biodiversity-Friendly Landscapes. *Trends Ecol. Evol.* **2021**, *36*, 919–930. [CrossRef]
11. Fernández, J.A.; Ayastuy, M.E.; Belladonna, D.P.; Comezaña, M.M.; Contreras, J.; de Maria Mourão, I.; Orden, L.; Rodríguez, R.A. Current Trends in Organic Vegetable Crop Production: Practices and Techniques. *Horticulturae* **2022**, *8*, 893. [CrossRef]
12. Carrascosa, Á.; Pascual, J.A.; Ros, M.; Petropoulos, S.; del Mar Alguacil, M. The Effect of Fertilization Regime on Growth Parameters of *Sonchus Oleraceus* and Two Genotypes of *Portulaca Oleracea*. *Biol. Life Sci. Forum* **2022**, *16*, 7. [CrossRef]
13. Molina, M.; Tardío, J.; Aceituno-Mata, L.; Morales, R.; Reyes-García, V.; Pardo-De-Santayana, M. Weeds and Food Diversity: Natural Yield Assessment and Future Alternatives for Traditionally Consumed Wild Vegetables. *J. Ethnobiol.* **2014**, *34*, 44–67. [CrossRef]
14. Paschoalinotto, B.H.; Polyzos, N.; Compocholi, M.; Rouphael, Y.; Alexopoulos, A.; Dias, M.I.; Barros, L.; Petropoulos, S.A. Domestication of Wild Edible Species: The Response of *Scolymus hispanicus* Plants to Different Fertigation Regimes. *Horticulturae* **2023**, *9*, 103. [CrossRef]
15. Carrascosa, A.; Pascual, J.A.; Ros, M.; Petropoulos, S.A.; Alguacil, M. Agronomical Practices and Management for Commercial Cultivation of *Portulaca oleracea* as a Crop: A Review. *Plants* **2023**, *12*, 1246. [CrossRef]
16. Montoya-García, C.O.; García-Mateos, R.; Becerra-Martínez, E.; Toledo-Aguilar, R.; Volke-Haller, V.H.; Jesús Magdaleno-Villar, J. Bioactive Compounds of Purslane (*Portulaca oleracea* L.) According to the Production System: A Review. *Sci. Hortic.* **2023**, *308*, 111584. [CrossRef]
17. Petropoulos, S.A.; Fernandes, Â.; Dias, M.I.; Vasilakoglou, I.B.; Petrotos, K.; Barros, L.; Ferreira, I.C.F.R. Nutritional Value, Chemical Composition and Cytotoxic Properties of Common Purslane (*Portulaca oleracea* L.) in Relation to Harvesting Stage and Plant Part. *Antioxidants* **2019**, *8*, 293. [CrossRef]
18. Zhou, Y.-X.; Xin, H.-L.; Rahman, K.; Wang, S.-J.; Peng, C.; Zhang, H. *Portulaca oleracea* L.: A Review of Phytochemistry and Pharmacological Effects. *BioMed Res. Int.* **2015**, *2015*, 925631. [CrossRef]
19. Kumar, A.; Sreedharan, S.; Kashyap, A.K.; Singh, P.; Ramchiary, N. A Review on Bioactive Phytochemicals and Ethnopharmacological Potential of Purslane (*Portulaca oleracea* L.). *Heliyon* **2022**, *8*, e08669. [CrossRef]
20. Botella, M.Á.; Hellín, P.; Hernández, V.; Dabauza, M.; Robledo, A.; Sánchez, A.; Fenoll, J.; Flores, P. Chemical Composition of Wild Collected and Cultivated Edible Plants (*Sonchus oleraceus* L. and *Sonchus tenerrimus* L.). *Plants* **2024**, *13*, 269. [CrossRef]
21. Peerzada, A.M.; O'Donnell, C.; Adkins, S. Biology, Impact, and Management of Common Sowthistle (*Sonchus oleraceus* L.). *Acta Physiol. Plant.* **2019**, *41*, 1–17. [CrossRef]
22. Jain, S.; Singh, G. Preliminary Phytochemical Sereening and in Vitro Antioxidant Activity of Extracts of Whole Plant of *Sonchus oleraceus* Asteraceae. *Res. J. Pharm. Sci.* **2014**, *3*, 1–12.

23. Alothman, E.A.; Awaad, A.S.; Safhi, A.A.; Almoqren, S.S.; El-Meligy, R.M.; Zain, Y.M.; Alasmary, F.A.; Alqasoumi, S.I. Evaluation of Anti-Ulcer and Ulcerative Colitis of *Sonchus oleraceus* L. *Saudi Pharm. J.* **2018**, *26*, 956–959. [CrossRef] [PubMed]
24. Saxena, A.; Kumar, J. Phytochemical Screening, Metal-Binding Studies and Applications of Floral Extract of *Sonchus oleraceus* as a Corrosion Inhibitor. *J. Bio- Tribo-Corrosion* **2020**, *6*, 55. [CrossRef]
25. Chen, L.; Teng, H.; Cao, H. Chlorogenic Acid and Caffeic Acid from *Sonchus oleraceus* Linn Synergistically Attenuate Insulin Resistance and Modulate Glucose Uptake in HepG2 Cells. *Food Chem. Toxicol.* **2019**, *127*, 182–187. [CrossRef]
26. Xia, D.-Z.; Yu, X.-F.; Zhu, Z.-Y.; Zou, Z.-D. Antioxidant and Antibacterial Activity of Six Edible Wild Plants (*Sonchus* spp.) in China. *Nat. Prod. Res.* **2011**, *25*, 1893–1901. [CrossRef]
27. Qiao, L.; Gao, M.; Zheng, J.; Zhang, J.; Lu, L.; Liu, X. Novel Browning Alleviation Technology for Fresh-Cut Products: Preservation Effect of the Combination of *Sonchus oleraceus* L. Extract and Ultrasound in Fresh-Cut Potatoes. *Food Chem.* **2021**, *348*, 129132. [CrossRef]
28. Khan, B.A.; Nadeem, M.A.; Anwar, S.; Maqbool, R.; Yasin, M.; Amin1, M.M.; Aziz5, A.; Asif, M.; Hayyat, M.S. A, The Phytotoxic Effects of *Sonchus oleraceus* L. on Emergence and Seedling Growth of *Echinocloa cruss-galli*. *J. Weed Sci. Res.* **2020**, *27*, 433–446. [CrossRef]
29. Bergstrand, K.-J. Organic Fertilizers in Greenhouse Production Systems—A Review. *Sci. Hortic.* **2022**, *295*, 110855. [CrossRef]
30. Richardson, A.D.; Duigan, S.P.; Berlyn, G.P. An Evaluation of Noninvasive Methods to Estimate Foliar Chlorophyll Content. *New Phytol.* **2002**, *153*, 185–194. [CrossRef]
31. Chrysargyris, A.; Panayiotou, C.; Tzortzakis, N. Nitrogen and Phosphorus Levels Affected Plant Growth, Essential Oil Composition and Antioxidant Status of Lavender Plant (*Lavandula angustifolia* Mill.). *Ind. Crops Prod.* **2016**, *83*, 577–586. [CrossRef]
32. Wojdyło, A.; Oszmiański, J.; Czemerys, R. Antioxidant Activity and Phenolic Compounds in 32 Selected Herbs. *Food Chem.* **2007**, *105*, 940–949. [CrossRef]
33. Číž, M.; Čížová, H.; Denev, P.; Kratchanova, M.; Slavov, A.; Lojek, A. Different Methods for Control and Comparison of the Antioxidant Properties of Vegetables. *Food Control* **2010**, *21*, 518–523. [CrossRef]
34. Goulas, V.; Manganaris, G.A. The Effect of Postharvest Ripening on Strawberry Bioactive Composition and Antioxidant Potential. *J. Sci. Food Agric.* **2011**, *91*, 1907–1914. [CrossRef]
35. Tzortzakis, N.G.; Tzanakaki, K.; Economakis, C.D. Effect of Origanum Oil and Vinegar on the Maintenance of Postharvest Quality of Tomato. *Food Nutr. Sci.* **2011**, *2*, 974–982. [CrossRef]
36. Xylia, P.; Fasko, K.G.; Chrysargyris, A.; Tzortzakis, N. Heat Treatment, Sodium Carbonate, Ascorbic Acid and Rosemary Essential Oil Application for the Preservation of Fresh *Rosmarinus officinalis* Quality. *Postharvest Biol. Technol.* **2022**, *187*, 111868. [CrossRef]
37. Chrysargyris, A.; Petropoulos, S.A.; Prvulovic, D.; Tzortzakis, N. Performance of Hydroponically Cultivated Geranium and Common Verbena under Salinity and High Electrical Conductivity Levels. *Agronomy* **2021**, *11*, 1237. [CrossRef]
38. Kazimierczak, R.; Hallmann, E.; Rembiałkowska, E. Effects of Organic and Conventional Production Systems on the Content of Bioactive Substances in Four Species of Medicinal Plants. *Biol. Agric. Hortic.* **2015**, *31*, 118–127. [CrossRef]
39. Nastou, E.; Thalassinos, G.; Polyzos, N.; Antoniadis, V.; Petropoulos, S.A. The Effect of Nitrogen Fertilization Rate on Growth and Physiological Parameters of Three Purslane Genotypes Grown in a Soilless Cultivation System. *Acta Hortic.* **2021**, *1321*, 125–132. [CrossRef]
40. Karkanis, A.; Asprogeraka, A.C.; Paouris, E.; Ntanasi, T.; Karavidas, I.; Rumbos, C.I.; Athanassiou, C.G.; Ntatsi, G. Yellow Mealworm Frass: A Promising Organic Fertilizer for Common Sowthistle (*Sonchus oleraceus* L.) and Bristly Oxtongue (*Helminthotheca echioides* (L.) Holub) Cultivation. *Heliyon* **2024**, *10*, e35508. [CrossRef]
41. Cardarelli, M.; El Chami, A.; Iovieno, P.; Rouphael, Y.; Bonini, P.; Colla, G. Organic Fertilizer Sources Distinctively Modulate Productivity, Quality, Mineral Composition, and Soil Enzyme Activity of Greenhouse Lettuce Grown in Degraded Soil. *Agronomy* **2023**, *13*, 194. [CrossRef]
42. Rodrigues, M.A.; Pereira, A.; Cabanas, J.E.; Dias, L.; Pires, J.; Arrobas, M. Crops Use-Efficiency of Nitrogen from Manures Permitted in Organic Farming. *Eur. J. Agron.* **2006**, *25*, 328–335. [CrossRef]
43. Molla, A.; Solomou, A.D.; Fountouli, A.; Chatzikirou, E.; Stamatakis, E.; Stamatakis, P.; Skoufogianni, E. Influence of an Organic Fertilizer on Agronomic Characteristics and Herbaceous Plant Diversity in a Greek Ecosystem: The Case of Cretan Dittany (*Origanum dictamnus* L.). *Nitrogen* **2024**, *5*, 426–438. [CrossRef]
44. Andrianto, H.; Suhardi; Faizal, A. Measurement of Chlorophyll Content to Determine Nutrition Deficiency in Plants: A Systematic Literature Review. In Proceedings of the 2017 International Conference on Information Technology Systems and Innovation (ICITSI), Bandung, Indonesia, 23–24 October 2017; IEEE: Piscataway, NJ, USA, 2017; pp. 392–397.
45. Wei, L.; Li, J.; Qu, K.; Chen, H.; Wang, M.; Xia, S.; Cai, H.; Long, X.E.; Miao, Y.; Liu, D. Organic Fertilizer Application Promotes the Soil Nitrogen Cycle and Plant Starch and Sucrose Metabolism to Improve the Yield of *Pinellia ternata*. *Sci. Rep.* **2024**, *14*, 12722. [CrossRef]
46. Ouzounidou, G.; Asfi, M.; Sotirakis, N.; Papadopoulou, P.; Gaitis, F. Olive Mill Wastewater Triggered Changes in Physiology and Nutritional Quality of Tomato (*Lycopersicon esculentum* Mill.) Depending on Growth Substrate. *J. Hazard. Mater.* **2008**, *158*, 523–530. [CrossRef]
47. Fontana, E.; Hoeberechts, J.; Nicola, S.; Cros, V.; Palmegiano, G.B.; Peiretti, P.G. Nitrogen Concentration and Nitrate/Ammonium Ratio Affect Yield and Change the Oxalic Acid Concentration and Fatty Acid Profile of Purslane (*Portulaca oleracea* L.) Grown in a Soilless Culture System. *J. Sci. Food Agric.* **2006**, *86*, 2417–2424. [CrossRef]

48. Martínez-Alcántara, B.; Martínez-Cuenca, M.R.; Bermejo, A.; Legaz, F.; Quiñones, A. Liquid Organic Fertilizers for Sustainable Agriculture: Nutrient Uptake of Organic versus Mineral Fertilizers in Citrus Trees. *PLoS ONE* **2016**, *11*, e0161619. [CrossRef]
49. Bakhtiari, M.; Mozafari, H.; Asl, K.K.; Sani, B.; Mirza, M. Plant Growth, Physiological, and Biochemical Responses of Medic Savory [*Satureja macrantha* (Makino) Kudô] to Bio-Organic and Inorganic Fertilizers. *J. Med. Plants By-Prod.* **2020**, *9*, 9–17. [CrossRef]
50. Chrysargyris, A.; Louka, S.; Petropoulos, S.A.; Tzortzakis, N. Soilless Cultivation of *Portulaca oleracea* Using Medicinal and Aromatic Plant Residues for Partial Peat Replacement. *Horticulturae* **2023**, *9*, 474. [CrossRef]
51. Serri, F.; Souri, M.K.; Rezapanah, M. Growth, Biochemical Quality and Antioxidant Capacity of Coriander Leaves under Organic and Inorganic Fertilization Programs. *Chem. Biol. Technol. Agric.* **2021**, *8*, 33. [CrossRef]
52. Lv, J.; Huang, H.; Yu, L.; Whent, M.; Niu, Y.; Shi, H.; Wang, T.T.Y.; Luthria, D.; Charles, D.; Yu, L.L. Phenolic Composition and Nutraceutical Properties of Organic and Conventional Cinnamon and Peppermint. *Food Chem.* **2012**, *132*, 1442–1450. [CrossRef]
53. Vera-Contreras, J.; Fuentealba, V.; Pinto-Morales, F.; Paz Jofré, M.; Rodriguez, P.; Rojas, C.; Vergara-Retamales, R.; Betancur, M.; Pinto-Poblete, A.; Rivas, Y. Effect of Organic Fertilization on Fatty Acid Production and Antidiabetic Activity of *Portulaca oleracea* L. *Rev. Macambira* **2024**, *8*, e081008. [CrossRef]
54. Sarwar, M.; Patra, J.K.; Ali, A.; Maqbool, M.; Arshad, M.I. Effect of Compost and NPK Fertilizer on Improving Biochemical and Antioxidant Properties of *Moringa oleifera*. *S. Afr. J. Bot.* **2020**, *129*, 62–66. [CrossRef]
55. Sałata, A.; Nurzyńska-Wierdak, R.; Kalisz, A.; Kunicki, E.; Ibáñez-Asensio, S.; Moreno-Ramón, H. Effects of Organic Cropping on Phenolic Compounds and Antioxidant Capacity of Globe Artichoke Herbs. *Agronomy* **2022**, *12*, 192. [CrossRef]
56. Nørbæk, R.; Aaboer, D.B.F.; Bleeg, I.S.; Christensen, B.T.; Kondo, T.; Brandt, K. Flavone C-Glycoside, Phenolic Acid, and Nitrogen Contents in Leaves of Barley Subject to Organic Fertilization Treatments. *J. Agric. Food Chem.* **2003**, *51*, 809–813. [CrossRef] [PubMed]
57. Zhang, P.; Yang, F.; Zhang, H.; Liu, L.; Liu, X.; Chen, J.; Wang, X.; Wang, Y.; Li, C. Beneficial Effects of Biochar-Based Organic Fertilizer on Nitrogen Assimilation, Antioxidant Capacities, and Photosynthesis of Sugar Beet (*Beta vulgaris* L.) under Saline-Alkaline Stress. *Agronomy* **2020**, *10*, 1562. [CrossRef]
58. Qiu, X.; Wang, Y.; Hu, G.; Wang, Q.; Zhang, X.; Dong, Y. Effect of Different Fertilization Modes on Physiological Characteristics, Yield and Quality of Chinese Cabbage. *J. Plant Nutr.* **2013**, *36*, 948–962. [CrossRef]
59. Poozesh, V.; Amirahmadi, A. The Impact of Organic and Inorganic Fertilizers on Some Phytochemical and Antioxidant Properties in *Zingiber officinale* Rosc. *J. Chem. Health Risks* **2024**, *14*, 555–567. [CrossRef]
60. Krey, K.L.; Nabity, P.D.; Blubaugh, C.K.; Fu, Z.; Van Leuven, J.T.; Reganold, J.P.; Berim, A.; Gang, D.R.; Jensen, A.S.; Snyder, W.E. Organic Farming Sharpens Plant Defenses in the Field. *Front. Sustain. Food Syst.* **2020**, *4*, 97. [CrossRef]

horticulturae MDPI

Article

Olive Oil Quality Produced Under Conventional and Organic Farming Systems in a Multisite Two-Year Evaluation in Greece †

Petros Anargyrou Roussos *, Asimina-Georgia Karyda, Vassilios Kaibalis and Konstantinos Zontanos

Laboratory of Pomology, Department of Crop Science, Agricultural University of Athens, Iera Odos 75, 118 55 Athens, Greece; romina_karyda@hotmail.com (A.-G.K.); billakosltg@gmail.com (V.K.); stud118025@aua.gr (K.Z.)

* Correspondence: roussosp@aua.gr; Tel.: +30-210-5294596

† This paper is an extended version of our paper published in Roussos, P.A.; Karida, A.-G.; Kaibalis, V. Olive oil quality characteristics under organic and integrated management in messinia county Greece. In Proceedings of the XIV International Scientific Agriculture Symposium "AGROSYM 2023", Jahorina, Bosnia and Herzegovina, 5–8 October 2023.

Abstract: Six olive groves (three conventional and three organic) of "Koroneiki" olive cultivar were selected in two different provinces in southern Greece during a two-year trial (one province per year). At harvest, the oil was extracted from olive fruits using an Abenchor-type olive mill. Olive oil quality characteristics, including free acidity, ultraviolet absorbance indexes, peroxide value, free fatty acids, total and individual phenolic compounds, antioxidant capacity, and α-tocopherol content, were assessed. During the second year, olive oil samples were stored for 12 months, and their quality was re-evaluated. Few significant differences were observed between organically and conventionally produced olive oils, while the quality of both deteriorated significantly after storage (decrease in antioxidant capacity and major phenolic compounds, except for hydroxytyrosol and tyrosol, and increase in peroxide index value). Principal component analysis revealed that, during the first year, oils produced following organic and conventional practices could be distinguished; however, this distinction was not evident during the second year. Nevertheless, significant differentiation was found between oils produced under the same cultivation practice. In conclusion, the discrepancies found in the literature regarding organic versus conventional olive oil properties probably originate from the differences detected even among similar products (organic or conventional), including high-standard deviations.

Keywords: free fatty acids; functional properties; Koroneiki; phenolic compounds; tocopherol

1. Introduction

The olive tree is an emblematic species of the Mediterranean basin, holding significant social, economic, and environmental value for the countries in the region [1]. It is highly appreciated for several characteristics, including its drought resistance, adaptability to diverse pedoclimatic conditions, and, most notably, its uniquely valuable products: table olives and olive oil. The olive tree is considered one of the oldest plants cultivated for its oil [2]. Although in the modern world olive oil accounts for only a small portion of the total consumption of plant-derived oils, it holds immense economic importance for producing countries and consumers worldwide due to its nutritional value and exceptional sensory attributes [2,3]. Extra-virgin olive oil (EVOO) is a cornerstone of the Mediterranean diet, with its consumption steadily increasing worldwide thanks to its health-promoting properties. Studies have shown that it can reduce the risk of coronary heart disease, cancer, inflammation, diabetes, Alzheimer's disease, and metabolic disorders [4,5].

These beneficial properties are attributed to the unique composition of olive oil. It is rich in monounsaturated fatty acids (MUFAs), primarily oleic acid (C18:1), which accounts for 56% to 85% of its total fatty acid content. Additionally, it contains low levels of polyunsaturated fatty acids (PUFAs), such as linoleic and linolenic acids, and saturated fatty acids (SFAs), mainly palmitic and stearic acids [4,6]. Olive oil is also abundant in natural antioxidants, including phenolic compounds, tocopherols, squalene, and carotenoids, all of which contribute to its antioxidant properties and stability over time [5,7,8].

Phenolic compounds represent a diverse group of bioactive metabolites found in olive oil [4]. Secoiridoids, such as the aldehydic forms of oleuropein and ligstroside aglycone, oleocanthal, and oleacein, as well as free tyrosol and hydroxytyrosol and their respective bound forms, are among the most important phenolic compounds present in EVOO. These compounds significantly enhance its health-promoting properties, oxidative stability, and sensory attributes [6,7,9,10]. Flavonoids such as luteolin and apigenin have been also found in EVOO, both contributing to its antioxidant, as well as health-promoting properties [9,11–13]. Additionally, tocopherols, particularly α-tocopherol, the most abundant in olive oil, serve as potent antioxidant molecules [4,7].

The overall quality of olive oil is evaluated through various attributes, including free acidity, peroxide value, spectrophotometric indices, phenolic content, oxidative stability, nutritional value (fatty acid and tocopherol composition), and sensory characteristics [6,7,10,14]. These quality parameters result from a complex interplay of numerous factors, with cultivar being one of the most significant [9,15,16]. However, other factors, such as pedoclimatic conditions, altitude, cultivation practices (irrigation, fertilization, pest, and disease control), harvest timing, oil extraction techniques, and storage conditions, also play crucial roles in determining the quality of the final product [9,17,18].

In recent decades, there has been growing interest in the benefits of organic farming for both the environment and human health. This trend has been driven largely by increased consumer awareness of pesticide residues in food and concerns about environmental contamination and sustainability [3,4,19]. An improper use of phytosanitary products in conventional farming can lead to pesticide residues in food. Even with proper use, detectable levels of pesticides may remain within legally defined limits. Organic agriculture, on the other hand, is generally considered environmentally friendly due to reduced chemical pollution. While evidence suggests that organic farming may produce food with a higher nutritional quality, some studies have found no clear advantages over conventional methods [1,20–22]. Moreover, research on the impact of organic practices on the quality and nutritional attributes of olive oil remains limited compared to other food products [23,24].

Given that olive oil quality is influenced by numerous factors, the present study aimed to investigate the effects of organic versus conventional farming practices. The trial was conducted as a multisite two-year experiment using the same cultivar to minimize variability arising from factors such as cultivar differences, pedoclimatic conditions, farming practices, and seasonal variations.

2. Materials and Methods

2.1. Test Site Location—Plant Material

The trial was conducted in southern Greece, in the Messenia Prefecture, a region where olive cultivation is extensively practiced. The trial lasted for two growing seasons, during the year 2019–2020 in Pylia (Pila) province (36°57′0″ N, 21°44′0″ E, altitude 110 m) and during the year 2020–2021 in Raches province (37°27′86″ N, 21°71′61″ E, altitude 130 m) (Figure 1). Pylia province is characterized by typical Mediterranean climate, with an average annual rainfall of about 800 mm and mild winters. The average annual temperature is around 20 °C, with January averaging approximately 10.8 °C and July around 25.5 °C. The soils are primarily calcareous and clayey–loamy. Raches province has also a Mediterranean climate, with mild winters and warm summers, and an average annual rainfall of about 800 mm. The average annual temperature is around 21 °C, with January having an average temperature of approximately 14.9 °C and July around 26.7 °C. The soils are primarily clayey–loamy, with good drainage. At each location, three conventionally managed orchards were selected, each belonging to a different grower, along with three organically managed olive groves, each owned by a separate farmer (Figure 2). Cultivation practices were individually recorded for each olive grove. The cultivar used in all groves was "Koroneiki", the most renowned olive oil cultivar in Greece and one of the most important olive oil cultivars worldwide, owing to its adaptability to various pedoclimatic conditions and its high productivity and supreme oil quality. The trees, ranging in age from 30 to 50 years, were fully productive. In each orchard, eight trees (organized as four replications of two trees each) were selected at the end of September based on uniform growth and production.

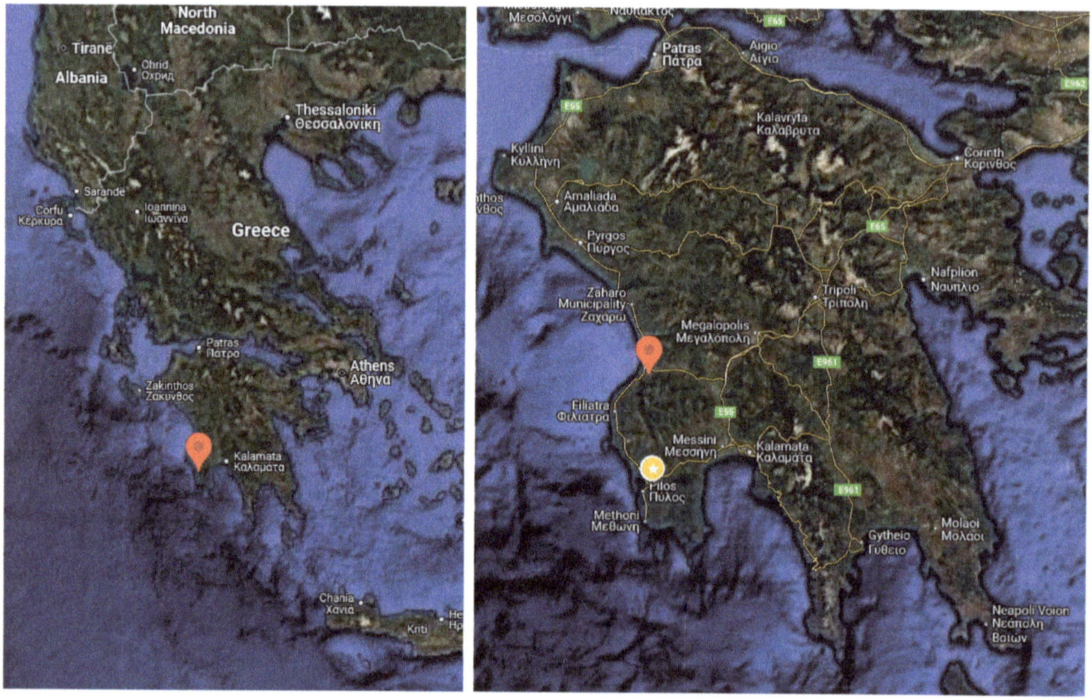

Figure 1. The location of the two provinces in Greece, in Messinia county (on the left). On the right a closer aspect of the area is depicted, where Pylia (yellow tag) and Raches (red tag) provinces are shown (produced by Google Maps).

At harvest, each tree was harvested separately, and the fruit weight was recorded. Approximately 1.5 kg of olives was randomly sampled from each plot and immediately transferred to the laboratory for processing.

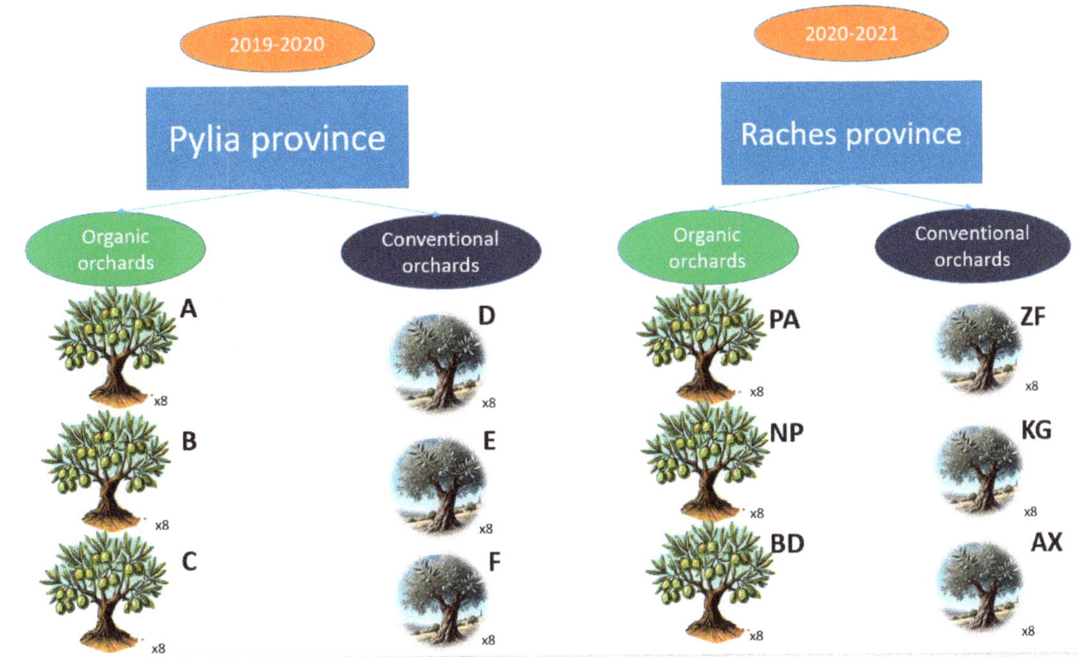

Figure 2. Schematic visualization of the trial. The letters indicate the farmers responsible for each olive grove (×8, indicates the number of trees used per orchard).

2.2. Olive Oil Extraction Procedure

In the laboratory, the maturity index of the fruits was determined using a sample of 100 randomly selected olive fruits, based on fruit skin and pulp color. After removing the leaves, the olives were crushed using a semi-industrial Abencor-type olive mill (Callis S.A., Athens, Greece). The crusher operated at 2800 rpm, and the resulting paste was transferred to a malaxer, where it was malaxed for 30 min at 50 rpm and 25–27 °C. Following malaxation, 50 g of the paste was transferred to a Falcon-type tube and centrifuged at 2500× g for 15 min to accurately determine the oil percentage in the paste. The remaining paste was centrifuged at 1400× g for 3–4 min using the semi-industrial centrifuge of the Abencor-type olive mill, and the oil produced was stored in filled bottles at 4 °C until analysis, which was performed within one week of extraction. During the second year, part of the olive oil produced from the Raches province trial was stored at room temperature in the dark to evaluate oil stability under different management practices over 12 months.

2.3. Olive Oil Analyses

The determination of olive oil free acidity, peroxide index, and ultraviolet absorption at 232 and 270 (K232, K270 respectively), as well as DK (Delta Kappa) was conducted according to the European Official Methods of Analysis 2016/1784.

2.4. Determination of Total Phenols

Phenolic compounds were extracted by adding 5 mL of hexane into 2.5 g of olive oil. After vortexing, 5 mL of methanol–water solution (60:40) was added and the mixture was vortexed for 2 min. The sample was then centrifuged at 2500× g for 10 min and the upper phase was discarded. The lower phase containing the phenolic fraction was transferred to another tube and used for the determination of the different phenol fractions. The total phenols were determined by the spectrophotometric Folin–Ciocalteu method, as described by Roussos et al. [25], and expressed in mg gallic acid equivalents (GAEs) kg^{-1} olive oil. The o-diphenols were determined according to Roussos et al. [25] and expressed as mg caffeic acid equivalents (CAEs) kg^{-1} olive oil, while the total flavonoids were determined according to Roussos et al. [25] and expressed as mg catechin equivalents (CtAEs) kg^{-1} olive oil.

Individual phenolic compounds were detected in the same phenolic extract by a high-performance liquid chromatography (HPLC), Shimadzu Nexera X2 (Shimadzu Europa GmbH, Albert-Hahn-Str. 6-10, 47269 Duisburg, Germany), equipped with a quaternary pump, a diode array detector operating at 260 nm, 280 nm, 325 nm, and 360 nm. A sample of 40 μL was injected in a column Luna LC Column 250 × 4.6 mm 5 μm C18 (2) 100 A (Phenomenex, 411 Madrid Avenue, Torrance, CA 90501-1430, USA) thermostated at 20 °C. The different phenols were eluted by a gradient program using two mobile phases, i.e., (A) methanol:acetonitrile (1:1) and (B) 1% v/v acetic acid in water. The flow rate was adjusted at 1 mL min^{-1} and the gradient profile was as follows: at time 0 min solvent A at 10% till time 5 min, at 10 min and till 35 min A at 20%, at 45 and till 55 min A at 25%, at 70 min A at 30%, at 100 min A at 40% and at 106 min till 123 min A at 70%. A total of eleven (11) phenolic compounds were detected based on their retention time and spectra similarity with the corresponding standards. The phenolic compounds detected were as follows: hydroxytyrosol, tyrosol, vannilic acid, caffeic acid, vannilin, p-coumaric acid, ferulic acid, oleacein, oleocanthal, luteolin, and apigenin. The corresponding standards were purchased from Sigma-Aldrich (Sigma-Aldrich, Inc., P.O. Box 14508, St. Louis, MO 63178, USA) and their concentration was expressed in mg kg^{-1} olive oil.

2.5. Antioxidant Capacity

The antioxidant capacity of the oils was determined in the same extract of phenolic compounds, based on two different assays, i.e., the diphenyl picryl hydrazyl (DPPH) and the ferric reducing antioxidant power (FRAP) assays, according to Roussos et al. [25]. The antioxidant capacity was expressed as μmol Trolox equivalent kg^{-1} olive oil.

2.6. a-Tocopherol Determination

The concentration of a-tocopherol was determined in the samples of the second year only. In 1 g of olive oil, 2 mL of hexane was added and the mixture was vortexed. The a-tocopherol was detected by the HPLC described above using a normal phase column (Lichrosorb 250 × 4.6 mm Si 60 5 μm) (Sigma-Aldrich, Inc., P.O. Box 14508, St. Louis, MO 63178, USA) thermostated at 20 °C using as eluent a mixture of 99% hexane and 1% 2-propanol. The flow rate was 0.8 mL min^{-1} and a-tocopherol was detected by a fluorescence detector at 290 nm excitation wavelength and 325 nm emission wavelength. Identification and quantification were performed using authentic standards.

2.7. Fatty Acids and Squalene Determination

Fatty acid composition, expressed as a percentage of corresponding methyl esters (FAMES), was determined by diluting 0.5 g of olive oil in 2 mL of GC-grade hexane and vigorously shaken, followed by the addition of 1 mL 0.2 N solution of KOH. The methyl-esters were analyzed by gas chromatography (Shimadzu GC-2030) (Shimadzu Europa GmbH, Albert-Hahn-Str. 6-10, 47269 Duisburg, Germany) equipped with an autosampler (AOC-20i+s), and a flame ionization detector (FID). The analysis was performed using a fused silica capillary column (Supelco SP®-2340 60 × 0.32 mm × 0.2 μm film thickness) (Sigma-Aldrich, Inc., P.O. Box 14508, St. Louis, MO 63178, USA). A split ratio of 80.0 was used and the injector temperature was set at 249 °C, while the detector temperature was set at 249 °C. Then, 10 μL of the sample was injected at an initial column temperature of 150 °C, which was held for 18 min. The temperature was then increased at a rate of 2 °C min^{-1} to 175 °C and held for 5 min, and was then raised at a rate of 5 °C min^{-1} to a final temperature of 200 °C, which was held for 22 min. The identified fatty acids were as follows: palmitic (C16:0), palmitoleic (C16:1), heptadecenoic (C17:1), stearic (C18:0), oleic (C18:1), linoleic (C18:2), linolenic (C18:3), arachidic (C20:0), and gadoleic (C20:1) acids. Squalene (Sigma-Aldrich, Inc., P.O. Box 14508, St. Louis, MO 63178, USA) concentration was determined by a five-point standard curve of the corresponding analytical standard and expressed as mg 100 g^{-1} of olive oil.

2.8. Statistical Analysis

The trial was designed as a pseudo-replicated trial (treatments, i.e., organic versus conventional cultivation practices were not in the same orchard) with four replicates of two trees in each orchard (i.e., four plots of two trees per orchard, three orchards per management system). The four plots within each orchard were randomly selected in each orchard's area (completely randomized in the orchard). Significant differences among cultivation practices for the same year (province) were determined based on the *t*-test at a = 0.05 after checking the normal distribution of the raw data using standard skewness, standard kurtosis, and the homogeneity of variances. The comparison of the cultivation practices used the data derived from all three orchards under the same practice. Principal component analysis of raw data (olive oil quality indexes, FAMES content, phenolic compounds concentration, squalene and a-tocopherol concentration, and antioxidant capacity as well as yield, oil percentage per fruit and oil quantity per tree at harvest only) was used to describe the characteristics and differences of each management practice as well as to assess possible differences among orchards under the same cultivation practice, using a small number of variables. A comparison of olive oil quality in time under the same management practice took also place, using *t*-test at a = 0.05 to assess the possible differences. The statistical software JMP 13.0 (SAS Institute) and Statgraphics Centurion XV (Statgraphics Technologies, Inc., The Plains, VA, USA) were used for the aforementioned analyses.

3. Results

3.1. Results at Harvest

Organic orchards in Pylia province presented significantly higher fruit production per tree compared to the conventional ones (Table 1). While the oil percentage per fruit and maturity index was similar under both organic and conventional management, the oil produced per tree was significantly higher under organic management.

Table 1. Effects of the cultivation management on olive fruit yield, olive oil percentage, and oil produced per tree at harvest.

Parameter	Organic	Conventional	Significance
		In Pylia province	
Yield (kg $tree^{-1}$)	35.64	25.65	*
Oil percentage per fruit	19.65	20.86	ns
Oil per tree (kg)	6.79	5.31	*
Maturity Index	2.8	3.1	ns
		In Raches province	
Yield (kg $tree^{-1}$)	29.10	40.94	**
Oil percentage per fruit	23.81	26.0	ns
Oil per tree (kg)	6.90	10.24	*
Maturity Index	3.2	3.5	ns

Asterisk indicates a significant difference between the means within the same row. ns, not significant, *, $p < 0.05$. **, $p < 0.01$.

Yield as well as oil produced per tree was significantly higher under conventional practice in Raches province, while no significant difference was determined regarding oil percentage per fruit and maturity index.

There were not any significant differences regarding the oil's free acidity, peroxide index, K232, and K270 in any of the two provinces, as can be seen in Table 2.

Table 2. Effects of the cultivation management on the oil's free acidity, peroxide index, K232, K270, at harvest.

Parameter	Organic	Conventional	Significance
		In Pylia province	
Free acidity (g oleic acid 100 g^{-1})	0.47	0.48	ns
Peroxides (meq O_2 kg^{-1})	13.75	10.83	ns
K232	1.41	1.43	ns
K270	0.111	0.121	ns
		In Raches province	
Free acidity (g oleic acid 100 g^{-1})	0.42	0.34	ns
Peroxides (meq O_2 kg^{-1})	12.91	9.58	ns
K232	1.64	1.92	ns
K270	0.17	0.18	ns

ns, not significant.

Total phenols and total o-diphenols were detected in higher concentrations in the oil produced under the conventional practice in Pylia province (Table 3). There were not any significant differences though regarding the total flavonoids' concentration and antioxidant capacity of the oils with either DPPH or FRAP assays. On the other hand, in Raches province, olive oils produced under conventional practice exhibited higher FRAP values than the corresponding organic ones, while no other significant difference was detected concerning total phenol concentration.

Table 3. Effects of cultivation management on olive total phenol, o-diphenol, and flavonoid concentration and antioxidant capacity at harvest.

Parameter	Organic	Conventional	Significance
		In Pylia province	
Total phenols (mg GAE kg^{-1})	374.1	418.9	ns
o-diphenols (mg CAE kg^{-1})	123.4	166.4	*
Total flavonoids (mg CtE kg^{-1})	124.1	187.8	*
FRAP (μmol Trolox equiv. kg^{-1})	1127.7	1442.4	ns
DPPH (μmol Trolox equiv. kg^{-1})	960.1	1210.6	ns
		In Raches province	
Total phenols (mg GAE kg^{-1})	283.8	330.5	ns
o-diphenols (mg CAE kg^{-1})	40.81	53.70	ns
Total flavonoids (mg CtE kg^{-1})	378.3	475.4	ns
FRAP (μmol Trolox equiv. kg^{-1})	1142.77	1431.1	*
DPPH (μmol Trolox equiv. kg^{-1})	1180.0	1400.2	ns

Asterisk indicates a significant difference between the means within the same row. ns, not significant, *, $p < 0.05$.

The concentration of the individual phenolic compounds detected was slightly affected by the management practice in Pylia province, as only vanillic acid was determined in higher concentration in organic olive oils, while p-coumaric acid and apigenin were found in higher concentrations in conventionally produced olive oils (Table 4). Only one significant difference was detected in oils produced in Raches province, as organically produced oils presented higher concentrations of hydroxytyrosol compared to the conventionally produced ones.

Table 4. Effects of cultivation management on the concentration of individual phenolic compounds detected in olive oils in mg kg^{-1} at harvest.

Parameter	Organic	Conventional	Significance
		In Pylia province	
Hydroxytyrosol	31.03	40.23	ns
Tyrosol	112.32	130.85	ns
Vanillic acid	21.49	17.67	**
Caffeic acid	0.52	0.57	ns
Vanillin	0.58	0.69	ns
p-coumaric acid	4.44	5.38	**
Ferulic acid	0.61	0.60	ns
Oleacein	12.23	12.11	ns
Oleocanthal	101.40	107.28	ns
Luteolin	42.40	52.39	ns
Apigenin	16.03	23.41	*
		In Raches province	
Hydroxytyrosol	5.69	4.14	*
Tyrosol	9.39	8.33	ns
Vanillic acid	0.48	0.42	ns
Caffeic acid	0.19	0.11	ns
Vanillin	0.15	0.15	ns
p-coumaric acid	0.64	0.45	ns
Ferulic acid	0.05	0.03	ns
Oleacein	37.46	39.19	ns
Oleocanthal	143.74	152.38	ns
Luteolin	15.20	14.02	ns
Apigenin	2.75	2.62	ns
a-tocopherol	81.38	85.60	ns

Asterisk indicates a significant difference between the means within the same row. ns, not significant, *, $p < 0.05$, **, $p < 0.01$.

The concentration of fatty acids (determined as FAMEs) differed among the olive oils produced under conventional and organic management (Table 5). Olive oils produced under conventional practice in Pylia province exhibited a higher content of palmitoleic (C16:1) and oleic (C18:1) acids, as well as a higher content of monounsaturated fatty acids (MUFAs), a higher ratio of MUFAs against polyunsaturated fatty acids (PUFAs), and oleic against linoleic (C18:2) acid. In contrast, organically produced olive oils exhibited higher content of C18:2, arachidic (C20), and gadoleic (C20:1) acids, as well as a higher content of PUFAs. There were not any significant differences concerning the content of the other fatty acids detected or squalene concentration. No significant difference was detected between organic and conventional olive oils regarding the content of free fatty acids and the concentration of squalene in oils produced in Raches province.

In Figure 3, it is obvious that there is a slight separation between organic and conventional management practices, taking into account all the raw data derived from sample analysis at harvest in Pylia province. Conventionally managed orchards can be found mainly on the positive side of principal component (PC) 1, while the organic ones are on the negative side of PC1.

Table 5. Effects of cultivation management on the content of free fatty acids (determined as FAMEs—%) and squalene concentration at harvest.

Parameter	Organic	Conventional	Significance
		In Pylia province	
C16	11.73	11.99	ns
C16:1	0.74	0.79	*
C17	0.025	0.024	ns
C18	2.53	2.53	ns
C18:1	77.94	78.35	*
C18:2	5.32	4.71	**
C20	0.46	0.44	**
C18:3	0.69	0.63	**
C20:1	0.33	0.31	**
C22	0.14	0.13	ns
C24	0.03	0.03	ns
SFA	14.9	15.2	ns
MUFA	79.06	79.49	**
PUFA	6.01	5.34	**
MUFA/PUFA	13.14	14.9	***
UFA/SFA	5.70	5.59	ns
C18:1/C18:2	14.66	16.65	***
Squalene (mg 100 g^{-1})	448.9	492.5	ns
		In Raches province	
C16	12.98	13.23	ns
C16:1	0.82	0.82	ns
C18	2.61	2.60	ns
C18:1	76.85	76.64	ns
C18:2	5.42	5.54	ns
C20	0.32	0.31	ns
C18:3	0.71	0.69	ns
C20:1	0.19	0.18	ns
SFA	15.91	16.14	ns
MUFA	77.85	77.68	ns
PUFA	6.13	6.23	ns
MUFA/PUFA	12.95	12.52	ns
UFA/SFA	5.28	5.20	ns
C18:1/C18:2	14.53	13.8	ns
Squalene (mg 100 g^{-1})	605.4	676.9	ns

Asterisk indicates a significant difference between the means within the same row. ns, not significant, *, $p < 0.05$, **, $p < 0.01$, ***, $p < 0.001$.

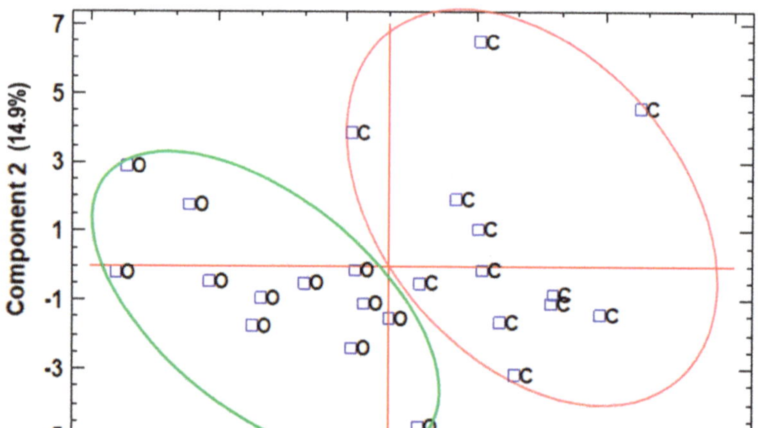

Figure 3. Scatterplot of the principal component analysis of the raw data at harvest in Pylia province. Abbreviations: O, organic management (denoted with the green eclipse), C, conventional management (denoted with the red eclipse).

Figure 4 presents the analysis of the PCA applied to the raw data per farmer in Pylia province, regarding only the conventionally managed orchards at harvest. There are significant differences among the three farmers, as they do not share common areas in the scatterplot.

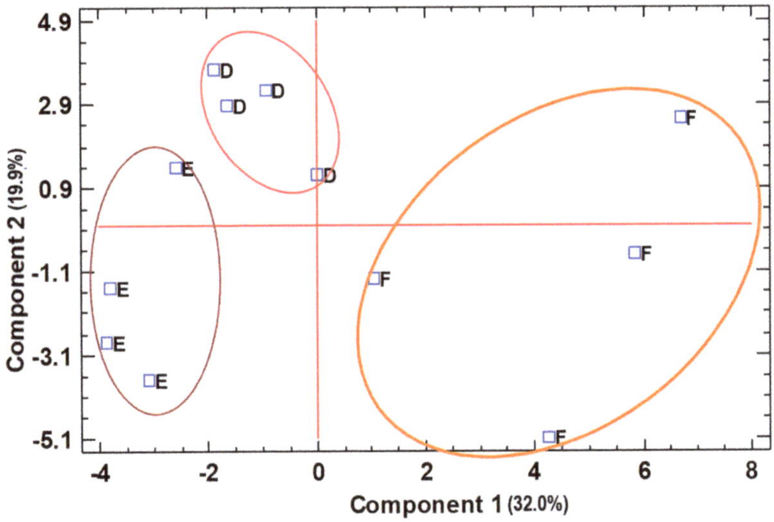

Figure 4. Scatterplot of the principal component analysis of the raw data of conventional orchards at harvest in Pylia province. The abbreviations (D, E, and F) represent the different olive groves (each grove is separately enclosed within one eclipse).

Figure 5 illustrates the raw data obtained from the analysis of the organic olive oil samples collected during harvest in the Pylia province. The data reveal a distinct separation among farmers: Farmer A is positioned on the negative side of PC1, whereas the other two farmers are located on the positive side of PC1.

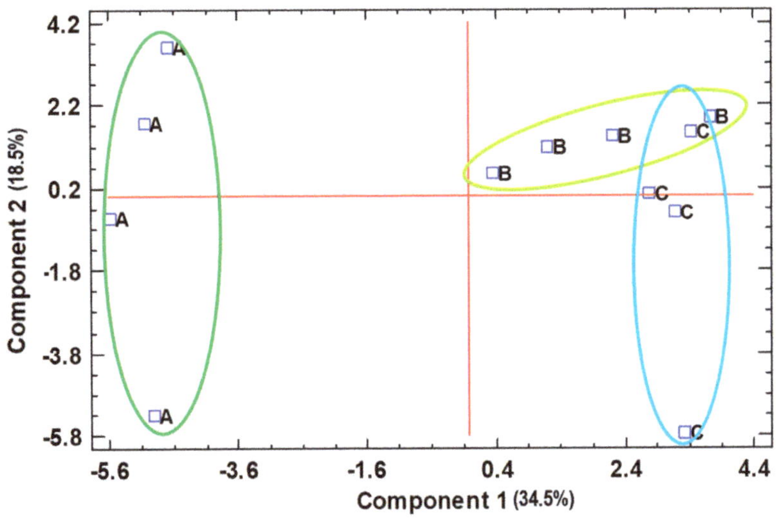

Figure 5. Scatterplot of the principal component analysis of the raw data of organic orchards at harvest in Pylia province. The abbreviations (A, B, and C) represent the different olive groves (each grove is separately enclosed within one eclipse).

Figure 6 shows that there is no distinct separation between organic and conventional management practices when considering the raw data from the sample analysis conducted at harvest in the Raches province.

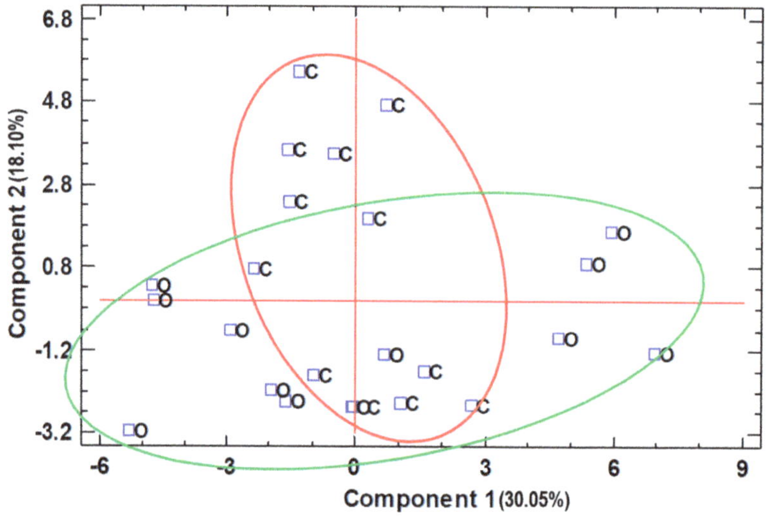

Figure 6. Scatterplot of the principal component analysis of the raw data at harvest in Raches province. Abbreviations: O, organic management (green eclipse), C, conventional management (red eclipse).

Figure 7 illustrates the PCA analysis applied to the raw data per farmer, focusing exclusively on conventionally managed orchards at harvest in Raches province. The results indicate that farmer AX is distinct from the other two, positioned on the positive side of PC1 and the negative side of PC2, while farmers KG and ZF occupy overlapping areas.

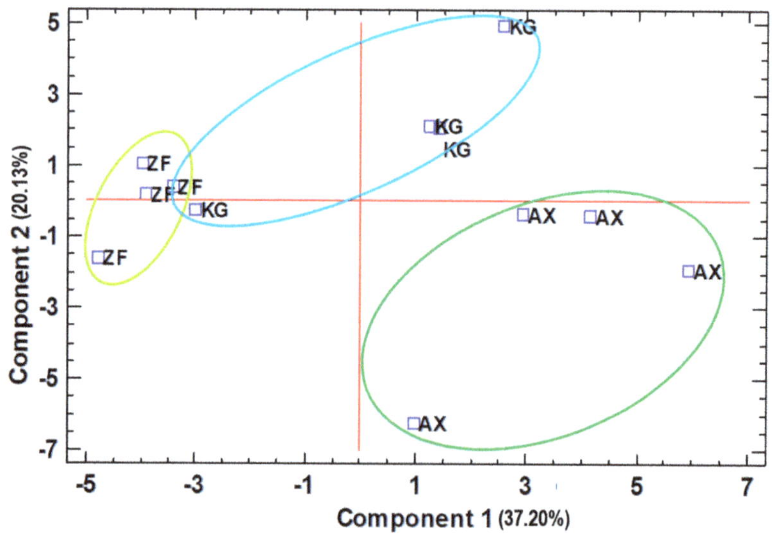

Figure 7. Scatterplot of the principal component analysis of the raw data of conventional orchards at harvest in Raches province. The abbreviations (ZF, KG, and AX) represent the initials of the farmers responsible for each olive grove (each grove is separately enclosed within one eclipse).

Figure 8 presents the raw data from the analysis of organic olive oil samples collected at harvest in Raches province. The results reveal a clear separation among farmers practicing organic management. Farmer PA is positioned on the positive side of PC2 and the negative side of PC1, farmer BD is located on the negative side of both PC2 and PC1, and farmer NP is situated on the positive side of PC1.

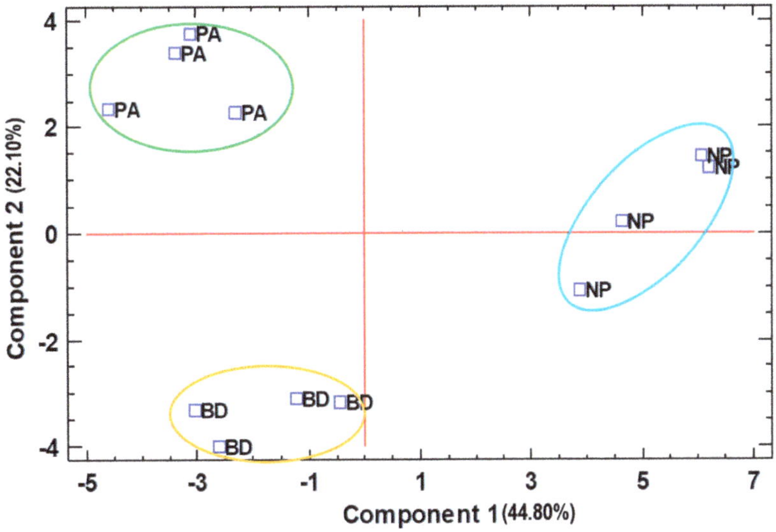

Figure 8. Scatterplot of the principal component analysis of the raw data of organic orchards at harvest in Raches province. The abbreviations (PA, NP, and BD) represent the initials of the farmers responsible for each olive grove (each grove is separately enclosed within one eclipse).

3.2. Olive Oil Quality 12 Months After Storage at Room Temperature (Derived from Raches Province Trial)

Twelve months after the oil was stored in the dark at room temperature, significant differences were detected regarding K232 and K270 indexes, with the conventionally produced oil presenting higher values (Table 6).

Table 6. Effects of the cultivation management on the oil's free acidity, peroxide index, K232, K270, twelve months after storage deriving from Raches province.

Parameter	Organic	Conventional	Significance
Free acidity (g oleic acid 100 g^{-1})	0.50	0.43	ns
Peroxides (meq O_2 kg^{-1})	16.67	17.91	ns
K232	1.09	1.96	**
K270	0.08	0.16	*

Asterisk indicates a significant difference between the means within the same row. ns, not significant, *, $p < 0.05$, **, $p < 0.01$.

On the other hand, conventionally produced oils presented higher antioxidant capacity based on the FRAP assay, as can be seen in Table 7.

Table 7. Effects of cultivation management on olive total phenol, o-diphenol, and flavonoid concentration and antioxidant capacity twelve months after storage derived from Raches province.

Parameter	Organic	Conventional	Significance
Total phenols (mg GAE kg^{-1})	249.8	283.7	ns
o-diphenols (mg CAE kg^{-1})	19.38	14.63	ns
Total flavonoids (mg CtE kg^{-1})	145.8	160.8	ns
FRAP (μmol Trolox equiv. kg^{-1})	1120.38	1297.3	*
DPPH (μmol Trolox equiv. kg^{-1})	750.8	831.0	ns

Asterisk indicates a significant difference between the means within the same row. ns, not significant, *, $p < 0.05$.

No significant differences were determined among organic and conventional oils twelve months after storage regarding the individual phenolic compound concentration (Table 8).

Table 8. Effects of cultivation management on the concentration of individual phenolic compounds detected in olive oils in mg kg^{-1} twelve months after storage derived from Raches province.

Parameter	Organic	Conventional	Significance
Hydroxytyrosol	14.05	12.94	ns
Tyrosol	51.24	53.97	ns
Vanillic acid	0.22	0.22	ns
Caffeic acid	0.14	0.07	ns
Vanillin	0.05	0.06	ns
p-coumaric acid	0.14	0.11	ns
Ferulic acid	0.03	0.03	ns
Oleacein	42.56	46.98	ns
Oleocanthal	129.85	127.23	ns
Luteolin	17.06	15.96	ns
Apigenin	2.61	2.53	ns
a-tocopherol	17.33	13.84	ns

ns, not significant.

Similarly, no significant differences were determined among organic and conventional oils twelve months after storage regarding the fatty acids content and squalene concentration (Table 9).

Table 9. Effects of cultivation management on the content of free fatty acids (determined as FAMEs—%) and squalene concentration twelve months after storage derived from Raches province.

Parameter	Organic	Conventional	Significance
C16	13.06	13.38	ns
C16:1	0.63	0.67	ns
C18	2.51	2.58	ns
C18:1	77.47	76.92	ns
C18:2	5.30	5.43	ns
C20	0.27	0.26	ns
C18:3	0.58	0.59	ns
C20:1	0.18	0.17	ns
SFA	15.82	16.19	ns
MUFA	78.29	77.77	ns
PUFA	5.88	6.03	ns
MUFA/PUFA	13.65	12.93	ns
UFA/SFA	5.32	5.17	ns
C18:1/C18:2	15.07	14.18	ns
Squalene (mg 100 g^{-1})	689.2	660.6	ns

ns, not significant.

The following table (Table 10) presents the significant changes observed during storage, as well as the percentage of change (positive—increase or negative—decrease) in the oils produced in Raches province under the two cultivation practices. Changes that were not significant under both cultivation practices are not presented. The concentration of o-diphenols was significantly reduced under storage, but the reduction was greater in the conventional oils. Total flavonoid concentration was also significantly reduced, similar to the antioxidant capacity measured by the DPPH assay. Free acidity increased under storage in both conventional and organic oils, as did the peroxide concentration. The UV absorbance indexes (K232 and K270) were significantly reduced only in organic oils during the storage period. Both C16:1 and C18:3 fatty acids were significantly reduced in both conventional and organic oils, presenting almost similar reduction percentages. On the other hand, the concentration of both hydroxytyrosol and tyrosol increased in storage, exhibiting almost similar percentages in both oils. Nonetheless, all four other phenolic compounds (vanillin acid, vanillin, p-coumaric acid, and oleocanthal) presented significant differences, with the most significant reduction being that of p-coumaric acid, reaching almost 80%. A-tocopherol presented a similar reduction too, losing almost 80% of its original concentration during storage.

Table 10. The percentage change and its significance during the twelve-month storage of the oils produced under the two cultivation practices in Raches province.

Parameter	Organic		Conventional	
	%	Significance	%	Significance
o-diphenols (mg CAE kg^{-1})	−52.5	ns	−72.7	***
Total flavonoids (mg CtE kg^{-1})	−61.4	***	−66.2	***
DPPH (μmol Trolox equiv. kg^{-1})	−36.4	***	−40.6	***
Free acidity (g oleic acid 100 g^{-1})	18.3	ns	26.10	*
Peroxides (meq O_2 kg^{-1})	29.0	ns	87.0	***
K232	−33.4	*	2.54	ns
K270	−52.2	***	−9.3	ns
C16:1	−22.2	***	−18.8	***
C18:3	−17.7	***	−13.3	*
Hydroxytyrosol	146.7	***	211.9	***
Tyrosol	445.6	***	547.9	***
Vanillic acid	−53.4	***	−48.5	***
Vanillin	−64.8	***	−59.7	***
p-coumaric acid	−78.6	***	−74.8	***
Oleocanthal	−9.70	*	−16.5	*
a-tocopherol	−78.7	***	−83.8	***

Asterisk indicates a significant difference between the means within the same row. ns, not significant, *, $p < 0.05$, ***, $p < 0.001$.

4. Discussion

The PCA analyses at harvest showed minimal significant differences between organically and conventionally produced olive oil. Similar findings have been reported by various researchers, suggesting that the observed differences are not substantial enough to clearly distinguish the two cultivation systems. Most researchers agree that the greatest difference lies in the zero-pesticide residues found in organic olive oil.

In the present experiment, the two practices resulted in similar but slightly differing olive oil quality in the trial conducted in Pylia. On the other hand, in the second year, the trial in Raches revealed no significant differences between the two systems concerning olive oil quality attributes. Based on our results, it can be inferred that the major reason for these non-significant differences is the variability in olive oil quality within each production system. This was demonstrated by the differences found between farmers managing their olive groves under the same principles of either organic or conventional practices.

In Pylia province, organic olive groves produced significantly higher yield and oil quantity per tree, as also reported by Volakakis et al. [26]. However, the opposite was observed in Raches province, underscoring the difficulty of determining which production system is superior based solely on yield. Similar discrepancies have been reported by other researchers too [1,21,23,26] driving the meta-analyses to conclude that there are generally no significant differences between the two systems [26]. This is especially true when organic olive groves are managed intensively with balanced nutrient inputs and under both irrigated as well as rain-fed conditions [1].

The oil content per fruit mass was not statistically different between the two systems in either region, suggesting that the biosynthetic pathway of olive oil is not affected by management practices, even when differences in fruit yield occur. Contrarily, Volakakis et al. [23] found a higher oil content per fruit in organically managed groves. This discrepancy could be attributed to several factors, such as cultivar, irrigation practices, fruit maturity at harvest, and pedoclimatic conditions.

None of the most commonly assessed quality attributes of the olive oil (i.e., free acidity, Kappa values, and peroxide index) differed between the two systems in either province. This aligns with findings from other studies [12,19,24,27], although conflicting reports exist, with some studies reporting higher or lower acidity in organic oils [3,17,19,28]. It appears that the oil acidity produced under either system largely depends on the cultivar assayed [19]. It must be noted here though that all oils produced were within the range of EVOO solely based on these parameters. Their values are significantly influenced by both pre-harvest (pest infestation, ripening stage etc) and post-harvest management (time passed between harvest and milling, malaxation conditions etc) [2,12,23,28,29]. As all these factors were carefully controlled and managed, and the milling conditions were consistent, it can be assumed that healthy fruits processed under favorable conditions preserved olive oil quality.

Phenolic compounds are a group of bioactive molecules of significant importance for human health [4,10,16]. Olive oil is considered a valuable source of such molecules and its role in the Mediterranean diet highlights its importance [27,30]. In this study, no significant differences in phenolic compounds were observed between conventionally and organically produced olive oil. However, in Pylia, o-diphenols and total flavonoids were higher in conventionally produced olive oil. Similarly, in Raches village the antioxidant capacity of the conventional olive oil (based on FRAP assay) was higher than the one of organic oil. While some studies report higher phenolic content in organic olive oil [12,13,15,18,19,31], others find no consistent or significant differences [3,6,27]. Lozano-Castellón et al. [16] reported higher tyrosol and hydroxytyrosol concentrations in conventional oils, while Jiménez et al. [24,32] observed higher phenol concentrations in conventionally produced oils from the Picual cultivar.

It is often expected that organic olive oil would contain higher phenolic compound concentrations, but this was not observed in our study. In Pylia, this discrepancy may be partly due to the higher fruit load in organically managed trees. Rosati et al. [3] and Inglese et al. [33] noted that low fruit load leads to earlier fruit coloration and higher polyphenol content, which may explain our findings. Furthermore, the variability among farms managed under the same practice, may also contribute to the lack of significant differences, as already mentioned by Rosati et al. [3]. The different microclimatic conditions of each orchard may have played a significant role in phenolic compound biosynthesis, justifying the differences found among orchards managed under the same practice. It is important to note that the oil (the fatty portion of the fruit) is only a small fraction of the constituents of the olive fruit. Therefore, any agronomic practice-induced changes in the fruit are often not detectable in the extracted oil [3]. This was further evidenced by the minimal differences observed in individual phenolic compound concentrations over both years, consistent with previous literature [9,27].

The phenolic profile of the oil samples included various classes of compounds, such as phenolic acids (vanillic acid, caffeic acid, p-coumaric acid, ferulic acid), phenolic alcohols (hydroxytyrosol, tyrosol), flavonoids (luteolin, apigenin), and secoiridoids (oleacein and oleocanthal), in line with previous studies [34–37]. Overall, the olive cultivar, the growing practices, the pedoclimatic conditions, and the technological factors during oil extraction collectively influence the phenolic composition of the olive products, making it difficult to identify a single dominant factor [13,15,17,32]. Similarly, α-tocopherol concentration did not differ between oils produced under the two practices in Raches, although previous studies have reported higher concentrations in organic oils [17,31].

Even though in Raches province the oils produced did not exhibit any significant difference regarding their fatty acid composition, this was not the same in oils from Pyllia. Conventionally produced oil presented a higher oleic acid and MUFA percentage, while organic oils exhibited higher PUFA percentages. No significant difference was observed regarding the squalene concentration. Rosati et al. [3] also reported increased PUFA concentrations in organic oils, attributing this to reduced nitrogen fertilization compared to conventional systems. Contrarily, Malheiro et al. [4] observed the opposite, while other studies found no differences [6,26,28], highlighting the complexity of agronomic practices' effects on fatty acid composition.

After twelve months of storage at room temperature, oils from Raches showed significant changes in major quality and functional attributes. Both systems exhibited similar degradation patterns, except for o-diphenol and peroxide concentrations, where conventional oils showed a larger change. It was clear that the quality of oils deteriorated significantly, based on most of the quality characteristics assayed. Similar quality degradation after storage has been reported by numerous researchers, attributed to various factors, with the most important ones being exposure to oxygen, high temperature, and light, inducing hydrolytic and oxidative phenomena [9,38,39]. These oxidation procedures result in phenol and tocopherol losses, followed by a decrease in the concentration of unsaturated fatty acids, similar to those recorded in the present study. Interestingly, only hydroxytyrosol and tyrosol concentrations increased during storage, likely due to their release from more complex compounds (bound phenols), as has been already described [7,9]. The decreases in antioxidant compounds such as phenolic compounds and a-tocopherol were expected, as these contribute to the antioxidant capacity and stability of the oil [7], protecting the fatty acids from oxidation and olive oil from degradation through oxidation and hydrolysis processes [6,7,39]. An average loss of 46% of the initial phenol concentration in the olive oil after 12 months of storage in the dark at 25 °C, as in the present experiment, has been also reported by others [40,41]. However, when their concentration falls below the limit for efficient protection of valuable olive biomolecules, the increase in free acidity and peroxides is expected [7].

The most significant result was the fact that the production system (organic or conventional) did not appear to influence olive oil stability or functional properties during storage. The fatty acid composition and the balance of antioxidants and pro-oxidants are the key factors affecting oxidation patterns [7]. Since in this study few differences were observed at harvest between the two systems, a similar stability behavior during storage is justifiable.

5. Conclusions

Olive oils produced under the two cultivation systems exhibited similar quality characteristics. All farmers managed their trees based on judicious use of inputs and protected the olive fruit from biotic damages, thus resulting in extra-virgin olive oil production. The individual farmer appears to play a significant role in the implementation of cultivation practices, which, in turn, affects the quality traits of the oil. This leads to variations even among oils produced within the same cultivation system (organic or conventional). In conclusion, when management practices lead to healthy fruit production and oil extraction occurs under optimal conditions, the oil quality is expected to be high, regardless of the management system used. However, farmers may influence the quality attributes of the oil produced by applying different fertilization and plant protection programs. In the future, attention should be focused on how such practices may affect desirable oil characteristics.

Supplementary Materials: The following supporting information can be downloaded at: https://www.mdpi.com/article/10.3390/horticulturae11020130/s1, Table S1: Pesticide and fertilizer applications in organic and conventional olive groves in Pyllia province; Table S2: Pesticide and fertilizer applications in organic and conventional olive groves in Raches province.

Author Contributions: Conceptualization, P.A.R.; methodology, P.A.R.; software, P.A.R. and A.-G.K.; validation, P.A.R.; formal analysis, A.-G.K., V.K. and K.Z.; investigation, P.A.R., A.-G.K., V.K. and K.Z.; resources, P.A.R.; data curation, P.A.R., A.-G.K., V.K. and K.Z.; writing—original draft preparation, P.A.R. and A.-G.K.; writing—review and editing, P.A.R.; visualization, P.A.R.; supervision, P.A.R. and A.-G.K.; project administration, P.A.R. and A.-G.K.; funding acquisition, P.A.R. All authors have read and agreed to the published version of the manuscript.

Funding: This research received no external funding.

Data Availability Statement: Data is contained within the article and Supplementary Materials.

Acknowledgments: We would like to thank all the farmers who gave us access to their farms and contributed with the olive fruits produced. This article is a revised and expanded version of a paper entitled "Olive oil quality characteristics under organic and integrated management in Messinia county Greece", which was presented at the XIV International Scientific Agriculture Symposium "AGROSYM 2023", Jahorina, Bosnia and Herzegovina, 5–8 October 2023.

Conflicts of Interest: The authors declare no conflicts of interest.

References

1. Pleguezuelo, C.R.R.; Zuazo, V.H.D.; Martínez, J.R.F.; Peinado, F.J.M.; Martín, F.M.; Tejero, I.F.G. Organic olive farming in Andalusia, Spain. A review. *Agron. Sustain. Dev.* **2018**, *38*, 20. [CrossRef]
2. Caporaso, N.; Boskou, D. Olive (*Olea europaea*). In *Oilseeds: Health Attributes and Food Applications*; Springer: Berlin/Heidelberg, Germany, 2021; pp. 211–252.
3. Rosati, A.; Cafiero, C.; Paoletti, A.; Alfei, B.; Caporali, S.; Casciani, L.; Valentini, M. Effect of agronomical practices on carpology, fruit and oil composition, and oil sensory properties, in olive (*Olea europaea* L.). *Food Chem.* **2014**, *159*, 236–243. [CrossRef] [PubMed]
4. Malheiro, R.; Casal, S.; Baptista, P.; Pereira, J.A. *How Agronomic Factors Affects Olive Oil Composition and Quality*; Nova Science Publishers: Hauppauge, NY, USA, 2014.
5. Beltrán, G.; Bucheli, M.E.; Aguilera, M.P.; Belaj, A.; Jimenez, A. Squalene in virgin olive oil: Screening of variability in olive cultivars. *Eur. J. Lipid Sci. Technol.* **2016**, *118*, 1250–1253. [CrossRef]
6. Lanza, B.; Ninfali, P. Antioxidants in extra virgin olive oil and table olives: Connections between agriculture and processing for health choices. *Antioxidants* **2020**, *9*, 41. [CrossRef] [PubMed]
7. Tsimidou, M.Z. Olive oil quality. In *Olive Oil*; Elsevier: Amsterdam, The Netherlands, 2006; pp. 93–111.
8. Mastralexi, A.; Tsimidou, M.Z. On the squalene content of CV Chondrolia Chalkidikis and Chalkidiki (Greece) virgin olive oil. *Molecules* **2021**, *26*, 6007. [CrossRef] [PubMed]
9. Kalogeropoulos, N.; Tsimidou, M.Z. Antioxidants in Greek virgin olive oils. *Antioxidants* **2014**, *3*, 387–413. [CrossRef]

10. Olmo-Cunillera, A.; Lozano-Castellón, J.; Pérez, M.; Miliarakis, E.; Tresserra-Rimbau, A.; Ninot, A.; Romero-Aroca, A.; Lamuela-Raventós, R.M.; Vallverdú-Queralt, A. Optimizing the Malaxation Conditions to Produce an Arbequina EVOO with High Content of Bioactive Compounds. *Antioxidants* **2021**, *10*, 1819. [CrossRef]
11. Jurado-Campos, N.; García-Nicolás, M.; Pastor-Belda, M.; Bußmann, T.; Arroyo-Manzanares, N.; Jiménez, B.; Viñas, P.; Arce, L. Exploration of the potential of different analytical techniques to authenticate organic vs. conventional olives and olive oils from two varieties using untargeted fingerprinting approaches. *Food Control* **2021**, *124*, 107828. [CrossRef]
12. Carrapiso, A.I.; Rubio, A.; Sánchez-Casas, J.; Martín, L.; Martínez-Cañas, M.; de Miguel, C. Effect of the organic production and the harvesting method on the chemical quality and the volatile compounds of virgin olive oil over the harvesting season. *Foods* **2020**, *9*, 1766. [CrossRef] [PubMed]
13. López-Yerena, A.; Lozano-Castellón, J.; Olmo-Cunillera, A.; Tresserra-Rimbau, A.; Quifer-Rada, P.; Jiménez, B.; Pérez, M.; Vallverdú-Queralt, A. Effects of organic and conventional growing systems on the phenolic profile of extra-virgin olive oil. *Molecules* **2019**, *24*, 1986. [CrossRef] [PubMed]
14. Inglese, P.; Famiani, F.; Galvano, F.; Servili, M.; Esposto, S.; Urbani, S. 3 factors affecting extra-virgin olive oil composition. *Hortic. Rev.* **2011**, *38*, 83.
15. Boskou, D. Olive fruit, table olives, and olive oil bioactive constituents. In *Olive and Olive Oil Bioactive Constituents*; Elsevier: Amsterdam, The Netherlands, 2015; pp. 1–30.
16. Lozano-Castellón, J.; López-Yerena, A.; Domínguez-López, I.; Siscart-Serra, A.; Fraga, N.; Sámano, S.; López-Sabater, C.; Lamuela-Raventós, R.M.; Vallverdú-Queralt, A.; Pérez, M. Extra virgin olive oil: A comprehensive review of efforts to ensure its authenticity, traceability, and safety. *Compr. Rev. Food Sci. Food Saf.* **2022**, *21*, 2639–2664. [CrossRef]
17. Ríos-Reina, R.; Camacho, F.; Morales, M.L.; Jiménez-Herrera, B.; Callejón, R.M. Influence of irrigation modalities (irrigation management and dryland), fruit ripening, and cultivation modality (organic and conventional) on quality and chemosensory profile of Hojiblanca and Picual extra virgin olive oils. *Eur. J. Lipid Sci. Technol.* **2021**, *123*, 2000375. [CrossRef]
18. Anastasopoulos, E.; Kalogeropoulos, N.; Kaliora, A.C.; Kountouri, A.; Andrikopoulos, N.K. The influence of ripening and crop year on quality indices, polyphenols, terpenic acids, squalene, fatty acid profile, and sterols in virgin olive oil (*Koroneiki* cv.) produced by organic versus non-organic cultivation method. *Int. J. Food Sci. Technol.* **2011**, *46*, 170–178. [CrossRef]
19. Jimenez, B.; Sánchez-Ortiz, A.; Lorenzo, M.L.; Rivas, A. Effect of organic cultivation of Picual and Hojiblanca olive varieties on the quality of virgin olive oil at four ripening stages. *Eur. J. Lipid Sci. Technol.* **2014**, *116*, 1634–1646. [CrossRef]
20. Panagodimou, E.; Roussis, I.; Papadas, C.T.; Bilalis, D.J. A Comparison of Cost in Organic and Conventional Olive Oil Production in Greece. *Bull. Univ. Agric. Sci. Vet. Med. Cluj-Napoca. Hortic.* **2019**, *76*, 127–130. [CrossRef]
21. Berg, H.; Maneas, G.; Salguero Engström, A. A Comparison between Organic and Conventional Olive Farming in Messenia, Greece. *Horticulturae* **2018**, *4*, 15. [CrossRef]
22. Kalogiouri, N.P.; Aalizadeh, R.; Thomaidis, N.S. Investigating the organic and conventional production type of olive oil with target and suspect screening by LC-QTOF-MS, a novel semi-quantification method using chemical similarity and advanced chemometrics. *Anal. Bioanal. Chem.* **2017**, *409*, 5413–5426. [CrossRef]
23. Volakakis, N.; Kabourakis, E.; Kiritsakis, A.; Rempelos, L.; Leifert, C. Effect of Production System (Organic versus Conventional) on Olive Fruit and Oil Yields and Oil Quality Parameters in the Messara Valley, Crete, Greece; Results from a 3-Year Farm Survey. *Agronomy* **2022**, *12*, 1484. [CrossRef]
24. Jimenez, B.; Sánchez-Ortiz, A.; Lorenzo, M.L.; Rivas, A. Effect of Agronomical Practices on the Nutritional Quality of Virgin Olive Oil at Different Ripening Stages. *J. Am. Oil Chem. Soc.* **2015**, *92*, 1491–1501. [CrossRef]
25. Roussos, P.A.; Karabi, A.; Anastasiou, L.; Assimakopoulou, A.; Gasparatos, D. Apricot tree nutrient uptake, fruit quality and phytochemical attributes, and soil fertility under organic and integrated management. *Appl. Sci.* **2023**, *13*, 2596. [CrossRef]
26. Volakakis, N.; Kabourakis, E.; Leifert, C. Conventional and organic cultivation and their effect on the functional composition of olive oil. In *Olives and Olive Oil as Functional Foods: Bioactivity, Chemistry and Processing*; Wiley: Oxford, UK, 2017; pp. 35–43.
27. Ninfali, P.; Bacchiocca, M.; Biagiotti, E.; Esposto, S.; Servili, M.; Rosati, A.; Montedoro, G. A 3-year study on quality, nutritional and organoleptic evaluation of organic and conventional extra-virgin olive oils. *J. Am. Oil Chem. Soc.* **2008**, *85*, 151–158. [CrossRef]
28. Luis, G.-G.D.; Ramón, A.-R.; Teresa, M.M. Chemical characterization of organic and non-organic virgin olive oils. *OCL* **2014**, *21*, D506.
29. Ghreishi Rad, S.A.; Jalili, M.; Ansari, F.; Rashidi Nodeh, H.; Rashidi, L. Maturity impact on physicochemical composition and polyphenol properties of extra virgin olive oils obtained from Manzanilla, Arbequina, and Koroneiki varieties in Iran. *Food Sci. Nutr.* **2023**, *11*, 5396–5408. [CrossRef]
30. López-Yerena, A.; Ninot, A.; Jiménez-Ruiz, N.; Lozano-Castellón, J.; Pérez, M.; Escribano-Ferrer, E.; Romero-Aroca, A.; Lamuela-Raventós, R.M.; Vallverdú-Queralt, A. Influence of the Ripening Stage and Extraction Conditions on the Phenolic Fingerprint of 'Corbella' Extra-Virgin Olive Oil. *Antioxidants* **2021**, *10*, 877. [CrossRef]
31. Gutiérrez, F.; Arnaud, T.; Albi, M.A. Influence of ecological cultivation on virgin olive oil quality. *J. Am. Oil Chem. Soc.* **1999**, *76*, 617–621. [CrossRef]

32. Jiménez, B.; Rivas, A.; Lorenzo, M.L.; Sánchez-Ortiz, A. Chemosensory characterization of virgin olive oils obtained from organic and conventional practices during fruit ripening. *Flavour Fragr. J.* **2017**, *32*, 294–304. [CrossRef]
33. Inglese, P.; Famiani, F.; Servili, M. I fattori di variabilità genetici, ambientali e colturali della composizione dell'olio di oliva. *Italus Hortus* **2009**, *16*, 67–81.
34. Pino, C.; Sepúlveda, B.; Tapia, F.; Saavedra, J.; García-González, D.L.; Romero, N. The Impact of Mild Frost Occurring at Different Harvesting Times on the Volatile and Phenolic Composition of Virgin Olive Oil. *Antioxidants* **2022**, *11*, 852. [CrossRef] [PubMed]
35. Faghim, J.; Mohamed, M.B.; Bagues, M.; Guasmi, F.; Triki, T.; Nagaz, K. Irrigation effects on phenolic profile and extra virgin olive oil quality of "Chemlali" variety grown in South Tunisia. *S. Afr. J. Bot.* **2021**, *141*, 322–329. [CrossRef]
36. Flamminii, F.; Gaggiotti, S.; Chiaudani, A.; Compagnone, D.; Cichelli, A. The Introduction of Allochthonous Olive Variety and Super High-Density System in the Abruzzo Region: A Study on Olive Oil Quality. *Foods* **2023**, *12*, 1292. [CrossRef]
37. Crizel, R.L.; Hoffmann, J.F.; Zandoná, G.P.; Lobo, P.M.S.; Jorge, R.O.; Chaves, F.C. Characterization of Extra Virgin Olive Oil from Southern Brazil. *Eur. J. Lipid Sci. Technol.* **2020**, *122*, 1900347. [CrossRef]
38. Velasco, J.; Dobarganes, C. Oxidative stability of virgin olive oil. *Eur. J. Lipid Sci. Technol.* **2002**, *104*, 661–676. [CrossRef]
39. Cinquanta, L.; Esti, M.; Notte, E.L. Evolution of phenolic compounds in virgin olive oil during storage. *J. Am. Oil Chem. Soc.* **1997**, *74*, 1259–1264. [CrossRef]
40. Diamantakos, P.; Ioannidis, K.; Papanikolaou, C.; Tsolakou, A.; Rigakou, A.; Melliou, E.; Magiatis, P. A New Definition of the Term "High-Phenolic Olive Oil" Based on Large Scale Statistical Data of Greek Olive Oils Analyzed by qNMR. *Molecules* **2021**, *26*, 1115. [CrossRef] [PubMed]
41. Roussos, P.A.; Karida, A.-G.; Kaibalis, V. Olive oil quality characteristics under organic and integrated management in messinia county Greece. In Proceedings of the XIV International Scientific Agriculture Symposium "AGROSYM 2023", Jahorina, Bosnia and Herzegovina, 5–8 October 2023; pp. 165–171.

horticulturae

Article

Effects of Organic and Conventional Cultivation on Composition and Characterization of Two Citrus Varieties 'Navelina' Orange and 'Clemenules' Mandarin Fruits in a Long-Term Study

Alfons Domínguez-Gento [1], Rosita Di Giorgi [2], María Dolores García-Martínez [2] and María Dolores Raigón [2,*]

[1] Organic Production and Innovation Service, Agricultural Experimental Station, Pda. Barraquet, s/n, 46740 Carcaixent, Spain; dominguez_alf@gva.es
[2] Departamento de Química, Instituto de Conservación y Mejora de la Agrobiodiversidad Valenciana, Universitat Politècnica de Valencia, Camino de Vera, s/n, 46022 Valencia, Spain; r.digiorgi1@studenti.unipi.it (R.D.G.); magarma8@qim.upv.es (M.D.G.-M.)
* Correspondence: mdraigon@qim.upv.es

Abstract: A transition towards an organic food system is taking place around the world. This process is favored by growing consumer demand, who associate organic crops with being healthier, tastier, and safer for the environment than conventional crops. Citrus is one of the most widely produced crops worldwide and has important socio-economic and cultural significance in the Mediterranean area. The aim of this work is twofold; on the one hand, it reveals the variability of a set of physical–chemical and nutritional quality parameters of two citrus fruit varieties, 'Navelina' oranges and 'Clemenules' mandarins, from organic and conventional production in a long-term study. On the other hand, taking advantage of the large number of results, a model is proposed that allows the successful differentiation of citrus fruits from organic and conventional production and a tool that allows predicting the production system of citrus fruits. The results suggest that organically produced citrus fruits do not generate differences in terms of external aspects, providing fruits that are acceptable to the market. Organic production techniques influenced the lower peel content and higher pulp and juice content in 'Navelina' orange and 'Clemenules' mandarin fruits and led to a greater ability to synthesize vitamin C in the juice, more essential oils in the skin, and higher seed numbers, although in all cases, the fruits can be classified as having low seed numbers. Two discriminating equations were obtained that use easy-to-measure parameters to successfully classify organic citrus fruits. The classification and prediction models obtained constitute useful tools to help in the control of the purity/authenticity of organic citrus fruits.

Keywords: vitamin C; citrus; comparison; Mediterranean conditions; physicochemical quality; nutritional

1. Introduction

Citrus fruits (*Citrus* L. from Rutaceae) are among the major fruit crops in international trade in terms of economic value and are in high demand worldwide for fresh consumption, juice production, the manufacture of goods such as jams and frozen concentrates, and as food additives for dishes and beverages [1,2]. In addition, citrus fruits are also widely used to produce essential oils and for the manufacture of liqueurs [3]. Annual citrus production is 158 million tons, with a cultivation area of more than 10.1 million hectares worldwide [4]. Citrus production has important economic, social, and cultural significance in the Mediterranean region; in fact, oranges are one of the largest agricultural export crops. Citrus orchards are especially important near Valencia, where they produce more than 70% of the total Spanish citrus crops [5].

The *Citrus* genus includes oranges, lemons, grapefruits, limes, mandarins, kumquats, tangelos, and others [6]. They represent an important part of our daily diet, not only due to their attractive colors, special aromas, and delicious flavors, but also their high nutritional value and health-promoting effects [7,8]. Evidence from several in vitro and in vivo studies has related citrus fruit consumption to a reduced risk of chronic diseases such as cancers, cardiovascular diseases, and diabetes [8,9]. These fruits contain a range of highly beneficial bioactive compounds, such as polyphenols, coumarins, limonoids, carotenoids, essential oils, minerals, and vitamins, which alone or in combination show antimicrobial and antioxidant properties and help boost the body's immune system [7,10–13]. Due to their high vitamin C concentration and the large per capita consumption volume, citrus fruits are an important source of ascorbic acid in the human diet [14,15]. As the main antioxidant compound, vitamin C concentration is an important nutritional quality indicator of citrus fruit and its derivative products. The current recommended daily allowance (RDA) for this vitamin intake is 80–95 mg/day to have a positive effect on the organism.

The nutritional value of food depends on many environmental factors, such as soil type, fertilization practices, and production system (organic vs. conventional farming), among others. Nowadays, environmentally friendly farming practices are essential to ensure future production. This is a current topic that has attracted public interest and has generated much discussion [16]. Conventional farming systems have focused mainly on yield rather than on the sustainable use of available resources [17]. Technological advances and economies of scale encouraged the use of chemical solutions to control pests and diseases and increase soil productivity [10]. The massive use of chemically synthesized fertilizers, heavy machinery, and intensive agricultural practices contribute to high greenhouse gas emissions [18]. In addition, these practices have led to a rapid decline in biodiversity, affecting important ecosystem functions, such as the regulation of herbivore populations by their natural enemies, plant pollination by insects, or the decomposition process of soil fauna, resulting in significant imbalances [19].

Nowadays, agriculture and the food industry are facing new trends in the development of organic farming practices [20,21]. These techniques offer an alternative to industrial farming and are favored by the growing consumer demand for organic fruit and vegetables [22]. This is because consumers associate organic products with healthier and tastier food and consider organic food to be more environmentally friendly than their conventionally grown counterparts [23]. One of the serious problems in the production of citrus fruits is the intensification of cultivation carried out in conventional systems, generating pollution mainly derived from the nitrogen fertilizers used on orange farms, which have caused many Spanish citrus-growing areas to have nitrate levels in their aquifers above the legal limit, exceeding 300 ppm in large areas, as well as concentrations of pesticides in drinking water [24,25].

Currently, an organic transition is taking place in agricultural soil management for citrus orchards [26]. Organic farming seeks solutions to avoid soil degradation and increase soil quality. Cerda et al. [27] reported that plant covers (crops and weeds) are valid measures to control extreme soil losses. Concerns for the environment, the welfare of future generations, the aspiration to produce high-quality products, higher economic profitability, and the will to reduce dependence on agrochemicals encourage citrus farmers in the Valencia Region toward the organic transition [28]. Most well-designed studies reported that organic farming has the potential to produce high-quality food, with some relevant increases in terms of health-promoting phytonutrients and certain vitamins and minerals, and lower levels of pesticide residues, which could be assumed to provide health benefits [29]. It has long been known that there is a clear link between the food we eat and our health [30]. Citrus-rich diets have been strongly associated with numerous health benefits and a reduced risk of disease [31].

The recent studies on the effects of applying organic methods in citrus crops focus on the impacts on the soil [32] and other environmental effects [33]; few papers study the effect on the quality of the fruits. Some previous work [34] found significantly higher vitamin C levels in red oranges and other fruits grown in organic conditions [35]. Recently, He et al. [36] indicated that the application of organic substance fertilizers, such as humic and fulvic acids, improved the quality of 'Eureka' lemon fruits. In particular, the edible rate and juice production increased, as well as the contents of vitamin C, total acid, total sugar, total soluble solids, and the number of seeds.

The objective of this study is twofold; on the one hand, it reveals the variability of a set of physical–chemical and nutritional quality parameters of two citrus fruit varieties, 'Navelina' oranges (*Citrus sinensis* L. Osbeck) and 'Clemenules' mandarins (*Citrus reticulata* Blanco), from organic and conventional production in a long-term study. On the other hand, taking advantage of the large number of results, a model is proposed that allows the successful differentiation of citrus fruits from organic and conventional production.

This study will contribute to increasing the knowledge of the effect of the organic production method on citrus fruit quality. It will also provide a prediction complex for classifying citrus fruits according to the production system (organic and conventional).

2. Materials and Methods

2.1. Experimental Materials

The fruits were grown in the Valencia area, based on two production systems (organic and conventional) and different production years. Several fruit quality parameters were analyzed, including physical (shape, fresh weight, diameter, height, peel color index, percentage of juice, peel, pulp, number of seeds, juice density, skin thickness, and albedo thickness) and chemical (total soluble solids, pH, total acidity, ripeness index, essential oil, and vitamin C concentration) characteristics. All parameters are related to quality and organoleptic attributes.

Fruits of the mandarin variety 'Clemenules' (a natural bud mutation of the variety Fina) and 'Navelina' orange were used as the plant material. The citrus fruits were harvested at their optimum ripening stage in the period between November and January during seven different seasons (1999–2000, 2001–2002, 2007–2008, 2008–2009, 2009–2010, 2010–2011, and 2020–2021). It was not possible to analyze fruits from all consecutive seasons since 1999, mainly due to resource problems and climatic conditions. However, there are a large number of study years that include possible seasonal variability. Plants were grown in the Central-East of the Valencian Community, in the province of Valencia, under open field conditions, both in organic (ORG) and conventional (CONV) systems, with similar edaphoclimatic and environmental conditions. The citrus plots are located inland, approximately 30 km from the Mediterranean coast, in an area known as La Casella (Ribera Alta region). With a semi-arid Mediterranean climate, the average annual temperature is 17.6 °C (July is the warmest month of the year, with average temperatures of 26.4 °C, and January is the coldest month, with average temperatures of 9.8 °C). The driest month is July, with 9 mm of precipitation, and most precipitation falls in October (63 mm); the annual accumulated rainfall is 460 mm.

The type of soil in the study area is 'entisols' [37]. In general, these soils are characterized by a loamy clay texture, and the main composition (10–30 cm deep) properties of organic and conventional soils are shown in Table 1. In general, organic soils are characterized by having a higher water content, higher organic matter content, higher enzymatic activity, and higher cation exchange capacity, due to the higher concentrations of calcium, potassium, and magnesium.

2.2. Experimental Method

The fruits were harvested in situ in the plots of origin, from healthy and mature trees that were initially older than 20 years. The variety 'Navelina' was grafted on Citrange troyer, on Citrange carrizo, and on bitter orange trees, depending on the plot. The variety 'Clemenules' was grafted on Citrange troyer, Citrange carrizo rootstock, particularly with loamy soils and drip irrigation systems. Organic farms are certified according to European regulations, specifically EEC Regulation 2092/91, Council Regulation (EC) 834/2007, and Regulation (EU) 2018/848 of the European Parliament and the Council. As it is a long-term study, the organic citrus plots have gone through different regulatory changes in the European Union, but the changes in regulations have not affected the traditional management systems that have been in place for more than twenty years.

Table 1. Composition characteristics of organic and conventional soils.

Parameter	Organic Soil	Conventional Soil
Humidity (%)	7.00 ± 0.23	5.88 ± 0.29
pH	7.75 ± 0.04	7.08 ± 0.05
Conductivity (dS m^{-1})	0.160 ± 0.006	0.122 ± 0.009
$CaCO_3$ (%)	1.79 ± 0.09	0.92 ± 0.06
Total N (%)	0.059 ± 0.005	0.041 ± 0.007
Organic matter (%)	1.28 ± 0.09	0.85 ± 0.27
C/N	12.92 ± 0.78	11.61 ± 2.13
Enzymatic activity * (μg PNP g^{-1} soil h^{-1})	186.15 ± 15.97	64.21 ± 10.47
K (mg kg^{-1})	113.74 ± 17.97	79.15 ± 10.58
P (mg kg^{-1})	54.93 ± 4.72	59.00 ± 1.81
Ca (mg kg^{-1})	1641.24 ± 153.39	664.20 ± 84.07
Mg (mg kg^{-1})	252.67 ± 28.62	149.20 ± 5.62
Na (mg kg^{-1})	51.57 ± 1.03	50.53 ± 0.37
Fe (mg kg^{-1})	5.85 ± 0.36	9.40 ± 0.44
Cu (mg kg^{-1})	0.71 ± 0.07	2.36 ± 0.43
Zn (mg kg^{-1})	3.81 ± 0.35	12.46 ± 1.79
Cation exchange capacity	10.78 ± 0.75	4.96 ± 0.48

* Determined as alkaline phosphatase (PNP = p-nitrophenol).

In all cases, five organic orchards and five conventional ones, located in the same production area, were sampled. The organic plots have live hedges in their perimeters that act as a natural border and as a refuge for auxiliary fauna. Organic soils have plant covers, mainly a mixture of vetch (100 kg ha^{-1}) and oats (80 kg ha^{-1}). Fertilization consisted of the contribution of between 15 and 20 tons of sheep manure per hectare and year; half was usually contributed at the beginning of September and the other half in March. Authorized biological control applications were made when the balance of the ecosystem was destabilized. The rest of the irrigation techniques and water doses, pruning, etc., were similar to conventional techniques. There are also no differences in the planting frames, with average densities of 6 m × 4 m. The conventional soils are characterized by not having vegetation covers (bare soils), using tillage and authorized herbicides. Conventional fertilization consisted of the application by hectare and year of 350 kg of nitrogen, 80 kg of P_2O_5, 140 kg of K_2O, and 180 kg of MgO, and applications of microelements, mainly 1 kg of iron, using inorganic sources of synthesis. Phytosanitary practices were carried out in the conventional plots when an imbalance occurred, using the authorized substances.

2.3. Sampling and Analytical Methodology

In each season, the harvesting was carried out manually, on the same day, taking fruits from the entire plot. Trees were randomly selected, eliminating those on the perimeter, and fruits were taken from inside and outside the canopy, from the four cardinal points. Approximately 100 kg of fruits per 1250 m^2 and year and from a whole experimental area were collected for analysis (15 sample trees).

The fruits were in a range of commercial ripeness recommended for local farmers, and this point varies from season to season depending on climatic parameters. In total, 1588 'Clemenules' fruits (794 ORG and 794 CONV) and 984 'Navelina' fruits (492 ORG and 492 CONV) were examined. The quality parameters analyzed for each fruit are presented in Figure 1. Each parameter analysis was performed in triplicate. The fresh citrus fruits studied were analyzed to assess their overall quality.

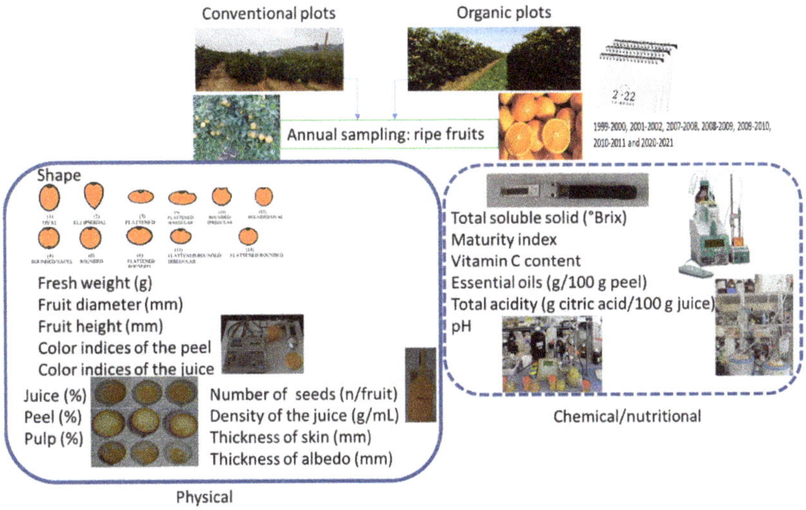

Figure 1. Scheme of several physical, chemical, and nutritional parameters evaluated for each fruit sample.

Representative fruits from each sample plot were individually weighed, measured, and used for juice extraction. The number of seeds was determined by counting the seeds extracted from the pulp of each fruit one by one. Fruit height (H), equatorial diameter (D), rind thickness (mm), skin thickness (ST), and albedo thickness (AT) were measured using a digital electronic slide gauge (model CD-15 DC; Mitutoyo (UK) Ltd., Telford, UK) within 0.01 mm accuracy. Weights of individual fruits (W), pulp, juice, and skin were obtained by weighing them on an analytical balance (CB-Junior, Cobos) with an accuracy of ±0.001 g; the content of pulp (P), skin (S), and juice (J) is expressed as a percentage. Peel color measurements in CieLab space were performed at two random points on the citrus skin and we averaged the values, and one point in the juice surface was measured using a colorimeter (Konica Minolta CR-300, Photo Imaging Inc., Mahwah, NJ, USA), obtaining the color components (L*, a*, b*), as well as the color index (CI) in skin and juice, which was calculated by the following equation: $CI = 1000 \times a/(L \times b)$, where L* indicates brightness, a* indicates chromaticity on a green (−) to red (+) axis, and b* indicates chromaticity on a blue (−) to yellow (+) axis; this index is widely used in the citrus industry as a maturation index [38].

Each fruit was manually squeezed and filtered to obtain juice, which was used to determine juice density (Dy) by the pycnometer method; total soluble solids content (TSS, °Brix) was determined by refractometry (method 932.12); and pH (method 981.12) and total acidity (TA) expressed in g anhydrous citric acid/100 mg juice (%) were determined (method 942.15) according to AOAC (1995) [39]. In addition, the TSS/TA ratio was calculated to define the ripeness index (RI) of citrus fruit [40]. Fruit shape was determined by visual observation, using a citrus chart, classifying them as oval, round, flattened, and ellipsoidal, according to a numerical scale ranging from 1 for an oval shape to 14 for a rounded/flattened shape. A frequency mosaic graph was used to identify the shape of the citrus fruit for the two varieties.

Vitamin C concentration (Vit C) in citrus juice was determined by potentiometric titration with chloramine-T, using automatic titration equipment (702 SM Titrino, Metrohm, Herisau, Switzerland) and expressed in mg ascorbic acid (AA) per 100 g of juice [41]. The essential oils are determined in the flavedo. The method applied for the determination of essential oil concentration (EOC) consisted of the extraction of these oils from citrus skin by the Clavenger system [42], and the results are expressed in g/100 g of peel (%).

2.4. Data Analysis

Data obtained were processed using Statgraphics® Centurion 18. For each citrus fruit variety, data were subjected to a two-factor analysis of variance (ANOVA) using a fixed effects model for cultivar season and production system. The significance of main effects and interactions was obtained from the ANOVA, and differences among main treatments and combinations were identified with Student–Newman–Keuls multiple-range tests. Discriminant analysis was also performed to classify the different observations. The main objective of the discriminant analysis was to calculate the linear combinations of the classifying fruits that maximize the difference between the production systems. Discriminant analysis was applied for explanatory and predictive purposes.

3. Results

3.1. Descriptive Results: Effect of Analysis of Variance

Tables 2 and 3 show the results of the physical and chemical quality parameters, respectively, of 'Navelina' fruit, produced by organic and conventional systems during the seven different seasons considered in the study. Tables 4 and 5 show the results of 'Clemenules' fruit for the same parameters. Some parameters were not analyzed in all the seasons due to the approach of the individual objectives in the corresponding season, but the existing data allow the statistical study to be carried out.

In the present study, the season had a significant influence on all physical and chemical quality parameters of 'Navelina' and 'Clemenules' fruits, except for the number of seeds for 'Navelina' fruits and skin thickness for 'Clemenules' fruits (Tables 2 and 4, respectively). This was mainly due to the influence of environmental factors, such as climatic conditions, rainfall frequency, air temperature, and daylight hours, among other factors. As reported by Lado et al. [43], climatic conditions are the most significant external factors affecting fruit quality. In general, the long-term study aims to minimize the effects of climatic incidences in favor of growing conditions. In this sense, shape, height, juice content, the number of seeds, juice color index, albedo thickness, total acidity, ripeness index, and vitamin C concentration in Navelina fruits are parameters that have the same behavior in all seasons; for example, for Navelina fruits, regardless of the season of the year, on average, the organic fruits had a higher total acidity value and contained a higher vitamin C content.

The significant differences detected in the physical quality parameters, for the cultivation system, in 'Navelina' fruit are CI of peel, skin, and pulp (%) (Table 2). Conventional oranges have a significantly higher peel color index and skin percentage (10.29 and 26.95%, respectively) than organic (9.97 and 25.17%, respectively), while the pulp percentage is significantly higher in organic 'Navelina' fruits (27.77%) than in conventional ones (26.05%).

Significant effects of cropping system x season are observed for several physical quality parameters: fresh weight, fruit diameter, skin CI, peel and pulp percentage, a* and b* (color parameters), and skin thickness (Table 2). In the parameters where no statistically significant interaction effects are observed, the farming system shows the same trend in all seasons.

For the chemical quality parameters of the 'Navelina' fruit samples, significant differences due to the cultivation system are observed for pH value, vitamin C concentration, and percentage of essential oils (Table 3). Organic 'Navelina' fruits had a significantly higher mean pH value (3.64) than conventional 'Navelina' fruits (3.58). Vitamin C concentration was significantly higher in 'Navelina' fruits grown under organic conditions with mean values of 63.70 mg per 100 g juice than their respective counterparts (57.51 mg per 100 g juice). Essential oil concentration was also significantly higher in organic 'Navelina' oranges (0.34 g per 100 g peel) than in conventional samples (0.28 g per 100 g peel). The growing method $\times$ season interaction significantly affects total soluble solids, pH, juice density, and essential oil content (Table 3). Organic 'Navelina' fruits showed a higher number of seeds per fruit (although in no case are values of the average number of seeds in fruits greater than 1) with differences to the limit to become significant (p-value = 0.0506).

Table 2. Physical quality characteristics of 'Navelina' fruits grown under organic (ORG) and conventional (CONV) conditions in the seven different seasons and probability (*p*-value) for the effects of cultivation method, season, and their interaction.

Factors [1]		Physical Quality Characteristics [2]												
		Shape	W (g/Fruit)	D (mm)	H (mm)	Skin CI	J (%)	S (%)	P (%)	L*	a*	b*	ST (mm)	AT (mm)
Cultivation system	ORG	6.60	222.82	75.25	78.53	9.97	47.07	25.17	27.77	65.06	35.09	56.47	4.87	2.18
	CONV	6.74	229.66	75.50	79.62	10.29	46.86	26.95	26.05	64.66	35.12	56.38	5.04	2.37
	p-value	0.7275	0.0868	0.5733	0.4197	0.0322	0.7221	0.0001	0.0027	0.2419	0.8824	0.7983	0.2907	0.0707
Season	1999–2000	1.40	250.09	77.90	83.32	8.19	43.00	26.83	30.16	62.22	34.53	61.67	5.09	1.60
	2001–2002						42.55	27.42	29.56				4.27	1.35
	2007–2008		263.21	80.70	79.66	7.17	50.09	26.88	23.04	68.93	34.32	69.71	5.55	
	2008–2009		183.71	69.70	73.82	13.90	46.65	26.47	26.87	55.47	42.22	55.30		
	2009–2010		236.83	76.26	85.39	14.23	49.44	25.20	25.36	55.67	38.65	50.81		
	2010–2011		199.45	72.12	77.15	11.39	44.29	26.52	29.19	65.24	39.72	56.20		
	2020–2021	11.93	224.18	75.56	75.12	5.87	52.75	23.08	24.16	81.62	21.20	44.86	4.91	3.87
	p-value	0.0000	0.0000	0.0000	0.0000	0.0000	0.0000	0.0000	0.0000	0.0000	0.0000	0.0000	0.0000	0.0000
Cultivation system × season	ORG × 1999–2000	1.52	230.47	76.52	82.35	7.08	44.08	26.00	29.92	62.50	33.80	61.13	5.29	1.57
	ORG × 2001–2002						43.93	25.16	30.94				3.80	1.24
	ORG × 2007–2008		254.65	79.63	79.34	7.34	49.06	27.59	23.35	68.74	35.13	69.96	5.67	
	ORG × 2008–2009		192.36	70.91	74.60	14.03	46.60	26.77	26.63	55.13	42.15	55.02		
	ORG × 2009–2010		221.64	75.46	80.78	14.89	48.87	24.81	26.32	55.90	39.59	49.24		
	ORG × 2010–2011		203.45	72.73	78.19	10.76	43.71	24.53	31.75	66.13	38.90	57.89		
	ORG × 2020–2021	11.68	234.38	76.22	75.95	5.68	53.25	21.29	25.45	81.95	20.97	45.57	4.73	3.74
	CONV × 1999–2000	1.28	269.71	79.29	84.29	9.30	41.92	27.66	30.41	61.94	35.27	62.21	4.88	1.65
	CONV × 2001–2002						41.17	29.67	28.19				4.76	1.46
	CONV × 2007–2008		271.76	81.78	79.99	7.01	51.12	26.16	22.72	69.12	33.52	69.45	5.44	
	CONV × 2008–2009		175.06	68.50	73.04	13.76	46.71	26.17	27.12	55.81	42.30	55.59		
	CONV × 2009–2010		252.02	77.06	90.00	13.57	50.00	25.59	24.41	55.44	37.71	52.38		
	CONV × 2010–2011		195.45	71.51	76.12	12.03	44.87	28.50	26.62	64.36	40.53	54.51		
	CONV × 2020–2021	12.17	213.99	74.89	74.27	6.06	52.26	24.87	22.87	81.29	21.42	44.15	5.08	4.00
	p-value	0.3378	0.0000	0.0005	0.0514	0.0000	0.0566	0.0000	0.0001	0.3461	0.0000	0.0000	0.0004	0.7610

[1] The *p*-values test the statistical significance of each of the factors. When *p*-values are less than 0.05, these factors have a statistically significant effect for each parameter with a 95% confidence level. [2] Unit fruit weight (W); equatorial fruit diameter (D); fruit height (H); skin color index (skin CI); juice content (J); skin content (S); pulp content (P); brightness (L*); chromaticity to red (a*); chromaticity to yellow (b*); skin thickness (ST); albedo thickness (AT).

Table 3. Chemical quality characteristics of 'Navelina' fruits grown under organic (ORG) and conventional (CONV) conditions in the seven different seasons and probability (*p*-value) for the effects of cultivation method, season, and their interaction.

Factors [1]		Chemical Quality Characteristics [2]								
		TSS (°Brix)	pH	TA (%)	RI	Vit C (mg AA/100 g)	EOC (%)	Dy (g/mL)	Juice CI	Seeds (n/Fruit)
Cultivation system	ORG	11.83	3.64	1.28	11.02	63.70	0.34	1.03	1.50	0.08
	CONV	12.08	3.58	1.36	10.43	57.51	0.28	1.04	1.29	0.00
	p-value	0.104	0.0358	0.1325	0.0713	0.0004	0.0011	0.054	0.733	0.0506
Season	1999–2000	11.42	3.73	0.83	8.94	47.11	0.32	1.05	2.58	0.12
	2001–2002	12.03	3.57	0.77	17.20	50.38	0.43	1.05	2.62	0
	2007–2008	11.66	3.54	1.11	10.72	62.42				
	2008–2009	11.15	3.82	1.60	7.42	59.89	0.35	1.02		
	2009–2010	12.52	3.71	2.98	4.79		0.25	1.01		
	2010–2011	12.30	3.52	1.00	12.41	65.74		1.04		
	2020–2021	12.58	3.37	0.95	13.61	78.10	0.21	1.06	1.02	0
	p-value	0.0000	0.0000	0.0000	0.0000	0.0000	0.0000	0.0000	0.0000	0.0838
Cultivation system × season	ORG × 1999–2000	11.04	3.66	0.85	8.84	49.46	0.39	1.05	3.44	0.24
	ORG × 2001–2002	12.31	3.51	0.79	16.51	56.85	0.51	1.05	2.31	0
	ORG × 2007–2008	11.45	3.62	0.96	11.94	63.27				
	ORG × 2008–2009	11.24	3.83	1.60	7.52	62.58	0.33	1.03		
	ORG × 2009–2010	11.76	3.85	2.90	5.13		0.25	0.98		
	ORG × 2010–2011	12.92	3.63	0.98	13.21	69.80		1.04		
	ORG × 2020–2021	12.06	3.42	0.89	13.99	80.27	0.21	1.05	1.25	0
	CONV × 1999–2000	11.80	3.80	0.80	9.04	44.77	0.26	1.05	1.73	0
	CONV × 2001–2002	11.74	3.63	0.76	17.88	43.91	0.35	1.05	2.93	0
	CONV × 2007–2008	11.87	3.47	1.25	9.51	61.57				
	CONV × 2008–2009	11.06	3.81	1.59	7.31	57.19	0.37	1.02		
	CONV × 2009–2010	13.28	3.57	3.07	4.46		0.24	1.04		
	CONV × 2010–2011	11.69	3.42	1.02	11.61	61.69		1.05		
	CONV × 2020–2021	13.1	3.33	1.00	13.23	75.94	0.20	1.06	0.78	0
	p-value	0.0000	0.0004	0.1867	0.1006	0.5005	0.0000	0.0001	0.2770	0.0838

[1] The *p*-values test the statistical significance of each of the factors. When *p*-values are less than 0.05, these factors have a statistically significant effect for each parameter with a 95% confidence level. [2] Total soluble solids content (TSS); total acidity (TA); ripeness index (RI); vitamin C concentration (Vit C); essential oils concentration (EOC); juice density (Dy); juice color index (juice CI); number of seeds (seeds).

Table 4. Physical quality characteristics of 'Clemenules' fruits grown under organic (ORG) and conventional (CONV) conditions in the seven different seasons and probability (*p*-value) for the effects of cultivation method, season, and their interaction.

Factors [1]		Physical Quality Characteristics [2]												
		Shape	**W (g/fruit)**	**D (mm)**	**H (mm)**	**Skin CI**	**J (%)**	**S (%)**	**P (%)**	**L***	**a***	**b***	**ST (mm)**	**AT (mm)**
Cultivation system	ORG	6.88	90.56	58.89	49.93	11.33	43.03	23.11	33.86	64.15	36.35	50.11	2.75	1.45
	CONV	6.80	81.87	56.13	47.30	10.13	40.55	26.81	32.65	65.50	34.18	52.04	2.28	1.30
	p-value	0.721	0.0001	0.0000	0.0000	0.0000	0.0041	0.0000	0.1502	0.0000	0.0000	0.0000	0.0136	0.2922
Season	1999–2000	2.76	70.82	55.06	45.41	10.01	38.62	26.84	34.53	60.66	35.56	58.98	2.59	1.48
	2001–2002	8.16	106.17	62.05	50.60	8.77	46.05	22.69	31.26				2.37	0.92
	2007–2008						41.32	28.78	29.91					
	2008–2009		74.46	53.79	47.23	15.04	37.74	23.19	39.07	53.90	41.87	52.38		
	2009–2010		77.88	55.01	48.65	12.83	38.84	27.21	33.95	66.73	39.49	46.89		
	2010–2011		81.16	56.57	50.05	12.23	39.41	23.86	36.74	62.80	40.85	54.69		
	2020–2021	9.59	106.80	62.59	49.75	5.50	50.54	22.12	27.34	80.04	18.57	42.45	2.59	1.73
	p-value	0.0000	0.0000	0.0000	0.0000	0.0000	0.0000	0.0000	0.0000	0.0000	0.0000	0.0000	0.0656	0.0000
Cultivation system × season	ORG × 1999–2000	2.52	82.92	59.23	50.02	10.94	41.63	26.03	32.33	59.59	37.30	57.41	3.17	1.68
	ORG × 2001–2002	8.46	119.36	64.83	50.60	9.23	43.87	20.57	35.56				2.53	0.94
	ORG × 2007–2008						45.80	26.11	28.09					
	ORG × 2008–2009		77.67	55.23	49.29	15.94	32.99	20.46	46.56	52.99	43.55	51.63		
	ORG × 2009–2010		77.67	55.17	49.07	12.71	40.27	27.39	32.34	67.22	39.77	47.42		
	ORG × 2010–2011		81.95	57.04	51.30	13.51	44.40	20.16	35.48	60.94	42.12	52.07		
	ORG × 2020–2021	9.64	103.81	61.82	49.28	5.68	52.29	21.04	26.68	80.00	19.01	42.02	2.54	1.74
	CONV × 1999–2000	3.0	58.72	50.89	40.81	9.06	35.62	27.65	36.73	61.73	33.81	60.54	2.0	1.27
	CONV × 2001–2002	7.85	92.98	59.27	50.59	8.31	48.23	24.81	26.95				2.21	0.90
	CONV × 2007–2008						36.83	31.44	31.72					
	CONV × 2008–2009		71.25	52.35	45.16	14.15	42.50	25.92	31.58	54.80	40.19	53.13		
	CONV × 2009–2010		78.09	54.84	48.22	12.97	37.41	27.04	35.55	66.24	39.21	46.37		
	CONV × 2010–2011		80.36	56.10	48.80	10.96	34.43	27.57	38.00	64.65	39.58	57.31		
	CONV × 2020–2021	9.54	109.79	63.35	50.21	5.32	48.80	23.20	27.99	80.09	18.13	42.88	2.63	1.71
	p-value	0.2002	0.0000	0.0000	0.0000	0.0000	0.0000	0.0000	0.0000	0.0000	0.0380	0.0000	0.0061	0.6561

[1] The *p*-values test the statistical significance of each of the factors. When *p*-values are less than 0.05, these factors have a statistically significant effect for each parameter with a 95% confidence level. [2] Unit fruit weight (W); equatorial fruit diameter (D); fruit height (H); skin color index (skin CI); juice content (J); skin content (S); pulp content (P); brightness (L*); chromaticity to red (a*); chromaticity to yellow (b*); skin thickness (ST); albedo thickness (AT).

Table 5. Chemical quality characteristics of 'Clemenules' fruits grown under organic (ORG) and conventional (CONV) conditions in the seven different seasons and probability (*p*-value) for the effects of cultivation method, season, and their interaction.

Factors [1]		Chemical Quality Characteristics [2]								
		TSS (°Brix)	pH	TA (%)	RI	Vit C (mg AA/100 g)	EOC (%)	Dy (g/mL)	Juice CI	Seeds (n/Fruit)
Cultivation system	ORG	12.00	3.82	1.05	13.91	55.58	0.36	1.05	3.33	12.34
	CONV	12.00	3.80	1.15	13.52	53.30	0.29	1.03	3.55	2.62
	p-value	0.9841	0.5200	0.0983	0.5033	0.3841	0.0002	0.1504	0.5795	0.0002
Season	1999–2000	11.32	4.22	0.48	16.36	46.08	0.40	1.06	5.88	12.66
	2001–2002	12.58	3.63	0.83	15.26	43.80	0.53	1.05	2.93	2.54
	2007–2008	10.77	3.78	0.76	14.25	55.47				
	2008–2009	12.77	4.01	1.29	10.28	58.96	0.29	1.03		
	2009–2010	12.31	3.71	2.84	5.17		0.23	1.01		
	2010–2011	12.58	3.72	0.73	17.90	60.28		1.04		
	2020–2021	11.68	3.61	0.76	16.79	62.06	0.18	1.06	1.52	7.32
	p-value	0.0019	0.0000	0.0000	0.0000	0.0000	0.0000	0.0000	0.0000	0.0000
Cultivation system × season	ORG × 1999–2000	10.16	4.20	0.43	14.30	47.13	0.24	1.04	6.03	24.33
	ORG × 2001–2002	13.0	3.69	0.79	16.53	51.31	0.80	1.06	2.74	3.89
	ORG × 2007–2008	10.9	3.79	0.79	13.81	55.46				
	ORG × 2008–2009	12.81	4.03	1.24	10.77	54.61	0.28	1.04		
	ORG × 2009–2010	12.52	3.73	2.71	6.17		0.28	1.06		
	ORG × 2010–2011	12.27	3.72	0.72	17.51	59.65		1.03		
	ORG × 2020–2021	12.31	3.61	0.69	18.28	65.33	0.18	1.05	1.23	8.79
	CONV × 1999–2000	12.47	4.24	0.54	18.43	45.03	0.56	1.07	5.74	1.0
	CONV × 2001–2002	12.17	3.56	0.87	13.98	36.29	0.26	1.05	3.11	1.18
	CONV × 2007–2008	10.65	3.78	0.73	14.70	55.49				
	CONV × 2008–2009	12.73	3.99	1.34	9.79	63.30	0.29	1.02		
	CONV × 2009–2010	12.1	3.69	2.98	4.17		0.18	0.96		
	CONV × 2010–2011	12.89	3.73	0.74	18.28	60.92		1.04		
	CONV × 2020–2021	11.04	3.61	0.83	15.30	58.79	0.18	1.06	1.82	5.66
	p-value	0.0688	0.9113	0.4327	0.0000	0.0000	0.0000	0.0000	0.5675	0.0308

[1] The *p*-values test the statistical significance of each of the factors. When *p*-values are less than 0.05, these factors have a statistically significant effect for each parameter with a 95% confidence level. [2] Total soluble solids content (TSS); total acidity (TA); ripeness index (RI); vitamin C concentration (Vit C); essential oils concentration (EOC); juice density (Dy); juice color index (juice CI); number of seeds (seeds).

Significant differences are observed in several physical quality parameters of 'Clemenules' fruits (Table 4). Particularly, when the effect of the cultivation method is considered, 'Clemenules' fruits grown in the organic cultivation method had significantly higher mean values of fresh weight (90.56 g) than those grown under the conventional system (81.87 g). The mean values of diameter and height were also significantly higher in the organic 'Clemenules' fruits (58.89 mm and 49.93 mm, respectively) compared to the mean values found in fruits grown under conventional practice (56.13 mm and 47.30 mm, respectively). The same trend is detected for the color index of the peel. This parameter was significantly higher in organically grown 'Clemenules' fruits (11.33) than conventionally grown 'Clemenules' samples (10.13), possibly due to the higher value of parameter a*. The percentage of juice was significantly higher in the organic fruits (43.03%) than in the conventional ones (40.55%). The cultivation method also affects the skin thickness, which was significantly higher in organic 'Clemenules' (2.75 mm) than in conventional ones (2.28 mm). However, the total skin content was higher in conventionally produced fruits. The cropping system × season interaction affects all physical parameters except shape and albedo thickness (Table 4).

The significant difference found in 'Clemenules' fruit, due to the growing system effect, for chemical and nutritional parameters, was in the concentration of essential oils in the skin and the number of seeds per fruit (Table 5). Organic 'Clemenules' fruit samples had a significantly higher content of essential oils, 0.36 g per 100 g of peel, compared to 0.29 g per 100 g in conventional peel fruits. The number of seeds per fruit was statistically higher in organic 'Clemenules' than in conventional ones (12 and 3 seeds, respectively). The vitamin C concentration was also higher in the organic fruits, although without statistical differences, with mean values of 55.58 mg per 100 g of juice compared to 53.30 mg per 100 g of juice for the conventional ones. Finally, significant differences for the cultivation system × season interaction found in 'Clemenules' samples were in the ripeness index, the vitamin C concentration, and the percentage of essential oils (Table 5). In the parameters where no statistically significant interaction effects are observed, the farming system shows the same trend in all seasons.

3.2. *Frequency Mosaic Graph: Fruit Shape Typing*

The mosaic graphs show the results of the contingency table (cross-frequency table), using rectangles whose area is proportional to the frequency of each of the class crossings. The height of the rectangles (ordinate axis) is proportional to the percentage distribution of the corresponding variable and the base of the rectangle (abscissa axis) is proportional to the percentage distribution of the other variables. Figure 2 shows the mosaic graph of the shape frequencies of 'Navelina' and 'Clemenules' citrus fruit grown under organic conditions, and Figure 3 represents the same graphical parameter for conventional oranges and mandarins.

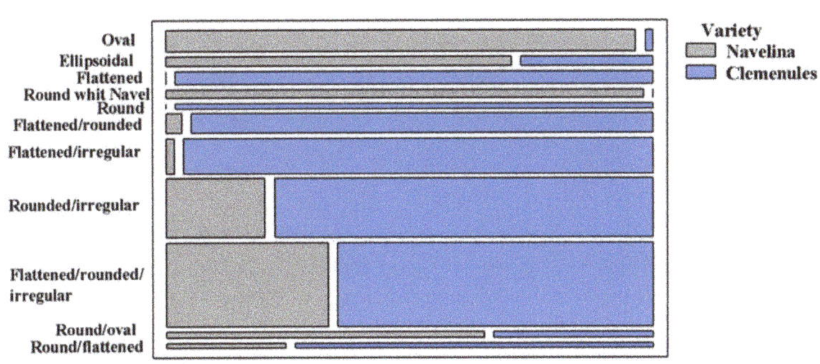

Figure 2. Mosaic graphic of frequencies of the shape of organic citrus 'Navelina' and 'Clemenules' varieties.

Figure 3. Mosaic graphic of frequencies of the shape of conventional citrus 'Navelina' and 'Clemenules' varieties.

A higher incidence of oval shape is observed for the citrus variety 'Navelina' compared to 'Clemenules', with a higher frequency of oval shape for conventional 'Navelina' fruits. The flattened shape is more prevalent for the 'Clemenules' variety, regardless of the cropping system. The mosaic graphs show that the irregularly rounded shape has a higher incidence in the fruits for the 'Clemenules' variety than in those of the 'Navelina' oranges, and in both, higher percentages are observed for the organic fruits. Conventional 'Clemenules' fruits showed a higher incidence of flattened/irregular shapes compared to 'Navelina' oranges. Figure 2 shows a higher frequency of flattened/rounded/irregular shapes for organic mandarins, while in Figure 3, the frequency is higher for conventional oranges.

3.3. *Typification of the Fruits According to the Production System*

Discriminant analysis is applied for explanatory and predictive purposes. With the explanatory use, the aim is to determine the contribution of each classifier variable in the correct classification of each individual (citrus) in the production system (organic and conventional). With the predictive application, it will be possible to determine in the future the group to which an individual belongs, for which the values taken by the classifying variables are known.

Two statistically significant (95% confidence level) and discriminant models have been obtained, which can help predict the production system to which the citrus fruits belong (Table 6). These models can be used to typify and predict the production system of the observations.

Table 6. Eigenvalues of the discriminant models in the classification of citrus fruits.

	Statistical Values					
Standardized Classifier Function	Eigenvalue	Canonical Correlation	Lambda of Wilks	Chi-Square	Freedom Degrees	p-Value
[1] Model 1 = −84.353 + 2.91167 × D − 0.498518 × W + 0.799272 × H	0.00407229	0.06368	0.99594	10.4181	3	0.0153
[2] Model 2 = −0.773399 × ST + 0.741901 × AT + 1.17359 × S	0.271197	0.46189	0.78666	186.088	3	0.0000

[1] Equatorial fruit diameter (D), unit fruit weight (W), fruit height (H). [2] Skin thickness (ST), albedo thickness, skin content (S).

To realize the two models, different analyses were carried out to arrive at an adequate classificatory function, with the main objective being that a reduced and simple number of parameters can be used for the successful classification of citrus fruits. A significant model (95% confidence level) was obtained using only physical parameters (unit weight, diameter, height, and the color parameters L*, a*, b*, CI). With the results of this preliminary study, the three parameters with the greatest weight in the classification function were selected and the analysis was performed again, where 2567 cases were used to develop model 1, which discriminates between organic and conventional systems, using three predictor variables. Model 1 allows 53.72% of the fruits to be classified correctly, using only the parameters of diameter, height, and unit weight of the citrus fruits. When the other parameters (physical, chemical, and nutritional) are used, a significant function is obtained that allows 79.47% of the cases to be classified correctly. Although the classification level is high, the resulting classification function is not valid because it requires a large number of parameters. The last analysis carried out includes only the parameters related to the skin measurement (skin thickness, albedo thickness, and skin content). Model 2 allows us to correctly classify fruits in 71.89% of the cases. A total of 779 cases were used to develop model 2, which discriminates between organic and conventional systems, using the three predictor variables, easily measurable from the skin of the fruits.

Table 7 shows that the classification of the observations (citrus fruits) based on Model 1 allows us to correctly classify, on average, 53.72% of the cases. Of the 1482 observations of organic fruits, 880 (59.38%) are correctly classified. Of the 1085 observations of conventional citrus fruits, 499 (45.99%) are correctly classified. Model 2, on average, allows 71.89% of the observations to be correctly classified.

Table 7. Classification of fruits by production system using the two models.

		Production System	Predicted Membership Group	
			Organic	Conventional
Model 1	Numerical count	Organic	880	602
		Conventional	586	499
	Percentage count	Organic	59.38%	40.62%
		Conventional	54.01%	45.99%
Model 2	Numerical count	Organic	323	103
		Conventional	116	237
	Percentage count	Organic	75.82%	24.18%
		Conventional	32.86%	67.14%

Figure 4 shows the scatter diagram of model 1 (left) and model 2 (right), according to the independent variables selected in each case. The highest level of separation between organic and conventional citrus fruits is observed with model 2.

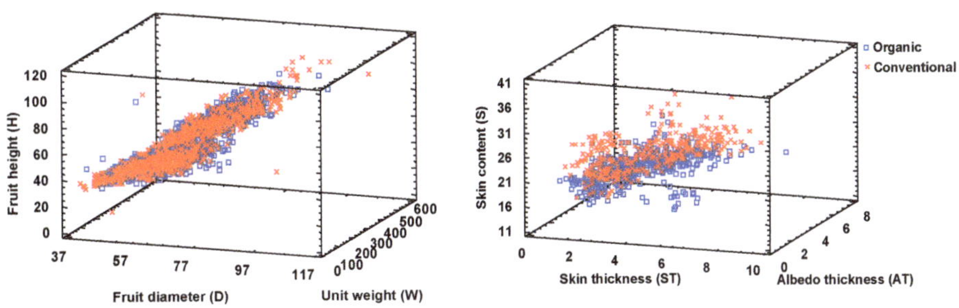

Figure 4. Scatterplot of model 1 (**left**) and model 2 (**right**) of the selected independent variables.

4. Discussion

Fruit quality is the result of complex interactions of genetic factors (plant crop and cultivar), environmental conditions, cultivation practices, harvesting time, and postharvest handling [10,43,44]. Our results suggest that there are considerable differences in various physical, chemical, and nutritional quality parameters in the citrus samples analyzed due to the different production systems. Citrus shape is strongly correlated to variety [45]. The results of this study show that 'Navelina' fruits are mainly oval in shape, regardless of farming system. For fruits of the 'Clemenules' variety, although they are oval and round in shape, in all cases, the main characteristic was the flattened shape. The higher percentage of irregular shapes in organic citrus fruits was expected to be an important index of biodiversity that characterizes organic vegetables and fruits. This can be a drawback for the marketing of fruits, but responsible organic consumers are aware of this variability in the fresh product [46,47].

The characteristic color of citrus fruits is mainly due to the presence of carotenoids, and their composition is influenced by several factors, including geographical origin, fruit ripeness, and, particularly, variety [48]. Color is one of the most remarkable features of the different citrus species and cultivars and a key factor of external and internal quality and consumer acceptance [11]. In fact, it is of great importance in evaluating commercial quality, and there have been attempts to quantify the sensation of color perceived by the brain through the human eye, although some fruit degreening in conventional practices by exogenous exposure to the ripening hormone ethylene could confuse the will of the consumer [49]. The skin color index shows significant differences in the case of both citrus varieties studied: in 'Clemenules' fruit, color values observed in organically produced fruit were higher, while in 'Navelina' oranges, the opposite was found, with higher values for conventional fruits, finding similarity with the results of other studies [50].

The varietal difference is observed in the size, being the fruits of 'Navelina' heavier than those of 'Clemenules'. The height of the fruits is significantly higher in citrus fruits of the 'Navelina' variety when the production system is organic. Fruit diameter is not significant, regardless of the variety. The cultivation system is not significant for the unit weight of the 'Navelina' fruits, with the weight ranging between 222.82 and 229.66 g/fruit. These grammages are within the caliber values for this variety [51]. Organically produced 'Clemenules' fruits are on average 9.6% heavier than conventional ones. This means that under organic production conditions, weights similar to or greater than those obtained with conventional methods can be obtained. Consequently, organic fruits reach the standards of commercial quality, related to the caliber.

We found significant differences in skin percentage in 'Navelina' and 'Clemenules' fruits, which are higher for conventional fruits. Similar results are reported [50] in the same varieties, but only in one growing season, where a lower skin percentage was found in citrus fruits produced on organic farms. In 'Navelina' oranges, the skin content of conventionally produced fruits is 26.95% compared to a value of 25.17% for organically produced fruits, which means that the amount of juice and pulp in conventionally produced 'Navelina' oranges is 73.05%, while in the case of organically produced 'Navelina' oranges, the pulp and juice content is 74.83%. In 'Clemenules' mandarins, the amount of juice and pulp from conventional production is 73.19%, while for organically produced mandarins, the pulp and juice content is 76.89%, which means that organic fruits are juicier than conventional ones. The skin levels of organic fruits are low, which means that organic fertilization invests more energy in the synthesis of juice and pulp, while chemical fertilization and, above all, excess nitrogen fertilization, concentrate on skin production. Some authors confirm that organic fertilizer treatments tend to decrease skin content and increase juice content in several citrus cultivars compared to conventional fertilizers [10,52–55], but variable seasonal conditions and variety may reverse the results [56]. When higher juice content occurs in conventional oranges, this effect could be attributed to the higher water concentration present in conventional fruits due to excess nitrogen fertilization, as reported by Neuhoff et al. [57].

The peel color, fruit size, and fruit shape contribute to the external appearance and influence consumers' attention and appreciation. Satsumas and Clementines are considered medium–small to medium-sized; common and Mediterranean mandarins are medium-sized. On the other hand, experience and market demands in the distribution of organic food mean that the fruits that reach the market are very homogeneous. Differences due to external attributes are only observed in markets with a higher degree of complicity with producers, and in this case, it is no longer a commercial problem. In addition, the consumption of irregular foods contributes to reducing food waste.

The higher number of seeds found in our work for the organic citrus fruits 'Navelina' and 'Clemenules' is consistent with what is reported in the literature [50]. The presence or absence of seeds in fruit is a determining factor of its commercial value. Citrus fruits which have a high number of seeds per fruit are generally not preferred for fresh consumption and are not suitable for the processing industry for juice extraction [44]. However, the presence of seeds in fruit is due to cross-pollination, favored mainly by bees. Their presence is an index of the increased biodiversity in organic fields, as the pesticide-free ecosystem favors the presence of pollinating fauna [58]. The importance of bee pollination for food crops has been widely recognized. In addition to improving the yield of some crops, bee pollination contributes to increasing the nutritional value, improving the quality, and prolonging the shelf life of many fruits and vegetables [59]. Regardless, the fruits of the present study can be classified as 'low-seeded' since according to commercial criteria, a 'low-seeded' variety should have no more than two seeds per fruit in a random sample of twenty-five fruits, with no fruit containing more than four seeds [60].

We found a significantly higher level of vitamin C in organic 'Navelina' oranges than in oranges grown under conventional practices. Likewise, in the case of 'Clemenules' fruits, although the result is not statistically significant, we found a higher vitamin C concentration in organic mandarins. These results are in agreement with those reported by other authors who observed a higher concentration of vitamin C in fruit yielded with an organic cultivation method compared to conventional cultivation systems, in citrus cultivars and varieties [10,11,29,55]. This vitamin is one of the most important antioxidant compounds in plant cells and one of the main contributors to the health benefits attributed to consumption. Recently, its importance in the Coronavirus disease 2019 (COVID-19) has also been highlighted due to its role in modulating the immune system and cellular defense against oxidative stress associated with infection [61]. In many crops, increased vitamin C concentration also prolongs the food's shelf-life [11]. In the case of oranges, studies have shown that the shelf life of organically grown oranges increases by five weeks compared to conventional oranges [62]. Vitamin C concentration, in different citrus species, is inversely correlated with the nitrogen supply [63]. Therefore, the lower vitamin C concentration in conventional fruits could be due to the nitrogen application from synthetic fertilizers immediately available in these crops, as the abuse of nitrogen fertilizers leads to an increase in the relative water content of the plant tissue, then a dilution effect, and induces an increase in the concentration of nitrate ions in plants, then simultaneously the decrease in ascorbic acid, which is an inhibitor in the formation of carcinogenic nitrous compounds [14,64].

The mean values of vitamin C concentration found in this work for 'Navelina' and 'Clemenules' fruits are included in the literature range, where the vitamin C content in commercial orange juices is between 35 and 74 mg per 100 mL of juice [60]. Furthermore, in our experiment, the vitamin C concentration was significantly higher in 'Navelina' fruit than in 'Clemenules' citrus, regardless of the production system. This is in agreement with other authors [14,50], who reported that the vitamin C concentration in oranges is generally higher than in mandarins. This confirms the influence of genetic factors on fruit quality [43,62]. Our long-term study concludes that, on average, 100 mL of organic 'Clemenules' juice provides 4.1% more vitamin C than its conventional counterpart. In the case of 'Navelina', organic production can increase the vitamin C content in citrus juice by almost 10%.

Another significant difference related to the cultivation method is highlighted for the pH parameter in citrus juice, which was higher in organic 'Navelina' fruits than in conventional ones (Table 2). The result of pH in juice from organic 'Navelina' fruits is inconsistent with that of Duarte et al. [10], who reported a higher mean pH value in organic fruits than those found in fruits grown with conventional practices, possibly due to the presence of other acids present in conventional fruits that influence the decrease in the pH value. There is a logical trend between the pH value and the total acidity of the juice, since juices with a higher pH value have lower total acidity content. There is also a slight relationship between acidity parameters and TSS, so that when juices are more acidic, TSS increases, decreasing the value of the ripeness index. Some authors have found no differences in the acidity levels and soluble solids content of the fruit when contrasting the effect of chemical and organic fertilizers [65]. Total soluble solids content and total acidity determine the organoleptic properties of fruits and the balance ratio.

Total juice acidity had the same trend in the two citrus varieties and in all seasons, with the highest acidity in fruits from organic systems. Fruit height, juice percentage, brightness, flesh thickness, ripeness index, vitamin C content, juice color index, and number of seeds of 'Navelina' fruits are parameters for which the cultivation system shows the same tendency in all seasons. In the case of 'Clemenules' fruits, the color parameter a*, thickness of the fruit flesh, soluble solids content, pH, and juice color index are parameters for which the cultivation system also shows the same trend in all seasons studied.

The exposure of plants to aggressive and stressful situations, such as pathogen attack, leads to a considerable increase in secondary metabolites as a defense mechanism in the absence of synthetic fertilizers and pesticides commonly used in the conventional growing system. Then, the absence of chemical resources that characterizes organic practices is related to a higher concentration of essential oils in organic fresh vegetables and fruits than in conventionally grown ones [47,66]. In addition to sugars and acids, volatile flavorings can also influence flavor perception. Our results about the concentration of essential oils in organic and conventional 'Navelina' and 'Clemenules' fruits are in line with previous statements and in agreement with other authors who indicate that citrus peels are an important source of essential oils [9]. These volatile compounds not only enhance the organoleptic profile of citrus fruit, due to their pleasant smell, but also possess excellent properties, such as antimicrobial, anti-inflammatory, antitumor, antioxidant, and neuroprotective activities. In fact, these characteristics have long been widely used in foods, cosmetics, and pharmaceuticals [7]. Our long-term study concluded that, on average, 100 g of peel from organic 'Navelina' fruits provide 17.65% more essential oils than their conventional counterparts. In the case of 'Clemenules' fruits, organic production can increase the essential oil content by almost 19.50%.

The discriminant function has been used to successfully verify the identity of organic foods [67]. It has been employed in the classification and differentiation of seven orange varieties using several chemometric techniques [68]. It has even been successfully used in the classification of organic and conventional citrus by detecting markers linked to nitrogen metabolism [55]. In both cases, the parameters chosen for citrus classification require complex analyses and/or have been performed with a small number of samples. Cuevas et al. [69] provided a methodology based on profiling stable isotope polyphenols, among other quality parameters, to classify citrus fruits using data fusion techniques. In this case, the number of organic fruits analyzed was 126 and the number of conventional fruits was 132. The present work amplifies the scientific gap in the classification of organic citrus fruits based on quality parameters obtained from seven study campaigns, and the result is two classification equations, both obtained with simple analytical parameters.

Consumers and farmers need tools that can ensure trust and prevent organic food fraud. For Danezis et al. [70], verifying the authenticity of organically produced products is also of interest to the scientific community and regulators. The models obtained in this work will successfully predict the production system and generate more confidence among consumers.

5. Conclusions

The long-term results of this work allow us to provide verified information on the global quality of organic 'Navelina' orange and 'Clemenules' mandarin fruits. Organic systems produce citrus fruits that meet market criteria standards in terms of color and size, providing a visual quality accepted by consumers. In line with the growing demand for environmentally friendly organic food products that guarantee a lower impact, the daily consumption of organic citrus fruits is recommended to improve human health due to the antioxidant properties of vitamin C present in concentrations 4.1% higher in 'Clemenules' organic and 10% higher in 'Navelina' organic fruits. In addition to the higher vitamin value, organic citrus fruits contain higher concentrations of essential oils that generate a more aromatic sensation, conveying the characteristic and authentic aroma of citrus fruits. The high number of results obtained in the study period has allowed us to obtain two discriminating equations that use easy-to-measure parameters to successfully classify organic citrus fruits. The classification and prediction models obtained can be used by both national organizations/fruit processors and international control authorities, and constitute useful tools to help predict the purity/authenticity of organic citrus fruits. The results obtained reinforce the idea of the high quality of organic food and are strongly supported by the long-term study.

Author Contributions: Conceptualization, A.D.-G. and M.D.R.; Data curation, M.D.R.; Formal analysis, R.D.G. and M.D.G.-M.; Investigation, M.D.G.-M. and M.D.R.; Methodology, A.D.-G. and M.D.R.; Project administration, M.D.R.; Resources, A.D.-G. and M.D.R.; Supervision, M.D.R.; Visualization, A.D.-G., R.D.G., M.D.G.-M. and M.D.R.; Writing—original draft, R.D.G. and M.D.R.; Writing—review and editing, M.D.R. All authors have read and agreed to the published version of the manuscript.

Funding: This research received no external funding.

Data Availability Statement: The data presented in this study are available on request from the corresponding author. The data are not publicly available due to the high volume of data.

Acknowledgments: The authors thank all the farmers and producers of citrus fruits who collaborated in a participatory way with the monitoring of the crops during all the citrus seasons and who work hard to maintain the cultivation traditions of the Mediterranean region.

Conflicts of Interest: The authors declare no conflict of interest.

References

1. Rafiq, S.; Kaul, R.; Sofi, S.A.; Bashir, N.; Nazir, F.; Nayik, G.A. Citrus peel as a source of functional ingredient: A review. *J. Saudi Soc. Agric. Sci.* **2018**, *17*, 351–358. [CrossRef]
2. Talón, M.; Caruso, M.; Gmitter, F.G., Jr. *The Genus Citrus*, 1st ed.; Elsevier: London, UK, 2020; p. 720.
3. Navarro-Martínez, P.; Noguera-Artiaga, L.; Sánchez-Bravo, P.; Rosas-Burgos, E.C.; Pérez-López, A.J.; Carbonell-Barrachina, Á.A. Development and characterization of liquors prepared with an underutilized citrus by-product, the peel. *Eur. Food Res. Technol.* **2019**, *245*, 41–50. [CrossRef]
4. FAO. 2022. Available online: https://www.fao.org/faostat/es/#data/QCL (accessed on 10 April 2023).
5. Di Prima, S.; Rodrigo-Comino, J.; Novara, A.; Iovino, M.; Pirastru, M.; Keesstra, S.; Cerdà, A. Soil physical quality of citrus orchards under tillage, herbicide, and organic managements. *Pedosphere* **2018**, *28*, 463–477. [CrossRef]
6. Czech, A.; Zarycka, E.; Yanovych, D.; Zasadna, Z.; Grzegorczyk, I.; Kłys, S. Mineral content of the pulp and peel of various citrus fruit cultivars. *Biol. Trace Elem. Res.* **2020**, *193*, 555–563. [CrossRef]
7. Karn, A.; Zhao, C.; Yang, F.; Cui, J.; Gao, Z.; Wang, M.; Wang, F.; Xiao, H.; Zheng, J. In-vivo biotransformation of citrus functional components and their effects on health. *Crit. Rev. Food Sci. Nutr.* **2020**, *61*, 756–776. [CrossRef]
8. Ma, G.; Zhang, L.; Sugiura, M.; Kato, M. Citrus and health. In *The Genus Citrus*; Woodhead Publishing: Cambridge, UK, 2020; pp. 495–511. [CrossRef]
9. Lv, X.; Zhao, S.; Ning, Z.; Zeng, H.; Shu, Y.; Tao, O.; Xiao, C.; Lu, C.; Liu, Y. Citrus fruits as a treasure trove of active natural metabolites that potentially provide benefits for human health. *Chem. Cent. J.* **2015**, *9*, 68. [CrossRef] [PubMed]
10. Duarte, A.; Caixeirinho, D.; Miguel, M.G.; Sustelo, V.; Nunes, C.; Mendes, M.; Marreiros, A. Vitamin C content of citrus from conventional *versus* organic farming systems. *Acta Hortic.* **2010**, *868*, 389–394. [CrossRef]
11. Zacarías-García, J.; Rey, F.; Gil, J.V.; Rodrigo, M.J.; Zacarías, L. Antioxidant capacity in fruit of Citrus cultivars with marked differences in pulp coloration: Contribution of carotenoids and vitamin C. *Food Sci. Technol. Int.* **2020**, *27*, 210–222. [CrossRef]
12. Patil, B.S.; Jayaprakasha, G.K.; Murthy, K.N.C. Beyond vitamin C: The diverse, complex health-promoting properties of citrus fruits. *Citrus Res. Technol.* **2017**, *38*, 107–121. [CrossRef]
13. Di Donna, L.; Bartella, L.; De Vero, L.; Gullo, M.; Giuffrè, A.M.; Zappia, C.; Capocasale, M.; Poiana, M.; D'Urso, S.; Caridi, A. Vinegar production from *Citrus bergamia* by-products and preservation of bioactive compounds. *Eur. Food Res. Technol.* **2020**, *246*, 1981–1990. [CrossRef]
14. Magwaza, L.S.; Mditshwa, A.; Tesfay, S.Z.; Opara, U.L. An overview of preharvest factors affecting vitamin C content of citrus fruit. *Sci. Hortic.* **2017**, *216*, 12–21. [CrossRef]

15. Costanzo, G.; Iesce, M.R.; Naviglio, D.; Ciaravolo, M.; Vitale, E.; Arena, C. Comparative studies on different citrus cultivars: A revaluation of waste mandarin components. *Antioxidants* **2020**, *9*, 517. [CrossRef] [PubMed]
16. De Castro, N.T.; de Alencar, E.R.; Zandonadi, R.P.; Han, H.; Raposo, A.; Ariza-Montes, A.; Araya-Castillo, L.; Botelho, R.B.A. Influence of Cooking Method on the Nutritional Quality of Organic and Conventional Brazilian Vegetables: A Study on Sodium, Potassium, and Carotenoids. *Foods* **2021**, *10*, 1782. [CrossRef]
17. Guilherme, R.; Reboredo, F.; Guerra, M.; Ressurreição, S.; Alvarenga, N. Elemental composition and some nutritional parameters of sweet pepper from organic and conventional agriculture. *Plants* **2020**, *9*, 863. [CrossRef]
18. Smith, L.G.; Kirk, G.J.; Jones, P.J.; Williams, A.G. The greenhouse gas impacts of converting food production in England and Wales to organic methods. *Nat. Commun.* **2019**, *10*, 4641. [CrossRef] [PubMed]
19. Domínguez Gento, A.; Ballester, R.; Raigón, M.D.; García, M.D.; Vercher, R.; Moscardó, E.; Calabuig, A. Efecto de cubiertas vegetales permanentes en la fertilidad del cultivo de cítricos ecológicos. In Proceedings of the IX Congreso SEAE, Lleida, Spain, 6–9 October 2010; Available online: http://www.agroecologia.net/recursos/publicaciones/publicaciones-online/2010/ix-congreso/cd-actas/p5-fertilidad-del-suelo-ii/5-6-efecto-dominguez.pdf (accessed on 23 April 2023).
20. Jeločnik, M.; Ion, R.A.; Jovanović, M.; Popescu, C.G. Has organic farming potential for development? Comparative Study in Romania and Serbia. *Procedia Econom. Bus. Adm.* **2015**, *22*, 268–276. [CrossRef]
21. Fernández, J.A.; Ayastuy, M.E.; Belladonna, D.P.; Comezaña, M.M.; Contreras, J.; de Maria Mourão, I.; Orden, L.; Rodríguez, R.A. Current trends in organic vegetable crop production: Practices and techniques. *Horticulturae* **2022**, *8*, 893. [CrossRef]
22. Ribes-Moya, A.M.; Adalid, A.M.; Raigón, M.D.; Hellín, P.; Fita, A.; Rodríguez-Burruezo, A. Variation in flavonoids in a collection of peppers (*Capsicum* sp.) under organic and conventional cultivation: Effect of the genotype, ripening stage, and growing system. *J. Sci. Food Agric.* **2020**, *100*, 2208–2223. [CrossRef]
23. Dall'Asta, M.; Angelino, D.; Pellegrini, N.; Martini, D. The nutritional quality of organic and conventional food products sold in Italy: Results from the food labelling of Italian products (FLIP) study. *Nutrients* **2020**, *12*, 1273. [CrossRef]
24. Raigón, M.D. Los cítricos: Reserva para el invierno. *Veritas* **2011**, *18*, 12–16. Available online: https://www.veritas.es/wp-content/uploads/2020/11/Estudio-naranjas-Dra.-Raigon.pdf (accessed on 23 April 2023).
25. Renau-Pruñonosa, A.; García-Menéndez, O.; Ibáñez, M.; Vázquez-Suñé, E.; Boix, C.; Ballesteros, B.B.; Hernández García, M.; Morell, I.; Hernández, F. Identification of aquifer recharge sources as the origin of emerging contaminants in intensive agricultural areas. La Plana de Castellón, Spain. *Water* **2020**, *12*, 731. [CrossRef]
26. Hondebrink, M.A.; Cammeraat, L.H.; Cerdà, A. The impact of agricultural management on selected soil properties in citrus orchards in Eastern Spain: A comparison between conventional and organic citrus orchards with drip and flood irrigation. *Sci. Total Environ.* **2017**, *581*, 153–160. [CrossRef]
27. Cerdà, A.; Rodrigo-Comino, J.; Giménez-Morera, A.; Keesstra, S.D. Hydrological and erosional impact and farmer's perception on catch crops and weeds in citrus organic farming in Canyoles river watershed, Eastern Spain. *Agric. Ecosyst. Environ.* **2018**, *258*, 49–58. [CrossRef]
28. Beltrán-Esteve, M.; Picazo Tadeo, A.J.; Reig Martínez, E. What makes a citrus farmer go 'organic'? Empirical evidence from Spanish citrus farming. *Span. J. Agric. Res.* **2012**, *10*, 901–910. [CrossRef]
29. Popa, M.E.; Mitelut, A.C.; Popa, E.E.; Stan, A.; Popa, V.I. Organic foods contribution to nutritional quality and value. *Trends Food Sci. Technol.* **2019**, *84*, 15–18. [CrossRef]
30. Martí, N.; Mena, P.; Cánovas, J.A.; Micol, V.; Saura, D. Vitamin C and the role of citrus juices as functional food. *Nat. Prod. Commun.* **2009**, *4*, 677–700. [CrossRef] [PubMed]
31. Chhikara, N.; Kour, R.; Jaglan, S.; Gupta, P.; Gat, Y.; Panghal, A. Citrus medica: Nutritional, phytochemical composition and health benefits—A review. *Food Funct.* **2018**, *9*, 1978–1992. [CrossRef]
32. Abobatta, W.F.; El-Azazy, A.M. Role of organic and biofertilizers in citrus orchards. *Aswan Univ. J. Environ. Stud.* **2020**, *1*, 13–27. [CrossRef]
33. Zhao, H.; Dong, Z.; Liu, B.; Xiong, H.; Guo, C.; Lakshmanan, P.; Wang, X.; Chen, X.; Shi, X.; Zhang, F.; et al. Can citrus production in China become carbon-neutral? A historical retrospect and prospect. *Agric. Ecosyst. Environ.* **2023**, *348*, 108412. [CrossRef]
34. Tarozzi, A.; Hrelia, S.; Angeloni, C.; Morroni, F.; Biagi, P.; Guardigli, M.; Cantelli-Forti, G.; Hrelia, P. Antioxidant effectiveness of organically and non-organically grown red oranges in cell culture systems. *Eur. J. Nutr.* **2006**, *45*, 152–158. [CrossRef]
35. Khalil, H.A.; Hassan, S.M. Ascorbic acid, β-carotene, total phenolic compound and microbiological quality of organic and conventional citrus and strawberry grown in Egypt. *Afr. J. Biotechnol.* **2015**, *14*, 272–277. [CrossRef]
36. He, X.; Zhang, H.; Li, J.; Yang, F.; Dai, W.; Xiang, C.; Zhang, M. The positive effects of Humic/Fulvic acid fertilizers on the quality of lemon fruits. *Agronomy* **2022**, *12*, 1919. [CrossRef]
37. United States Department of Agriculture (USDA). Soil Mechanics Level 1 Module 3. In *Soil Textural Classification Study Guide*; USDA Soil Conservation Service: Washington, DC, USA, 1987.
38. Vidal, A.; Talens, P.; Prats-Montalbán, J.M.; Cubero, S.; Albert, F.; Blasco, J. In-line estimation of the standard colour index of citrus fruits using a computer vision system developed for a mobile platform. *Food Bioprocess Technol.* **2013**, *6*, 3412–3419. [CrossRef]
39. AOAC. *Official Methods of Analysis*, 16th ed.; Association of Official Analytical Chemists: Washington, DC, USA, 1995.
40. Zhou, Y.; He, W.; Zheng, W.; Tan, Q.; Xie, Z.; Zheng, C.; Hu, C. Fruit sugar and organic acid were significantly related to fruit Mg of six citrus cultivars. *Food Chem.* **2018**, *259*, 278–285. [CrossRef]

41. Arya, S.P.; Mahajan, M.; Jain, P. Non-spectrophotometric methods for the determination of vitamin C. *Anal. Chim. Acta* **2000**, *417*, 1–14. [CrossRef]
42. Centro de Comercio Internacional UNCTAD/GATT (CCI). Aceites esenciales y oleorresinas: Estudio de distintos productores y de mercados importantes. In *Monografía Sobre Determinados Aceites Esenciales y Oleorresinas de Especias*; CCI, UNCTAD-GATT: Geneva, Switzerland, 1986; p. 222.
43. Lado, J.; Gambetta, G.; Zacarias, L. Key determinants of citrus fruit quality: Metabolites and main changes during maturation. *Sci. Hortic.* **2018**, *233*, 238–248. [CrossRef]
44. Aregay, N.; Belew, D.; Zenebe, A.; Haile, M.; Gebresamuel, G.; Girma, A. Tree age and harvesting season affected physico-chemical and bioactive compounds of elite type of Gunda Gundo orange (*Citrus* Spp.) in the Northern Ethiopia. *Int. J. Fruit Sci.* **2021**, *21*, 26–39. [CrossRef]
45. Goldenberg, L.; Yaniv, Y.; Kaplunov, T.; Doron-Faigenboim, A.; Porat, R.; Carmi, N. Genetic diversity among mandarins in fruit-quality traits. *J. Agric. Food Chem.* **2014**, *62*, 4938–4946. [CrossRef]
46. Lauterbach, J.; Bantle, C. For More Diversity, Better Taste and My Own Health. Exploring Organic Consumers Purchasing Motives for Heirloom Vegetable Varieties. *Sustainability* **2022**, *14*, 4068. [CrossRef]
47. Rahman, S.M.E.; Mele, M.A.; Lee, Y.T.; Islam, M.Z. Consumer preference, quality, and safety of organic and conventional fresh fruits, vegetables, and cereals. *Foods* **2021**, *10*, 105. [CrossRef]
48. Beltrán-González, F.; Pérez-López, A.J.; López-Nicolás, J.M.; Carbonell-Barrachina, Á.A. Effects of agricultural practices on instrumental colour, mineral content, carotenoid composition, and sensory quality of mandarin orange juice, cv. Hernandina. *J. Sci. Food Agric.* **2008**, *88*, 1731–1738. [CrossRef]
49. Matsumoto, H.; Ikoma, Y.; Kato, M.; Nakajima, N.; Hasegawa, Y. Effect of postharvest temperature and ethylene on carotenoid accumulation in the flavedo and juice sacs of satsuma mandarin (*Citrus unshiu* Marc) fruit. *J. Agric. Food Chem.* **2009**, *57*, 4724–4732. [CrossRef] [PubMed]
50. Raigón, M.D.; Domínguez Gento, A.; García Martínez, M.D. La calidad de los frutos cítricos de producción ecológica. In *Memorias del V Congreso Latinoamericano de Agroecología-SOCLA*; Facultad de Ciencias Agrarias y Forestales: La Plata, Argentina, 2015; pp. 688–693.
51. Barkley, N.A.; Roose, M.L.; Krueger, R.R.; Federici, C.T. Assessing genetic diversity and population structure in a citrus germplasm collection utilizing simple sequence repeat markers (SSRs). *Theor. Appl. Genet.* **2006**, *112*, 1519–1531. [CrossRef] [PubMed]
52. Roussos, P.A. Phytochemicals and antioxidant capacity of orange (*Citrus sinensis* (L.) Osbeck cv. Salustiana) juice produced under organic and integrated farming system in Greece. *Sci. Hortic.* **2011**, *129*, 253–258. [CrossRef]
53. Sánchez-Bravo, P.; Noguera-Artiaga, L.; Martínez-Tomé, J.; Hernández, F.; Sendra, E. Effect of Organic and Conventional Production on the Quality of Lemon "Fino 49". *Agronomy* **2022**, *12*, 980. [CrossRef]
54. Carranca, C.; Brunetto, G.; Tagliavini, M. Nitrogen nutrition of fruit trees to reconcile productivity and environmental concerns. *Plants* **2018**, *7*, 4. [CrossRef]
55. Rapisarda, P.; Calabretta, M.L.; Romano, G.; Intrigliolo, F. Nitrogen metabolism components as a tool to discriminate between organic and conventional citrus fruits. *J. Agric. Food Chem.* **2005**, *53*, 2664–2669. [CrossRef]
56. Li, Z.; Zhang, R.; Xia, S.; Wang, L.; Liu, C.; Zhang, R.; Fan, Z.; Cheng, F.; Liu, Y. Interactions between N, P and K fertilizers affect the environment and the yield and quality of satsumas. *Glob. Ecol. Conserv.* **2019**, *19*, e00663. [CrossRef]
57. Neuhoff, D.; Vlatschkov, V.; Raigón, D. Comparison of the quality of conventionally and organically grown oranges in Spain. In Proceedings of the 17th IFOAM Organic World Congress, Organic is Life-Knowledge for Tomorrow, Namyangju, Republic of Korea, 28 September–1 October 2011; pp. 487–490.
58. Cuevas, E.; Blancas, J.; Caballero, J.; Hinojosa-Díaz, I.A.; Martínez-Ballesté, A. Agricultural management and local knowledge: Key factors for the conservation of socio-ecosystems in the face of the pollinator world crisis. *Bot. Sci.* **2021**, *99*, 305–320. [CrossRef]
59. Patel, V.; Pauli, N.; Biggs, E.; Barbour, L.; Boruff, B. Why bees are critical for achieving sustainable development. *Ambio* **2021**, *50*, 49–59. [CrossRef]
60. Goldenberg, L.; Yaniv, Y.; Porat, R.; Carmi, N. Mandarin fruit quality: A review. *J. Sci. Food Agric.* **2018**, *98*, 18–26. [CrossRef] [PubMed]
61. Bellavite, P.; Donzelli, A. Hesperidin and SARS-CoV-2: New light on the healthy function of citrus fruits. *Antioxidants* **2020**, *9*, 742. [CrossRef] [PubMed]
62. Cuevas, F.J.; Pradas, I.; Moreno, J.M. Efecto del manejo ecológico en la calidad general y bio-funcional de dos variedades de naranjas. In Proceedings of the XI Congreso de la Sociedad Española de Agricultura Ecológica, Vitoria-Gasteiz, Spain, 1–4 October 2014; p. 676.
63. Mozafar, A. Nitrogen fertilizers and the amount of vitamins in plants: A review. *J. Plant. Nutr.* **1993**, *16*, 2479–2506. [CrossRef]
64. Raigón, M.D.; Rodriguez-Burruezo, A.; Prohens, J. Effects of organic and conventional cultivation methods on composition of eggplant fruits. *J. Agric. Food Chem.* **2010**, *58*, 6833–6840. [CrossRef] [PubMed]
65. Abd El-Migeed, M.M.; Saleh, M.M.S.; Mostafa, E.A. The beneficial effect of minimizing mineral nitrogen fertilization on Washington navel orange trees by using organic and biofertilizers. *World J. Agric. Sci.* **2007**, *3*, 80–85.
66. Mditshwa, A.; Magwaza, L.S.; Tesfay, S.Z.; Mbili, N. Postharvest quality and composition of organically and conventionally produced fruits: A review. *Sci. Hortic.* **2017**, *216*, 148–159. [CrossRef]

67. Capuano, E.; Boerrigter-Eenling, R.; van der Veer, G.; van Ruth, S.M. Analytical authentication of organic products: An overview of markers. *J. Sci. Food Agric.* **2013**, *93*, 12–28. [CrossRef]
68. Vavoura, M.V.; Karabagias, I.K.; Kosma, I.S.; Badeka, A.V.; Kontominas, M.G. Characterization and Differentiation of Fresh Orange Juice Variety Based on Conventional Physicochemical Parameters, Flavonoids, and Volatile Compounds Using Chemometrics. *Molecules* **2022**, *27*, 6166. [CrossRef]
69. Cuevas, F.J.; Pereira-Caro, G.; Muñoz-Redondo, J.M.; Ruiz-Moreno, M.J.; Montenegro, J.C.; Moreno-Rojas, J.M. A holistic approach to authenticate organic sweet oranges (*Citrus sinensis* L. cv Osbeck) using different techniques and data fusion. *Food Control* **2019**, *104*, 63–73. [CrossRef]
70. Danezis, G.P.; Tsagkaris, A.S.; Brusic, V.; Georgiou, C.A. Food authentication: State of the art and prospects. *Curr. Opin. Food Sci.* **2016**, *10*, 22–31. [CrossRef]

horticulturae

MDPI

Article

An Updated Isotopic Database of Fertilizers Used in Intensive Organic Farming: A Case Study on Protein Hydrolyzed Derivatives and Chelated Nutrients

José Manuel Muñoz-Redondo, Francisco Julián Cuevas, José Carlos Montenegro, José Luis Ordóñez-Díaz and José Manuel Moreno-Rojas *

Department of Agro-Industry and Food Quality, Andalusian Institute of Agricultural and Fisheries Research and Training (IFAPA), Alameda del Obispo, Avda Menéndez Pidal, s/n. 14004 Córdoba, Spain; josem.munoz.redondo@juntadeandalucia.es (J.M.M.-R.); fjulian.cuevas@juntadeandalucia.es (F.J.C.); josec.montenegro@juntadeandalucia.es (J.C.M.)

* Correspondence: josem.moreno.rojas@juntadeandalucia.es; Tel.: +34-671-53-27-58

Abstract: The global demand for organic food products has rapidly increased over the last years, becoming an emerging niche market targeting the high-income segment. The higher retailing price for organic food products may increase the risk of fraudulent practices at the different stages of the food supply chain, and consequently, substantial control is needed. Currently, the authentication of organic food products, such as those of plant origin, remains a key challenge in analytical chemistry. While stable isotopes have emerged as a powerful tool for this purpose, most studies have focused on crops, missing the agricultural inputs used for fertilization that influence the isotopic values of the crops. In this study, we aimed to isotopically characterize commonly used fertilizers, soil conditioners, and micronutrient fertilizers in intensive organic agriculture in the largest organic production region in the world (Almería, Spain). Our goal was to clarify the limitations that nitrogen isotopic fingerprinting presents for the fertilizer input industry and to characterize the organic inputs. The conventional fertilizers analyzed in this study showed low $\delta^{15}N$ values compared to their organic counterparts, except for some plant-based fertilizers, protein hydrolyzed fertilizers, and chelated nutrients. Both protein hydrolyzed fertilizers and micronutrient fertilizers presented a wide range of variability in their $\delta^{15}N$ values, including some very low or even negative values, more similar to those of conventional fertilizers. The results of this study highlight the challenges of authenticating organic foods in agriculture when using nitrogen isotope analysis.

Keywords: fertilizers; nitrogen stable isotopes; organic food authentication; soil fertility assessment; agricultural practices

1. Introduction

The organic food industry has experienced substantial growth, with an estimated global market of EUR 106.4 billion in 2019 [1]. This expansion is largely driven by consumer demand for quality, sustainability, and food safety [2], making the consumers willing to pay more. However, the industry faces important challenges due to the lower agricultural yields of the organic farming systems compared to the conventional ones, implying higher retail prices. To address this challenge, the election of fertilizers in organic production has caught increasing attention as a solution to this yield gap. In this sense, the use of any agricultural input must be authorized by the appropriate inspection bodies, and regulatory frameworks are responsible for regulating the use of agricultural inputs in both organic and conventional farming. Examples include the European Council Regulation (EC) in Europe, the National Organic Standards Board (NOSB) in the United States, and the National Accreditation Services for Conformity Assessment in China. Currently, the European regulations address the harmonization of conditions for inorganic, organic, and waste-based fertilizers within the internal European Union market. Nonetheless, some innovative organic fertilizers, such as protein hydrolyzed fertilizers (PHFs), have not been thoroughly addressed due to their rapid emergence in the market and their diverse origins. Consequently, individual national laws are required for each member state, making access to a single market more difficult due to the differing rules.

Protein hydrolyzed fertilizers have gained great attention among organic producers due to their valuable nutrients for plants, which are related with increases in soil fertility. Those characteristics have boosted a strong development of these agricultural inputs in organic management. PHFs are mainly produced by the denaturing and hydrolysis of various protein sources, such as plants, animals, or microorganism derivatives, leading to a wide diversity of origins and methods of synthesis. This varying composition makes it difficult to standardize their use by regulatory frameworks. Thus, PHFs were classified in the European laws as fertilizers and soil conditioners, as well as plant protection products. This latter classification makes their usage controversial, since no requirements for nitrogen concentration are needed. The same applies to the other class of organic fertilizers—micronutrients, which also do not require a nitrogen declaration. Therefore, more research is still required to clarify the origin of these fertilizers so that their legal status can be clearly defined and inspection bodies can ensure their appropriate use in organic crops.

Numerous analytical techniques have emerged over the recent years to address the differentiation between organically and conventionally grown crops [3], some of them based on the isotopic fingerprinting of the fertilizers used. For the authentication of organic foodstuffs, techniques based on isotope-ratio mass spectrometry have demonstrated their potential [4]. In this regard, the isotopic ratio of nitrogen of organic products has been related with the inputs used in agricultural systems [5], linking the soil and plant isotopic values of nitrogen [6] to fertilizers [7]. Particularly, several studies assessed the nitrogen isotopic values of fertilizers through correlations with nitrogen signatures found in fruits and vegetables in different agricultural systems [5]. This made it possible to discriminate between organic and conventional products, but it depended on different parameters that in many cases could not be controlled by the farmers, such as physiological processes. A previous study [8] stated that, although the nitrogen isotopic composition of the fertilizers used in conventional and organic systems were different, the variable impact can be reflected in the crop. This hypothesis was based on the broad range of environmental conditions and agricultural practices existing in the world and the fluctuating nitrogen cycle in different ecosystems. Therefore, it is necessary to gain an in-depth understanding of all the mechanisms influencing the isotopic signature of the products cultivated under different water and fertilizer regimes. Another important issue regarding these studies is the recent emergence of soil conditioners. To date, these agricultural inputs have not been isotopically characterized, but their nutrient content directly influences the absorption and bioavailability of the nitrogen supplied [9]. Soils have little capacity to retain oxidized forms of nitrogen and ammonium, and therefore, the accumulation of these types of fertilizers in soils is small. For this reason, most of the organic nitrogen fertilizers are associated with organic matter.

Considering the aforementioned, the aim of this study was to isotopically characterize the most commonly used fertilizers, soil conditioners, and micronutrient fertilizers in intensive organic agriculture in the largest organic production region in the world (Almería, Spain) to try to clarify the policy constraints that nitrogen isotopic fingerprinting presents for the fertilizer input industry. This will help create new databases to characterize organic inputs.

2. Materials and Methods

2.1. Samples

A total of 205 fertilizers (both liquid or solid) authorized for conventional (n = 54) and organic (n = 151) cultivation systems were obtained from different commercial fertilizer suppliers. We selected fertilizers that are commonly used in the region of Almeria, aiming to include a heterogeneous sample. This approach involved the selection of a variety of fertilizers with different compositions and suppliers. We consulted with local farmers and agricultural extension officers to identify the most widely used types of organic fertilizers. Detailed information about the origin and number of fertilizers by origin is shown in Table 1. These fertilizers are used in the systems under intensive conditions like the ones occurring in the south of Spain, including organic farming systems.

Table 1. Summary of the descriptive statistics for the fertilizers evaluated.

		Synthetic fertilizers $\delta^{15}N$ (‰)					
Composition		**Origin**	**Mean**	**SD**	**Max**	**Min**	n
Ammonia		Air	0.40	0.63	1.20	−0.30	8
Ammonia + nitrate		Air	0.35	0.88	1.80	−0.50	16
Ammonia + urea		Air	0.47	1.15	2.00	−1.20	12
Ammonia + nitrate + urea		Air	0.47	0.29	0.80	0.30	6
Nitrate		Air	−0.30	0.00	−0.30	−0.30	4
Urea		Air	−0.23	0.68	0.30	−1.20	8
		Conventional organic fertilizers $\delta^{15}N$ (‰)					
Category	**Type**	**Subtype**	**Mean**	**SD**	**Max**	**Min**	n
Terrestrial	Animal	Blood	5.24	1.56	7.00	3.40	5
		By-products	7.97	4.32	12.31	1.88	7
		Horn	4.65	4.33	5.40	3.90	2
		Manure and slurry	11.21	3.14	14.40	5.15	8
	Plant	By-products	2.71	3.83	11.27	−1.63	23
Marine	Seaweed		3.67	2.62	6.10	0.90	3
	Animal		2.30	0.57	2.70	1.90	2
		Organic fertilizers $\delta^{15}N$ (‰) (declaration of nitrogen content not required)					
PHF			3.41	3.10	10.35	−2.22	81
Micronutrients			1.09	3.12	6.79	−2.34	20

Both organic and inorganic fertilizers were processed similarly based on whether they were solid or liquid. All the liquid samples were homogenized by a vortex shaker for 1 min and sampled, and an aliquot of 2 mL was transferred to a vial and preserved in a freezer (−80 °C). All the liquid samples were weighed using 0.5 mg of Chemosorb© (nitrogen-free adsorber pad) and introduced into tin capsules. The Chemosorb© retained the fertilizer preventing losses of compounds due to the volatilization of the liquid samples, improving the repeatability of the results. Meanwhile, the solid samples were homogenized by grinding and milling in a ball mill for 60 s or more to obtain a fine powder. The organic fertilizers used in this approach were authorized by the European Union Council Regulation (Regulation (EC) No. 2018/848) and certification bodies.

2.2. Instruments and Standards

The choice of standards was based on the signal compatibility with the different types of fertilizers being analyzed (organic and synthetic). To achieve this, a series of preliminary tests were conducted to analyze the signal responses of various organic and inorganic fertilizers. Based on these tests, the standards used for synthetic fertilizers were IAEA-N1 (0.43‰), IAEA-N2 (20.30‰), IAEA-NO_3 (4.70‰), and Urea Isotopic Working Standard (−0.73‰) provided by the International Atomic Energy Agency (IAEA, Vienna, Austria) and IVA Analysentechnik e. K., (Dusseldorf, Germany). These standards are well established for covering a broad range of $\delta^{15}N$ values relevant to synthetic fertilizers [10]. To account for the signal responses observed in our tests, we added two additional standards for organic fertilizers: flour (sorghum flour, 1.58‰) and protein (casein, 5.94‰) standards provided by Elemental Microanalysis and IVA Analysentechnik e. K., respectively.

A Delta V Advantage Isotope Ratio Mass Spectrometer from Thermo Fisher Scientific (Bremen, Germany) equipped with a Flash EA elemental analyzer was used for the analysis of stable isotope ratios. The instrument was calibrated with the following international reference standards: IAEA-N1 (ammonium sulfate, $\delta^{15}N$ = 0.43‰), IAEA-NO_3 (potassium nitrate, $\delta^{15}N$ value of −4.70‰), and IAEA-N2 (potassium nitrate, $\delta^{15}N$ value of 20.30‰). All the reference materials were certified by the International Atomic Energy Agency (IAEA, Austria) and used within the ranges 250–280 μg (except sorghum flour; 700–1000 μg) to achieve a signal intensity between 2700 and 6000 mV. The long-term standard deviation of their values was lower than 0.15‰.

2.3. Determination of 15N/14N Isotope Ratio by EA-IRMS

The samples were weighed into tin capsules (3.3 × 5 mm, IVA Analysentechnik e. K., Dusseldorf, Germany).

The carrier and reference gases (He) were circulated at 140 and 200 mL min^{-1}, respectively. The temperatures of the elemental analyzer were set at 65 and 1020 °C and 650 °C for the oven and the reduction and oxidation reactors, respectively. The results of the nitrogen isotope ratio ($\delta^{15}N$) analyses were reported as the per mile (‰) enrichment relative to the international standard (air nitrogen) for the nitrogen isotope ratio according to Equation (1):

$$\delta^{15}\text{Nsample}(‰) = [(R_{sample} - R_{standard})/R_{standard}] \times 1000 \quad (1)$$

where R_{sample} and $R_{standard}$ denote the ratio between the heavier and the lighter isotope ($^{15}N/^{14}N$) in the sample and reference material, respectively.

The fertilizers were analyzed in duplicate with variability below 0.3‰ for $\delta^{15}N$ measurements as previously reported [7].

In the same analysis, the percentage of nitrogen was measured by the elemental analyzer. All the commercial fertilizers presenting a declared nitrogen content were checked, with no significant deviation compared to the nitrogen declared. The box and whisker plot was constructed using SPSS© software (v. 17.0, IBM, Chicago, IL, USA).

2.4. Statistical Analyses

An analysis of variance (ANOVA) was conducted to determine the differences among the various classes of fertilizers. To further compare the means of the different groups, the Tukey's honestly significant difference (Tukey HSD) test was used as a post hoc analysis. This allowed us to determine which groups exhibited statistically significant differences in their means. The differences were considered statistically significant at a p-value < 0.05.

3. Results and Discussion

3.1. Synthetic Fertilizers

Synthetic fertilizers are produced from the fixation of atmospheric nitrogen into ammonium through a chemical reaction known as the Haber–Bosch process [11]. This process does not cause significant N isotope fractionation, although slight fractionation may occur in subsequent processes such as the transformation of NH_4^+ into NO_3 [12]). Therefore, synthetic fertilizers are expected to display $\delta^{15}N$ values close to that of the atmospheric nitrogen (0‰). In this sense, these fertilizers displayed homogeneous $\delta^{15}N$ values (Figure 1) that ranged between −1.2 and 2.0‰, with a mean value of 0.26‰ and a standard deviation of 0.8‰. These results were in accordance with previous studies that showed a similar range of variation [5,7]. When assessing the results of inorganic and organic fertilizers, a lower range of variation was observed in the inorganic fertilizers (Figure 1). An analysis of variance (ANOVA) was performed to compare the two types of fertilizers, obtaining statistically significant differences at a p-value < 0.001. The mean value for organic fertilizers was 3.69‰, in comparison with 0.26‰ obtained for inorganic fertilizers. These results were as expected, since it has already been demonstrated that inorganic fertilizers show lower $\delta^{15}N$ values than their organic counterparts [7].

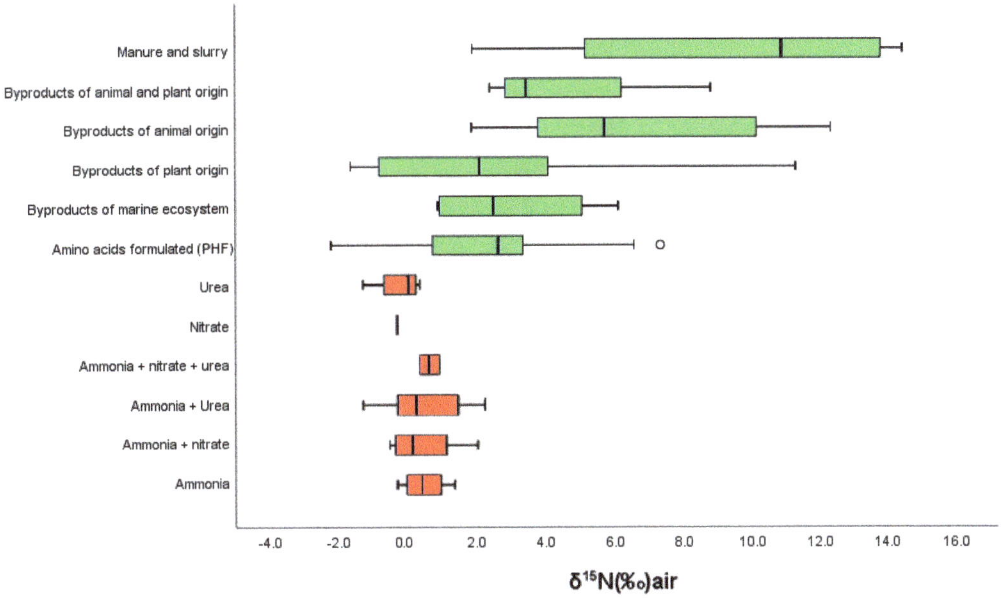

Figure 1. Box and whisker plot for the distribution of $\delta^{15}N$ values of conventional and organic fertilizers used in intensive conventional (red) and intensive organic (green) agricultural systems.

3.2. Organic Fertilizers

The wide diversity of organic fertilizers makes their classification complex, although two main categories can be distinguished according to their ecosystem origin: marine and terrestrial. These categories were further categorized in different subtypes, as shown in Table 1. The $\delta^{15}N$ (‰) value for terrestrial ecosystems ranged between a minimum of −1.63‰ and a maximum of 14.40‰. The lower values belonged to plant by-products and the maximum ones to animal-based fertilizers. The plant by-products showed high variability in relation to the different ways in which nitrogen is incorporated into the plants. The plant $^{15}N/^{14}N$ ratio is an integrator of the nitrogen cycle between soil and air, which can be affected by several processes, such as plant fertilizer sources, symbiotic atmospheric fixation, mycorrhizal symbioses, or fractionation during uptake [13]. Within the terrestrial ecosystem, the animal samples were more enriched in ^{15}N compared to the plants. The highest $\delta^{15}N$ values were detected in manure and slurry samples (Figure 1) in agreement with the previous literature [6,7]. Moreover, the $^{15}N/^{14}N$ ratio of an animal-based fertilizer is highly dependent on the trophic level of the animal from which it derives. Additionally, Choi et al. [14] reported that the isotopic composition of fresh animal excrement is not usually different from the $\delta^{15}N$ of synthetic fertilizers. Thus, the main fractionation vector in manure is maturity, due to its exposure to the atmosphere and the speedy loss of $^{14}NH_3$ via volatilization [15]. These authors reported that the $\delta^{15}N$ of total N of cattle manure compost increased rapidly within 10 days of composting, but further increases were not observed thereafter in the following 90 days of composting, suggesting that the ^{15}N enrichment of composted manure is a fast process occurring in the early stage of composting and that it is affected by the bedding (sawdust, crop residues, etc.) material [15]. The mean values for the blood samples reported in our study were about 5.24‰ (Table 1). The animal by-products category was mainly characterized by meat powder samples with a high variability, values of $\delta^{15}N$ ranging from 1.88‰ to 12.31‰. Moreover, the mean value for fertilizers based on horns was slightly slower (4.65‰) than that reported by other authors [7] (around 6‰).

Our results for marine fertilizers were similar to the $\delta^{15}N$ values previously obtained in the literature [7]. The values of this type of organic inputs varied from 0.9‰ to 6.10‰, displaying slightly higher values than inorganic fertilizers (Figure 1). Seaweed cannot fix nitrogen and first take up nitrogen as nitrate and ammonium. These seaweeds extracts are growing in popularity among farmers in intensive agriculture due to their enhanced abiotic stress tolerance, growth, nutritional quality, and postharvest quality, which have been recently reviewed [16]. One of the main advantages of these fertilizers is the fact that they are available as a liquid extract or in soluble powder forms. This fact makes it possible to mix the extracts with irrigation water that can be easily applied by drip irrigation in intensive greenhouses.

3.2.1. Protein Hydrolyzed Fertilizers (PHFs)

Protein hydrolyzed fertilizers are allowed to be used in organic farming, as shown in Figure 2A. The European regulation allows their use either as fertilizers, soil conditioners, or nutrients, where the % N needs to be declared, or as plant protection products, where the % N does not need to be declared. This lack of requirement for declaring the nitrogen content in plant protection products may make their use controversial, highlighting the need for in-depth research to provide greater clarity in regulatory standards.

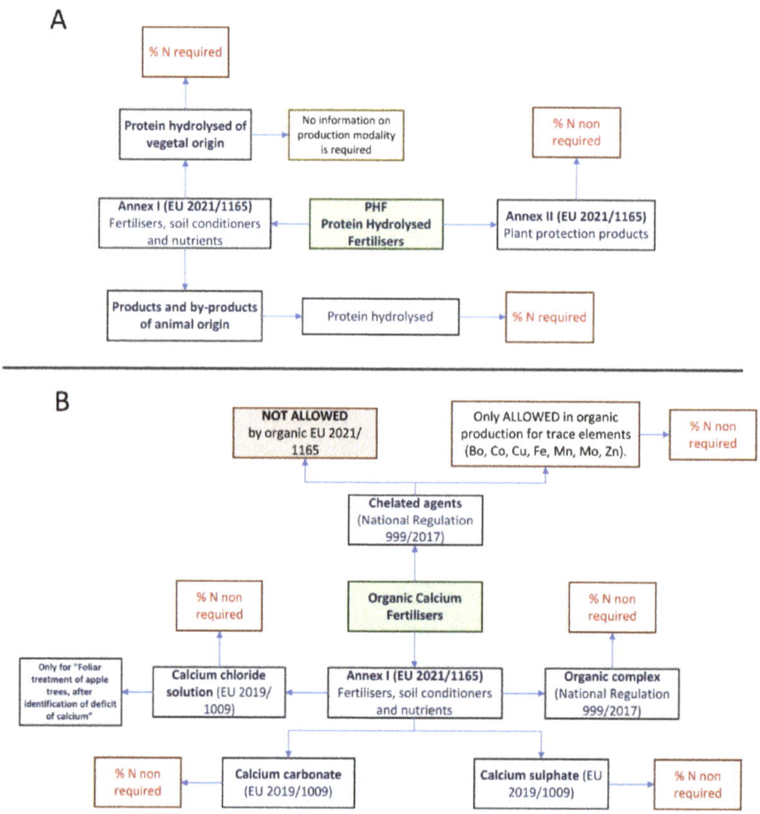

Figure 2. Authorized formulations of protein hydrolyzed fertilizers (**A**) and micronutrients in organic farming (**B**).

The PHFs analyzed in this study showed a high variability in their $\delta^{15}N$ values due to their different origins and the production techniques used, with a mean $\delta^{15}N$ value of 3.41‰ and a distance between the maximum (10.35‰) and minimum (−2.22‰) values of 12.57‰ (Table 1). This was explained by the low (even negative) $\delta^{15}N$ values of several PHFs, which were in the same range as the synthetic fertilizers (Tables 1 and 2). Low $\delta^{15}N$ values were predominantly found in PHFs derived from leguminous plants. This could be attributed to the capacity of leguminous plants to fix atmospheric nitrogen (which has $\delta^{15}N$ values close to 0), minimizing the isotopic fractionation [13] and resulting in lower $\delta^{15}N$ values in their derived fertilizers. From the analysis of variance (ANOVA), we observed a clear differentiation of PHFs according to their origin at a p-value < 0.001 (Table 2). The hydrolyzed fertilizers from vegetal material clearly displayed lower $\delta^{15}N$ values (with a mean of 2.25‰) than the animal-origin counterparts (6.27‰). Figure 3 illustrates the distribution of these values, revealing a separation between animal- and plant-derived PHFs at $\delta^{15}N$ values of approximately 3.8‰. Consequently, depending on the management practices employed, the application of plant-based PHFs with low isotopic values could influence organic crop cultivars by modifying the $\delta^{15}N$ values in cultivated plants and fruits. This reduction could result in isotopic signatures typically observed in conventional crops. These results, combined with the fact that regulations do not require the declaration of the nitrogen content in these types of fertilizers when used as plant protection products, could make their use controversial under the current regulatory framework when using $\delta^{15}N$ to certify the organic origin of products. Consequently, this study highlights the need for research on the impact of PHFs with low $\delta^{15}N$ values on the isotopic values of the crops.

Table 2. Mean and deviation for the $\delta^{15}N$ and % N of PHFs according to their hydrolysis type and origin. Results for the analysis of variance (ANOVA) are also shown.

Factor	PHF	$\delta^{15}N$ Mean (‰)	$\delta^{15}N$ SD (‰)	p-Value (ANOVA)	% N Mean (‰)	% N SD (‰)	p-Value (ANOVA)
Origin	Animal (N = 15)	6.27	3.08	<0.001	7.33	4.18	<0.05
	Vegetal (N = 30)	2.25	2.63		4.59	4.24	
	Unknown (N = 36)	3.18	2.78	-	6.65	4.73	-
Hydrolysis	Acid (N = 10)	3.83	3.35	ns	8.13	7.00	ns
	Enzymatic (N = 51)	2.89	2.42		5.58	4.23	
	Unknown (N = 20)	4.51	4.21	-	6.06	3.71	-

ns = not significant

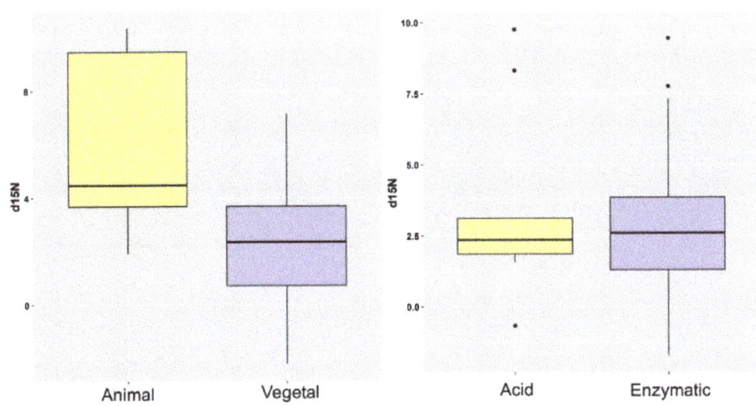

Figure 3. Box and whisker plot for the distribution of $\delta^{15}N$ values of PHFs from different origins (plant or animal; **left**) and hydrolysis methods (acid or enzymatic; **right**).

Significant differences were also found in the % N between PHFs from vegetal and animal origin, with the latter showing a higher percentage of nitrogen (Table 2) and highlighting the influence of the source material on the final PHF composition. Additionally, different hydrolysis processes, either acid or enzymatic, can be used for production of organic fertilizers, which may potentially cause different fractionation patterns. However, no statistically significant differences were found in the $\delta^{15}N$ values between the fertilizers obtained under these production techniques, as shown in Table 2 and Figure 3.

3.2.2. Micronutrients

Another type of fertilizer permitted in organic farming is micronutrients, which include calcium and trace element formulations as shown in Figure 2B. These do not require a declaration of the nitrogen content either, although some of them have shown high total nitrogen levels and low $\delta^{15}N$ values (Table 1). Chelated micronutrients are fertilizers with essential trace elements chemically bonded with a chelating agent. This enhances the bioavailability and solubility of the micronutrients, making them more accessible to plants. Examples of chelated micronutrients include chelated forms of iron, zinc, manganese, and copper, among others. The chelating agent helps prevent the micronutrients from binding with other substances in the soil, allowing the plants to absorb them more effectively. These properties make them widely used in agriculture. Moreover, the addition of certain carbon and nitrogen substrates can influence soil organic matter decomposition and N_2O emissions [17], which may have implications for the effectiveness and environmental impact of using these types of fertilizers in organic farming.

Among the different nutrients, calcium is essential for crops, especially in greenhouses in Almería, where 'blossom-end rot' due to low calcium levels in nutrient solutions [18] is a common issue for tomatoes and peppers. Farmers and previous research also recognized the importance of other trace elements (B, Zn, Mo, etc.) in organic farming nutrition to avoid deficiencies and promote plant growth [19].

Twenty commercial micronutrient fertilizers were characterized, of which five had a total nitrogen content below 0.5%, eleven ranged between 1 and 5%, and four displayed values above 5% (Table 3). The highest nitrogen content was observed in formulations with chelated calcium and trace elements. The $\delta^{15}N$ values of the micronutrients analyzed ranged from −2.34 to 9.48‰, averaging 2.4‰, which is more similar to values observed in conventional fertilizers. In fact, half of the fertilizers analyzed had $\delta^{15}N$ values below 0‰. Consequently, no statistically significant differences (p-value > 0.05) were found from an analysis of variance (ANOVA) comparing these fertilizers with the conventional ones. Similar to the PHFs, certain fertilizers allowed in organic farming exhibited $\delta^{15}N$ values and high percentages of total nitrogen, which did not need to be declared according to regulations. This raises the need for further research to determine whether these fertilizers could impact the authentication of organic products when $\delta^{15}N$ is used for verification.

Table 3. Categories, declared elements, $\delta^{15}N$ and $\delta^{13}C$ values, and percentage of nitrogen in chelated nutrients.

Category	Declared Elements	$\delta^{15}N$	$\delta^{13}C$	%N
Chelated calcium	Ca, EDTA	−0.57	−26.91	9.78
Trace element	Fe, Mn, Zn, B, Cu, EDTA	−0.57	−27.20	8.28
Chelated calcium	Ca	0.25	−30.01	7.42
Trace element	Zn, EDTA	−1.14	−31.01	6.88
Trace element	Bo	−2.34	−30.25	4.99
Organic complex	Ca	1.11	−30.41	4.89
Trace element	Zn, Fe, B, Mo	4.15	−30.01	4.82
Trace element	Zn, B, Mo	−1.87	−26.78	3.43
Calcium chloride	Ca, Mat.Org	3.36	−19.33	2.48
Trace element	B, Ma, Mn	9.02	−26.06	1.97
Calcium carbonate	Ca, Mg	3.22	−22.84	1.91

Table 3. *Cont.*

Category	Declared Elements	$\delta^{15}N$	$\delta^{13}C$	%N
Calcium carbonate	Ca, Mg	2.76	-	1.78
Trace element	Zn, B, Mo	−0.62	−27.41	1.45
Trace element + chelated calcium	Ca, Mg, Fe	9.48	−24.62	1.29
Trace element	Zn, Fe, B Mo	−0.88	−24.63	1.12
Calcium chloride	Ca	7.73	−7.31	0.43
Calcium sulfate	Ca, Mg	−0.21	-	0.10
Calcium chloride	Ca	6.79	−3.34	0.09
Calcium chloride	Ca	−0.88	−17.65	0.04
Trace element + calcium chloride	Zn, Ca	−0.19	-	0.03

4. Conclusions

This study reflects the importance of deepen research into the limitations of stable isotopes for the traceability of organic and conventional foodstuffs. The conventional fertilizers analyzed in this study showed low $\delta^{15}N$ values compared to their organic counterparts, except for some plant-based fertilizers, protein hydrolyzed fertilizers, and chelated nutrients. Both protein hydrolyzed fertilizers and micronutrient fertilizers presented a wide range of variability in their $\delta^{15}N$ values, including some low or even negative values. Additionally, several of both types of fertilizers showed a high nitrogen content, which does not need to be declared according to regulations (in the case of chelating agents and protein hydrolyzed fertilizers when used as plant protection products). Consequently, the use of these fertilizers may present challenges in authenticating organic products and the likely impact of their $\delta^{15}N$ values. Therefore, based on the results of this study, it is essential to conduct further research on how factors affecting the fertilizers' isotopic composition impact crops and the final food products.

Author Contributions: Conceptualization, J.M.M.-R. (José Manuel Moreno-Rojas); methodology, J.M.M.-R. (José Manuel Moreno-Rojas), J.M.M.-R. (José Manuel Muñoz-Redondo), F.J.C. and J.C.M.; software, J.M.M.-R. (José Manuel Muñoz-Redondo) and F.J.C.; formal analysis, J.M.M.-R. (José Manuel Muñoz-Redondo); investigation, J.M.M.-R. (José Manuel Moreno-Rojas), J.M.M.-R. (José Manuel Muñoz-Redondo), F.J.C., J.L.O.-D. and J.C.M.; resources, J.M.M.-R. (José Manuel Moreno-Rojas); data curation, J.M.M.-R. (José Manuel Moreno-Rojas), J.M.M.-R. (José Manuel Muñoz-Redondo), F.J.C. and J.C.M.; writing—original draft preparation, J.M.M.-R. (José Manuel Muñoz-Redondo), J.M.M.-R. (José Manuel Moreno-Rojas), and F.J.C.; writing—review and editing, J.M.M.-R. (José Manuel Moreno-Rojas), J.L.O.-D. and J.M.M.-R. (José Manuel Muñoz-Redondo); supervision, J.M.M.-R. (José Manuel Moreno-Rojas); project administration, J.M.M.-R. (José Manuel Moreno-Rojas); funding acquisition, J.M.M.-R. (José Manuel Moreno-Rojas). All authors have read and agreed to the published version of the manuscript.

Funding: This work was funded by the Andalusian Institute of Agricultural and Fisheries Research and Training (IFAPA, Spain) through the projects 'Evaluación de la trazabilidad de nuevos insumos derivados de hidrolizados de proteínas (DHP) utilizados en agricultura ecológica respecto al origen de los productos de partida, metodologías de síntesis y su integridad (FERTI-TRACE)' (PP.AVA.AVA2019.045) and 'Viabilidad de la relación de isótopos estables de nitrógeno ($^{15}N/^{14}N$) como metodología para la caracterización de la producción ecológica frente a la convencional' (PR.PEI.IDF2016.01), and the European Rural Development Fund (ERDF, EU). J.M.M.-R. (José Manuel Muñoz-Redondo) and J.C.M. received grants through a research contract at IFAPA funded through the projects PP.AVA.AVA2019.045 and PR.PEI.IDF2016.01, respectively.

Data Availability Statement: Data are contained within the article.

Conflicts of Interest: The authors declare no conflicts of interest.

References

1. Willer, H.; Trávnícek, J.; Meier, C.; Schlatter, B. *The World of Organic Agriculture. Statistics and Emerging Trends*; IFOAM: Bonn, Germany, 2021; p. 340.
2. Baron, J.; Jay-Russell, M.T. Assessment of Current Practices of Organic Farmers Regarding Biological Soil Amendments of Animal Origin in a Multi-Regional US Study. *Food Prot. Trends* **2018**, *38*, 347–362.
3. Capuano, E.; Boerrigter-Eenling, R.; van der Veer, G.; van Ruth, S.M. Analytical Authentication of Organic Products: An Overview of Markers. *J. Sci. Food Agric.* **2013**, *93*, 12–28. [CrossRef] [PubMed]
4. Laursen, K.H.; Schjoerring, J.K.; Kelly, S.D.; Husted, S. Authentication of Organically Grown Plants–Advantages and Limitations of Atomic Spectroscopy for Multi-Element and Stable Isotope Analysis. *TrAC Trends Anal. Chem.* **2014**, *59*, 73–82. [CrossRef]
5. Inácio, C.T.; Chalk, P.M.; Magalhães, A.M. Principles and Limitations of Stable Isotopes in Differentiating Organic and Conventional Foodstuffs: 1. Plant Products. *Crit. Rev. Food Sci. Nutr.* **2015**, *55*, 1206–1218. [CrossRef] [PubMed]
6. Chalk, P.M.; Magalhães, A.M.; Inácio, C.T. Towards an Understanding of the Dynamics of Compost N in the Soil-Plant-Atmosphere System Using 15N Tracer. *Plant Soil* **2013**, *362*, 373–388. [CrossRef]
7. Bateman, A.S.; Kelly, S.D. Fertilizer Nitrogen Isotope Signatures. *Isotopes Environ. Health Stud.* **2007**, *43*, 237–247. [CrossRef] [PubMed]
8. Bateman, A.S.; Kelly, S.D.; Woolfe, M. Nitrogen Isotope Composition of Organically and Conventionally Grown Crops. *J. Agric. Food Chem.* **2007**, *55*, 2664–2670. [CrossRef] [PubMed]
9. Fageria, N.K.; Baligar, V.C. Enhancing Nitrogen Use Efficiency in Crop Plants. *Adv. Agron.* **2005**, *88*, 97–185.
10. Gentile, N.; Rossi, M.J.; Delémont, O.; Siegwolf, R.T.W. δ15N Measurement of Organic and Inorganic Substances by EA-IRMS: A Speciation-Dependent Procedure. *Anal. Bioanal. Chem.* **2013**, *405*, 159–176. [CrossRef] [PubMed]
11. Schlesinger, W.H. On the Fate of Anthropogenic Nitrogen. *Proc. Natl. Acad. Sci. USA* **2009**, *106*, 203–208. [CrossRef] [PubMed]
12. Freyer, H.D.; Aly, A.I.M. Nitrogen-15 Variations in Fertilizer Nitrogen†. *J. Environ. Qual.* 1947; 3, 405–406. [CrossRef]
13. Robinson, D. δ15N as an Integrator of the Nitrogen Cycle. *Trends Ecol. Evol.* **2001**, *16*, 153–162. [CrossRef] [PubMed]
14. Choi, W.-J.; Kwak, J.-H.; Lim, S.-S.; Park, H.-J.; Chang, S.X.; Lee, S.-M.; Arshad, M.A.; Yun, S.-I.; Kim, H.-Y. Synthetic Fertilizer and Livestock Manure Differently Affect δ15N in the Agricultural Landscape: A Review. *Agric. Ecosyst. Environ.* **2017**, *237*, 1–15. [CrossRef]
15. Kim, Y.-J.; Choi, W.-J.; Lim, S.-S.; Kwak, J.-H.; Chang, S.X.; Kim, H.-Y.; Yoon, K.-S.; Ro, H.-M. Changes in Nitrogen Isotopic Compositions during Composting of Cattle Feedlot Manure: Effects of Bedding Material Type. *Bioresour. Technol.* **2008**, *99*, 5452–5458. [CrossRef] [PubMed]
16. Battacharyya, D.; Babgohari, M.Z.; Rathor, P.; Prithiviraj, B. Seaweed Extracts as Biostimulants in Horticulture. *Sci. Hortic.* **2015**, *196*, 39–48. [CrossRef]
17. Li, Y.; Moinet, G.Y.; Clough, T.J.; Whitehead, D. Organic Matter Contributions to Nitrous Oxide Emissions Following Nitrate Addition Are Not Proportional to Substrate-Induced Soil Carbon Priming. *Sci. Total Environ.* **2022**, *851*, 158274. [CrossRef] [PubMed]
18. Saure, M.C. Why Calcium Deficiency Is Not the Cause of Blossom-End Rot in Tomato and Pepper Fruit—A Reappraisal. *Sci. Hortic.* **2014**, *174*, 151–154. [CrossRef]
19. Wyszkowski, M.; Brodowska, M.S. Content of Trace Elements in Soil Fertilized with Potassium and Nitrogen. *Agriculture* **2020**, *10*, 398. [CrossRef]

horticulturae

MDPI

Communication

Innovative Tools for the Nitrogen Fertilization Traceability of Organic Farming Products

Simona Fabroni [1,*], Luana Bontempo [2], Gabriele Campanelli [3], Stefano Canali [4] and Francesco Montemurro [5]

1 Council for Agricultural Research and Economics (CREA)—Research Center for Olive, Fruit and Citrus Crops, Corso Savoia 190, 95024 Acireale, Italy
2 Edmund Mach Foundation, Traceability Unit, Via E. Mach 1, 38098 San Michele all'Adige, Italy; luana.bontempo@fmach.it
3 Council for Agricultural Research and Economics (CREA)—Research Center for Vegetable and Ornamental Crops, Via Salaria 1, 63030 Monsampolo del Tronto, Italy; gabriele.campanelli@crea.gov.it
4 Council for Agricultural Research and Economics (CREA)—Research Center for Agriculture and Environment, Via della Navicella 4, 00184 Roma, Italy; stefano.canali@crea.gov.it
5 Council for Agricultural Research and Economics (CREA)—Research Center for Agriculture and Environment, Via Celso Ulpiani 5, 70125 Bari, Italy; francesco.montemurro@crea.gov.it
* Correspondence: simona.fabroni@crea.gov.it

Abstract: In the last decades, consumers have become increasingly interested in organic products, and they strongly demand reliability in the traceability of the organic products they buy and eat. Several research methods have been developed in the last decades to study inexperienced and reliable systems and to assess the authenticity of products obtained using organic cultivation practices. The monitoring of some chemical compounds, originating from primary and/or secondary metabolism, in horticultural organic and conventional products has shown the diversity generated using the two production approaches. The difference in fertilization practices has been also shown to have an effect on the isotopic distribution of some elements, with particular reference to nitrogen. An integrated system is proposed to evaluate the validity of organic goods using collected isotopic data and other chemical and biological parameters. This approach is intended to be coupled to the application of chemometric multivariate analysis on quality and nutraceutical parameters combined with isotopic data. Indeed, this will give the opportunity to discriminate organic from conventional products based on different isotopic signatures, due to the different nitrogenous sources, combined with the qualitative profile of the crops, which are significantly affected by the different agronomic treatments. The main perspectives of the presented integrated approach, based on the combined use of chemometric and analytical tools, are linked to the feasibility of applying a reliable system for traceability. This will authenticate productions obtained using organic fertilizers (organic agriculture) with respect to those obtained with the use of synthetic fertilizers (conventional agriculture), protect and valorize virtuous farmers and support political stakeholders and decision-makers to counteract food fraud.

Keywords: organic horticultural practices and production; chemical metabolites; isotopic ratios; multivariate analysis; chemometrics

1. Background

In recent years, a growing interest in organic products has been recorded as they are considered more sustainable, safer, and healthier than conventional ones by consumers. As a matter of fact, the traceability of organic products and the definition of their authenticity have been subjects of relevant interest for many years. Several studies have been performed to investigate the authenticity of products obtained using organic cultivation methods [1,2]. They focused on the identification of new quality "markers" that allow differentiating, from field to fork, an organic product from a conventional one. In particular, the monitoring of some chemical compounds, originating from primary and/or secondary metabolism, in horticultural organic and conventional products has shown the diversity generated using the two production approaches [3,4]. Indeed, based on biochemical and transcriptional approaches, crucial evidence has been provided about the physiological changes that can occur in organic and conventional cultivated crops [5]. Moreover, it is widely recognized that the diversity in fertilization practices for the two cultivation systems influences the isotopic distribution of some elements present in fruit and vegetable crops, with particular reference to nitrogen (N) [6,7]. Starting from these backgrounds and the knowledge shared so far within the scientific community, the main aim of the present communication is to propose an innovative approach for the N fertilization traceability of organic farming products.

2. The Up-to-Date Literature

Based on the Sustainable Development Goals introduced by the United Nations [8] and the targets of the European Green Deal [9], it is known that one of the main worldwide shared goals is to increase the percentage of organically managed farms by 2030. The focused target is to have 25% of the whole European cultivated area devoted to organic agriculture. In this context, it has been widely discussed how organic agriculture is more environmentally sustainable with respect to conventional farming for different reasons, including lower nutrient losses, reduced fossil energy consumption, and the ban on synthetic pesticides and fertilizers [10]. One of the most important aspects of organic agriculture is to manage N fertilization since synthetic fertilizers are not allowed. In fact, N is a crucial element for plant growth, and it has been estimated that about half of the amount of N provided to the soil is lost [11]; therefore, there is a need to synchronize the amount of N demand and the N supply when different fertilization strategies are used, specifically when organic and conventional practices are used [12]. Furthermore, consumers are increasingly interested in organic food, and they strongly demand reliability in the traceability of the organic products they usually buy and eat. The actual European regulation (EU Reg. 2018/848) establishes that official controls are to be performed in accordance with Regulation (EU) 2017/625 (art. 14) [13]. Specifically, this regulation establishes that official control methods and techniques shall include the examination of documents and traceability records that may be relevant to an assessment for compliance with the rules of organic management. In fact, on the majority of organic farms, the sole traceability mechanism in place is based on documentary evidence that agricultural management complies with European regulations. Furthermore, multiresidue analyses are carried out only upon the specific request from certification bodies or distribution buyers, and relative samplings are randomly performed [14,15]. Indeed, this might not be enough to certify the authenticity of organic products as farmers could fraudulently use fertigation to supply mineral fertilizers to the soil or apply pesticides, observing large shortage intervals, and all these behaviors would not be easily detected [16,17]. So, there is a need for an objective traceability of organic management based on scientifically collected data [18].

The possibility of differentiating two samples on the basis of isotopic analysis is linked to the isotopic fractionation that occurs in raw materials. Isotopic ratios of natural elements such as nitrogen, carbon, oxygen, hydrogen, and sulfur have been conveniently used for food authenticity purposes, as their fractionation occurs in response to very different biogeochemical phenomena [19]. It has been demonstrated that isotopic analysis can be reliably used to check for food authenticity [20–22]. Specifically for organic products, a deviation from the natural isotopic distribution of N happens during the production and maturation of organic fertilizers. Indeed, enrichment in the lighter isotope (^{14}N) occurs due to its preferential volatilization. As a consequence, the N isotopic ratio [$\delta(^{15}N) = {}^{15}N/{}^{14}N$]

$$\delta^i\left(E_{sample/standard}\right) = \frac{R(i_E/j_E)\ sample}{R(i_E/j_E)\ standard} - 1$$

is increased in organic fertilizers and amendments of both vegetable and animal origin [6,7]. In the aforementioned equation, $\delta(^{15}N)$ is expressed with respect to the international standard air (atmospheric N_2), and i_E/j_E is the isotope ratio between the heavier and lighter N isotopes. Air has more or less the same N isotopic composition throughout the world ($\delta(^{15}N)$ = 0‰ by definition and internationally recognized), and industrially produced mineral fertilizers typically have $\delta(^{15}N)$ close to 0‰ because their N directly derives from atmospheric nitrogen, and there is basically no isotope fractionation during their industrial synthesis [6,7]. Conversely, organic fertilizers allowed in organic farming have positive values of $\delta(^{15}N)$ (ranging from +1‰ to +37‰); therefore, they are easily distinguishable from synthetic fertilizers. It has been demonstrated that these differences are reflected in the final vegetable product, thereby confirming the great potential for N isotopic analysis to check the authenticity of organic products [23–33]. At the same time, it must be stressed that it has been widely argued that $\delta(^{15}N)$ may not be discriminating for all crops or in every management condition, e.g., when organic fertilizers are contemporarily used in conventional practice, and vice versa, or for crops which are characterized by low N requirements [33]. In addition, it is also worth noting that few studies have been reported on the effect of using agricultural service crops (such as N-fixing plants (legumes) or not N-fixing plants) for green manure in organic management on the isotopic signature of the final product [34]. As a matter of fact, N-fixing plants fix nitrogen from the air (having $\delta(^{15}N)$ values close to 0‰) in root nodules, thus the N isotopic signature of the final products could reflect this root uptake, putatively leading to misclassification of authentic organic products.

Several researchers have reported that organically grown fruit is richer in antioxidants when compared with conventional counterparts [3,35–37]. For almost two decades, it has been postulated that N deficiency in organic products may induce the production of carbon-rich phenylpropanoids, thus explaining the higher healthy value of organic products. Recent research [4] has reported results that conflict with this theory, showing that long-term organic fields are associated with no relevant or significant difference in N and carbon (C) content or in the C:N ratio between organic and conventional products. The same study reported that pest attacks were preferentially set on conventional plants with respect to organic ones, but no evidence suggesting plant nutrient unbalance was directly linked with this result. Consistently, the study demonstrated that organic soil management boosted salicylic acid build-up, which resulted in discouraging plant–insect interactions. Indeed, salicylic acid accumulation was not associated with N deficiency in the organic plants, but it depended on adaptations in soil microbial communities associated with long-term organic management. Thus, based on this recent experimental evidence reporting how soil microbiota influences plant resistance to pests, it may also be postulated that polyphenolics and other antioxidants may accumulate in organic plants due to functional shifts in soil microflora. Furthermore, based on a biochemical and molecular approach, it was recently demonstrated [5] that different fertilization practices (organic or conventional) induced differential accumulation of free amino acids, with higher concentrations of proline and contemporarily lower concentrations of glutamate in fruits from organically managed orange trees. In addition, the upregulation of glutamate dehydrogenase gene expression was recorded, thus identifying a possible adaptive response in common orange plants to organic fertilizers. It has been suggested that this adaptive response may putatively be based on a metabolism switch aimed at ensuring turnover of the tricarboxylic acid cycle to continuously supply carbon skeletons for biosynthetic demands, as a regulatory phenomenon. Other studies recently reported the possibility of discriminating between organic and conventional crops based on the differential metabolome associated with the two productions [38,39].

3. Methodology

In order to achieve the goal of implementing a reliable system for tracing the authenticity of organic products, different treatments should be used and compared. One possibility is to incorporate conventional, organic, and mixed N fertilization options in the experimental design, comparing them at varying fertilizer percentages. It is also relevant to include the use of agronomic techniques for the management of soil fertility, such as the use of agroecology strategies [40]. The variations in $\delta(^{15}N)$ induced using the combination of different fertilizations together with the use of green manure or cover crops exhibiting an agroecological function in the organic production system should be quantified. This would allow for evaluating how soil management based on the agroecological approach affects the $\delta(^{15}N)$ in the final product, in comparison with organic management based on the full substitution and/or conventional agriculture approaches. At the same time, the variations in physicochemical and nutraceutical attributes induced using the different treatments should be evaluated. Once collected, the obtained data should be treated with a chemometric multivariate approach able to display the data in a way that chemical interpretation of the system is allowed. This may involve transforming the original variable into new variables that are functions of the original data and verifying that a sample can be assigned to previously defined groups formed using samples of known origin ('organic' or 'not organic') on the basis of their compositional similarity with groups. This system will be able to discriminate, at a high confidence level, between organic and conventional products. Finally, the use of the implemented database should be validated on a multiyear basis (Figure 1).

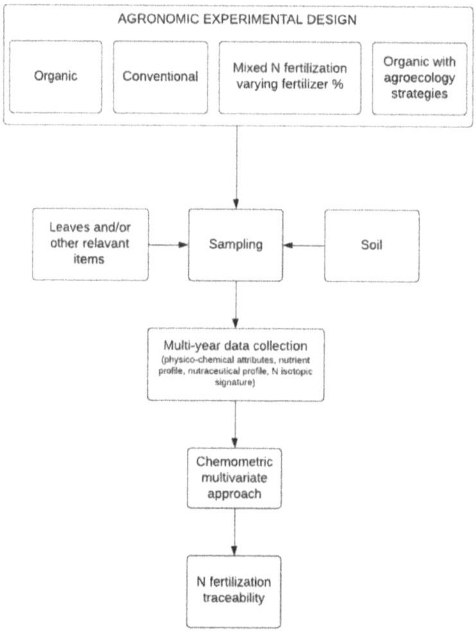

Figure 1. Methodology flow chart.

4. Perspectives

The main aim of the present communication is to provide insights and general suggestions for the application of a feasible methodology for the implementation of an investigation system based on a combined chemical and chemometric approach. In particular, the main perspectives for the application of this kind of approach are linked to the possibility to: (i) allow, with the acquisition of relevant chemical and biochemical attributes and isotopic data, one to discern between productions obtained with mineral fertilizers, typically applied in conventional agriculture and not allowed in organic agriculture, and productions obtained with organic agriculture practice; (ii) improve the general understanding of the factors influencing the quality of organic products and their differentiation from conventional ones; (iii) allow the traceability of organic products from field to fork which, in turn, would contribute to protect and reward virtuous farmers; (iv) valorize the environmental externalities linked to the application of organic farming practices; and (v) support policymakers in identifying methodologies for contrasting food fraud.

Author Contributions: Conceptualization, S.F., L.B., G.C., S.C. and F.M.; methodology, S.F., L.B., G.C., S.C. and F.M.; resources, S.F., L.B., G.C., S.C. and F.M.; writing—original draft preparation, S.F., L.B. and F.M.; writing—review and editing, S.F., L.B., G.C., S.C. and F.M.; visualization, S.F., L.B., G.C., S.C. and F.M.; supervision, S.F., L.B. and F.M.; project administration, S.F.; funding acquisition, S.F. All authors have read and agreed to the published version of the manuscript.

Funding: This research was funded by the Italian Ministry of Agriculture, Food Sovereignty and Forestry, grant number 93173/2017, 'Application of innovative methods for the traceability of organic farming products'—INNOVABIO project and by the European Commission Research Executive Agency, grant agreement (GA) number 101084527, 'Supporting UPtake Integrated Pest Management and lOw-Risk pesTicide Use'—SUPPORT project.

Institutional Review Board Statement: Not applicable.

Informed Consent Statement: Not applicable.

Data Availability Statement: Not applicable.

Conflicts of Interest: The authors declare no conflict of interest.

References

1. Miller, H.I. Buying 'Organic' to Get 'Authenticity'? Or Safer and More Nutritious Food? Think Again. And Again. *Mo. Med.* **2019**, *116*, 8–11.
2. Prache, S.; Lebret, B.; Baéza, E.; Martin, B.; Gautron, J.; Feidt, C.; Médale, F.; Corraze, G.; Raulet, M.; Lefèvre, F.; et al. Review: Quality and authentication of organic animal products in Europe. *Animal* **2022**, *16*, 100405. [CrossRef]
3. Chebrolu, K.K.; Jayaprakasha, G.K.; Jifon, J.; Patil, B.S. Production system and storage temperature influence grapefruit vitamin C, limonoids, and carotenoids. *J. Agric. Food Chem.* **2012**, *60*, 7096–7103. [CrossRef]
4. Blundell, R.; Schmidt, J.E.; Igwe, A.; Cheung, A.L.; Vannette, R.L.; Gaudin, A.C.; Casteel, C.L. Organic management promotes natural pest control through altered plant resistance to insects. *Nat. Plants* **2020**, *6*, 483–491. [CrossRef]
5. Fabroni, S.; Amenta, M.; Rapisarda, S.; Torrisi, B.; Licciardello, C. Amino acid metabolism and expression of genes involved in nitrogen assimilation in common oranges cv. Valencia Late. *Biol. Plant.* **2022**, *66*, 155–162. [CrossRef]
6. Bateman, A.S.; Kelly, S.D.; Jickells, T.D. Nitrogen Isotope Relationships between Crops and Fertilizer: Implications for Using Nitrogen Isotope Analysis as an Indicator of Agricultural Regime. *J. Agric. Food Chem.* **2005**, *53*, 5760–5765. [CrossRef]
7. Muñoz-Redondo, J.M.; Montenegro, J.C.; Moreno-Rojas, J.M. Using Nitrogen Stable Isotopes to Authenticate Organically and Conventionally Grown Vegetables: A New Tracking Framework. *Agronomy* **2023**, *13*, 131. [CrossRef]
8. United Nations. Available online: https://www.un.org/sustainabledevelopment/ (accessed on 15 May 2023).
9. European Commission. Available online: https://commission.europa.eu/strategy-and-policy/priorities-2019-2024/european-green-deal_en (accessed on 15 May 2023).
10. Reganold, J.P.; Wachter, J.M. Organic agriculture in the twenty-first century. *Nat. Plants* **2016**, *2*, 15221. [CrossRef]
11. Diacono, M.; Montemurro, F. Towards a Better Understanding of Agronomic Efficiency of Nitrogen: Assessment and Improvement Strategies. *Agronomy* **2016**, *6*, 31. [CrossRef]
12. Persiani, A.; Diacono, M.; Monteforte, A.; Montemurro, F. Agronomic performance, energy analysis and carbon balance comparing different fertilization strategies in horticulture under Mediterranean conditions. *Environ. Sci. Pollut. Res.* **2019**, *26*, 19250–19260. [CrossRef]
13. Regulation (EU) 2018/848 of the European Parliament and of the Council of 30 May 2018 on Organic Production and Labelling of Organic Products and Repealing Council Regulation (EC) No 834/2007. Available online: https://eur-lex.europa.eu/legal-content/EN/TXT/?uri=CELEX%3A32018R0848 (accessed on 15 May 2023).
14. Ambrus, A. Reliability of measurements of pesticide residues in food. *Accredit. Qual. Assur.* **2004**, *9*, 288–304. [CrossRef]
15. Kazimierczak, R.; Średnicka-Tober, D.; Golba, J.; Nowacka, A.; Hołodyńska-Kulas, A.; Kopczyńska, K.; Góralska-Walczak, R.; Gnusowski, B. Evaluation of Pesticide Residues Occurrence in Random Samples of Organic Fruits and Vegetables Marketed in Poland. *Foods* **2022**, *11*, 1963. [CrossRef] [PubMed]
16. Rapisarda, P.; Calabretta, M.L.; Romano, G.; Intrigliolo, F. Nitrogen Metabolism Components as a Tool to Discriminate between Organic and Conventional Citrus Fruits. *J. Agric. Food Chem.* **2005**, *53*, 2664–2669. [CrossRef] [PubMed]

17. Tan, S.; Xie, D.; Ni, J.; Chen, F.; Ni, C.; Shao, J.; Zhu, D.; Wang, S.; Lei, P.; Zhao, G.; et al. Characteristics and influencing factors of chemical fertilizer and pesticide applications by farmers in hilly and mountainous areas of Southwest, China. *Ecol. Indic.* **2022**, *143*, 109346. [CrossRef]
18. Codex Alimentarius Commission, CAC/GL 60-Principles for Traceability/Product Tracing as a Tool within a Food Inspection and Certification System. 2006. Available online: www.codexalimentarius.net/input/download/standards/.../CXG_060e.pdf (accessed on 15 May 2023).
19. Picó, Y.; Barceló, D. Isotopic Mass Spectrometry in Food and Environmental Chemistry. Mass Spectrometry in Food and Environmental Chemistry. In *The Handbook of Environmental Chemistry*; Picó, Y., Campo, J., Eds.; Springer: Cham, Switzerland, 2022; Volume 119. [CrossRef]
20. Zhi, L.; Yuan, W.; Yudi, H.; Wei, L.; Bin, L.; Guiyuan, M. Multi-stable isotope and multi-element origin traceability of rice from the main producing regions in Asia: A long-term investigation during 2017–2020. *Food Chem.* **2023**, *412*, 135417. [CrossRef]
21. Bontempo, L.; Perini, M.; Pianezze, S.; Horacek, M.; Roßmann, A.; Kelly, S.D.; Thomas, F.; Heinrich, K.; Schlicht, C.; Schellenberg, A.; et al. Characterization of Beef Coming from Different European Countries through Stable Isotope (H, C, N, and S) Ratio Analysis. *Molecules* **2023**, *28*, 2856. [CrossRef]
22. Popîrdă, A.; Luchian, C.E.; Colibaba, L.C.; Focea, E.C.; Nicolas, S.; Noret, L.; Cioroiu, I.B.; Gougeon, R.; Cotea, V.V. Carbon-Isotope Ratio (δ13C) and Phenolic-Compounds Analysis in Authenticity Studies of Wines from Dealu Mare and Cotnari Regions (Romania). *Agronomy* **2022**, *12*, 2286. [CrossRef]
23. Flores, P.; Fenoll, J.; Hellin, P. The Feasibility of Using δ15N and δ13C Values for Discriminating between conventionally and Organically Fertilized Pepper (*Capsicum annuum* L.). *J. Agric. Food Chem.* **2007**, *55*, 5740–5745. [CrossRef]
24. Bateman, A.S.; Kelly, S.D.; Woolfe, M. Nitrogen Isotope Composition of Organically and Conventionally Grown Crops. *J. Agric. Food Chem.* **2007**, *55*, 2664–2670. [CrossRef]
25. Rapisarda, P.; Camin, F.; Fabroni, S.; Perini, M.; Torrisi, B.; Intrigliolo, F. Influence of Different Organic Fertilizers on Quality Parameters and the δ15N, δ13C, δ2H, δ34S, and δ18O Values of Orange Fruit (*Citrus sinensis* L. Osbeck). *J. Agric. Food Chem.* **2010**, *58*, 3502–3506. [CrossRef]
26. Camin, F.; Perini, M.; Bontempo, L.; Fabroni, S.; Faedi, W.; Magnani, S.; Baruzzi, G.; Bonoli, M.; Tabilio, M.R.; Musmeci, S.; et al. Potential isotopic and chemical markers for characterizing organic fruits. *Food Chem.* **2011**, *125*, 1072–1082. [CrossRef]
27. Hayashi, N.; Ujihara, T.; Tanaka, E.; Kishi, Y.; Ogwa, H.; Matsuo, H. Annual Variation of Natural 15N Abundance in Tea Leaves and Its Practicality as an Organic Tea Indicator. *J. Agric. Food Chem.* **2011**, *59*, 10317–10321. [CrossRef]
28. Carter, J.F.; Chesson, L.A. *Food Forensics: Stable Isotopes as a Guide to Authenticity and Origin*; CRC Press: Boca Raton, FL, USA, 2017; pp. 273–302.
29. Novak, V.; Adler, J.; Husted, S.; Fromberg, A.; Laursen, K.H. Authenticity Testing of Organically Grown Vegetables by Stable Isotope Ratio Analysis of Oxygen in Plant-Derived Sulphate. *Food Chem.* **2019**, *291*, 59–67. [CrossRef]
30. Gatzert, X.; Chun, K.P.; Boner, M.; Hermanowski, R.; Mäder, R.; Breuer, L.; Gattinger, A.; Orlowski, N. Assessment of multiple stable isotopes for tracking regional and organic authenticity of plant products in Hesse, Germany. *Isot. Environ. Health Stud.* **2021**, *57*, 281–300. [CrossRef]
31. Bontempo, L.; van Leeuwen, K.A.; Paolini, M.; Holst Laursen, K.; Micheloni, C.; Prenzler, P.D.; Ryan, D.; Camin, F. Bulk and compound-specific stable isotope ratio analysis for authenticity testing of organically grown tomatoes. *Food Chem.* **2020**, *318*, 126426. [CrossRef] [PubMed]
32. Wassenaar, L.I.; Kelly, S.D.; Douence, C.; Islam, M.; Monteiro, L.; Abrahim, A.; Rinke, P. Assessment of rapid low-cost isotope (δ15N, δ18O) analyses of nitrate in fruit extracts by Ti(III) reduction to differentiate organic from conventional production. *Rapid Commun. Mass Spectrom.* **2022**, *36*, e9259. [CrossRef] [PubMed]
33. Inácio, C.T.; Chalk, P.M.; Magalhães, A.M. Principles and limitations of stable isotopes in differentiating organic and conventional foodstuffs: 1. Plant products. *Crit. Rev. Food Sci. Nutr.* **2015**, *55*, 1206–1218. [CrossRef]
34. Yun, S.-I.; Lim, S.-S.; Lee, G.-S.; Lee, S.-M.; Kim, H.-Y.; Ro, H.-M.; Choi, W.-J. Natural 15N abundance of paddy rice (*Oryza sativa* L.) grown with synthetic fertilizer, livestock manure compost, and hairy vetch. *Biol. Fertil. Soils* **2011**, *47*, 607–617. [CrossRef]
35. Lester, G.E.; Manthey, J.A.; Buslig, B.S. Organic vs. conventionally grown Rio Red whole grapefruit and juice: Comparison of production inputs, market quality, consumer acceptance, and human health-bioactive compounds. *J. Agric. Food Chem.* **2007**, *55*, 4474–4480. [CrossRef]
36. Navarro, P.; Pérez-López, A.J.; Mercader, M.T.; Carbonell-Barrachina, A.A.; Gabaldon, J.A. Antioxidant activity, color, carotenoids composition, minerals, vitamin C and sensory quality of organic and conventional mandarin juice, cv. Orogrande. *Food Sci. Technol. Int.* **2011**, *17*, 241–248. [CrossRef]
37. Esch, J.R.; Friend, J.R.; Kariuki, J.K. Determination of the vitamin c content of conventionally and organically grown fruits by cyclic voltammetry. *Int. J. Electrochem. Sci.* **2010**, *5*, 1464–1474.
38. Mihailova, A.; Kelly, S.D.; Chevallier, O.P.; Elliott, C.T.; Maestroni, B.M.; Cannavan, A. High-Resolution Mass Spectrometry-Based Metabolomics for the Discrimination between Organic and Conventional Crops: A Review. *Trends. Food Sci. Technol.* **2021**, *110*, 142–154. [CrossRef]

39. Cubero-Leon, E.; De Rudder, O.; Maquet, A. Metabolomics for Organic Food Authentication: Results from a Long-Term Field Study in Carrots. *Food Chem.* **2018**, *239*, 760–770. [CrossRef] [PubMed]
40. Canali, S.; Diacono, M.; Campanelli, G.; Montemurro, F. Organic no-till with roller crimpers: Agro-ecosystem services and applications in organic Mediterranean vegetable productions. *Sustain. Agric. Res.* **2015**, *4*, 70. [CrossRef]

horticulturae

MDPI

Article

Innovative Tools for Nitrogen Fertilization Traceability in Organic Farming Products: A Fennel Case Study

Francesco Montemurro [1], Mariangela Diacono [1], Angelo Fiore [1], Luana Bontempo [2], Alberto Roncone [2], Margherita Amenta [3], Gabriele Ballistreri [3], Nicolina Timpanaro [3], Biagio Torrisi [3] and Simona Fabroni [3,*]

1 Council for Agricultural Research and Economics (CREA), Research Center for Agriculture and Environment, Via Celso Ulpiani, 5, 70125 Bari, Italy; francesco.montemurro@crea.gov.it (F.M.); mariangela.diacono@crea.gov.it (M.D.); angelo.fiore@crea.gov.it (A.F.)

2 Traceability Unit, Edmund Mach Foundation, Via E. Mach 1, 38098 San Michele All'Adige, Italy; luana.bontempo@fmach.it (L.B.); alberto.roncone@fmach.it (A.R.)

3 Council for Agricultural Research and Economics (CREA), Research Center for Olive, Fruit and Citrus Crops, Corso Savoia 190, 95024 Acireale, Italy; margherita.amenta@crea.gov.it (M.A.); gabriele.ballistreri@crea.gov.it (G.B.); nicolina.timpanaro@crea.gov.it (N.T.); biagiofrancesco.torrisi@crea.gov.it (B.T.)

* Correspondence: simona.fabroni@crea.gov.it

Abstract: Food fraud is a matter of great interest, particularly when organic productions are involved. Therefore, policymakers and institutions are asked to introduce new effective official control methods, not only based on documentary compliance to EU regulations. Lately, an integrated approach based on the use of isotopic, chemical and biological data treated by chemometrics has been suggested to authenticate organic products. The present research aims to validate the beforementioned integrated approach for the reliable classification of organically and conventionally grown fennel productions obtained by applying different treatments at diverse percentages of mineral fertilizers. Moreover, to the best of our knowledge, for the first time, an N-fixing plant (alfalfa), as an agroecological service crop in intercropping fennel, was introduced. Physicochemical parameters (fresh weight, total soluble solids, total acidity, consistency, Commission Internationale de l'Eclairage (CIE) lightness (L*), green-red (a*), blue-yellow (b*) color components), total phenolic content, in vitro antioxidant activity, total and inorganic N, and stable isotope ratios ($^{15}N/^{14}N$, $^{13}C/^{12}C$, and $^{34}S/^{32}S$) analyses were performed over a two-year field trial and further multivariate discriminant analysis was performed. The physicochemical parameters were not affected by the differential growing practice while antioxidant activity showed higher values in the organic treatments. The conventional treatments increased the NO_3^- values over 440% compared to the organic ones and the lowest values were found when alfalfa was used for intercropping. $\delta(^{15}N)$ of fennel, $\delta(^{34}S)$ of fennel and $\delta(^{15}N)$ of leaves showed great discriminatory potential, with respect to the applied agronomic practice. However, these parameters alone were not sufficient as biomarkers for differentiating cultivation methods. Instead, the proposed innovative tool demonstrated that only a comprehensive multivariate approach was able to achieve a correct classification of grouped cases. Furthermore, the discrimination between organic and conventional horticultural products showed feasible results, even when N-fixing plants were incorporated in organic practice, thus avoiding misinterpretations solely based on the isotopic N fingerprint. These promising findings highlight the potential of this method and suggest that further research should extend its application to other horticultural crops to confirm and expand upon the current results.

Keywords: organic and conventional fertilization; metabolites; isotopic ratios; multivariate analysis; agroecological practices

https://doi.org/10.3390/horticulturae11030329

1. Introduction

Nowadays, the conversion of cropland from conventional to organic agriculture is constantly increasing, sustained by both consumers' high sensitivity to sustainability issues and institutional policy targets introduced by the European Green Deal strategic initiatives package [1] and the United Nations' sustainable development goals [2]. In this context, farmers are encouraged to implement the conversion to organic farming, also taking into account that the quality of organic products is more profitable on the markets, compared to that of conventional ones. Data from the Italian utilized agricultural area (UAA) in 2023, showed, thanks to the support offered by the Common Agriculture Policy (CAP) [3], a significant growth reaching 2.5 million hectares (+4.5% compared to 2022, equal to +106 thousand hectares) [4]. This increase shortens the gap from the target of 25% of organic UAA to be achieved by 2030, as set by the European Commission within the Farm to Fork Strategy [1]. In the last year, the Italian organic horticultural crop production has recorded an increase of 1% respect to the previous year [4]. Taking into consideration the great efforts and challenges to be faced by the farmers where organic horticultural crop production is concerned, this growth represents a fairly interesting perspective in terms of the continuity of the whole supply chain.

The conversion to organic farming requires the adoption of sustainable practices aimed at eliminating the use of synthetic fertilizers and pesticides. Thereby, with the aim of increasing the fertility of the soil and preventing pests and pathogens attacks, organic farmers are required to apply only the allowed organic fertilizers and plant protection products, while implementing sustainable measures to avoid crop production losses. In organic vegetable rotations, the introduction of cover crops, planted either before or concurrent with a cash crop, may provide many beneficial ecosystem services, which could both sustain agricultural production and promote environmental protection [5]. Different cover crop species have been evaluated as a tool to improve agroecosystem resilience to control the weeds, pest occurrence, preserve in-field biodiversity, and therefore, to increase crop production [6]. The implementation of these agroecological practices requires redesigning existing cropping systems, which are typically relatively undiversified. One of these practices is intercropping, which consists of coexistence of two or more crop species simultaneously in the same field with different spatial arrangements [7]. In sustainable agricultural systems, and, in particular, in organic vegetable production, the intercropping can provide many beneficial ecosystem services and could be used in many horticultural systems to reduce the nitrogen (N) supply of cash crops. Specifically, when leguminous cover crops are introduced, atmospheric N fixation enhances nitrogen-use efficiency. The use of self-reseeding annual legumes, such as alfalfa, may play a crucial role in Mediterranean organic farming by reducing the need for external agricultural inputs [8].

However, the effort made by responsible farmers to carry out high-yield, safe and environmentally compatible horticultural organic productions may clash with the incorrect behavior of some deceitful producers who adopt fraudulent agronomic practices not allowed in organic farming. According to current European Regulation (EU) 2018/848 on the organic production and labeling of organic products [9], official controls must be carried out in accordance with Regulation (EU) 2017/625 [10]. This regulation establishes that official control methods must include an examination of documents, traceability records and other records which may be relevant for the assessment of compliance with the rules. However, to ensure high yields and avoid phytosanitary threats, fraudulent farmers adopt unfair practices and produce fictitious farm registers to the control bodies. That is why the possibility to unequivocally trace the origin of organic products is of great relevance to reassure the consumers who consciously choose to buy organic products. Organic trade associations and consumer federations are thus pressingly asking for the implementation of a traceability system based on reliable and reproducible analytical tools. Since organic authenticity has increasingly been a matter of interest, over the last two decades, many studies have proposed analytical methods to discriminate organic vs. conventional farming products on the basis of isotopic fractionation of the N source used as fertilizer [11–13]. The possibility of differentiating two samples on the basis of isotopic analysis is related to the N isotopic fractionation that happens in raw materials, i.e., the deflection from the natural isotopic allotment of N, the N isotopic ratio ($^{15}N/^{14}N$), as a consequence of natural biogeochemical processes. Specifically, the production and maturation of organic fertilizers cause a preferential volatilization of the lighter isotope; therefore, the resulting N isotopic ratio in organic fertilizers turns to be significantly higher respect to synthetic ones. It has been widely demonstrated that this difference is also reflected in the harvested crops [14–16]. However, some controversial studies have also reported that N isotopic ratio may not be discriminatory for all the crops and management conditions. Indeed, it has been demonstrated that when organic fertilizers are applied in conventional agriculture, or when crops with low N requirement are concerned, the N isotopic ratio alone may not allow a reliable discrimination [17,18]. Moreover, specifically targeting horticultural crop productions, the application of N-fixing plants as cover crops or agroecological service crops in organic agriculture could have an impact on the N isotopic signature of the harvested crop leading to the misclassification of real organic horticultural products.

Very recently, an integrated approach based on the use of isotopic, chemical and biological data treated by chemometrics has been suggested by the authors to authenticate organic products [19]. This approach has been further tested and validated for the cauliflower crop in different productive and environmental conditions [20]. In this study, different treatments were carried out and compared, including organic, conventional, and mixed ones at different percentages of mineral fertilizers, including the application of a non-N-fixing plant, the horseradish (*Raphanus sativus* L.), as an agroecological service crop. However, there is a lack of knowledge about the validation of this approach to different crops, environments and while applying N-fixing plants as cover crops.

Based on what has been stated so far, the present research aims are to test and validate the beforementioned integrated approach as a reliable analytical tool to improve organic fennel authenticity. To achieve a reliable classification of organically and conventionally grown fennel productions, different treatments representing diverse percentages of mineral fertilizers were applied. Moreover, to the best of our knowledge, the first time a N-fixing plant (alfalfa) was introduced was as an agroecological service crop in intercropping fennel. The proposed validated tool will help to protect the efforts made by responsible organic farmers while preventing food frauds committed by some deceitful organic producers who adopt fraudulent agronomic practices, including the use of synthetic fertilizers, which is not allowed in organic farming.

2. Materials and Methods

2.1. Field Experiment Setup, Experimental Design, Crop Production and Plant Material Determinations

A two-year experimental field trial was carried out at the Council for Agricultural Research and Economics in the Research Centre for Agriculture and Environment, in the experimental farm "Campo 7" (located in Metaponto (MT)—South Italy, lat. 40°24′ N; long 16°48′ E and 8 m.a.s.l.). The soil of the experimental farm is classified as Vertic Cambisols according to WRB classification. It contains 42, 39, and 19% of clay, loam, and sand, respectively. It is also characterized by about 25 g·kg^{-1} of organic matter, 759 mg·kg^{-1} of exchangeable potassium, 31.1 mg·kg^{-1} of Olsen phosphorus, pH 7.8 and 0.48 mS·cm^{-1} of electrical conductivity. We decided to investigate the presence of N content only in the bulbs and not in the soils, since fennel is consumed as fresh vegetable, and the nitrate level is a limiting factor for consumer health. Therefore, we focused on the crop production and N content in bulbs.

In the study, here reported, two years of fennel cultivation (Aurelio/Augusto F1 cv—Cycle: 150 days) are presented, during the 2018 and 2019 cropping seasons. According to the Köppen–Geiger classification, the area falls under the Csa climate, characterized by hot, dry summers and mild, wet winters. The average monthly temperatures (September to December) during each cropping cycle (16.9 and 15.9 °C in the first and second cycle, respectively) were quite close to averages of the long-term value (15.3). The cumulative rainfall recorded during the fennel field experiment was 206.2 and 342.4 mm in the first and second cycle, respectively, compared to 244.8 mm in the long period.

Six fertilization strategies, placed on three field strips of the experimental field were compared as follows: (1) "Organic" as pelleted animal manure (3% N) applied at transplanting in a soil managed without synthetic fertilization in the last several years; (2) "Mix-organic" with 1/3 N by pelleted animal manure at transplanting + 2/3 as synthetic N (i.e., UREA) one month after transplanting, in a soil without synthetic fertilization for several years; (3) "Conventional" with only synthetic N fertilization, in a soil in which chemical fertilization has always been managed using conventional techniques, by using UREA as 1/2 N at the transplanting + 1/2 one month after transplanting; (4) "Mix conventional-a" carried out in a soil in which chemical fertilization has always been managed using conventional techniques. Fertilization was performed at the transplanting with 1/3 N by pelleted animal manure + 2/3 with synthetic N one month after transplanting; (5) "Mix conventional-b" carried out in a soil in which chemical fertilization has always been managed using conventional techniques. Fertilization was performed at the transplanting with 2/3 pelleted animal manure + 1/3 with synthetic N one month after transplanting; (6) "Organic + agro-ecological practices (Organic + AEP)" with N applied at transplanting and only as organic fertilizer of vegetable origin (Vegand: 4–1.4–0). This latter treatment was a part of a long-term filed experiment [21], in which different agroecological practices are being tested. In particular, in fennel cropping system the alfalfa (*Medicago sativa* L.) was used as intercropping to sustain the cash crop and to increase N utilization efficiency of the entire system. The total amount of fertilizer (organic plus mineral) was 120 kg·ha^{-1} for each elementary plot, which is the normal amount of fertilizer applied for this crop by the organic producers. Each elementary plot consisted of 57.6 m^2, replicated 4 times. Based on the percentage of organic and mineral fertilizers (pellet, Vegand, and UREA) applied for each treatment, we calculated the right amount of organic and/or mineral fertilizers distributed in each treatment in one solution (at transplanting) or two times (transplanting and one month after), according to the fertilizer strategies adopted and above described.

The fennel crops were transplanted on 06th and 10th September 2018 and 2019, respectively, with a spacing of 0.30 m per row and 0.80 m between rows (about 240 plants for each treatment). To avoid the fertilizer contamination among the treatments we separated the plot along the row for 1 m. Moreover, we collected the plants for the analyses in the center of each elementary plot to be sure that there was any possible contamination among fertilizer treatments. The harvest occurred at the commercial maturity on 28 November 2018 and 4 December 2019, respectively. Pathogens control was the same in all N fertilization strategies, according to the European regulation for organic productions. In particular, even if two of the three fields were not conducted according to organic production, we decided to apply the same procedure for crop husbandry at all treatments. Finally, we adopted the same agronomical practices (no irrigation, weed control, etc.) at all treatments even if they were cultivated in different fields, in order to attribute the difference only at the fertilizer strategies.

Every year in June and at the fennel harvest (late November and early December in 2018 and 2019, respectively), one soil sample for each elementary plot (6 treatments × 4 replicates = 24 samples) was collected. To have a more representative sample we collected five sub-samples of 0–30 cm from each plot randomly along the diagonal of each plot, and they were pulled into one sample for the chemical analyses. We did this operation manually with a hand auger. At the fennel harvest, we collected the plants samples in the middle of each elementary plot. Each plant was divided into leaves and fennel marketable yield for each elementary plot (24 samples).

The physicochemical parameters (fresh weight, total soluble solids, total acidity, consistency, CIE L*, a*, b* color components) were determined on fresh fennel bulb samples, while total phenolic content, in vitro antioxidant activity, total and inorganic N, and stable isotope ratio ($^{15}N/^{14}N$, $^{13}C/^{12}C$, and $^{34}S/^{32}S$) analyses were performed on freeze-dried fennel bulbs. Stable isotope ratio analyses were also performed on fertilizers ($^{15}N/^{14}N$), soil samples ($^{15}N/^{14}N$ and $^{34}S/^{32}S$) and leaves samples ($^{15}N/^{14}N$, $^{13}C/^{12}C$, and $^{34}S/^{32}S$). All the samples (soil, leaves, and bulbs) were analyzed, with four experimental replicates for each treatment for each year of experimentation. Each experimental replicate was analyzed in triplicate.

2.2. Physicochemical Parameters

A representative sample of four bulbs was collected at commercial maturity from each of the six treatments and was used for determining the physicochemical parameters. Fennel plant fresh weight (g) for each treatment was expressed as average value (of four measurements). The total acidity (TA) was determined by automatic titration (Mettler Toledo, Greifensee, Switzerland, DL 25) of the fennel bulb juice with a 0.1 M NaOH alkaline solution and expressed as % of oxalic acid. The pH was measured with a pH meter (Mettler Toledo, Greifensee, Switzerland, DL 25). A digital refractometer (ATAGO, Minato-ku, Tokyo, Japan, RX-5000) was used to determine the total soluble solids (TSS) which were expressed as °Brix of the fennel bulb juice. The firmness was measured at two opposite points per fennel bulb using a texture analyzer (Zwick/Roell (Ulm, Germany), DO-FB0.5 TS model 2002) and expressed in N. Finally, the color analysis was evaluated on four equidistant points of the equatorial part of the fennel bulb and it was expressed as CIE L*, a*, b* values using a colorimeter (Minolta, Chiyoda-ku, Tokyo, Japan, CR-300).

2.3. In Vitro Antioxidant Activity

The in vitro antioxidant activity was measured through two assays based on spectrophotometric and spectrofluorimetric methods. It was evaluated on freeze-dried fennel bulb samples (500 mg) obtained from the six different agronomic plots. Samples were extracted in 100 mL of an hydroalcoholic solution (ethanol:water, 80:20 *v*/*v*) at 1% citric acid kept stirring at 120 rpm for 24 h at room temperature. The extraction procedure was repeated again on the residue obtained by filtration. Then, the two extracts were combined and further distilled under vacuum in a rotavapor lab system to remove ethanol. The concentrated extracts were then used for the determination of the antioxidant activity by Folin–Ciocalteu Reagent (FCR) and Oxygen Radical Absorbance Capacity (ORAC) assays. Both assays were used as they are based on two different reaction mechanisms, an electron-transfer (ET) and an hydrogen atom transfer (HAT) reaction, respectively [22]. Therefore, they might be representative of the antioxidant capacity exerted by different classes of compounds present in the raw sample.

2.3.1. Folin–Ciocalteu Reagent (FCR)

The FCR colorimetric method was applied as previously described [20] with some modifications. This colorimetric assay was used to determine the total phenolic content as a measure of the total antioxidant capacity. Diluted samples (1:20, *v*/*v*) of the concentrated extracts were mixed with 5 mL of FC commercial reagent (previously diluted with water 1:10 *v*/*v*) and 4 mL of a 7.5% sodium carbonate solution. After reacting for 2 h at room temperature in the dark, the absorbance was spectrophotometrically measured at 765 nm, and the concentration of total phenolic was expressed as mg (±) gallic acid equivalents (mg GAE·g^{-1} DW).

2.3.2. Oxygen Radical Absorbance Capacity (ORAC)

The ORAC assay was performed on appropriately diluted samples (1:50, *v*/*v*) of the concentrated aqueous extracts as previously described [19] on a Wallac 1420 Victor III 96-well plate reader (Perkin Elmer, Shelton, DC, USA). The assay was carried out using fluorescein (116 nM) as the target molecule for the free radical attack from the AAPH (153 mM) peroxyl radical generator, Trolox (1 μM) as the control standard and phosphate buffer (pH 7.0) as the blank. Results were expressed as μmoles of Trolox equivalents (μmol TE·g^{-1} DW).

2.4. Total and Inorganic N on Fennel Bulb Samples

To determine total N (mg·kg^{-1}) after dry matter digestion, the Kjeldahl method was adopted, using the AutoKjeldahl Unit K-370 (Buchi, Flawil, Switzerland) [14,23].

Nitrate and ammonium were determined as follows: 25.0 g of ground and freeze-dried sample were transferred into a 200 mL glass flask, added with 100 mL of distilled water, homogenized by stirring in a water bath, at a temperature of (+80 °C) for 20 min. Once cool, the sample was filtered and 5.0 mL of the first filtrate was discarded, whereas the subsequent filtrate was collected. The filtrate was used for nitrate and ammonium determinations [24]. The concentration was performed by a continuous flow colorimetry auto-analyzer (Flow SyS Systea, Anagni, Italy).

2.5. Stable Isotope Ratio Analysis

Freeze-dried fennel bulb samples were subjected to $^{13}C/^{12}C$, $^{15}N/^{14}N$, and $^{34}S/^{32}S$ analyses using an isotope ratio mass spectrometer (Delta plus XP Thermo Finnigan, Bremen, Germany) equipped with an elemental analyzer (EA Flash 1112 Thermo Finnigan, Bremen, Germany). The values of the isotopic ratios were expressed in δ‰ and calibrated against international reference materials including NBS-22 oil (IAEA) for $\delta^{13}C$ and glycine (USGS 64) for $\delta^{15}N$ measurement. Hair USGS 42 ($\delta^{34}S$ = +7.84 ± 0.25‰), USGS 43 ($\delta^{34}S$ = +10.46 ± 0.22‰), and barium sulfates IAEA-SO-5 and NBS 127 (IAEA) were used for $^{34}S/^{32}S$ calibration.

As far as soil and fertilizers samples are concerned, the analysis of $\delta^{15}N$ was carried out by weighing 25 mg for soil samples and 1 to 7 mg for fertilizer samples, based on their elemental composition, into tin capsules and measuring them using an EA Flash 1112 (Thermo Finnigan). The isotopic values, expressed in delta per mil, were calculated relative to a laboratory working standard (casein, $\delta^{15}N$ = +7.38), which was calibrated against international reference materials: IAEA-NO-3 (IAEA—International Atomic Energy Agency, Vienna, Austria, $\delta^{15}N$ = +4.7‰) and L-glutamic acid USGS 40 ($\delta^{13}C$ = −26.39‰ and $\delta^{15}N$ = −4.52‰) (U.S. Geological Survey, Reston, VA, USA). Sulfur values were measured using a Vario Cube (Elementar Analysensysteme GmbH, Langen-Selbold, Germany), with 20 mg of sample weighed into tin capsules. The values were calculated relative to the international reference material (USGS 42, Tibetan hair, $\delta^{34}S$ = 7.84‰) and an internal working standard (flour, $\delta^{34}S$ = 0.41‰).

The CNS values of fennel leaves were obtained using a Vario Cube (Elementar Analysensysteme GmbH, Langen-Selbold, Germany) by weighing 2 mg of sample into tin capsules. Isotopic ratios were determined relative to one internal standard (casein $\delta^{13}C$ = −21.98‰, $\delta^{15}N$ = 7.38‰) for C and N, and using two points for sulfur: one international reference material (USGS 42, Tibetan hair $\delta^{34}S$ = 7.84‰) and an internal reference material (flour, $\delta^{34}S$ = 0.41‰). The internal reference standards were calibrated against internationally recognized materials, as follows:

- For $\delta^{13}C$: fuel oil NBS-22 ($\delta^{13}C$ = −30.03‰), sucrose IAEA-CH-6 ($\delta^{13}C$ = −10.45‰), and L-glutamic acid USGS 40 ($\delta^{13}C$ = −26.39‰ and $\delta^{15}N$ = −4.52‰).
- For $\delta^{15}N$: L-glutamic acid USGS 40 and potassium nitrate IAEA-NO-3 ($\delta^{15}N$ = +4.7‰).
- For $\delta^{34}S$: barium sulfate IAEA-SO-5 ($\delta^{34}S$ = +0.5‰) and NBS 127 ($\delta^{34}S$ = +20.3‰).

The maximum acceptable standard deviations for repeatability were set at 0.1‰ for $\delta^{13}C$, 0.2‰ $\delta^{15}N$, and 0.3‰ $\delta^{34}S$.

The stable isotope composition of a substance is commonly represented using the delta (δ) notation, with values expressed in parts per thousand (‰) or in the SI unit Urey (1 mUr = 1‰), as recommended by the IUPAC. This notation describes the variations in the molar ratio (R) of the heavy to light isotopes of an element, or the ratio of a rare to a common stable isotope relative to a defined reference standard. The calculation follows this equation:

$$\delta_{ref}(^{i}E/^{j}E, sample) = \left[\frac{R(^{i}E/^{j}E,\ \ sample)}{R(^{i}E/^{j}E,\ \ ref)}\right] - 1$$

where *ref* represents the international measurement standard, *sample* refers to the analyzed material, and *iE/jE* denotes the isotope ratio between the heavier and lighter isotopes. The delta values are multiplied by 1000 and are typically expressed in "per mil" (‰) units. Alternatively, following the International System of Units (SI), they can be reported as "milliUrey" (mUr), as recommended by Coplen [25].

2.6. Statistical and Chemometric Multivariate Analysis

The statistical differences in the mean values of all the determined parameters were determined through variance analysis (ANOVA) and the treatment means separation was performed by Tukey's post hoc test (STATSOFT 6.0 software—Statsoft Italia srl, Vigonza, Italy). The analysis of variance was first carried out considering two fixed factors ("years" and "fertilization treatments"). Considering that the factor "years" and the interaction between the two factors were not significant ($p > 0.01$) for most of the investigated parameters, in this study we present the average values for each treatment in the two years of experimentation. The discriminating ability of the implemented system was tested by a multivariate statistical tool, the linear discriminant analysis (LDA) by using the IBM SPSS Statistics (version 20) software (IBM corporation; Armonk, NY, USA) with the within-class covariance matrix as classification model. The aim of the chemometric multivariate analysis was to determine whether all the determined variables could discriminate among the samples belonging to the six different treatments including three theses whose fertilization was entirely ("Organic"; "Organic + AEP") or mainly ("Mix-Organic") organic and three theses whose fertilization was entirely ("Conventional") or mainly conventional ("Mix-Conventional-a"; "Mix-Conventional-b"). Therefore, a linear discriminant analysis was performed with the aim of investigating differences between groups on the basis of the attributes of the analyzed variables for each case. The performed discriminant analysis created a number of discriminant functions which minimized the possibility of misclassifying cases into their respective groups or categories, indicating which attributes contributed most to group separation. The number of the generated discriminant functions is always one less than the number of the groups; thus, in our study five discriminant functions were produced.

3. Results and Discussion

3.1. Physicochemical and Qualitative Fennel Parameters

The results of the physicochemical and qualitative fennel parameters are reported in Table 1. The values for each parameter are reported as mean of the two years of experimentation. No significant differences were found among treatments for all the parameters; they were not affected by climate conditions which were comparable in the two cropping cycles. Similar results were previously reported for fennel cv. 'Giulio' and 'Helvia' in a study aimed at comparing conventional and organic vegetable farming systems in a long-term experimental field [26]. Even if two years are not enough to draw general conclusion, it is possible to confirm that slight meteorological differences did not modify physicochemical and qualitative fennel performance. The plant fresh weight values recorded at harvest for the six treatments were consistent with the study by Atta-Aly for sweet fennel var. 'Dolce' [27]. The different types of N fertilization did not affect the plant fresh weight at harvest, thus indicating that the supplied mineral fertilizer ("Conventional") the organic amendments ("Organic" and "Organic + AEP" treatments), and their coformulations ("Mix-organic", "Mix conventional-a" and "Mix conventional-b" treatments) fulfilled the fennel plants' N requirements. For the agroecological point of view, this result indicated that it is possible to substitute totally or partially mineral fertilizers with organic ones without reducing fennel performances [28]. Indeed, during the two years of trial, in all the treatments the plants were thick and compact, with a rounded head already in the early stages of growth. As far as pH, TA and TSS are concerned, the recorded values were in agreement with the levels reported for these parameters in previous studies on fennel var 'Azoricum' cv. 'Augusto' and 'Tiziano' [29] and 'Clio' [30]. The pH values ranged from 6.42 ± 0.20 of the "Organic" treatment to 6.78 ± 0.06 of the "Mix conventional-a" treatment while TA showed values ranging from $0.28 \pm 0.01\%$ of oxalic acid of the "Mix conventional-b" treatment to $0.36 \pm 0.02\%$ of oxalic acid of the "Organic" treatment. TSS varied from 5.07 ± 0.23 °Brix of the "Organic + AEP" treatment to 5.68 ± 0.41 °Brix of the "Mix-organic". The CIE L* a* b* values of the fennel samples for each treatment were in accordance with the values reported in previous studies [28, 29] putting in evidence that no differences on the color coordinates of the fennel bulbs were produced as a result of the different type and percentage of fertilizer distributed. Firmness was not affected by the growing practice and its levels showed no statistically significant differences among treatments, showing values ranging from 74.59 ± 2.53 N of the "Conventional" treatment to 81.10 ± 1.53 N of the "Organic" one, showing that, once again, this agronomical practice cannot produce any effect to the genetical and environmental sources of variations. The results have highlighted how the differential fertilization and agronomic practice applied in the six different treatments did not produce any significant difference on the main physicochemical parameters, demonstrating that fennel plants' N requirements were equally fulfilled.

Table 1. Physicochemical and qualitative fennel bulb parameters. The data reported are average values ± standard errors ($p \leq 0.01$)., relating to four experimental replicates for each treatment for two years of experimentation (N = 8).

Treatment	Plant Fresh Weight (g FW)	pH	Total Acidity (% Oxalic Acid)	Total Soluble Solids (°Brix)	L*	a*	b*	Firmness (N)
organic	979.51 ± 46.38	6.42 ± 0.20	0.36 ± 0.02	5.38 ± 0.32	76.91 ± 2.19	−2.64 ± 0.14	13.33 ± 0.99	81.10 ± 1.53
mix-organic	984.43 ± 36.91	6.69 ± 0.04	0.36 ± 0.03	5.68 ± 0.41	76.66 ± 2.38	−2.52 ± 0.22	14.03 ± 1.15	80.38 ± 1.61
conventional	1072.08 ± 71.30	6.69 ± 0.05	0.31 ± 0.01	5.32 ± 0.37	81.62 ± 0.35	−2.79 ± 0.10	13.99 ± 0.98	74.59 ± 2.53
Mix-conventional-a	936.67 ± 33.04	6.78 ± 0.06	0.34 ± 0.03	5.59 ± 0.38	79.74 ± 1.04	−2.88 ± 0.20	13.72 ± 1.08	79.99 ± 1.76
Mix-conv.-b	1020.77 ± 63.09	6.64 ± 0.03	0.28 ± 0.01	5.20 ± 0.34	80.26 ± 0.90	−2.71 ± 0.12	13.71 ± 1.05	78.21 ± 3.17
organic + AEP	1131.86 ± 63.51	6.48 ± 0.11	0.34 ± 0.02	5.07 ± 0.23	79.35 ± 0.74	−3.00 ± 0.20	13.42 ± 0.81	79.34 ± 1.82

3.2. *In Vitro Antioxidant Activity*

The total polyphenolic content measured through FCR assay and expressed as mg of gallic acid equivalents g-1 DW, showed (Figure 1) statistically significant differences among the six treatments ($p \leq 0.01$). Higher values were found in the "Organic" treatment (18.17 ± 0.28 mg GAE·g^{-1} DW), intermediate values in the Mix-organic" and "Organic + AEP" treatments (16.70 ± 0.30 mg GAE·g^{-1} DW and 15.75 ± 0.68 mg GAE·g^{-1} DW, respectively) and lower ones in the three conventional treatments (15.71 ± 0.46, 15.40 ± 0.41, and 14.98 ± 0.68 mg GAE·g^{-1} DW, for "Mix conventional-b", "Mix conventional-a", and "Conventional", respectively). The recorded total polyphenolic values were in accordance with the values reported for fennel cv. 'Chiarino' by Ninfali et al. in a previous work [31]. The differences found among the six treatments are in accordance with what previously reported by Salama et al. [32] who showed that organic and bio-organic fertilization produced an increase in total polyphenolic content in 'Dolce' and 'Zefa fino' fennel compared with conventional fertilization. As for ORAC assay, significant differences ($p \leq 0.01$) were found among the treatments which partially reflected those found for total polyphenolic content. Indeed, higher values were recorded in the "Organic" (2785.39 ± 21.94 ORAC units) and "Mix-organic" (2743.04 ± 85.17 ORAC units) treatments while the lowest values were found in "Mix conventional-b" (2207.99 ± 82.31 ORAC units) and "Organic + AEP" (2378.03 ± 46.34 ORAC units). The ORAC assay's sensitivity to different types of antioxidants and the specific conditions of the assay may account for this variation respect to the FCR assay. It is widely recognized that ORAC units are an expression of the antioxidant quality of raw food. Indeed, a daily intake of raw vegetables providing at least 3000–5000 ORAC units has been previously recommended to prevent the incipience of free radicals at a physiological level [33]. Our results show a relevant contribution of fennel in terms of antioxidant capacity, in accordance with another research [31]. The higher accumulation of secondary metabolites exerting relevant antioxidant capacities in organic fertilization treatments, compared to conventional ones, has also been reported for other horticultural products, such as cauliflower [19]. Moreover, a systematic review by Baranski et al. [34], based on an extensive dataset of 343 peer-reviewed publications, confirmed that organic crops have higher concentrations of polyphenols and a higher antioxidant activity. For over a decade, it has been suggested that the accumulation of secondary metabolites in organic products is linked to a C:N ratio unbalance. This imbalance is based on a transient nitrogen deficiency in organic management which induces the biosynthesis of carbon-rich phenylpropanoids [35]. In addition, it has been recently suggested that the organic farming practices promote a metabolism shift that ensures a continuous supply of carbon skeletons for the biosynthetic demands of organic plants [36]. Our results have shown that organic management produced a significant increase in total phenolics and ORAC units supporting the accumulation of carbon-rich phenylpropanoids as a consequence of the organic farming practice.

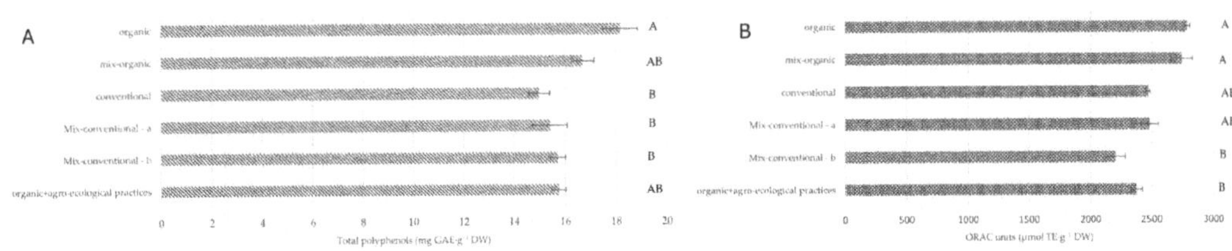

Figure 1. Total polyphenol content (**A**) and ORAC units (**B**) in fennel bulbs (four experimental replicates for each treatment for two years of experimentation; N = 8) ± standard errors. Different letters indicate statistically significant differences ($p \leq 0.01$).

3.3. Total and Inorganic Nitrogen on Fennel Bulb Samples

In Table 2 the average concentrations of total N, total inorganic N, NO_3^- and NH_4^+ in fennel bulbs of the six different treatments are reported. Both the total N and total inorganic N showed significantly ($p \leq 0.01$) higher values in the "Organic + AEP" treatment, showing that the alfalfa consociation contributed to the N availability for the cash crop. As expected, for both these parameters the significantly lowest values ($p \leq 0.01$) were detected in the "Organic" treatment. Synthetic N sources of the "Conventional", "Mix conventional-a" and "Mix conventional-b" treatments produced a significant increase ($p \leq 0.01$) in nitrate content compared to the organic treatments. In particular, as mean value, the conventional treatments increased the NO_3^- values over 440% compared to the organic ones, without increasing the fennel growth and production (as shown in Table 1). Being that fennel is consumed as a fresh vegetable, the nitrate level is a limiting factor for consumer health. It has been recommended that fresh vegetables with nitrate levels exceeding 1000 mg·kg^{-1} of fresh weight may constitute a risk for human health [37]. Our results confirmed that nitrate levels in the different treatments were far below this limit; therefore, fennel can be considered safe for consumption as fresh product.

Table 2. Total N, total inorganic, nitrate, and ammonium N concentrations in fennel bulb. The data reported are average values ± standard errors, relating to 4 experimental replicates for each treatment for two years of experimentation (N = 8).

Treatment	Total Nitrogen (%)	Total Inorganic Nitrogen (mg·kg^{-1} DW)	NO_3^- (mg·kg^{-1} DW)	NH_4 (mg·kg^{-1} DW)
organic	1.72 ± 0.04 D	480.90 ± 4.34 C	51.03 ± 4.65 B	429.87 ± 7.41 C
mix-organic	2.13 ± 0.04 C	596.77 ± 29.29 BC	49.70 ± 3.99 B	547.07 ± 25.79 BC
conventional	2.40 ± 0.06 BC	774.18 ± 37.05 AB	236.48 ± 30.99 A	537.70 ± 16.40 BC
Mix-conventional-a	2.28 ± 0.07 C	843.00 ± 54.81 BC	227.35 ± 47.81 A	615.65 ± 25.89 B
Mix-conventional-b	2.69 ± 0.08 AB	929.54 ± 32.41 A	305.12 ± 28.58 A	624.42 ± 12.87 B
organic + agro-ecological practices	2.75 ± 0.05 A	844.96 ± 56.53 A	41.57 ± 2.45 B	803.39 ± 55.72 A

Different letters indicate statistically significant differences ($p \leq 0.01$).

Furthermore, the "Organic + AEP" treatment was part of a long-term experiment in which agroecological practices were studied [21]. These practices likely contributed to increase the efficiency of nutrient use, by retaining excess N after the harvest of commercial crops and enhancing its availability for subsequent commercial crops [38,39]. This is evident from the higher content of total N in the fennel bulb. On the other hand, the nitrate content was lower in this treatment since the mineralization is modulated. Furthermore, the presence of alfalfa (*Medicago sativa* L.) in the soil may have determined an increase in organic carbon, thus improving the availability of N in the soil as the soil organic matter mineralized [40]. The presence of alfalfa as cover crop may have also increased the soil microbial biomass pool [41,42] with notable positive effects also on the cash crop.

3.4. *Stable Isotope Ratios Analysis*

3.4.1. Analyses of $^{15}N/^{14}N$ on Fertilizers

The fertilizers used in this study exhibited $\delta(^{15}N)$ values (Table 3) that align well with the ranges previously reported by Bateman and Kelly [11,17] for various types of fertilizers. Specifically, synthetic nitrogen fertilizers tend to have stable nitrogen isotope values close to zero, typically ranging between −2 and 2‰. In this study, the measured values were 0.0‰ and 0.5‰, consistent with expectations. The nitrogen in these fertilizers originates from atmospheric nitrogen ($\delta(^{15}N)$ atm = 0‰), with minimal fractionation occurring during their production. Conversely, as expected, organic fertilizers displayed noticeably higher $\delta(^{15}N)$ values. In particular, the animal-derived fertilizer showed the highest values (8.6‰), while the plant-derived fertilizer had comparatively lower values (5.0‰). These differences are attributable to the source materials and their respective nitrogen cycles before application.

Table 3. $\delta(^{15}N)$ values determined in the fertilizer samples.

$\delta(^{15}N)$ (‰ vs. AIR)	$\delta(^{13}C)$ (‰ vs. V-PDB)	$\delta(^{34}S)$ (‰ vs. V-CDT)	Type of Fertilizer
8.6	−25.5	0.6	Pelleted animal manure
5.0	−22.6	4.5	Vegand
0.0	−43.1	4.7	Urea (one month after transplanting)
0.5	−16.0	1.5	Urea (at transplanting)

3.4.2. Analyses of $^{15}N/^{14}N$ $^{13}C/^{12}C$ and $^{34}S/^{32}S$ on Fennel Bulb Samples

The Box and Whisker plots reported in Figure 2 are referred to the results of the stable isotope ratio analysis of $^{15}N/^{14}N$ [$\delta(^{15}N)$], $^{13}C/^{12}C$ [$\delta(^{13}C)$] and $^{34}S/^{32}S$ [$\delta(^{34}S)$] in fennel bulb samples of the six different treatments.

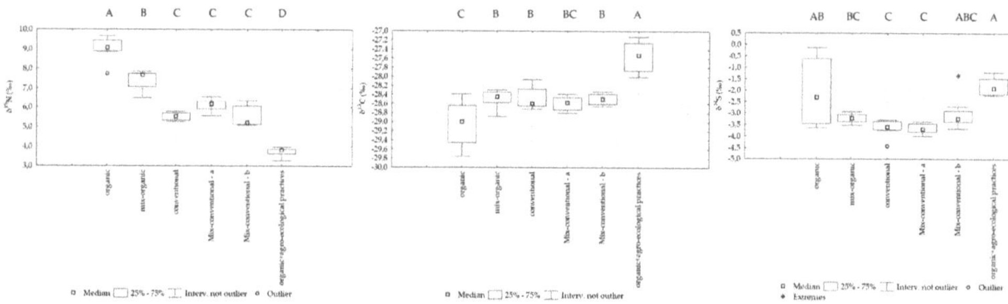

Figure 2. Boxplot of $^{15}N/^{14}N$ [$\delta(^{15}N)$], $^{13}C/^{12}C$ [$\delta(^{13}C)$] and $^{34}S/^{32}S$ [$\delta(^{34}S)$] in fennel bulb samples (four experimental replicates for each treatment for two years of experimentation; N = 8). Different letters indicate statistically significant differences ($p \leq 0.01$).

As regards the $\delta(^{15}N)$ values, our results showed a clear effect of the fertilization on its relative values in the different treatments. The $\delta(^{15}N)$ highest values (9.03 ± 0.19‰, $p \leq 0.01$) was recorded in the "Organic treatment". The "Mix-organic" treatment showed significantly lower values (7.42 ± 0.18‰, $p \leq 0.01$) with respect to the "Organic" plot while the three conventional treatment showed even lower values, in particular "Conventional", "Mix conventional-a" and "Mix conventional-b" were equal to 5.52 ± 0.07‰, 6.11 ± 0.11‰ and 5.54 ± 0.18‰, respectively. The lowest values were recorded in the "Organic + AEP" treatment (3.73 ± 0.07‰). As previously widely assessed [11,12,18,20,43], the N isotope ratio ($^{15}N/^{14}N$) can be recognized as a reliable indicator of the applied agricultural regime, due to the favored volatilization of NH_4^+ having the lighter N isotope during the generation of organic manures which causes significant increases in its value. At the same time, it has been postulated that if green manure based on N-fixing plants (i.e., leguminous crops) is applied in crop rotation in the organic management production, the reliability of the discrimination can result significantly affected [18,20]. The results of our study highlighted that the alfalfa (*Medicago sativa* L.), used to sustain the cash crop and to increase N utilization efficiency, produced a significant effect on the fractionation of the N isotope which reflected on significantly lower levels of its ratio ($^{15}N/^{14}N$). Therefore, although the "Organic + AEP" plot was fertilized only with an organic fertilizer of vegetable origin, the $\delta(^{15}N)$ analysis produced values compatible with a synthetic mineral fertilization, thus putatively leading to a misclassification of the samples collected in this plot, if $\delta(^{15}N)$ alone was considered as indicator of the growing practice. It could also be considered that this latter treatment was carried out in an experimental field in which agroecological practices have been adopted for several years, and these procedures are encouraged by policymakers, which would increase environmental production sustainability [44]. Therefore, if we consider only the $\delta(^{15}N)$ value as a tool for N fertilization traceability we will possibly have a misunderstanding reading of results, with wrong legislation rules.

The $\delta(^{13}C)$ values of fennel showed significant differences ($p \leq 0.01$) among the treatments. The highest values were recorded in the "Organic + AEP" ($-27.55 \pm 0.11‰$) samples, intermediate ones in "Mix-organic" ($-28.49 \pm 0.07‰$), "Conventional" ($-28.48 \pm 0.08‰$), "Mix conventional-a" ($-28.59 \pm 0.05‰$) and "Mix conventional-b" ($-28.50 \pm 0.04‰$) treatments, whereas the lowest values were found in "Organic" ($-29.04 \pm 0.17‰$). Our results are in agreement with previous studies that have shown significant lower $\delta(^{13}C)$ values in organic onions and cabbages [45] and in organic peaches [17] due to the higher microbiological activity in the organic soil resulting in respiratory CO_2 with lower $\delta(^{13}C)$. It has also been previously demonstrated that with high N availability, as in the conventional plots, the photosynthetic rate increases, thus reducing the discrimination of the enzyme RuBisCo against $^{13}CO_2$. This determines higher $\delta(^{13}C)$ values in conventional plots in comparison to the organic ones [17,43]. Indeed, our results confirm this finding as the "Organic" treatment showed the lowest $\delta(^{13}C)$ values. In addition, although being an organically grown treatment, the "Organic + AEP" thesis showed the highest $\delta(^{13}C)$ values, probably due to the higher N availability for the cash crop due to the consociation of alfalfa as intercropping which further increased the rate of the photosynthesis. In line with this evidence, the "Organic + AEP" treatment showed higher total N, total inorganic N and NH_4^+ contents, as already shown in Section 3.3.

The different agronomic practices applied produced effects on the $\delta(^{34}S)$ values. Significantly lower values ($p \leq 0.01$) were found for the "Conventional" ($-3.64 \pm 0.12‰$) and "Mix conventional-a" ($-3.67 \pm 0.07‰$) treatments compared to "Organic" ($-2.06 \pm 0.50‰$) and "Organic + AEP" ($-1.81 \pm 0.13‰$). It is known that sulfur isotope ratio may provide supplementary information for the applied agricultural practice, as it is strictly related to the soil and the sulfate fertilizer employed [46,47]. Our results demonstrate that the soil of the conventional plots, which is different from the soil of the organic ones, produced useful insights into the discrimination of the agricultural regime.

3.4.3. Analyses of $^{15}N/^{14}N$ and $^{34}S/^{32}S$ on Soils

Figure 3 presents the boxplots of $\delta(^{15}N)$ and $\delta(^{34}S)$ values in the soil across the six fertilization treatments analyzed in this study. The impact of conventional, organic, and mixed fertilization strategies was statistically evaluated. Significant differences ($p \leq 0.01$) were observed for $\delta(^{34}S)$ between "Organic" and "Mix-conventional-b" treatments, despite partial overlap between the two groups. This finding aligns with previous studies [46], which suggest that, although fertilizer type influences $\delta(^{34}S)$ values, this parameter is more sensitive to geographical origin than to the agricultural system. The consistency in $\delta(^{34}S)$ variability across the six tested treatments supports this hypothesis, as these similarities are likely driven by geological soil characteristics, microbial activity, and proximity to the sea [46]. On the other hand, the ANOVA test identified more pronounced statistically significant differences ($p \leq 0.01$) in $\delta(^{15}N)$ values, based on agricultural practices. A tendency for higher $\delta(^{15}N)$ values was observed in organic and mixed-organic treatments, in soils with a longer history of organic management, which resulted in statistically significant differences compared to the lowest values recorded in the mixed-conventional treatment b. Conversely, conventional, mixed-conventional a, and agroecological treatments displayed $\delta(^{15}N)$ values that were not statistically different from each other and from the previous theses, falling within an intermediate range. This $\delta(^{15}N)$ pattern reflects the lasting influence of fertilization practices implemented in recent years, leaving a detectable isotopic imprint on the soil. Consequently, soils with a long-term history of organic management tend to exhibit higher $\delta(^{15}N)$ values than those under conventional farming. Soils managed through agroecological practices showed intermediate $\delta(^{15}N)$ values, which, while distinct, were not entirely separable from those of organically and conventionally managed soils.

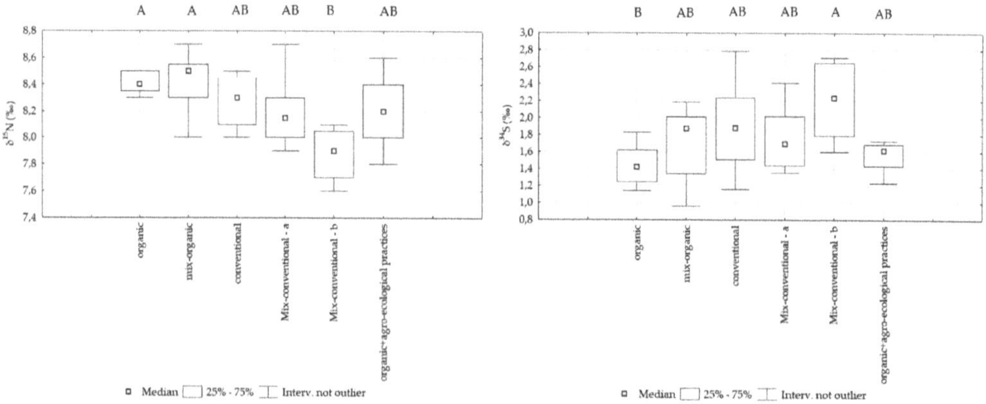

Figure 3. Boxplot of $^{15}N/^{14}N$ [$\delta(^{15}N)$] and $^{34}S/^{32}S$ [$\delta(^{34}S)$] in the soil samples (four experimental replicates for each treatment for two years of experimentation; N = 8). Different letters indicate statistically significant differences ($p \leq 0.01$).

3.4.4. Analyses of $^{15}N/^{14}N$ $^{13}C/^{12}C$, and $^{34}S/^{32}S$ on Leaves

Figure 4 shows the distribution of $\delta(^{15}N)$, $\delta(^{13}C)$ and $\delta(^{34}S)$ values in fennel leaves according to the different fertilization strategies applied. A trend toward lower $\delta(^{13}C)$ values was observed in the conventional and conventional-mixed treatments. However, across all samples, $\delta(^{13}C)$ remained within a narrow range around −29.0‰, which is characteristic of plants following a C3 photosynthetic pathway. Statistically significant differences were found between the conventional and conventional-mixed treatments compared to the agroecological treatment. No significant differences were observed between the organic and organic-mixed treatments, as their $\delta(^{13}C)$ values overlapped. The $\delta(^{15}N)$ pattern clearly reflected the fertilization practices with animal manure used during plant growth. Leaves from organically managed soils reached the highest $\delta(^{15}N)$ values, while lower values were recorded in the other treatments. Notably, no significant differences in $\delta(^{15}N)$ were observed between the agroecological and conventional treatments, mainly in relation to the "Organic + AEP" theses which implied the use of a leguminous crop (*Medicago sativa* L.) as a cover crop. In contrast, $\delta(^{34}S)$ values revealed a distinct pattern associated with the type of fertilizer applied. The agroecological treatment displayed statistically significant differences in $\delta(^{34}S)$ compared to the conventional and conventional-mixed treatments, indicating a clear sulfur isotopic distinction based on fertilization strategy. In this specific case, then, the combination of δ(^{15}N) and δ(^{34}S) can led to a complete traceability of the organic and conventional theses. The isotopic values of δ(^{13}C), δ(^{15}N), and δ(^{34}S) reflect the relative isotopic ratios of fennel bulbs and can serve as primary markers for distinguishing between organic and conventional fennel in traceability studies.

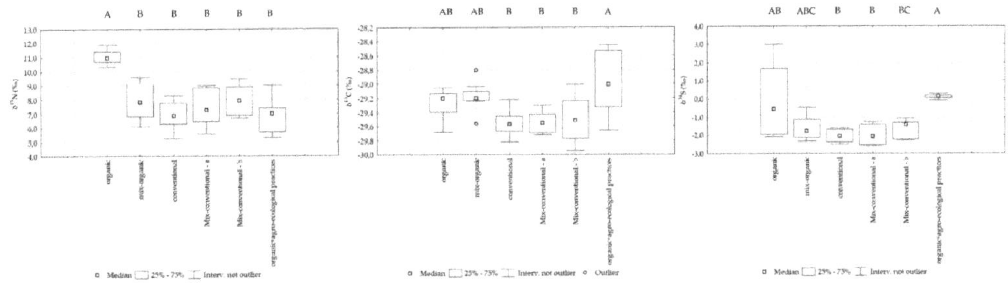

Figure 4. Boxplot of $^{15}N/^{14}N$ [δ(^{15}N)], $^{13}C/^{12}C$ [δ(^{13}C)] and $^{34}S/^{32}S$ [δ(^{34}S)] in fennel leaves (four experimental replicates for each treatment for two years of experimentation; N = 8). Different letters indicate statistically significant differences ($p \leq 0.01$).

3.5. Chemometric Multivariate Analysis

Table 4 reports on the eigenvalues of the canonical discriminant functions and the relative percentages of variance and cumulated variance of the model. The higher canonical correlation coefficients, which are an expression of the multiple correlation between the variables and the discriminant functions, were recorded by the first two discriminant functions, 0.966 and 0.958, respectively. Indeed, the first two functions expressed the 96.2% of the cumulated variance of the model, 87.8% and 8.3% for the first and the second, respectively. Table 5 provides an index of the importance of each variable for the group separation where the sign (+/−) indicates the direction of the relationship. Considering the values of the standardized coefficients of the first canonical discriminant function, TSS (1.953), NO_3^- (1.646), δ(^{15}N) of fennel (1.596), δ(^{34}S) of fennel (−1.424) and δ(^{15}N) of leaves (1.384) stand out as those that strongly predict allocation of the samples to a group. All the other variables were less successful as predictors. As a matter of fact, our results have highlighted that the variables which mostly contributed to the discrimination of fennel samples among the six different fertilization strategies groups were the differential accumulation of TSS and nitrates and the isotopic N and S signature in fennel bulbs and leaves. The original classification results showed a perfect group prediction as all the cases lied on the diagonal and 100.0% of original grouped cases were correctly classified (Table 6). In order to evaluate the power of the discriminant functions, a cross-validation was performed. It means that a leave-one-out classification was performed and the developed discriminant functions were used to further categorize the case that was left out. The procedure is automatically repeated in turn with each case left out. It was previously indicated that the overall predictive accuracy of the discriminant functions can be considered as acceptable if it is 25% larger than that due to chance [48]. In our study, the cross-validated classification showed that overall 77.1% of cases were correctly classified. We had six different groups with equal sizes so the right classification of a jack-knifed case could be achieved with a 1/6 chance (~17%). Thus, the 77.1% of correct classification for the overall cross-validated grouped cases was a very high percentage achieved, much larger than that due to chance and this confirms the great predictive accuracy of our implemented model. Figure 5 shows the combined-groups plot in which the first two discriminant functions represent the cartesian axes. As it can be noted, with respect to the first canonical discriminant function which alone displays the 87.8% of the variance, treatments are well discriminated and only a limited overlapping of the cases from treatments "Conventional" and "Mix-Conventional-b" can be detected. As far as function 2 (accounting for further 8.3% of the variance) is concerned, the three theses whose fertilization was entirely ("Organic"; "Organic + AEP") or mainly ("Mix-Organic") organic are well discriminated with respect to the three theses whose fertilization was entirely ("Conventional") or mainly conventional ("Mix-Conventional-a"; "Mix-Conventional-b").

Table 4. Eigenvalues of the canonical discriminant functions engendered by LDA.

Function	Eigenvalue	% Variance	% Cumulated Variance	Canonical Correlation
1	116.378	87.8	87.8	0.996
2	11.031	8.3	96.2	0.958
3	2.897	2.2	98.4	0.862
4	1.569	1.2	99.5	0.781
5	0.608	0.5	100	0.615

Table 5. Standardized coefficients of the canonical discriminant functions.

	Function				
	1	2	3	4	5
Plant_fresh_weight	0.349	0.008	0.267	−0.464	−0.002
pH	0.741	−0.294	0.723	0.130	−0.201
total_acidity	0.591	−0.101	0.784	0.067	−0.345
TSS	1.953	−0.844	0.796	0.584	−0.899
L*	0.819	−0.388	0.288	−0.327	0.029
a*	−0.318	0.319	0.155	−0.016	0.319
b*	−1.198	−0.366	0.190	0.499	−0.291
consistency	−0.217	0.550	−0.343	−0.007	−0.247
Total_polyphenols	0.208	0.375	−0.326	0.633	−0.293
ORAC_units	−0.127	0.359	0.609	0.393	−0.128
total_N	−0.941	0.264	−0.330	0.734	0.525
total_inorganic_N	−0.756	0.160	0.055	0.438	−0.820
NO_3	1.646	−0.840	−0.409	−0.911	0.706
delta_^{15}N_fennel	1.596	0.011	−0.168	0.304	0.350
delta_^{13}C_fennel	−0.315	0.004	−0.120	−0.234	0.393
delta_^{34}S_fennel	−1.424	0.032	0.303	0.990	0.493
delta_^{15}N_soil	−0.165	0.888	0.017	−0.805	1.084
delta_^{34}S_soil	−0.020	−0.316	−0.367	0.451	0.611
delta_^{15}N_leaves	1.384	−0.079	−0.694	−0.251	−0.383
delta_^{13}C_leaves	0.097	0.580	−0.226	0.649	0.041
delta_^{34}S_leaves	0.236	0.660	−0.865	−1.957	−0.041

Table 6. LDA classification results.

Thesis			Predicted Group Membership						Tot.
			Organic	Mix-Organic	Conventional	Mix-Conventional-a	Mix-Conventional-b	Organic + AEP	
Original	Count	Organic	8	0	0	0	0	0	8
		Mix-Organic	0	8	0	0	0	0	8
		Conventional	0	0	8	0	0	0	8
		Mix-Conventional-a	0	0	0	8	0	0	8
		Mix-Conventional-b	0	0	0	0	8	0	8
		Organic + AEP	0	0	0	0	0	8	8
	%	Organic	100.0	0.0	0.0	0.0	0.0	0.0	100.0
		Mix-Organic	0.0	100.0	0.0	0.0	0.0	0.0	100.0
		Conventional	0.0	0.0	100.0	0.0	0.0	0.0	100.0
		Mix-Conventional-a	0.0	0.0	0.0	100.0	0.0	0.0	100.0
		Mix-Conventional-b	0.0	0.0	0.0	0.0	100.0	0.0	100.0
		Organic + AEP	0.0	0.0	0.0	0.0	0.0	100.0	100.0

Table 6. *Cont.*

Thesis			Predicted Group Membership						Tot.
			Organic	Mix-Organic	Conventional	Mix-Conventional-a	Mix-Conventional-b	Organic + AEP	
Cross-validated [a]	Count	Organic	8	0	0	0	0	0	8
		Mix-Organic	0	5	0	3	0	0	8
		Conventional	0	0	6	1	1	0	8
		Mix-Conventional-a	0	0	2	5	1	0	8
		Mix-Conventional-b	0	0	3	0	5	0	8
		Organic + AEP	0	0	0	0	0	8	8
	%	Organic	100.0	0.0	0.0	0.0	0.0	0.0	100.0
		Mix-Organic	0.0	62.5	0.0	37.5	0.0	0.0	100.0
		Conventional	0.0	0.0	75.0	12.5	12.5	0.0	100.0
		Mix-Conventional-a	0.0	0.0	25.0	62.5	12.5	0.0	100.0
		Mix-Conventional-b	0.0	0.0	37.5	0.0	62.5	0.0	100.0
		Organic + AEP	0.0	0.0	0.0	0.0	0.0	100.0	100.0

100.0% of original grouped cases correctly classified. 77.1% of cross-validated grouped cases correctly classified. [a]. Cross validation is performed only for those cases in the analysis. In cross-validation, each case is classified by the functions derived from all cases other than that case.

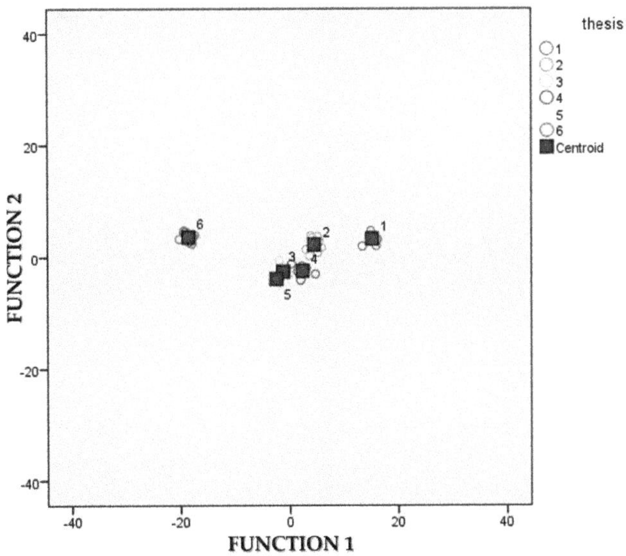

Figure 5. Graphical representation of the first two canonical discriminant functions. Treatments: (1) "Organic"; (2) "Mix-Organic"; (3) "Conventional"; (4) "Mix-Conventional-a"; (5) "Mix-Conventional-b"; and (6) "Organic + Agroecological practices".

Our results demonstrated that the joint use of physicochemical, nutritional and isotopic variables combined with their multivariate chemometric treatment can overcome the misclassification of a fennel sample of unknown agronomic origin (organic or conventional), also in the cases where mixed fertilization strategies are applied, thereby confirming what previously reported for cauliflower [19]. In addition, to the best of our knowledge, the presented innovative tool demonstrated for the first time that the discrimination between organic and conventional horticultural products is feasible even when N-fixing plants are applied in organic practice, thereby avoiding misinterpretations univocally based on the isotopic N fingerprint.

4. Conclusions

The different types of N fertilization did not affect the plant fresh weight at harvest, thus indicating that the supplied mineral fertilizer ("Conventional") the organic amendments ("Organic" and "Organic + AEP" treatments) and their coformulations ("Mix-organic", "Mix conventional-a" and "Mix conventional-b" treatments) fulfilled the fennel plants' N requirements. In addition, the conventional treatments increased the NO_3^- values over 440% compared to the organic ones, without increasing the fennel growth and production with possible negative effects on human health, if fertilizers doses might be increased.

Given the urgent need for innovative tools to trace N fertilization in organic farming and assist stakeholders, retailers, and policymakers in preventing food fraud, we propose a multivariate approach applied to an open-field fennel cultivation case study. Different agronomic treatments were tested and compared, including organic, conventional, and mixed fertilization strategies with varying proportions of mineral fertilizers. Furthermore, agroecological service crops were introduced as a sustainable alternative to traditional organic and conventional methods using a leguminous crop for the fertilization of the soil. The findings of our study indicated that multivariate linear discriminant analysis (LDA), including the investigated parameters as variables, achieved the correct classification of the grouped cases from the six different agronomic treatments, showing a great predictive accuracy of the implemented model. This demonstrated that even if none of the parameters alone may be considered as a reliable biomarker of the agronomic practice, a comprehensive multivariate approach can contribute to reliably trace the organic origin of the crop. Key contributors to the classification model included total soluble solids (TSS), nitrate (NO_3^-), $\delta(^{15}N)$ of fennel, $\delta(^{34}S)$ of fennel, and $\delta(^{15}N)$ of leaves. To the best of our knowledge, the presented innovative tool demonstrated for the first time that the discrimination between organic and conventional horticultural products is feasible even when N-fixing plants are applied in organic practice, thereby avoiding misinterpretations based univocally on the isotopic N fingerprint. These promising results highlight the potential of this method and suggest that further research should extend its application to other horticultural crops to confirm and expand upon the current findings.

Author Contributions: Conceptualization, L.B., F.M. and S.F.; methodology, G.B., M.A., L.B., F.M., A.F., M.D., N.T., B.T. and S.F.; software, S.F.; validation, L.B., F.M. and S.F.; formal analysis, L.B., F.M. and S.F.; investigation, G.B., M.A., L.B., F.M., A.F., M.D., N.T., B.T. and S.F.; resources, S.F.; data curation, L.B., A.R. and S.F.; writing—original draft preparation, L.B., F.M., M.D. and S.F.; writing—review and editing, G.B., M.A., L.B., A.R., F.M., A.F., M.D., N.T., B.T. and S.F.; visualization, L.B., F.M. and S.F.; supervision, S.F.; project administration, S.F.; funding acquisition, S.F. All authors have read and agreed to the published version of the manuscript.

Funding: This research was funded by Italian Ministry of Agriculture, Food Sovereignty and Forestry, grant number 93173/2017, 'Application of innovative methods for the traceability of organic farming products'—INNOVABIO project.

Data Availability Statement: Data are available from the authors.

Acknowledgments: The authors acknowledge Rosalba Scazzarriello, Marco Favale and Francesco Rinaldi for their field management and technical assistance.

Conflicts of Interest: The authors declare no conflicts of interest.

References

1. European Commission. Available online: https://commission.europa.eu/strategy-and-policy/priorities-2019-2024/europeangreen-deal_en (accessed on 10 February 2025).
2. United Nations. Available online: https://www.un.org/sustainabledevelopment/ (accessed on 10 February 2025).

3. COMMISSION IMPLEMENTING DECISION Approving the Amendment of the 2023–2027 CAP Strategic Plan of Italy for Union Support Financed by the European Agricultural Guarantee Fund and the European Agricultural Fund for Rural Development CCI: 2023IT06AFSP001. Available online: https://ec.europa.eu/transparency/documents-register/detail?ref=C(2023)6990&lang=en (accessed on 10 February 2025).
4. ISMEA Database 'Bio in Cifre'. Available online: https://www.sinab.it/sites/default/files/BIo%20in%20cifre%202024_0.pdf (accessed on 10 February 2025).
5. Blackshaw, R.E.; Moyer, J.R.; Huang, H.C. Beneficial effects of cover crops on soil health and crop management. In *Recent Research Developments in Soil Science*; Pandalai, S.G., Ed.; Research Signpost: Trivandrum, India, 2005; pp. 15–35.
6. Wezel, A.; Casagrande, M.; Celette, F.; Vian, J.-F.; Ferrer, A.; Peigné, J. Agroecological practices for sustainable agriculture. A review. *Agron. Sustain. Dev.* **2014**, *34*, 1–20. [CrossRef]
7. Diacono, M.; Persiani, A.; Testani, E.; Montemurro, F. Sustainability of agro-ecological practices in organic horticultural: Yield, energy-use and carbon footprint. *Agroecol. Sustain. Food Syst.* **2020**, *44*, 726–746. [CrossRef]
8. Masiunas, J.B. Production of vegetables using cover crop and living mulches: A review. *J. Veg. Crop Prod.* **1998**, *4*, 11–31.
9. Regulation (EU). *2018/848 of the European Parliament and of the Council of 30 May 2018 on Organic Production and Labelling of Organic Products and Repealing Council Regulation (EC) No 834/2007*; EU: Maastricht, The Netherlands, 2018. Available online: https://eur-lex.europa.eu/legal-content/EN/TXT/?uri=CELEX:02018R0848-20230221 (accessed on 10 February 2025).
10. Regulation (EU). *2017/625 of the European Parliament and of the Council of 15 March 2017 on Official Controls and Other Official Activities Performed to Ensure the Application of Food and Feed Law, Rules on Animal Health and Welfare, Plant Health and Plant Protection Products, Amending Regulations (EC) No 999/2001, (EC) No 396/2005, (EC) No 1069/2009, (EC) No 1107/2009, (EU) No 1151/2012, (EU) No 652/2014, (EU) 2016/429 and (EU) 2016/2031 of the European Parliament and of the Council, Council Regulations (EC) No 1/2005 and (EC) No 1099/2009 and Council Directives 98/58/EC, 1999/74/EC, 2007/43/EC, 2008/119/EC and 2008/120/EC, and Repealing Regulations (EC) No 854/2004 and (EC) No 882/2004 of the European Parliament and of the Council, Council Directives 89/608/EEC, 89/662/EEC, 90/425/EEC, 91/496/EEC, 96/23/EC, 96/93/EC and 97/78/EC and Council Decision 92/438/EEC (Official Controls Regulation)*; EU: Maastricht, The Netherlands, 2017. Available online: https://eur-lex.europa.eu/legal-content/EN/TXT/?uri=CELEX:02017R0625-20220128 (accessed on 10 February 2025).
11. Bateman, A.S.; Kelly, S.D.; Jickells, T.D. Nitrogen Isotope Relationships between Crops and Fertilizer: Implications for Using Nitrogen Isotope Analysis as an Indicator of Agricultural Regime. *J. Agric. Food Chem.* **2005**, *53*, 5760–5765. [CrossRef]
12. Rapisarda, P.; Calabretta, M.L.; Romano, G.; Intrigliolo, F. Nitrogen Metabolism Components as a Tool to Discriminate between Organic and Conventional Citrus Fruits. *J. Agric. Food Chem.* **2005**, *53*, 2664–2669. [CrossRef]
13. Wassenaar, L.I.; Kelly, S.D.; Douence, C.; Islam, M.; Monteiro, L.; Abrahim, A.; Rinke, P. Assessment of rapid low-cost isotope (δ15N, δ18O) analyses of nitrate in fruit extracts by Ti(III) reduction to differentiate organic from conventional production. *Rapid Commun. Mass. Spectrom.* **2022**, *36*, e9259. [CrossRef]
14. Inácio, C.T.; Chalk, P.M.; Magalhães, A.M.T. Principles and Limitations of Stable Isotopes in Differentiating Organic and Conventional Foodstuffs: 1. Plant Products. *Crit. Rev. Food Sci. Nutr.* **2015**, *55*, 1206–1218. [CrossRef]
15. Bontempo, L.; van Leeuwen, K.A.; Paolini, M.; Holst Laursen, K.; Micheloni, C.; Prenzler, P.D.; Ryan, D.; Camin, F. Bulk and compound-specific stable isotope ratio analysis for authenticity testing of organically grown tomatoes. *Food Chem.* **2020**, *318*, 126426. [CrossRef]
16. Gatzert, X.; Chun, K.P.; Boner, M.; Hermanowski, R.; Mäder, R.; Breuer, L.; Gattinger, A.; Orlowski, N. Assessment of multiple stable isotopes for tracking regional and organic authenticity of plant products in Hesse, Germany. *Isot. Environ. Health Stud.* **2021**, *57*, 281–300. [CrossRef]
17. Bateman, A.S.; Kelly, S.D.; Woolfe, M. Nitrogen Isotope Composition of Organically and Conventionally Grown Crops. *J. Agric. Food Chem.* **2007**, *55*, 2664–2670. [CrossRef]
18. Laursen, K.H.; Mihailova, A.; Kelly, S.D.; Epov, V.N.; Bérail, S.; Schjoerring, J.K.; Donard, O.F.X.; Larsen, E.H.; Pedentchouk, N.; Marca-Bell, A.D.; et al. Is it really organic?—Multi-isotopic analysis as a tool to discriminate between organic and conventional plants. *Food Chem.* **2013**, *141*, 2812–2820. [CrossRef] [PubMed]
19. Fabroni, S.; Bontempo, L.; Campanelli, G.; Canali, S.; Montemurro, F. Innovative Tools for the Nitrogen Fertilization Traceability of Organic Farming Products. *Horticulturae* **2023**, *9*, 723. [CrossRef]
20. Campanelli, G.; Amenta, M.; Bontempo, L.; Leteo, F.; Montemurro, F.; Platani, C.; Timpanaro, N.; Torrisi, B.; Fabroni, S. Innovative Tools for Nitrogen Fertilization Traceability in Organic Farming Products: A Cauliflower Case Study. *Horticulturae* **2024**, *10*, 94. [CrossRef]
21. Persiani, A.; Diacono, M.; Montemurro, F. The impact of long-term organic horticultural systems on energy outputs and carbon storages in relation to extreme rainfall events. *Eur. J. Agron.* **2024**, *161*, 127398. [CrossRef]
22. Huang, D.; Ou, B.; Prior, R.L. The Chemistry behind Antioxidant Capacity Assays. *J. Agric. Food Chem.* **2005**, *53*, 1841–1856. [CrossRef]

23. Torrisi, B.; Giuffrida, A.; Stagno, F.; Intrigliolo, F.; Canali, S.; Trinchera, A.; Pompili, L.; Nisini, L. Application of organic fertiliseres in a orange orchard in Southern italy: Efects on soil fertility trees nutritional status, yield and quality of fruits. *Proc. Int. Soc. Citric.* **2004**, 2, 649–653.
24. Nerdy, N.; De Lux Putra, E. Spectrophotometric method for determination of nitrite and nitrate levels in broccoli and cauliflower with different fertilization treatment. *Orient. J. Chem.* **2018**, *34*, 2983–2991. [CrossRef]
25. Coplen, T.B. Guidelines and Recommended Terms for Expression of Stable-Isotope-Ratio and Gas-Ratio Measurement Results. *Rapid Commun. Mass. Spectrom.* **2011**, *25*, 2538–2560. [CrossRef]
26. Campanelli, G.; Canali, S. Crop Production and Environmental Effects in Conventional and Organic Vegetable Farming Systems: The Case of a Long-Term Experiment in Mediterranean Conditions (Central Italy). *J. Sustain. Agric.* **2012**, *36*, 599–619. [CrossRef]
27. Atta-Aly, M.A. Fennel swollen base yield and quality as affected by variety and source of nitrogen fertilizer. *Sci. Hortic.* **2001**, *88*, 191–202. [CrossRef]
28. Persiani, A.; Diacono, M.; Monteforte, A.; Montemurro, F. Agronomic performance, energy analysis and carbon balance comparing different fertilization strategies in horticulture under Mediterranean conditions. *Environ. Sci. Pollut. Res.* **2019**, *26*, 19250–19260. [CrossRef] [PubMed]
29. Valentino, M.; Sequino, G.; De Filippis, F.; Di Monaco, R.; Cavella, S.; Torrieri, E. The effect of edible coating based on sodium caseinate and propyl gallate on the shelf life of minimally processed fennel during storage. *Appl. Food Res.* **2024**, *4*, 100462. [CrossRef]
30. Artés, F.; Escalona, V.H.; Artés-Hdez, F. Modified Atmosphere Packaging of Fennel. *J. Food Sci.* **2002**, *67*, 1550–1554. [CrossRef]
31. Ninfali, P.; Mea, G.; Giorgini, S.; Rocchi, M.; Bacchiocca, M. Antioxidant capacity of vegetables, spices and dressings relevant to nutrition. *Br. J. Nutr.* **2005**, *93*, 257–266. [CrossRef]
32. Salama, Z.A.; El Baz, F.K.; Gaafar, A.A.; Zaki, M.F. Antioxidant activities of phenolics, flavonoids and vitamin C in two cultivars of fennel (*Foeniculum vulgare* Mill.) in responses to organic and bio-organic fertilizers. *J. Saudi Soc. Agric. Sci.* **2015**, *14*, 91–99. [CrossRef]
33. Prior, R.L.; Joseph, J.A.; Cao, G.; Shukitt-Hale, B. *Can Foods Forestall Aging? February Issue*; Agricultural Research, USDA Magazine: Washington, DC, USA, 1999; pp. 15–17.
34. Barański, M.; Srednicka-Tober, D.; Volakakis, N.; Seal, C.; Sanderson, R.; Stewart, G.B.; Benbrook, C.; Biavati, B.; Markellou, E.; Giotis, C.; et al. Higher antioxidant and lower cadmium concentrations and lower incidence of pesticide residues in organically grown crops: A systematic literature review and meta-analyses. *Br. J. Nutr.* **2014**, *112*, 794–811. [CrossRef]
35. Brandt, K.; Leifert, C.; Sanderson, R.; Seal, C.J. Agroecosystem Management and Nutritional Quality of Plant Foods: The Case of Organic Fruits and Vegetables. *Crit. Rev. Plant Sci.* **2011**, *30*, 177–197. [CrossRef]
36. Fabroni, S.; Amenta, M.; Rapisarda, S.; Torrisi, B.; Licciardello, C. Amino acid metabolism and expression of genes involved in nitrogen assimilation in common oranges cv. Valencia Late. *Biol. Plant.* **2022**, *66*, 155–162. [CrossRef]
37. Karłowski, K. Azotany w warzywach--propozycje limitowania w polsce [Nitrates in vegetables--proposals for their limitation in Poland]. *Rocz Panstw Zakl Hig.* **1990**, *41*, 1–9.
38. Tonitto, C.; David, M.B.; Drinkwater, L.E. Replacing bare fallows with cover crops in fertilizer-intensive cropping systems: A meta-analysis of crop yield and N dynamics. *Agric. Ecosyst. Environ.* **2006**, *112*, 58–72. [CrossRef]
39. Brennan, E.B.; Boyd, N.S. Winter Cover Crop Seeding Rate and Variety Affects during Eight Years of Organic Vegetables: II. Cover Crop Nitrogen Accumulation. *Agron. J.* **2012**, *104*, 799–806. [CrossRef]
40. Hu, S.; Grunwald, N.J.; van Bruggen, A.H.C.; Gamble, G.R.; Drinkwater, L.E.; Shennan, C.; Demment, M.W. Short-Term Effects of Cover Crop Incorporation on Soil Carbon Pools and Nitrogen Availability. *Soil Sci. Soc. Am. J.* **1997**, *61*, 901–911. [CrossRef]
41. Brennan, E.B.; Acosta-Martinez, V. Cover cropping frequency is the main driver of soil microbial changes during six years of organic vegetable production. *Soil Biol. Biochem.* **2017**, *109*, 188–204. [CrossRef]
42. Dos Santos Cordeiro, C.F.; Rodrigues, D.R.; Rocha, C.H.; Araujo, F.F.; Echer, F.R. Glomalin and microbial activity affected by cover crops and nitrogen management in sandy soil with cotton cultivation. *Appl. Soil Ecol.* **2021**, *167*, 104026. [CrossRef]
43. Högberg, P.; Johnnisson, C.; Högberg, M.; Högbom, L.; Näsholm, T.; Hällgren, J.E. Measurements of abundances of 15N and 13C as tools in retrospective studies of N balances and water stress in forests: A discussion of preliminary results. *Plant Soil* **1995**, *168*, 125–133. [CrossRef]
44. Diacono, M.; Trinchera, A.; Montemurro, F. An overview on agroecology and organic agriculture strategies for sustainable crop production. *Agronomy* **2021**, *11*, 223. [CrossRef]
45. Georgi, M.; Voerkelius, S.; Rossmann, A.; Graßmann, J.; Schnitzler, W.H. Multielement Isotope Ratios of Vegetables from Integrated and Organic Production. *Plant Soil* **2005**, *275*, 93–100. [CrossRef]
46. Kelly, S.; Heaton, K.; Hoogewerff, J. Tracing the geographical origin of food: The application of multi-element and multi-isotope analysis. *Trends Food Sci. Technol.* **2005**, *16*, 555–567. [CrossRef]

47. Chung, I.M.; Han, J.; Kong, W.; Kim, J.; An, M.; Lee, J.; An, Y.; Jung, M.Y.; Kim, S. Regional discrimination of Agaricus bisporusmushroom using the natural stable isotope ratios. *Food Chem.* **2018**, *264*, 92–100. [CrossRef]
48. Leech, N.; Barrett, K.; Morgan, G.A. *SPSS for Intermediate Statistics: Use and Interpretation*, 3rd ed.; Routledge: London, UK, 2008. [CrossRef]

horticulturae

MDPI

Article

Assessment of Potentially Toxic Elements in Four Melon Fruit Varieties Grown in the Ganges and Yamuna River Basin

Mohssen Elbagory [1], Mohamed A. Abd El-Aziz [2], Alaa El-Dein Omara [3], Sami Abou Fayssal [4,5] and Vinod Kumar [6,*]

1 Health Specialties, Basic Sciences and Applications Unit, Applied College, King Khalid University, Abha 61421, Mohayil Asir, Saudi Arabia; mhmohammad@kku.edu.sa
2 Soil Improvement and Conservation Research Department, Soils, Water, and Environment Research Institute (SWERI), Agriculture Research Center (ARC), Giza 12112, Egypt; m.abdelaziz6275@gmail.com
3 Soil Microbiology Research Department, Soils, Water, and Environment Research Institute (SWERI), Agriculture Research Center (ARC), Giza 12112, Egypt; alaa.omara@yahoo.com
4 Department of Agronomy, Faculty of Agronomy, University of Forestry, 10 Kliment Ohridski Blvd, 1797 Sofia, Bulgaria; samiaboufayssal95@gmail.com
5 Department of Plant Production, Faculty of Agriculture, Lebanese University, Beirut 1302, Lebanon
6 Agro-Ecology and Pollution Research Laboratory, Department of Zoology and Environmental Science, Gurukula Kangri (Deemed to be University), Haridwar 249404, India
* Correspondence: drvksorwal@gkv.ac.in

Abstract: The present study aimed to investigate the occurrence of eight potentially toxic elements (PTEs) in selected varieties of watermelon (*Citrullus lanatus* var. *Arka Shyama* and *Crimson Sweet*) and muskmelon (*Cucumis melo* var. *Cantaloupe* and *Kajri*) grown near riverbanks in the Yamuna and Ganga River basins of Northern India. For this purpose, samples of melon fruits were collected from ten sampling sites from May to June 2024 and analyzed using ICP-OES. The results showed that the levels of PTEs varied significantly across the sampling sites, with muskmelons exhibiting slightly higher concentrations compared to watermelons. Specifically, the concentration (mg/kg dry weight) ranges for the watermelon varieties were Cd (0.05–0.20), Cr (0.40–1.10), Cu (1.50–4.90), Pb (0.01–0.11), As (0.01–0.08), Fe (80.00–120.00), Mn (9.00–15.80), and Zn (5.00–18.00). For muskmelons, the ranges were Cd (0.05–0.23), Cr (0.40–1.00), Cu (2.40–4.80), Pb (0.01–0.08), As (0.02–0.08), Fe (80.00–120.00 g), Mn (9.00–15.00), and Zn (8.00–18.00). In particular, the variability coefficients (CV%) indicated differential contamination in *Crimson Sweet*. On the other hand, Principal Component Analysis (PCA) and Hierarchical Cluster Analysis (HCA) tools facilitated the identification of sites of significant contamination and their respective interactions. The health risk studies using the health risk index (HRI), dietary intake modeling (DIM), and the target hazard quotient (THQ) also revealed no significant health risk of eight PTEs in melon fruits. Therefore, this study provides valuable insights into the biomonitoring of PTE contamination in widely consumed summer fruits of Northern India and the subsequent health risk assessment.

Keywords: fruit contamination; health risk; potentially toxic elements; riverbank cultivation; summer fruits

1. Introduction

Melons (*Cucumis* and *Citrullus* spp.; Cucurbitaceae) are tender trailing vines belonging to the gourd family [1]. They mainly grow in warm regions and are native to Central Asia. Melons are well known for their round, hard yellow/green skin, sweet flesh, and musky-scented edible fruit with numerous seeds [2]. In 2024, global melon production has been estimated at 29 million metric tons, with China as the leading country (14.6 million metric tons; 2.0% yearly growth rate), followed by India (1.4 million metric tons; 2.65% yearly growth rate), and Kazakhstan (1.1 million metric tons; 5.5% yearly growth rate) [3]. Melon fruits are consumed raw, mixed in salads, used to make luscious salsa with honeydew or avocado, skewered, pickled, blended to prepare refreshing drinks and soda, grilled to caramelize their sugar, and served as a sweet and tangy chilled soup or as a frozen dessert. Melon seeds can be roasted to make snacks or incorporated in baking practices to add a nutty flavor, mixed in salads, incorporated in soups and stews, smoothies and shakes, and served as grain dishes (with rice and quinoa), desserts (cakes, cookies, and energy bars), and dips and spreads (with hummus, tahina, or pesto). On the other hand, melon fruit holds several nutritional and medicinal compounds. For instance, it is a rich source of folic acid, and vitamins C and E, and is low in fat and sodium [4]. Melon fruit holds significant amounts of β-carotene, carotenoids, and lycopene responsible for the regulation of body metabolism [1]. Palmitessa et al. [5]. also outlined the richness of *Cucumis melo* in polyphenols; such dietary compounds can help reduce risks of cancer and cardiovascular diseases [6] and, with selenium and vitamin E compounds, protect body cells and tissues from oxidative damage [7]. Thus, it contributes to a healthy diet rich in antioxidants. Melon is also rich in fiber, protein, mono- and poly-unsaturated fatty acids. Such fatty acid intake helps reduce cholesterol and cardiovascular diseases [8,9]. Fiber intake promotes nutrient absorption and regulates the rate of food digestion by producing beneficial microbial metabolites [10], while protein can aid in muscle damage repair and stimulate growth [11].

Melon is considered a prolific bioaccumulator of potentially toxic elements (PTEs) [12,13]. However, a high bioaccumulation of PTEs in melon can lead to reduced size and number of leaves due to loss of turgor hung down from petioles [14]. Mgbemena et al. [15]. reported the following decreasing order of PTEs up taken by melon vine: zinc (Zn) > lead (Pb) > iron (Fe) > cadmium (Cd) > chromium (Cr) > nickel (Ni), with strong positive correlations between these elements found in soil and bioaccumulated in melon seeds. Another study found that melon tends to bioaccumulate PTEs (arsenic (As), Cd, copper (Cu), Pb, and Zn) in the following decreasing order: leaves > stalk/root > fruit [16]. Melon fruits collected from the cement industry and nearby areas in Lahore, Pakistan, revealed a significant level of cobalt (Co) (3.9 mg/kg) with a bioaccumulation factor (BAF) of 2.286 and a target hazard quotient (THQ) > 1 [17]. Such results are alarming as they depict a high health risk associated with the consumption of melon contaminated by Co; the latter can lead to cardiovascular, endocrine, and neurological deficits [18]. Leaves of *Cucumeropsis manii* (West African melon) showed the following decreasing order of bioaccumulated PTEs: manganese (Mn) > Pb > Co > Zn > Cu > Cd > Ni with Cd surpassing the maximum permissible level (0.02 mg/kg) set by the World Health Organization (WHO) [19,20]. Increased Cd uptake by melon induces increases in fruit size, sweetness, and antioxidant enzyme activities while reducing anthocyanin and chlorophyll contents [21]. However, the consumption of Cd-exposed food may be associated with cancer, cardiovascular diseases, chronic kidney disease, diabetes, and osteoporosis [22,23]. Another study found that PTE bioaccumulation was higher in melon fruit tissue than in its pericarp [24]. These authors adopted weighted gene co-expression network analysis (WGCNA) and qRT-PCR to detect specific modules correlated with bioaccumulated melon PTEs. Consequently, they identified five candidate genes responsible for PTEs bioaccumulation in melon peel and pulp. Brecchia et al. [25]. mentioned that the early season cultivation of *Cucumis melo* was strongly associated with reduced melon uptake of Cu and Pb. They also outlined a high correlation between PTEs and stem and leaf calcium (Ca), mostly associated with increased abiotic stress.

Arka Shyama is an early (65–70 days to harvest), newly introduced icebox segment watermelon owing greenish black rind, oblong fruit shape, dark red color with crispy and sweet (total soluble solids (TSS): 12%) flesh. It is an open-pollinated variety with high productivity (14.8 tons/ha), average fruit weight of 2–3 kg, average three fruits/vine, low number of seeds per fruit (219), and good lycopene content (6.40 mg/100 g). *Crimson Sweet* is an indeterminate watermelon variety with an 85-day harvest period, sweet, firm, and bright red flesh. Round *Crimson Sweet* fruits weigh around 11 kgs each, have light and dark green stripes, and exceptionally sweet and luscious flesh. Previous research showed that *Crimson Sweet* has the potential to bioaccumulate PTEs in the following decreasing order: Fe > Co > Zn > Mn > Ni > Cd > Cu > Pb in leaf > stem > root > seed > pulp [26]. *Cantaloupe* is one of the most popular muskmelons, with high amounts of antioxidants (mainly α and β-carotene, and carotenoids), minerals (Ca), magnesium (Mg), Mn, phosphorus (P), potassium (K), sodium (Na), vitamins (A–E, and K), and a sweet refreshing taste [27,28]. It also encloses good amounts of amino acids (mainly lysine, leucine, isoleucine, aspartic and glutamic acids, and alanine) and fatty acids (mainly palmitic and stearic acids) [28]. *Kajri* muskmelon, commonly known as Delhi melon, is a Punjab-native (Northern India) short-season (70–75 days) honeydew cultivar with a thin red-orange rind, green vertical stripes, slightly oblate shape, an interior green flesh, and an overall weight of 0.9–1.4 kg. Each *Kajri* vine produces 6–9 melons with 60–70 seeds per fruit. It is a rich source of vitamins A, B, C, and K, Ca, Fe, K, fiber, and folate [29].

The Ganga and Yamuna River Basin has been known for decades as an extensive melon production region. However, numerous reports mentioned the latter's sharp fall as a consequence of sewage and chemical water dumping; melons die after barely sprouting. Moreover, large pesticide applications have been increasing in the last few years within and around the basin region. These factors have naturally induced higher PTE contamination of surviving melon vines. Therefore, the current study aimed, for the first time, to assess the PTEs bioaccumulation in four melon fruit varieties (watermelon: *Arka Shyama* and *Crimson Sweet*; muskmelon: *Cantaloupe* and *Kajri*) grown in the Ganga and Yamuna River Basin in Northern India. The findings of the current study will eventually help determine the suitability of human consumption of grown melons in the studied region with a continuous need for monitoring.

2. Materials and Methods

2.1. Description of Study Area, Sampling Design, and Fruit Collection

The present study was conducted in Saharanpur and Haridwar districts of North India (29°28′18.85″ N to 78°01′10.47″ E; 29.471903 N/78.019576 E), covering an area of approximately 5500 km^2 (Figure 1). The region lies in the Ganga–Yamuna River basin, characterized by fertile alluvial soils and extensive riverbed cultivation. Farmers predominantly grow summer crops such as watermelon, muskmelon, cucumber, and Armenian cucumber during the summer months. For this, field sampling was conducted from May to June 2024, targeting ten sites (L1–L10), with L1–L6 in Saharanpur and L7–L10 in Haridwar districts, respectively. The sites L1 to L4 are located on the riverbank of the Yamuna River, L5 to L8 are situated on the banks of the Hindon and Solani tributaries, and L8 to L9 are located on the riverbank of the Ganges. These sites were selected to ensure representation of typical cultivation practices and agro-environmental conditions. The average temperature, humidity, and total rainfall of selected regions during the study period were as follows: 28.5 °C, 60%, and 243.00 mm, respectively [30]. At each site, five mature fruit samples of each crop variety of watermelon (*Arka Shyama* and *Crimson Sweet*) and muskmelon (*Cantaloupe* and *Kajri*) were randomly collected from different plants (Figure 2). Fruits of each variety were pooled separately to obtain composite samples, resulting in 50 samples per variety and

200 samples in total across all varieties. Collected samples were immediately transported under cooled conditions to prevent degradation. Each sample was analyzed three times to derive the mean value.

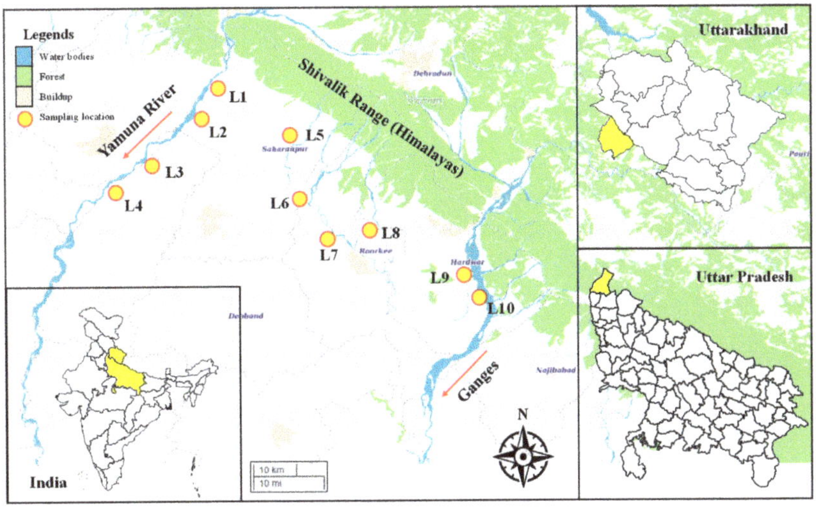

Figure 1. Map of the study area (Source: Bhuvan: https://bhuvan.nrsc.gov.in/ (accessed on 2 January 2025) and Power-user).

A: Arka Shyama B: Crimson Sweet C: Cantaloupe D: Kajri

Figure 2. Photograph of four melon fruit types (Watermelon: *Arka Shyama* and *Crimson Sweet*; Muskmelon: *Cantaloupe* and *Kajri*).

2.2. PTEs Analysis

Watermelon and muskmelon fruit samples collected from the selected sites were analyzed for eight PTEs (Cd, Cr, Cu, Pb, As, Fe, Mn, and Zn) using inductively coupled plasma optical emission spectroscopy (ICP-OES, 7300 DV, Perkin Elmer, Boston, MA, USA). For this, the edible portion of ripened fruit (pulp) was separated and oven-dried at 60 °C until constant weight was achieved. The samples were ground to a fine powder using a mechanical grinder. A total of 2 g of dried fruit sample powder was dissolved in 20 mL of a diacid mixture prepared using a 3:1 ratio of HNO_3 and $HClO_4$, respectively [31]. The mixture was self-digested overnight (12 h) and then contents were digested using a hot plate at 150 °C for 1 h until a clear solution was obtained. For this, the mixture was digested in a closed system to prevent any loss of sample. The digestion was performed in a fume hood with a reflux condenser to maintain a closed environment. The contents were further cooled, and the final volume was adjusted to 50 mL by the addition of 3% HNO_3 solution.

This content was filtered through Whatman no. 41 filter paper and finally used for metal element analysis by ICP-OES, as adopted by AL-Huquail [32]. Analytical measurements were validated using certified reagent materials (CRM; BCR679, Merck, Mumbai, India) and standard reference materials (SRM; 1000 mg/L stock, Merck, Mumbai, India), as per the manufacturer's instruction.

2.3. Data Analysis and Statistics

The datasets were analyzed using Microsoft Excel (Version 365, Microsoft Corp., Redmond, WA, USA) and OriginPro (Version 2023, OriginLab, Northampton, MA, USA) software packages. The data were processed using principal component analysis (PCA) and hierarchically clustered heatmap tools. The level of statistical significance was $p < 0.05$. To study health risks in relation to dietary intake of melon fruits contaminated with PTEs, three different non-carcinogenic risk indices i.e., health risk index (HRI), dietary intake modeling (DIM), and target hazard quotient (THQ), were used, as previously reported by Abou Fayssal et al. [33]. The following equations were used to compute HRI, DIM, and THQ values:

$$\text{HRI} = \text{DIM}/\text{RfD} \quad (1)$$

$$\text{DIM} = \text{SS} \times (\text{HMc}/\text{Bw}) \quad (2)$$

$$\text{THQ} = 10^{-3} \times (\text{Ef} \times \text{Ed} \times \text{Ir} \times \text{HMc}/\text{Bw} \times \text{Cp} \times \text{RfD}) \quad (3)$$

where RfD, SS, HMc, and Bw are oral reference doses, the serving of melon fruit (300 g per day dried weight), PTEs concentration in melon fruit (mg/kg), and body weight of consumer (70 kg), respectively, and 10^{-3} is the conversion factor, Ef is the exposure frequency, Ed is the exposure duration (365 days), Ir represents the melon ingestion rate (0.30 kg/day), Cp is the consumption period (25,550 days for an adult), and Rd is the reference dose (Cd: 1.0×10^{-3}; Cr: 5.0×10^{-3}; Cu: 4.0×10^{-2}; Pb: 4.0×10^{-4}, As: 3.0×10^{-4}, Fe: 7.0×10^{-1}; Mn: 1.4×10^{-2}, and Zn: 3.0×10^{-1}).

3. Results and Discussion

3.1. Concentration of PTEs in Melon Fruits

As depicted in Tables 1 and 2, the PTEs analysis of selected watermelon and muskmelon varieties showed significant variations in their concentrations across ten sampling locations. In watermelon, Cd ranged from 0.05 to 0.20 mg/kg, Cr from 0.40 to 1.10 mg/kg, Cu from 1.50 to 4.90 mg/kg, Pb from 0.01 to 0.11 mg/kg, As from 0.01 to 0.08 mg/kg, Fe from 80.00 to 120.00 mg/kg, Mn from 9.00 to 15.80 mg/kg, and Zn from 5.00 to 18.00 mg/kg. On the other hand, muskmelon showed slightly higher ranges: Cd 0.05–0.23 mg/kg, Cr 0.40–1.00 mg/kg, Cu 2.40–4.80 mg/kg, Pb 0.01–0.08 mg/kg, As 0.02–0.08 mg/kg, Fe 80.00–120.00 mg/kg, Mn 9.00–15.00 mg/kg, and Zn 8.00–18.00 mg/kg. Watermelon varieties exhibited high Fe and Mn levels, with mean concentrations exceeding 94 mg/kg and 13 mg/kg, respectively, as compared to muskmelon varieties. Cu in *Crimson Sweet* was slightly higher (3.64 mg/kg) compared to *Arka Shyama* (3.40 mg/kg). The variability coefficients (CV%) for Cd, As, and Pb indicated differential patterns of contamination, particularly in *Crimson Sweet*.

Table 1. Concentration (mg/kg) of eight PTEs in two varieties of watermelon collected from the Yamuna and Ganga River basins.

Watermelon Variety	Location	Potentially Toxic Elements							
		Cd	Cr	Cu	Pb	As	Fe	Mn	Zn
Arka Shyama	L1	0.09 ± 0.02	0.60 ± 0.07	2.20 ± 0.15	0.03 ± 0.01	0.01 ± 0.00	95.20 ± 2.55	11.50 ± 0.10	10.00 ± 0.51
	L2	0.18 ± 0.01	0.95 ± 0.11	3.80 ± 0.20	0.05 ± 0.02	0.02 ± 0.01	105.00 ± 3.38	15.40 ± 0.05	16.50 ± 0.10
	L3	0.20 ± 0.02	0.80 ± 0.09	4.50 ± 0.12	0.06 ± 0.02	0.03 ± 0.01	110.50 ± 4.04	14.00 ± 0.10	18.00 ± 0.21
	L4	0.13 ± 0.00	0.90 ± 0.12	1.50 ± 0.10	0.04 ± 0.01	0.01 ± 0.00	97.00 ± 2.16	13.80 ± 0.05	17.50 ± 0.27
	L5	0.10 ± 0.02	0.65 ± 0.05	4.80 ± 0.30	0.05 ± 0.02	0.02 ± 0.01	93.50 ± 3.30	12.10 ± 0.09	9.50 ± 0.32
	L6	0.06 ± 0.01	0.70 ± 0.04	2.10 ± 0.17	0.07 ± 0.02	0.03 ± 0.01	100.80 ± 3.05	14.30 ± 0.20	15.50 ± 0.19
	L7	0.15 ± 0.02	0.85 ± 0.05	4.90 ± 0.19	0.05 ± 0.02	0.01 ± 0.00	101.40 ± 4.70	14.20 ± 0.03	13.00 ± 0.08
	L8	0.11 ± 0.02	0.88 ± 0.07	4.40 ± 0.25	0.03 ± 0.01	0.03 ± 0.01	96.80 ± 3.03	15.00 ± 0.10	17.00 ± 0.07
	L9	0.19 ± 0.02	0.60 ± 0.11	2.80 ± 0.21	0.04 ± 0.01	0.03 ± 0.01	105.20 ± 4.20	13.60 ± 0.06	16.00 ± 0.12
	L10	0.05 ± 0.01	0.92 ± 0.02	3.00 ± 0.25	0.03 ± 0.01	0.02 ± 0.01	99.00 ± 3.09	12.00 ± 0.07	8.00 ± 0.10
	Mean	0.13	0.79	3.40	0.05	0.02	100.44	13.59	14.10
	SD	0.05	0.14	1.24	0.02	0.01	5.25	1.31	3.70
	CV (%)	42.08	17.30	36.47	34.47	44.03	5.23	9.65	26.21
Crimson Sweet	L1	0.08 ± 0.03	0.50 ± 0.05	3.20 ± 0.20	0.02 ± 0.01	0.01 ± 0.00	85.00 ± 2.50	9.00 ± 1.09	5.00 ± 0.32
	L2	0.20 ± 0.02	1.10 ± 0.10	3.90 ± 0.30	0.07 ± 0.02	0.05 ± 0.01	105.00 ± 3.50	15.00 ± 1.58	7.00 ± 0.54
	L3	0.05 ± 0.01	0.55 ± 0.05	4.10 ± 0.25	0.09 ± 0.02	0.06 ± 0.02	120.00 ± 4.00	12.50 ± 1.05	12.00 ± 1.08
	L4	0.23 ± 0.02	1.00 ± 0.10	3.80 ± 0.30	0.05 ± 0.01	0.02 ± 0.01	98.00 ± 3.03	15.50 ± 1.51	10.00 ± 0.81
	L5	0.13 ± 0.02	0.85 ± 0.10	3.50 ± 0.25	0.04 ± 0.01	0.03 ± 0.01	90.00 ± 2.55	11.80 ± 1.04	10.50 ± 1.05
	L6	0.25 ± 0.01	0.40 ± 0.05	3.60 ± 0.20	0.01 ± 0.00	0.07 ± 0.02	88.00 ± 2.54	15.80 ± 1.58	15.00 ± 1.23
	L7	0.10 ± 0.02	0.90 ± 0.10	3.30 ± 0.20	0.01 ± 0.00	0.02 ± 0.01	80.00 ± 2.50	12.50 ± 1.27	16.00 ± 1.07
	L8	0.09 ± 0.03	0.45 ± 0.05	4.20 ± 0.25	0.06 ± 0.02	0.08 ± 0.02	92.50 ± 3.00	10.20 ± 1.05	12.50 ± 1.03
	L9	0.15 ± 0.02	0.80 ± 0.10	3.70 ± 0.20	0.01 ± 0.00	0.06 ± 0.02	87.00 ± 2.54	14.80 ± 1.20	9.00 ± 0.58
	L10	0.06 ± 0.00	1.05 ± 0.10	3.10 ± 0.20	0.11 ± 0.02	0.03 ± 0.01	96.00 ± 3.51	15.60 ± 1.23	6.00 ± 0.52
	Mean	0.13	0.76	3.64	0.05	0.04	94.15	13.27	10.30
	SD	0.07	0.26	0.37	0.04	0.02	11.53	2.43	3.68
	CV (%)	53.67	34.65	10.21	74.49	57.43	12.25	18.31	35.69

L1–L10: Sampling locations; SD: standard deviation; CV: coefficient of variance.

Table 2. Concentration (mg/kg) of eight PTEs in two varieties of muskmelon collected from the Yamuna and Ganga River basins.

Muskmelon Variety	Location	Potentially Toxic Elements							
		Cd	Cr	Cu	Pb	As	Fe	Mn	Zn
Cantaloupe	L1	0.08 ± 0.02	0.60 ± 0.10	2.40 ± 0.05	0.01 ± 0.00	0.02 ± 0.01	80.00 ± 5.20	11.50 ± 0.52	8.00 ± 0.28
	L2	0.18 ± 0.05	0.90 ± 0.05	3.90 ± 0.14	0.03 ± 0.01	0.05 ± 0.01	105.00 ± 4.03	14.80 ± 0.08	16.00 ± 0.32
	L3	0.10 ± 0.03	0.65 ± 0.11	4.50 ± 0.22	0.04 ± 0.01	0.06 ± 0.02	95.00 ± 2.92	12.50 ± 0.11	18.00 ± 0.10
	L4	0.20 ± 0.07	0.75 ± 0.10	4.10 ± 0.10	0.03 ± 0.01	0.03 ± 0.01	88.00 ± 4.81	13.60 ± 0.25	12.50 ± 0.11
	L5	0.05 ± 0.03	0.80 ± 0.05	3.00 ± 0.05	0.04 ± 0.01	0.04 ± 0.01	100.00 ± 3.06	13.00 ± 0.07	14.00 ± 0.81
	L6	0.15 ± 0.05	0.55 ± 0.10	4.80 ± 0.07	0.02 ± 0.01	0.07 ± 0.03	105.50 ± 5.12	15.00 ± 0.14	17.50 ± 0.32
	L7	0.11 ± 0.04	0.70 ± 0.14	3.50 ± 0.06	0.05 ± 0.01	0.03 ± 0.01	110.00 ± 7.54	12.00 ± 0.05	11.00 ± 0.07
	L8	0.14 ± 0.05	0.95 ± 0.05	4.20 ± 0.03	0.02 ± 0.01	0.08 ± 0.02	90.00 ± 0.36	14.50 ± 0.19	16.50 ± 0.27
	L9	0.19 ± 0.05	0.85 ± 0.09	2.70 ± 0.12	0.04 ± 0.01	0.07 ± 0.02	95.80 ± 5.07	13.70 ± 0.03	15.80 ± 0.51
	L10	0.09 ± 0.04	0.88 ± 0.05	3.10 ± 0.05	0.06 ± 0.02	0.04 ± 0.01	99.00 ± 4.84	14.00 ± 0.07	10.00 ± 0.02
	Mean	0.13	0.76	3.62	0.03	0.05	96.83	13.46	13.93
	SD	0.05	0.14	0.80	0.02	0.02	9.08	1.19	3.41
	CV (%)	39.76	17.76	22.21	45.01	42.02	9.38	8.83	24.51
Kajri	L1	0.09 ± 0.01	0.70 ± 0.11	3.40 ± 0.12	0.01 ± 0.00	0.02 ± 0.01	85.00 ± 1.07	10.00 ± 0.34	9.00 ± 0.08
	L2	0.20 ± 0.02	1.00 ± 0.08	4.00 ± 0.07	0.06 ± 0.02	0.04 ± 0.02	115.00 ± 2.90	14.50 ± 0.02	18.00 ± 0.10
	L3	0.06 ± 0.01	0.55 ± 0.15	3.70 ± 0.02	0.05 ± 0.02	0.06 ± 0.03	120.00 ± 1.51	11.80 ± 0.06	15.00 ± 0.12
	L4	0.15 ± 0.04	0.75 ± 0.12	3.80 ± 0.05	0.03 ± 0.01	0.03 ± 0.01	95.00 ± 3.20	14.30 ± 0.10	17.00 ± 0.50
	L5	0.11 ± 0.03	0.85 ± 0.05	4.10 ± 0.04	0.04 ± 0.01	0.05 ± 0.02	105.00 ± 6.75	13.50 ± 0.07	12.00 ± 0.14
	L6	0.23 ± 0.05	0.50 ± 0.02	3.90 ± 0.12	0.02 ± 0.01	0.07 ± 0.03	110.50 ± 5.00	9.00 ± 0.30	14.00 ± 0.30
	L7	0.13 ± 0.01	0.72 ± 0.06	3.50 ± 0.09	0.04 ± 0.02	0.03 ± 0.01	100.10 ± 3.94	12.60 ± 0.40	13.50 ± 0.63
	L8	0.10 ± 0.02	0.60 ± 0.09	3.80 ± 0.04	0.02 ± 0.01	0.07 ± 0.02	90.00 ± 1.08	13.00 ± 0.14	15.50 ± 0.10
	L9	0.15 ± 0.03	0.80 ± 0.04	3.90 ± 0.06	0.04 ± 0.01	0.06 ± 0.02	118.00 ± 2.44	11.00 ± 0.29	17.00 ± 0.06
	L10	0.08 ± 0.01	0.88 ± 0.07	3.30 ± 0.13	0.08 ± 0.02	0.04 ± 0.01	95.00 ± 6.08	15.00 ± 0.70	10.00 ± 0.40
	Mean	0.13	0.74	3.74	0.04	0.05	103.35	12.47	14.10
	SD	0.05	0.16	0.26	0.02	0.02	12.23	2.00	3.03
	CV (%)	41.47	21.18	7.04	53.85	38.84	11.84	16.03	21.46

L1–L10: Sampling locations; SD: standard deviation; CV: coefficient of variance.

The comparative analysis of PTE concentrations in watermelons and muskmelons showed significant differences in PTE concentration. Both fruits exhibit higher Fe levels, though muskmelon varieties (*Cantaloupe* and *Kajri*) have slightly higher mean Fe concentrations compared to watermelons. Cu levels in muskmelons were higher than in watermelons. While, As and Cd had similar mean values across both fruits, variability was more pronounced in muskmelons for Pb and Cr. The accumulation of PTEs across watermelon and muskmelon varieties grown near riverbanks can be attributed to several physiological and soil–chemical mechanisms. Out of the studied PTEs, Cd may accumulate in fruits and vegetables due to its similarity to Ca, which results in competitive uptake at plant root sites [34]. On the other hand, Cr is found in the trivalent form, which becomes bioavailable through reduction–oxidation reactions in soils [35]. Similarly, Cu and Zn are essentially required for plant growth and are actively transported into plant tissues through membrane-bound proteins, but their excessive concentrations may result in disrupted homeostasis mechanisms [36]. Also, Pb is a non-essential and toxic element that enters plant roots via passive diffusion or endocytosis, binding with cell wall components and sequestering in vacuoles [37]. As a toxic element, As also mimics phosphate ions, which makes their uptake possible through phosphate transporters [38]. Nevertheless, Fe and Mn are essential elements required for plant growth and are actively absorbed by plant root systems for growth and physiologic processes [39]. However, the determined concentrations of Cd, Pb, and As in some watermelon samples exceed the permissible limits set by WHO/FAO regulations [40]. Specifically, Cd levels in certain samples surpass the 0.05 mg/kg threshold, while Pb concentrations approach the 0.11 mg/kg at the L10 limit which indicates potential health risks requiring further assessment.

Previously, several reports have shown the detection of harmful PTEs in watermelon and muskmelon varieties. Soil properties, stress levels, and irrigation water sources also play an important role in PTE accumulation in watermelon as reported by Khanjani et al. [41] and Gabrijel et al. [42]. In a study, Kolo et al. [43]. reported the occurrence of seven PTEs (Cd, Pb, Cu, Ni, Mn, Fe, Zn, and Co) in Sugar Baby and *Crimson Sweet* varieties of watermelon and found that leaf tissues showed the highest concentrations as compared to pulp, seed, and root. A similar research study was published by Oladele et al. [44]. for the detection of Ni, Cr, Cu, Cd, and Pb in three major markets in Akure (Isinkan, Shasha, and Oja-Oba) of Nigeria. They found that the levels of Cr and Ni were above WHO's maximum permissible limits which indicated a health risk. Therefore, the occurrence of these PTEs in watermelon and muskmelon fruits collected from the Yamuna and Ganga River basins might be due to their presence in riverbank soils and their uptake by the plant's root systems.

3.2. PCA and HCA Analysis Results

Figure 3a–d shows the clustered heatmaps and PCA biplots of PTEs distribution at ten sampling sites for two watermelon varieties, i.e., *Arka Shyama* and *Crimson Sweet*. Herein, the heatmap showed a distinct spatial variability in PTE concentrations across the sites, except for Fe and Pb. Specifically for *Arka Shyama*, PTE concentrations exhibit stratified clustering patterns while for *Crimson Sweet* it was a more heterogeneous distribution. The sampling sites showed distinct distribution patterns of all PTEs with the highest similarities between L2–L3 and L5–L7 sites for *Arka Shyama* and between L3–L8 and L4–L5 for *Crimson Sweet* varieties. On the other hand, PCA biplots were also helpful in reducing the dimensionality of the data and identifying dominant contributors to variability. In this case, the data were transformed into two principal components (PC1 and PC2) having a variance of 77.25 and 18.40% for *Arka Shyama* and 87.53 and 8.69% for *Crimson Sweet* varieties, respectively. It was observed that Fe and Zn showed the most dominant distributions on L3 and L6 sites, as indicated by the vector lengths in Figure 3b,d.

It was also reported that most PTEs strongly were associated with specific sampling sites. Similarly, Figure 4a,b showed the clustered heatmaps and PCA biplots of PTEs distribution in *Cantaloupe* and *Kajri* muskmelon varieties. The clustered heatmap showed no significant variation in Fe and Cr concentration across selected sampling sites for both muskmelon varieties while distinct variations were found for Cd, Cu, Pb, and As, which might be due to varying pollution levels at their corresponding sites. Herein, the highest similarities were recorded at L2–L9 and L7–L10 for *Cantaloupe* and L5–L9 and L4–L7 for *Kajri* varieties, respectively. In contrast, the PCA biplots showed similar patterns for Fe and Zn distributions (Figure 4b,d). The calculated variances of PC1 and PC2 were 87.30

and 11.43 for *Cantaloupe* and 93.74 and 4.28 for *Kajri* varieties, respectively. The distinct variations in PTE concentrations might be due to localized soil contamination in these regions [45].

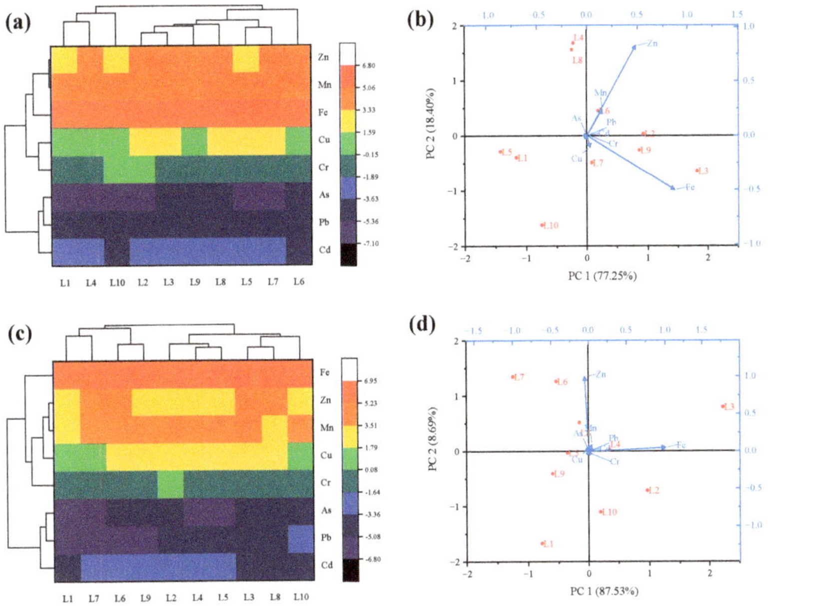

Figure 3. Clustered heatmap matrices and principal component biplots of eight PTEs at ten sampling sites for two varieties of watermelon: (**a**,**b**) *Arka Shyama*; (**c**,**d**) *Crimson Sweet*.

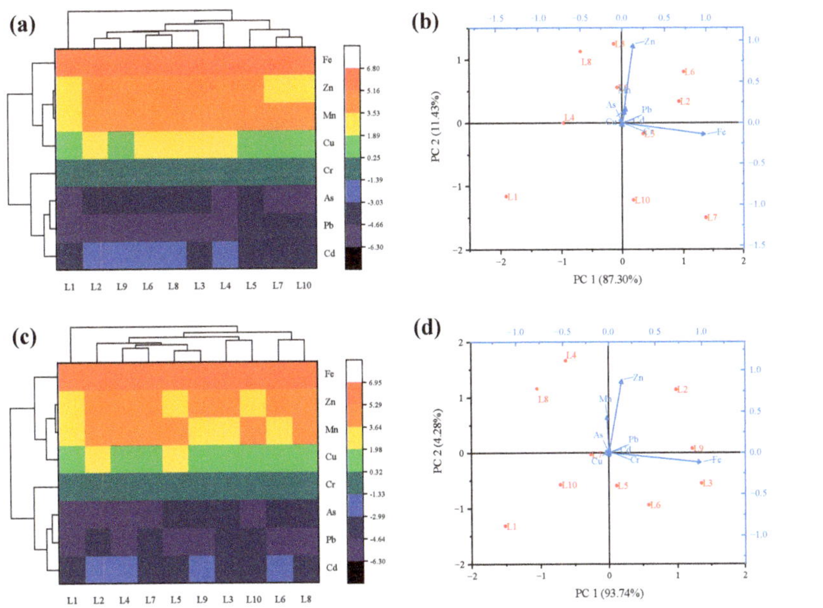

Figure 4. Clustered heatmap matrices and principal component biplots of eight PTEs at ten sampling sites for two varieties of muskmelon (**a**,**b**): *Cantaloupe*; (**c**,**d**): *Kajri*.

Clustered heat maps visualized the distribution and relationships between PTE concentrations across different sampling sites and fruit varieties. In this study, hierarchical clustering groups were found at similar sites based on PTE profiles, revealing patterns of spatial variability. Distinct clusters indicate sites with comparable PTE levels, while isolated clusters highlight sites with unique contamination patterns. On the other hand, PCA reduces the dimensionality of the data while retaining most of the variability. In this case, PCA transforms the PTE data into principal components that capture the highest variance. Together, HCA and PCA analyses have been used by several researchers to reduce the complexity of large datasets and understand PTE occurrence in fruits and vegetables. Out of them, Zhang et al. [46]. employed both HCA and PCA tools to understand the occurrence of PTEs in wastewater irrigation vegetables across China and found that these tools help to identify the most polluted sites. Similarly, Alves et al. [47]. also reported that PCA and HCA were useful in studying the mineral composition of native and non-native Amazonian fruits.

3.3. *Health Risk of PTEs*

In this study, a health risk assessment of PTEs analyzed in selected watermelon and muskmelon varieties was performed using selected indices i.e., DIM, HRI, and THQ (Table 3). In this, DIM represents the estimated daily ingestion of PTEs through food consumption and is expressed as mg/kg/day. This helps in studying the long-term effects of PTEs exposed to consumers [48]. It was observed that, for watermelon and muskmelon, the highest DIM values were recorded for Fe (0.4035–0.4429) followed by Zn (0.0441–0.0604) while the lowest values were recorded for Pb and As (0.0001–0.0002), respectively. Also, the *Arka Shyamza* (watermelon) and *Kajri* (muskmelon) varieties showed higher DIM values for most PTEs. In contrast, the HRI is defined as the ratio of the DIM and the oral reference dose (RfD) of PTEs [49]. A value exceeding 1 indicates a potential health risk if a contaminated food item is consumed. This study showed that the HRI of both watermelon and muskmelon did not exceed the threshold value (1). However, Cd, Cr, Pb, and Fe approached the level of 0.5, which was higher than other PTEs studied. In this case, watermelon varieties showed higher HRI values as compared to muskmelon. In addition, the THQ is a non-carcinogenic risk index tool that evaluates the risk of exposure in relation to safe reference doses. A THQ value greater than 1 indicates potential health risks [32]. The THQ values (<0.0007) for all PTEs remained below 1, which indicated negligible non-carcinogenic risks at current exposure levels. Therefore, the results indicate no significant health risks associated with PTE exposure across watermelon and muskmelon varieties.

It is evident that the presence of Cd, Pb, and As in food poses serious health risks. Chronic exposure to Cd can lead to kidney damage, respiratory issues, and weakened bones [50]. Pb primarily affects the nervous system, impairing cognitive development in children, while As is associated with skin lesions, cancer, and cardiovascular diseases when consumed over time [51]. Previous studies have shown the applicability of health risk indicators in the successful understanding of potential health hazards associated with the consumption of contaminated fruits. Out of them, Al-Huqail et al. [32]. reported HRI, DIM, and THQ of eight PTEs in lychee (*Litchi chinensis* Sonn., Sapindaceae) fruit collected from the Saharanpur region of northern India. Similarly, Chen et al. [52]. also reported the health risk of Cr pollution in selected irrigated crops in South Asian regions and utilized HRI, DIM, and THQ.

Table 3. Calculated values of selected health risk indices for eight PTEs at ten sampling sites for selected varieties of watermelon and muskmelon.

Index	Fruit	Varieties	Potentially Toxic Elements							
			Cd	Cr	Cu	Pb	As	Fe	Mn	Zn
DIM	Watermelon	*Arka Shyama*	0.0005	0.0034	0.0146	0.0002	0.0001	0.4305	0.0582	0.0604
		Crimson Sweet	0.0006	0.0033	0.0156	0.0002	0.0002	0.4035	0.0569	0.0441
	Muskmelon	*Cantaloupe*	0.0005	0.0033	0.0155	0.0001	0.0002	0.4150	0.0577	0.0597
		Kajri	0.0006	0.0032	0.0160	0.0002	0.0002	0.4429	0.0534	0.0604
HRI	Watermelon	*Arka Shyama*	0.5357	0.6729	0.3643	0.5357	0.2821	0.6149	0.4160	0.2014
		Crimson Sweet	0.5679	0.6514	0.3900	0.5357	0.5714	0.5764	0.4062	0.1471
	Muskmelon	*Cantaloupe*	0.5486	0.6540	0.3879	0.3214	0.6714	0.5928	0.4120	0.1990
		Kajri	0.5507	0.6300	0.4007	0.4286	0.6429	0.6328	0.3817	0.2014
THQ	Watermelon	*Arka Shyama*	0.0005	0.0007	0.0004	0.0005	0.0003	0.0006	0.0004	0.0002
		Crimson Sweet	0.0006	0.0007	0.0004	0.0005	0.0006	0.0006	0.0004	0.0001
	Muskmelon	*Cantaloupe*	0.0006	0.0007	0.0004	0.0003	0.0007	0.0006	0.0004	0.0002
		Kajri	0.0006	0.0006	0.0004	0.0004	0.0007	0.0006	0.0004	0.0002

HRI: health risk index; DIM: dietary intake modeling (mg/kg/day); THQ: target hazard quotient.

4. Conclusions

The present study showed that PTE accumulation in selected watermelon and muskmelon varieties is significantly influenced by location. In general, the watermelon varieties indicated higher levels of Fe and Mn, while muskmelons had slightly higher Cu concentrations. The concentration of PTEs like Cd, Pb, and As were also varying at levels of contamination. The results suggest that the presence of PTEs in fruits from the river basins may be linked to soil contamination and plant uptake mechanisms. It is suggested to consider the pollution prevention and management of agricultural lands near riverbanks to prevent crop contamination. Future research should focus on the long-term monitoring of PTE accumulation in fruits and explore mitigation strategies such as phytoremediation or soil amendment techniques. Additionally, assessing the bioavailability of PTEs in other crops growing in the region is highly recommended.

Author Contributions: Conceptualization, M.E. and V.K.; Data curation, M.E., M.A.A.E.-A., A.E.-D.O. and S.A.F.; Formal analysis, V.K.; Funding acquisition, M.E.; Investigation, V.K.; Methodology, M.E. and V.K.; Project administration, M.E.; Resources, V.K.; Software, M.A.A.E.-A., A.E.-D.O. and S.A.F.; Supervision, V.K.; Validation, M.E., M.A.A.E.-A., A.E.-D.O. and S.A.F.; Visualization, M.A.A.E.-A., A.E.-D.O. and S.A.F.; Writing—original draft, S.A.F. and V.K.; Writing—review and editing, M.E., M.A.A.E.-A. and A.E.-D.O. All authors have read and agreed to the published version of the manuscript.

Funding: The Deanship of Research and Graduate Studies at King Khalid University funded this work through Large Research Project under grant number RGP2/118/45.

Data Availability Statement: The original contributions presented in this study are included in the article. Further inquiries can be directed to the corresponding author.

Acknowledgments: The authors extend their appreciation to the Deanship of Research and Graduate Studies at King Khalid University funded this work through Large Research Project under Grant number RGP2/118/45.

Conflicts of Interest: The authors declare no conflicts of interest.

References

1. Silva, M.A.; Albuquerque, T.G.; Alves, R.C.; Oliveira, M.B.P.P.; Costa, H.S. Melon (*Cucumis melo* L.) by-Products: Potential Food Ingredients for Novel Functional Foods? *Trends Food Sci. Technol.* **2020**, *98*, 181–189. [CrossRef]

2. Pitrat, M.; Chauvet, M.; Foury, C. Diversity, History and Production of Cultivated Cucurbits. *Acta Hortic.* **1999**, *492*, 21–28. [CrossRef]
3. ReporLinker Global Melons Production by Country. Available online: https://www.reportlinker.com/dataset/a0495cccd8f317070b99bf8305fdec829bb55eb7 (accessed on 10 November 2024).
4. Pulela, B.L.; Maboko, M.M.; Soundy, P.; Amoo, S.O. Cultivar and Postharvest Storage Duration Influence Fruit Quality, Nutritional and Phytochemical Profiles of Soilless-Grown Cantaloupe and Honeydew Melons. *Plants* **2022**, *11*, 2136. [CrossRef]
5. Palmitessa, O.D.; Durante, M.; Somma, A.; Mita, G.; D'imperio, M.; Serio, F.; Santamaria, P. Nutraceutical Profile of "Carosello" (*Cucumis melo* L.) Grown in an Out-of-Season Cycle under LEDs. *Antioxidants* **2022**, *11*, 777. [CrossRef]
6. Rudrapal, M.; Khairnar, S.J.; Khan, J.; Dukhyil, A.B.; Ansari, M.A.; Alomary, M.N.; Alshabrmi, F.M.; Palai, S.; Deb, P.K.; Devi, R. Dietary Polyphenols and Their Role in Oxidative Stress-Induced Human Diseases: Insights Into Protective Effects, Antioxidant Potentials and Mechanism(s) of Action. *Front. Pharmacol.* **2022**, *13*, 806470. [CrossRef]
7. Cory, H.; Passarelli, S.; Szeto, J.; Tamez, M.; Mattei, J. The Role of Polyphenols in Human Health and Food Systems: A Mini-Review. *Front. Nutr.* **2018**, *5*, 87. [CrossRef] [PubMed]
8. Schwingshackl, L.; Hoffmann, G. Monounsaturated Fatty Acids and Risk of Cardiovascular Disease: Synopsis of the Evidence Available from Systematic Reviews and Meta-Analyses. *Nutrients* **2012**, *4*, 1989–2007. [CrossRef] [PubMed]
9. Telle-Hansen, V.H.; Gaundal, L.; Bastani, N.; Rud, I.; Byfuglien, M.G.; Gjøvaag, T.; Retterstøl, K.; Holven, K.B.; Ulven, S.M.; Myhrstad, M.C.W. Replacing Saturated Fatty Acids with Polyunsaturated Fatty Acids Increases the Abundance of Lachnospiraceae and Is Associated with Reduced Total Cholesterol Levels—A Randomized Controlled Trial in Healthy Individuals. *Lipids Health Dis.* **2022**, *21*, 92. [CrossRef] [PubMed]
10. Fu, J.; Zheng, Y.; Gao, Y.; Xu, W. Dietary Fiber Intake and Gut Microbiota in Human Health. *Microorganisms* **2022**, *10*, 2507. [CrossRef] [PubMed]
11. Stokes, T.; Hector, A.J.; Morton, R.W.; McGlory, C.; Phillips, S.M. Recent Perspectives Regarding the Role of Dietary Protein for the Promotion of Muscle Hypertrophy with Resistance Exercise Training. *Nutrients* **2018**, *10*, 180. [CrossRef]
12. Dovlatabadi, A.; Estiri, E.H.; Najafi, M.L.; Ghorbani, A.; Rezaei, H.; Behmanesh, M.; Miri, M. Bioaccumulation and health risk assessment of exposure to potentially toxic elements by consuming agricultural products irrigated with wastewater effluents. *Environ. Res.* **2022**, *205*, 112479. [CrossRef] [PubMed]
13. Irias Zelaya, C.R.; Gadelha, J.C.; Hernandez, F.F.F.; Ortiz Escobar, M.E. Heavy Metals Behavior in the Presence of Organic Acids and Proteins in the Rhizosphere and Plant Tissues of Yellow Melon Var. Natal (*Cucumis melo* L.) Cultivated in a Quartzarenic Neosol. *Water Air Soil Pollut.* **2020**, *231*, 572. [CrossRef]
14. Shahid, M.; Dumat, C.; Khalid, S.; Niazi, N.K.; Antunes, P.M.C. Cadmium Bioavailability, Uptake, Toxicity and Detoxification in Soil-Plant System. In *Reviews of Environmental Contamination and Toxicology*; Springer: Cham, Switzerland, 2017; Volume 241, ISBN 978-3-319-46945-4.
15. Mgbemena, N.M.; Amako, F.N.; Awa, F.C. Distribution of Heavy Metals in Melon Seeds and Soils along Ikot Ekpene Road, Umuahia, Nigeria. *J. Chem. Soc. Niger.* **2019**, *44*, 1341–1350.
16. Zhou, H.; Yang, W.T.; Zhou, X.; Liu, L.; Gu, J.F.; Wang, W.L.; Zou, J.L.; Tian, T.; Peng, P.Q.; Liao, B.H. Accumulation of Heavy Metals in Vegetable Species Planted in Contaminated Soils and the Health Risk Assessment. *Int. J. Environ. Res. Public Health* **2016**, *13*, 289. [CrossRef] [PubMed]
17. Yaqub, G.; Khan, A.; Zishan Ahmad, M.; Irshad, U. Determination of Concentration of Heavy Metals in Fruits, Vegetables, Groundwater, and Soil Samples of the Cement Industry and Nearby Communities and Assessment of Associated Health Risks. *J. Food Qual.* **2021**, *2021*, 3354867. [CrossRef]
18. Leyssens, L.; Vinck, B.; Van Der Straeten, C.; Wuyts, F.; Maes, L. Cobalt Toxicity in Humans—A Review of the Potential Sources and Systemic Health Effects. *Toxicology* **2017**, *387*, 43–56. [CrossRef] [PubMed]
19. Ohiagu, F.O.; Lele, K.C.; Chikezie, P.C.; Verla, A.W.; Enyoh, C.E. Bioaccumulation and Health Risk Assessment of Heavy Metals in Musa Paradisiaca, Zea Mays, Cucumeropsis Manii and Manihot Esculenta Cultivated in Onne, Rivers State, Nigeria. *Environ. Anal. Health Toxicol.* **2020**, *35*, e2020011. [CrossRef] [PubMed]
20. Zhang, W.; Zhu, G.; Zhao, L.; Wang, L.; Qiu, D.; Ye, L.; Lu, S.; Lin, X. Quantifying the Changes in Solute Transport Caused by Human Influence on River Connectivity in Inland River Basins. *Catena* **2024**, *246*, 108360. [CrossRef]
21. Wang, Y.; Lei, Z.; Ye, R.; Zhou, W.; Zhou, Y.; Zou, Z.; Li, J.; Yi, L.; Dai, Z. Effects of Cadmium on Physiochemistry and Bioactive Substances of Muskmelon (*Cucumis melo* L.). *Molecules* **2022**, *27*, 2913. [CrossRef] [PubMed]
22. Chunhabundit, R. Cadmium Exposure and Potential Health Risk from Foods in Contaminated Area, Thailand. *Toxicol. Res.* **2016**, *32*, 65–72. [CrossRef]
23. Han, H.; Sha, R.; Dai, J.; Wang, Z.; Mao, J.; Cai, M. Garlic Origin Traceability and Identification Based on Fusion of Multi-Source Heterogeneous Spectral Information. *Foods* **2024**, *13*, 1016. [CrossRef]
24. Shen, Q.; Wu, X.; Tao, Y.; Yan, G.; Wang, X.; Cao, S.; Wang, C.; He, W. Mining Candidate Genes Related to Heavy Metals in Mature Melon (*Cucumis melo* L.) Peel and Pulp Using WGCNA. *Genes* **2022**, *13*, 1767. [CrossRef]

25. Brecchia, M.; Buscaroli, E.; Mazzon, M.; Blasioli, S.; Braschi, I. Effect of the Growing Season, Trichoderma, and Clinoptilolite Application on Potentially Toxic Elements Uptake by *Cucumis melo* L. *HortScience* **2022**, *57*, 1548–1557. [CrossRef]
26. Kagu, B.M.; Jatau, A.M. Levels of Selected Heavy Metals and Physicochemical Parameters in Soils and Watermelon Samples from Mainok Farmland, Kaga LGA, Borno State, Nigeria. *ChemSearch J.* **2023**, *14*, 45–53.
27. Medeiros, I.; de Oliveira, G.L.R.; de Queiroz, J.L.C.; de Carvalho Gomes, C.; de Carvalho, F.M.C.; de Souza Lima, M.C.J.; Serquiz, A.C.; de Andrade Santos, P.P.; da Silva Camillo, C.; Maciel, B.L.L.; et al. Safety and Bioactive Potential of Nanoparticles Containing Cantaloupe Melon (*Cucumis melo* L.) Carotenoids in an Experimental Model of Chronic Inflammation. *Biotechnol. Rep.* **2020**, *28*, e00567. [CrossRef] [PubMed]
28. Ali, S.; Rahman, A.U.; Ali, E.; Karabulut, F.; Ali, S.; Ahmad, R.; Fadl, M.E.; AbdelRahman, M.A.E.; Ahmed, M.A.A.; Scopa, A. Insights on the Nutritional Profiling of Cantaloupe (*Cucumis melo* L.) via 1-Naphthalene Acetic Acid. *Plants* **2023**, *12*, 2969. [CrossRef] [PubMed]
29. Yousuf, B.; Srivastava, A.K. A Novel Approach for Quality Maintenance and Shelf Life Extension of Fresh-Cut Kajari Melon: Effect of Treatments with Honey and Soy Protein Isolate. *Lwt* **2017**, *79*, 568–578. [CrossRef]
30. IMD Local Weather Report and Forecast for: Saharanpur, India Meteorological Department. Available online: https://city.imd.gov.in/citywx/localwx.php (accessed on 12 December 2024).
31. USEPA (United States Environmental Protection Agency). *EPA Method. 3050B Acid. Digestion of Sediments, Sludges and Soils, SW-846 Revision 2*; USEPA: Washington, DC, USA, 1996.
32. AL-Huqail, A.A.; Singh, R.; Širić, I.; Kumar, P.; Abou Fayssal, S.; Kumar, V.; Bachheti, R.K.; Andabaka, Ž.; Goala, M.; Eid, E.M. Occurrence and Health Risk Assessment of Heavy Metals in Lychee (*Litchi chinensis* Sonn., Sapindaceae) Fruit Samples. *Horticulturae* **2023**, *9*, 989. [CrossRef]
33. Abou Fayssal, S.; Kumar, P.; Popescu, S.M.; Sardar, H.; Ahmad, R.; Gupta, D.; Gaur, S.K.; Alharby, H.F.; Al-Ghamdi, A.G. Health Risk Assessment of Heavy Metals in Saffron (*Crocus sativus* L.) Cultivated in Domestic Wastewater and Lake Water Irrigated Soils. *Heliyon* **2024**, *10*, e27138. [CrossRef]
34. Rizwan, M.; Ali, S.; Adrees, M.; Ibrahim, M.; Tsang, D.C.W.; Zia-ur-Rehman, M.; Zahir, Z.A.; Rinklebe, J.; Tack, F.M.G.; Ok, Y.S. A Critical Review on Effects, Tolerance Mechanisms and Management of Cadmium in Vegetables. *Chemosphere* **2017**, *182*, 90–105. [CrossRef]
35. Pattnaik, B.K.; Sahu, C. Chromium Dynamics in Soil-Plant System. In *Response of Field Crops to Abiotic Stress: Current Status and Future Prospects*; CRC Press: Boca Raton, FL, USA, 2022; pp. 55–65. ISBN 9781003258063.
36. Mir, A.R.; Pichtel, J.; Hayat, S. Copper: Uptake, Toxicity and Tolerance in Plants and Management of Cu-Contaminated Soil. *BioMetals* **2021**, *34*, 737–759. [CrossRef] [PubMed]
37. Jagota, N.; Singh, S.; Kaur, H.; Kaur, R.; Sharma, A. Oxidative Stress in Lead Toxicity in Plants and Its Amelioration. In *Sustainable Nexus Approaches: Nature Switzerland*; Mitigation, L.T., Ed.; Springer: Cham, Switzerland, 2024; pp. 299–333.
38. Garbinski, L.D.; Rosen, B.P.; Chen, J. Pathways of Arsenic Uptake and Efflux. *Environ. Int.* **2019**, *126*, 585–597. [CrossRef]
39. Vatansever, R.; Ozyigit, I.I.; Filiz, E. Essential and Beneficial Trace Elements in Plants, and Their Transport in Roots: A Review. *Appl. Biochem. Biotechnol.* **2017**, *181*, 464–482. [CrossRef]
40. Åkesson, M.T.; Point, C.C.; di Caracalla, V.D.T. *Joint FAO/WHO Food Standards Programme CODEX Committee on Contaminants in Foods*; WHO: Geneva, Switzerland, 2015.
41. Khanjani, M.J.; Maghsoudi Moud, A.A.; Saffari, V.R.; Hashamipor, S.M.; Soltanizadeh, M. Statistical Analysis of Heavy Metals Concentration in Watermelon Plants Irrigated by Wastewater. In *AIP Conference Proceedings*; American Institute of Physics: College Park, MD, USA, 2007; Volume 971, pp. 197–201.
42. Gabrijel, O.; Davor, R.; Zed, R.; Marija, R.; Monika, Z. Cadmium Accumulation by Muskmelon under Salt Stress in Contaminated Organic Soil. *Sci. Total Environ.* **2009**, *407*, 2175–2182. [CrossRef]
43. Kolo, B.G.; Shugaba, A.; Kagu, M. Determination of Some Heavy Metals in Variety of Watermelon Samples Cultivated in Kaga Local Government Area, Borno State, Nigeria. In *Economic and Social. Development: Book of Proceedings*; Varazdin Development and Entrepreneurship Agency: Varazdin, Croatia, 2020; pp. 377–386.
44. Oladele, O.O.; Aladesanmi, O.C. Level of Heavy Metals in Sliced Watermelon Fruits in Selected Markets in Akure, Nigeria. *Bull. Natl. Res. Cent.* **2020**, *44*, 102. [CrossRef]
45. Kumar, P.; Alhag, S.K.; Al-Shahari, E.A.; Al-Fakeh, M.S.; Abou Fayssal, S.; Bachheti, R.K.; Širić, I.; Eid, E.M. Impact of Irrigation with Contaminated River Water on Growth, Yield, and Heavy Metals Accumulation in Planted Armenian Cucumber (*Cucumis melo* Var. Flexuosus (L.) Naudin.). *Water Air Soil Pollut.* **2025**, *236*, 1. [CrossRef]
46. Zhang, X.; Wang, X.; Cao, X.; Xiao, G.; Miao, H. Heavy Element Contents of Vegetables and Health-Risk Assessment in China. *Sci. Total Environ.* **2022**, *828*, 154552. [CrossRef]
47. Alves, B.S.F.; Pereira Junior, J.B.; Carvalho, F.I.M.; Dantas Filho, H.A.; Fernandes Dantas, K.G. Mineral Composition of Amazonian Fruits by Flame Atomic Absorption Spectrometry Using Multivariate Analysis. *Biol. Trace Elem. Res.* **2019**, *189*, 259–266. [CrossRef]

48. Shehu, I.; Malsiu, A.; Bajraktari, N. Assessment of Potentially Toxic Element Concentrations in Soil And Vegetables and Impact on Human Health Through TF, EDI, and HRI Indicators: Case Study Anadrinia Region (Kosovo). *Biol. Trace Elem. Res.* **2023**, *201*, 479–492. [CrossRef] [PubMed]
49. Chaoua, S.; Boussaa, S.; El Gharmali, A.; Boumezzough, A. Impact of Irrigation with Wastewater on Accumulation of Heavy Metals in Soil and Crops in the Region of Marrakech in Morocco. *J. Saudi Soc. Agric. Sci.* **2019**, *18*, 429–436. [CrossRef]
50. Fatima, G.; Raza, A.M.; Hadi, N.; Nigam, N.; Mahdi, A.A. Cadmium in Human Diseases: It's More than Just a Mere Metal. *Indian J. Clin. Biochem.* **2019**, *34*, 371–378. [CrossRef] [PubMed]
51. Karri, V.; Schuhmacher, M.; Kumar, V. Heavy Metals (Pb, Cd, As and MeHg) as Risk Factors for Cognitive Dysfunction: A General Review of Metal Mixture Mechanism in Brain. *Environ. Toxicol. Pharmacol.* **2016**, *48*, 203–213. [CrossRef] [PubMed]
52. Chen, F.; Ma, J.; Akhtar, S.; Khan, Z.I.; Ahmad, K.; Ashfaq, A.; Nawaz, H.; Nadeem, M. Assessment of Chromium Toxicity and Potential Health Implications of Agriculturally Diversely Irrigated Food Crops in the Semi-Arid Regions of South Asia. *Agric. Water Manag.* **2022**, *272*, 107833. [CrossRef]

horticulturae

MDPI

Article

Cu Concentration in Vegetable-Cultivated Soil: Comparison Between Open Fields and Protected Environments in Brazil

Odair J. Schmitt [1], Jerônimo L. Andriolo [1], Tadeu L. Tiecher [2], Lessandro De Conti [3], Cledimar R. Lourenzi [4], Renan Costa Beber Vieira [5], Douglas R. Kaiser [5], Carina Marchezan [6], Aparecida Miranda Corrêa [6], Isley C. Bicalho Silva [6], Tatiane Chassot [5], Mario S. Wolski [5], Letícia Morsch [4], Edicarla Trentin [6], Moreno Toselli [7], Eleną Baldi [7], Paulo Ademar Avelar Ferreira [6] and Gustavo Brunetto [6,*]

1 Plant Science Department, Federal University of Santa Maria (UFSM), Santa Maria 97105-900, RS, Brazil; odair.schmitt@uffs.edu.br (O.J.S.); jeronimoandriolo@gmail.com (J.L.A.)
2 Federal Institute Rio Grande do Sul, Campus Restinga, Porto Alegre 91791-508, RS, Brazil; tadeu.t@hotmail.com
3 Federal Institute Farroupilha, Campus Santo Augusto, Santo Augusto 98590-000, RS, Brazil; lessandrodeconti@gmail.com
4 Rural Engineering Department, Center of Agricultural Science, Federal University of Santa Catarina (UFSC), Florianópolis 88034-001, SC, Brazil; lourenzicr@gmail.com (C.R.L.); lehmorsch@gmail.com (L.M.)
5 Department of Agronomy, Federal University of Fronteira Sul–Campus Cerro Largo (UFFS), Cerro Largo 97900-000, RS, Brazil; renan.vieira@uffs.edu.br (R.C.B.V.); douglasrodrigokaiser@gmail.com (D.R.K.); tatianechassot@yahoo.com.br (T.C.); mario.wolski@uffs.edu.br (M.S.W.)
6 Soil Science Department, Federal University of Santa Maria (UFSM), Santa Maria 97105-900, RS, Brazil; marchezancarina@yahoo.com.br (C.M.); aparecida.correa@acad.ufsm.br (A.M.C.); isleybicalho@yahoo.com.br (I.C.B.S.); edicarlatrentin@gmail.com (E.T.); ferreira.aap@gmail.com (P.A.A.F.)
7 Department of Agricultural and Food Sciences, University of Bologna, Viale Fanin, 46 40127 Bologna, Italy; moreno.toselli@unibo.it (M.T.); elena.baldi7@unibo.it (E.B.)
* Correspondence: brunetto.gustavo@gmail.com; Tel.: +55-55-99984-0302

Abstract: High copper (Cu) concentration in soils used for vegetable production is an agricultural, social and environmental issue. The excess of Cu in soil can cause toxicity in plants, reducing growth and yields. Moreover, Cu can also be absorbed and accumulated in edible organs, increasing risks for human health. This study aimed to compare Cu fractions in soils: (1) non-cultivated, natural soil (NC), (2) soils cultivated with open field crops (FCs) and (3) soils cultivated in controlled environments (CEs). The survey was carried out on 25 sites in Rio Grande do Sul (Southern Brazil), with each site containing the three types of soils described above. From these sites, the four with the highest soil Cu concentrations were selected to compare soluble (Cu-$CaCl_2$), available (Cu-EDTA) and total (Cu-EPA) Cu. Both total and available soil Cu concentration in soil solution, in CE and FC areas were higher than in NC. At sites 23R1 (Oxisol) and 11R1 (Molisol) the soluble and available Cu content was higher in the FC environment than in CEs, and the same was observed for the total Cu content at sites 23R1 (Oxisol) and 9R1 (Molisol). Some FC soils showed total Cu concentrations higher than the limits established by Brazilian environmental legislation, being, as a consequence, potentially contaminated. Soil monitoring and cropping practices must be adopted to reduce the Cu content in soils of vegetable crops in this region.

Keywords: heavy metals; environmental contamination; Cu-soil solution; available Cu; total Cu

1. Introduction

Vegetable production stands out by the intensive use of chemical and organic agricultural inputs, mainly of fertilizers [1,2]. Indiscriminate agricultural fertilizer use is mainly associated with: (i) inexistence and/or lack of research indicating mineral nutrient requirements during the growth and development of more than 50 species of vegetables cultivated in Brazil [3]; (ii) short development cycles of vegetables, generally between

30 and 150 days, associated with concerns on high yield obtainment; and (iii) the use of several intensive cropping systems, such as cultivation in an open field, in a protected environment and soilless. This generates several variables and specificities to be considered in the formulation, recommendation and management of the crop fertilization.

In agricultural soils, copper (Cu) concentration is usually naturally low, but may increase due to intensive application via chemical and/or organic fertilization, since high fertilizer rates are used in vegetable cultivation [4,5]. Also, Cu product use in phytosanitary treatments, especially fungicides, can contribute to increased soil Cu [6]. In Brazil, there are 28 products formulated with Cu, which are recommended for at least one vegetable species [7]. This way, frequent use of these products, associated with successive crops in the same area and high rates of fertilization, can cause Cu accumulation in the soil [8–10], especially since Cu is a micronutrient and, therefore, absorbed by plants in small quantities [11].

Moreover, Cu can be absorbed and accumulated in edible plant organs at levels that are considered toxic [12], representing a risk to consumer health due to its cumulative potential in the human organism [13], as well as reducing crop productivity and causing economic losses to producers. Although this study does not evaluate Cu levels in vegetables, the literature indicates that increased soil Cu can lead to absorption by plants, causing accumulation that may negatively affect plant growth and pose health risks when consumed. It is important to note that the available Cu fraction is the result of several physicochemical reactions between the element and soil constituents, mainly clay, iron (Fe) and aluminum (Al) oxides, organic matter content, pH and the cation exchange capacity (CEC) in soil [14,15]. Therefore, Cu availability can vary according to soil classes and the management practices adopted in production systems.

In this context, it is essential to monitor soil Cu availability to ensure safe and sustainable vegetable production. Studies on heavy metal presence in soils cultivated with vegetables remain limited. This study aims to address this gap by diagnosing Cu concentration and availability in soils used for open-field and protected-environment vegetable cultivation. Out of the initial 25 sampling sites, four were selected for detailed analysis due to their higher Cu concentrations, representing higher-risk scenarios. Protected environments refer to covered cultivation areas, such as greenhouses or plastic tunnels, where climatic conditions are partially controlled.

2. Materials and Methods

2.1. Soil Sampling Sites

Sites where soil samples were collected were defined with the aim of covering the five regions classified and presented in Ordinance No. 85, issued by the State Foundation for Environmental Protection and published on 5 September 2014. This ordinance establishes soil Quality Reference Values (QRVs) for nine chemical elements, among them Cu, naturally present in the different geomorphological/geological provinces of Rio Grande do Sul state, Southern Brazil.

Preliminary selection of sampling sites was carried out by using satellite images (Google Earth Pro, version 7.3.3.) and the choice was made according to the following criteria: (i) existence of two cropping systems, i.e., field and greenhouse cultivation; (ii) successive cultivation of vegetables for at least 5 years; and (iii) the presence of a non-anthropized area (uncultivated field or native forest) close to cultivated areas (used as a control). Sampled sites identification was performed through the collection and annotation of geographical coordinates, from a portable GPS navigator, as shown in Table 1 and Figure 1.

Table 1. Identification, municipalities, soil class and geographic coordinates of the 25 experimental sites.

Identification	Municipalities	Soil Class (1)	Geographic Coordinates
1R3	Santa Maria	Alfisol	29°41′27.67″ S; 53°42′34.38″ O
2R3	Santa Maria	Alfisol	29°38′51.81″ S; 53°40′36.51″ O
3R1	**Cruz Alta**	**Oxisol**	**28°36′29.69″ S; 53°37′43.50″ O**
4R4	Cruz Alta	Oxisol	28°34′49.23″ S; 53°32′28.35″ O
5R2	Caçapava do Sul	Entisol	30°33′18.41″ S; 53°28′40.26″ O
6R4	Caçapava do Sul	Alfisol	30°35′42.87″ S; 53°22′39.61″ O
7R5	Turuçu	Alfisol	31°26′27.23″ S; 52°13′40.39″ O
8R3	São Vicente do Sul	Alfisol	29°41′1.18″ S; 54°41′12.78″ O
9R1	**Feliz**	**Molisol**	**29°31′3.96″ S; 51°19′30.71″ O**
10R4	Cacequi	Alfisol	29°51′55.86″ S; 54°49′47.29″ O
11R1	**Bom Princípio**	**Molisol**	**29°30′39.29″ S; 51°21′21.17″ O**
12R1	Cruzeiro do Sul	Oxisol	29°32′59.30″ S; 52° 3′36.62″ O
13R1	Venâncio Aires	Alfisol	29°34′14.05″ S; 52°12′2.29″ O
14R3	Santa Cruz do Sul	Alfisol	29°40′42.11″ S; 52°19′16.01″ O
15R2	São Sepé	Alfisol	30°11′4.21″ S; 53°36′47.98″ O
16R3	São Gabriel	Alfisol	30°21′7.08″ S; 54°16′57.43″ O
17R1	Erechim	Oxisol	27°39′56.73″ S; 52°18′41.99″ O
18R1	Caxias do Sul	Inceptsoil	29°13′18.26″ S; 51° 1′13.13″ O
19R2	Canguçu	Entisol	31°25′44.40″ S; 52°40′3.07″ O
20R5	Camaquã	Alfisol	30°49′6.77″ S; 51°46′44.64″ O
21R1	Passo Fundo	Oxisol	28°14′30.07″ S; 52°33′20.33″ O
22R4	Tabaí	Alfisol	29°41′31.80″ S; 51°42′18.08″ O
23R1	**Santa Rosa**	**Oxisol**	**27°46′56.37″ S; 54°27′29.19″ O**
24R1	Itaqui	Alfisol	29° 7′47.54″ S; 56°33′48.01″ O
25R1	Júlio de Castilhos	Alfisol	29° 9′54.77″ S; 53°39′37.96″ O

In identification, the "R" followed by the numbers (1, 2, 3, 4, 5) represents the geomorphological region in which the state is inserted. See Figure 1. (1) The soils were classified according to the Soil Survey Staff (2014).

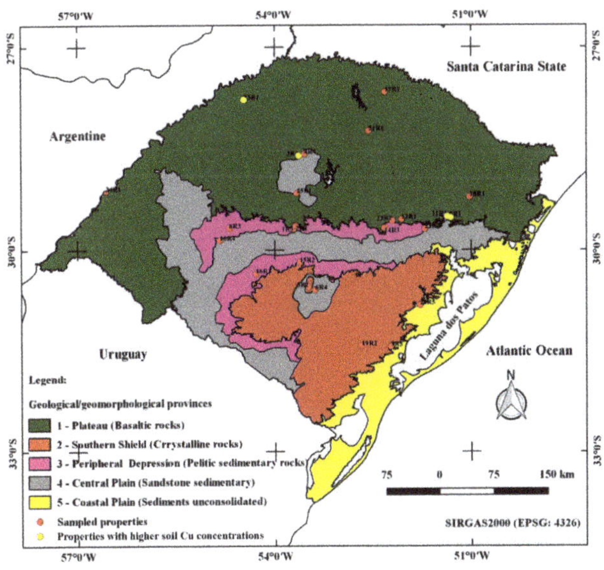

Figure 1. Georeferenced location of sampled properties across the five geomorphological/geological provinces according to FEPAM Ordinance No. 85/2014. Sites sampled for this study are marked in red, while those selected for discussion due to higher soil Cu concentrations are shown in yellow. More information about the sampling sites can be found in Table S1. Created by the author. Source: Geological Map [16], adapted from FEPAM [17].

2.2. Soil Sampling Procedure

Twenty-five sites were chosen and, within the same site, soil samples were collected in three different environments: non-cultivated, natural soil (NC), cultivated with open field crops (FCs) and cultivated in controlled environments (CEs), at the depth of 0.0–0.20 m according to the methodology proposed by CONAMA [18] and recommendations from CQFS RS/SC [19]. In all the areas, 15 subsamples were collected with a cutting blade and mixed together to create a composite sample, used for analysis. In areas under FCs and CEs, soil samples were collected between planting lines, while in native or forest areas, soil sampling was performed according to a zigzag design. Samples were placed in plastic bags, properly labeled and sealed. Subsequently, they were air dried, ground, sieved with 2 mm mesh and stored.

2.3. Chemical Analysis of Soil

Soil samples collected in the CE environment were extracted by Mehlich-1 [20] to evaluate Cu concentration. Based on these results, the four sites with the highest Cu concentrations were selected (sites 23R1, 3R1, 11R1 and 9R1) since they were those with the highest potential for environmental contamination or toxicity to plants. The sites have 31, 24, 28 and 23 years of cultivation with vegetables, respectively. The samples deriving from FCs, CPs and NCs of the four soils were analyzed for clay, organic matter, pH in water, CEC (pH 7.0), Cu-$CaCl_2$, Cu-EDTA and Cu-EPA.

Soluble Cu was extracted by $CaCl_2$ (0.01 mol L^{-1}), according to Novozamsky et al. [21]; briefly, 1.0 g of dry soil was mixed with 10 mL of $CaCl_2$ (0.01 mol L^{-1}) and stirred for 2 hours. Samples were then centrifuged for 15 min at 3200 rpm, filtered (Whatman No 1 filter paper, Germany, medium weight with a particle retention capacity of approximately 11 μm) and stored since analysis.

Available Cu (Cu-EDTA) was extracted with EDTA (Na_2-EDTA 0.01 mol L^{-1}/1.0 mol L^{-1} ammonium acetate, pH adjusted to 7.0 [22]. Total Cu concentration (Cu-EPA) was determined according to USEPA [23], by oxidation and acid attack using concentrated HNO_3, H_2O_2 and HCl under heating. In all extraction methods, Cu concentration was determined by an atomic absorption spectrophotometer (EEA; Perkin Elmer, Analyst 200, Shelton, CT, USA).

2.4. Statistical Analysis

Copper concentration determined by different extraction ($CaCl_2$, EDTA and EPA) was compared by orthogonal contrasts at 1% error probability, using the SISVAR software version 5.2 [24].

3. Results

3.1. Physical-Chemical Characteristics in Soils

The four sites selected for a more detailed analysis (23R1, 3R1, 11R1 and 9R1) were classified and inserted in geomorphological region 1 (Figure 1), whose source material is Plateau volcanic rocks [17]. Soils from 23R1 and 3R1 sites were classified as Oxisol (Table 1), while soils from the 11R1 and 9R1 sites were classified as Molisol (Tables 1 and 2). These sites were chosen due to their significantly higher Cu concentrations compared to the other sites. Sites 23R1 and 3R1 showed average clay contents, which were higher than the values observed in 11R1 and 9R1. The clay contents in CEs were lower than the values observed in FCs in all sites (Table 2). The organic matter contents ranged from 12 g kg^{-1} to 31 g kg^{-1} among the sites. In the four sites, the organic matter contents in the CE environment were higher than those observed in FCs (Table 2).

Table 2. Soil clay, organic matter (O.M.), pH, cation exchange capacity ($CEC_{pH7.0}$) and Cu evaluated in samples referring to the following environments: non-cultivated, natural soil (NC); cultivated with open field crops (FCs) and cultivated in controlled environments (CEs) from the four selected properties for further determination of Cu contents by $CaCl_2$, EDTA and EPA.

Environment	Clay	O.M.	pH	$CEC_{pH\ 7.0}$	Cu
	(g kg^{-1})		(1:1)		(mg dm^{-3})
23R1					
NC	690	31	5.7	13.1	7.40
FC	790	22	5.5	11.5	10.38
CE	470	31	5.3	17.0	60.00
3R1					
NC	360	18	5.3	7.9	1.08
FC	600	24	4.8	10.1	11.32
CE	470	30	5.3	22.6	13.90
11R1					
NC	170	12	4.6	6.9	1.54
FC	140	17	6.3	6.0	7.47
CE	90	29	5.9	8.4	10.00
9R1					
NC	320	23	5.3	16.2	3.00
FC	200	17	7.2	14.6	2.67
CE	140	28	7.4	12.9	12.5

The pH values in water were similar in soils from sites 23R1 and 3R1 (Table 2). But the pH values in water in the soils of the 11R1 and 9R1 sites in the FC and CE environments were higher than the values verified in the NC environment (Table 2). The highest CEC (pH 7.0) values were observed in the CE environment at sites 23R1, 3R1 and 11R1. In the soil of the 9R1 site, the highest CEC (pH 7.0) values were observed in the NC environment (Table 2). The lowest Cu extracted by Mehlich-1 (Cu Mehlich-1) at sites 23R1, 3R1 and 11R1 was observed in the sequence NC < FC < CE (Table 2). At the 9R1 site, the lowest Cu-Mehlich-1 content was observed in the FC < NC < CE sequence (Table 2).

3.2. *Cu Availability in Soils*

The Cu fractions extracted by $CaCl_2$, EDTA and EPA in the soil of the NC environment were different from those observed in FCs and CEs in the four sites (Table 3). In the 3R1, 11R1 and 9R1 sites, the Cu-$CaCl_2$, Cu-EDTA and Cu-EPA contents in the FC and CE environments were higher than the values verified in the NC sites. At the 23R1 site, Cu contents in the FC environment did not differ from the NCs when the element was extracted by $CaCl_2$ and EPA (Table 3).

At sites 23R1 and 11R1, Cu-$CaCl_2$ contents in the CE environment were approximately 68% and 14% higher than those observed in the FCs (Table 3), while the Cu-EDTA contents at the 23R1 and 11R1 sites were approximately 79% and 14% higher in the CE environment compared to the FCs (Table 3). The Cu-EPA contents in the 23R1 and 9R1 sites were approximately 55% and 11% higher in the CE environment in relation to the FCs, but the Cu-EPA contents did not differ in the CE and FC environments for the other sites (Table 3).

Table 3. Orthogonal contrasts of Cu-$CaCl_2$, Cu-EDTA and Cu-EPA present in non-cultivated, natural soil (NC); cultivated with open field crops (FCs) and cultivated in controlled environments (CEs) from the four states (23R1, 3R1, 11R1 and 9R1).

Environment	23R1	3R1	11R1	9R1
		Cu-$CaCl_2$ (mg kg^{-1})		
NC	0.14	0.34	0.60	0.76
FC	0.18	0.55	0.73	0.93
CE	0.57	0.64	0.85	0.99
		Orthogonal Contrast F Test		
NC × FC e CE	**	**	**	**
NC × FC	ns	**	**	**
NC × CE	**	**	**	**
FC × CE	**	ns	**	ns
		Cu-EDTA (mg kg^{-1})		
NC	18.65	1.79	2.06	7.81
FC	21.70	18.23	10.67	10.88
CE	105.41	18.72	12.42	11.66
		Orthogonal Contrast F Test		
NC × FC e CE	**	**	**	**
NC × FC	**	**	**	**
NC × CE	**	**	**	**
FC × CE	**	ns	**	ns
		Cu-EPA (mg kg^{-1})		
NC	110.12	18.97	10.90	22.57
FC	116.17	116.80	21.33	28.10
CE	260.23	117.60	22.70	31.60
		Orthogonal Contrast F Test		
NC × FC e CE	**	**	**	**
NC × FC	ns	**	**	**
NC × CE	**	**	**	**
FC × CE	**	ns	ns	**

** Significant at 0.001 significance level. ns: not significant.

4. Discussion

Based on Cu concentrations extracted by Mehlich-1 it is possible to observe (Table 2) that in environments that were not previously cultivated (NC), in the four sites, the Cu concentration is higher than 0.5 mg/dm^3 and therefore classified in the class of "high" availability according to reference levels from CQFS RS/SC [19]. This is probably associated with the material source in these soils, which are basaltic rocks, and have high Cu concentrations in their composition [25,26]. Consequently, Cu additions would not be necessary for cultivating the majority of vegetables, as the native soil Cu is sufficient to meet crop needs.

In the two cultivated soils (FCs and CEs) with vegetable crops, Cu concentrations were higher than in native soils. The increase may be associated with the use of chemical and organic fertilizers, and fungicides that contain Cu in the composition, whose most common active principles are Cu oxychloride and Cu sulfate, which are highly effective and low cost [12]. The Cu accumulation in soil can occur since application of these fungicides in vegetable cultivation areas was listed as routine by farmers, being carried out weekly in the four sites studied.

Vegetable cultivation in the CE environment accumulated the highest Cu in the soil, compared to cultivation in FCs. This can be explained by the higher binding of Cu to functional groups present in soil organic matter [27,28]. In fact, a trend towards higher organic matter values in the four sites of the CE environment was observed (Table 2). The electronic configuration of Cu, $[Ar]3d^{10}4s^1$, confers high adsorption affinity to carboxylic and phenolic groups of organic matter [29]. Over time, some of the soil organic matter can be decomposed, which contributes to higher Cu availability [30]. In addition, protected cultivation has less water flow in the soil profile, reducing the likelihood of soluble Cu movement and surface runoff [27]. These factors contribute to the maintenance of higher concentrations of Cu in the soil. On the other hand, the higher organic matter content in CE may be contributing to the higher values of $CEC_{(pH7.0)}$ in sites 23R1, 3R1 and 11R1. In higher CEC environments, a higher Cu adsorption and lower metal availability is expected, as this is an important factor in soil Cu dynamics, since it acts directly on ion concentrations present in soil solution and in sites [14,30].

In the FC and CE environments, the 23R1 and 3R1 sites were classified as Oxisol, and have the highest Cu-EPA and the lowest Cu-$CaCl_2$ soil concentration. On the other hand, sites 11R1 and 9R1, classified as Molisol, had the lowest Cu-EPA and the highest Cu-$CaCl_2$ concentration. According to Brunetto et al. [31], Cu concentration in soil solution (for example, extracted with $CaCl_2$) is generally low, approximately 2% of total Cu content. In our study, the Cu extracted by both the $CaCl_2$ and the EPA method showed percentages varying from 0.1 to 1.7%, in Oxisol (23R1 e 3R1) and from 3.1 to 5.5% in Molisols (11R1 e 9R1). The Cu present in the soil solution is the result of adsorption/desorption, complexing and precipitation reactions, which affect metal distribution between soil solid and liquid fractions, being responsible for its mobility and bioavailability [32]. Higher concentrations of Cu in the solution can increase toxicity in plants [12], but also the potential for water contamination [33].

The Cu-EDTA contents were lower in soils at sites 11R1 and 9R1 in both cultivated environments. In these areas, the highest pH values were also observed. This can be attributed to recent applications of soil acidity correctives, such as limestone [34]. The free ion Cu^{2+} predominates in acidic soils and with the increase in pH, the deprotonation of the functional groups of minerals and organic matter occurs, which increases the CEC of the soil and consequently the adsorption of Cu, thus decreasing its availability [9,35]. In addition, an increase of soil anions such as carbonates (CO_3^{-2}) and hydroxyls (OH^-) promotes Cu precipitation, thus reducing its mobility [14,30].

Total Cu content (Cu-EPA) in soil is the reference for assessing the potential of environmental contamination by heavy metals in several countries such as Brazil [18], United States of America [36], Australia and New Zealand [37] and also in the European continent [38]. In Brazilian environmental legislation, total Cu concentrations values were stipulated, as reference, in areas exploited by agriculture, for prevention (PV) and investigation value (IV), being 60 and 200 mg kg^{-1}, respectively [18]. However, state legislations have different values and parameters. For Rio Grande do Sul state QRVs, they were determined according to the soils from the five geomorphological regions, ranging from 9 to 203 mg Cu kg^{-1} in agricultural soils, with VI = QRV = PV corresponding to 203 mg Cu kg^{-1} in agricultural soils from region 1 [17] in which the four sites evaluated in the present study are inserted. Considering these references, although they show differences, it is possible to observe that the soils sampled in NCs, FCs and CEs of 23R1 and FCs and CEs of 3R1 exceeded the respective QRVs according to the legislation of CONAMA and CETESB. Only the total Cu content from soil sampled in CP of 23R1 (Table 3), 260 mg Cu kg^{-1}, exceeds CONAMA and FEPAM investigation values, which indicates the need for practices reducing Cu content in soils of these areas [18].

The high Cu concentrations observed in soils of some areas of this study can hinder plant growth and production of these crops. The adoption of conservation practices, such as minimum cultivation and direct planting of vegetables, could minimize Cu transfer via erosion and leaching. However, this could lead to an increased Cu concentration in the surface layer (0–20 cm) that is the most exploited by most vegetables root system [3]. On the other hand, reduced tillage contributes to limiting OM mineralization, thus increasing Cu complexing [31,39], leading to decreased Cu availability.

5. Conclusions

The Cu concentration in soils cultivated with vegetables increased in comparison to native soil as a result of agricultural management. This has led to alterations in Cu concentration in the soil solution, in the available fraction and total Cu in both open field (FC) and under protected cultivation (CE) areas. This increase in Cu contents in more labile forms favors the metal bioavailability in soil, which can cause damage to plants and soil microorganisms, as well as increase of environmental contamination risks. In addition, soil from CE environment of site 23R1 presented a total Cu content with high contamination potential, which surpassed the quality referential value and constant intervention value in environmental legislation of Rio Grande do Sul state, Southern Brazil, deserving attention and monitoring, as well as the adoption of measures to reduce or stabilize the content of this metal in soil. This emphasizes the importance of conducting future studies involving the collection and determination of Cu contents in vegetables cultivated in soils with high Cu levels. Such research will help elucidate the occurrence of phytotoxicity and prevent risks associated with consumer health regarding vegetables grown in these soils.

Additionally, it is important to discuss the limitations of this study, including potential biases in sampling and analysis methods. Future research should aim to incorporate a larger sample size and diverse geographical locations to enhance the robustness of the findings. Addressing these limitations could provide a more comprehensive understanding of Cu dynamics in agricultural systems and inform better management practices.

Supplementary Materials: The following supporting information can be downloaded at: https://www.mdpi.com/article/10.3390/horticulturae10111183/s1, title; Table S1: Chemical, physical and mineralogical properties evaluated in samples from the protected cultivation environment (CE) from the 25 sampled sites and extractor/methodology used; Table S2: Identification, municipalities, chemical, physical and mineralogical properties evaluated at 25 sample sites of the protected cultivation environment (CE).

Author Contributions: Conceptualization, O.J.S., J.L.A. and G.B.; methodology, C.M., A.M.C. and L.M.; investigation, O.J.S., R.C.B.V., D.R.K., T.C., M.S.W., G.B., T.L.T., L.D.C. and C.R.L.; writing—review and editing, C.M., I.C.B.S., L.M., E.T., A.M.C., M.T., E.B. and G.B.; visualization, C.M.; supervision, G.B. and J.L.A.; project administration, P.A.A.F. and G.B. All authors have read and agreed to the published version of the manuscript.

Funding: We are grateful to Conselho Nacional de Desenvolvimento Científico e Tecnológico—CNPq (Brazilian National Council for Scientific and Technological Development) for the research productivity scholarship, granted to the second author.

Data Availability Statement: The raw data supporting the conclusions of this article will be made available by the authors on reques.

Conflicts of Interest: The authors declare that they have no conflicts of interest.

References

1. Franco, T.F.; Lima, E.S.A.; Amaral Sobrinho, N.M.B.; Carmo, M.G.F.; Breda, F.A.D.F. Enrichment and bioavailability of toxic elements in intensive vegetable production areas. *Rev. Caatinga* **2020**, *33*, 124–134. [CrossRef]
2. Silva, L.S.; Galindo, I.C.L.; Nascimento, C.W.A.; Gomes, R.P.; Campos, M.C.C.; Freitas, L.; Oliveira, I.A. Heavy metal contents in Latosols cultivated with vegetable crops. *Pesqui. Agropecu. Trop.* **2016**, *46*, 391–400. [CrossRef]
3. Filgueira, F.A.R. *Novo Manual de Olericultura: Agrotecnologia Moderna na Produção e Comercialização de Hortaliças*; UFV: Viçosa, Brazil, 2012.
4. Silva, L.S.; de Luna Galindo, I.C.; Gomes, R.P.; Filla, V.A.; Campos, M.C.C.; de Freitas, L.; de Oliveira, I.A.; Marques, K.P.P.; de Souza, E.D.; Cazzeta, J.O. Investigation of heavy metal accumulation in soil, water and plants in areas with intensive horticulture. *Aust. J. Crop Sci.* **2019**, *13*, 192–198. [CrossRef]
5. Zhen, H.; Jia, L.; Huang, C.; Qiao, Y.; Li, J.; Li, H.; Wan, Y. Long-term effects of intensive application of manure on heavy metal pollution risk in protected-field vegetable production. *Environ. Pollut.* **2020**, *263*, 114552. [CrossRef]
6. Fernández-Calviño, D.; Soler-Rovira, P.; Polo, A.; Arias-Estévez, M.; Plaza, C. Influence of humified organic matter on copper behavior in acid polluted soils. *Environ. Pollut.* **2010**, *158*, 3634–3641. [CrossRef]

7. Brasil. Sistema de Agrotóxicos Fitossanitários (AGROFIT). 2018. Available online: http://agrofit.agricultura.gov.br/agrofit_cons/principal_agrofit_cons (accessed on 12 July 2024).
8. Wang, J.; Guo, W.; Chen, X.; Shi, Y. Effect of cultivation ages on Cu accumulation in Greenhouse Soils in North China. *IOP Conf. Ser. Earth Environ. Sci.* **2017**, *94*, 012071. [CrossRef]
9. Yang, L.; Huang, B.; Hu, W.; Chen, Y.; Mao, M.; Yao, L. The impact of greenhouse vegetable farming duration and soil types on phytoavailability of heavy metals and their health risk in eastern China. *Chemosphere* **2014**, *103*, 121–130. [CrossRef]
10. Zhang, H.Z.; Li, H.; Wang, Z.; Zhou, L.D. Accumulation Characteristics of Copper and Cadmium in Greenhouse Vegetable Soils in Tongzhou District of Beijing. *Procedia Env. Sci.* **2011**, *10*, 289–294. [CrossRef]
11. Kabata-Pendias, A. *Trace Elements in Soils and Plants*; CRC Press: Boca Raton, FL, USA, 2011.
12. Tiecher, T.L.; Soriani, H.H.; Tiecher, T.; Ceretta, C.A.; Nicoloso, F.T.; Tarouco, C.P.; Clasen, B.E.; De Conti, L.; Tassinari, A.; Melo, G.W.B.; et al. The interaction of high copper and zinc doses in acid soil changes the physiological state and development of the root system in young grapevines (*Vitis vinifera*). *Ecotoxicol. Environ. Saf.* **2018**, *148*, 985–994. [CrossRef]
13. Mendoza, J.C.; Garrido, T.R.; Quilodrán, C.R.; Segovia, M.C.; Parada, J.A. Evaluation of the bioaccessible gastric and intestinal fractions of heavy metals in contaminated soils by means of a simple bioaccessibility extraction test. *Chemosphere* **2017**, *176*, 81–88. [CrossRef]
14. Brunetto, G.; de Melo, G.W.B.; Terzano, R.; Del Buono, D.; Astolfi, S.; Tomasi, N.; Pii, Y.; Mimmo, T.; Cesco, S. Copper accumulation in vineyard soils: Rhizosphere processes and agronomic practices to limit its toxicity. *Chemosphere* **2016**, *162*, 293–307. [CrossRef] [PubMed]
15. Kumar, V.; Pandita, S.; Sidhu, G.P.S.; Sharma, A.; Khanna, K.; Kaur, P.; Bali, A.S.; Setia, R. Copper bioavailability, uptake, toxicity and tolerance in plants: A comprehensive review. *Chemosphere* **2021**, *262*, 127810. [CrossRef] [PubMed]
16. CPRM—Comissão de Pesquisa em Recursos Minerais. Mapa Geológico do Estado do Rio Grande do Sul. Projeto Geologia do Brasil ao Milionésimo. 2006. Available online: http://www.cprm.gov.br/publique/media/geologia_basica/cartografia_regional/mapa_rio_grande_sul.pdf (accessed on 7 May 2024).
17. FEPAM—Fundação Estadual de Proteção Ambiental. Portaria FEPAM nº 85/2014. 2014. Available online: https://www.fepam.rs.gov.br/upload/arquivos/202301/27114351-portaria085-2014.pdf (accessed on 21 March 2024).
18. CONAMA—Conselho Nacional do Meio Ambiente. Resolução CONAMA no 420 de 28/12/2009. 2009. Available online: https://www.normasbrasil.com.br/norma/resolucao-420-2009_110436.html (accessed on 21 March 2024).
19. CQFS RS/SC—Comissão de Química e Fertilidade do Solo. *Manual de Adubação e de Calagem Para os Estados do Rio Grande do Sul e de Santa Catarina*; SBCS/Núcleo Regional Sul: Porto Alegre, Brazil, 2016.
20. Tedesco, M.J.; Gianello, C.; Biassani, C.A.; Bohnen, H.; Volkweiss, S.J. Análise de solo, plantas e outros materiais. In *Boletim Técnico*; Universidade Federal do Rio Grande do Sul: Porto Alegre, Brazil, 1995; p. 174.
21. Novozamsky, I.; Lexmond, T.M.; Houba, V.J.G. A Single Extraction Procedure of Soil for Evaluation of Uptake of Some Heavy Metals by Plants. *Int. J. Environ. Anal. Chem.* **1993**, *51*, 47–58. [CrossRef]
22. Chaignon, V.; Hinsinger, P. A Biotest for Evaluating Copper Bioavailability to Plants in a Contaminated Soil. *J. Environ. Qual.* **2003**, *32*, 824–833. [CrossRef]
23. USEPA. (n.d.). EPA Method 3050B: Acid Digestion of Sediments, Sludges, and Soils. Available online: https://www.epa.gov/homeland-security-research/epa-method-3050b-acid-digestion-sediments-sludges-and-soils (accessed on 28 March 2024).
24. Ferreira, D.F. SISVAR: Um programa para análises e ensino de estatística. *Rev. Symp.* **2008**, *6*, 36–41.
25. Althaus, D.; Gianello, C.; Tedesco, M.J.; Justin, K. Natural Fertility and Metals Contents in Soils of Rio Grande do Sul (Brazil). *Rev. Bras. Ciênc. Solo* **2018**, *42*, e0160418. [CrossRef]
26. Zinn, Y.L.; de Faria, J.A.; Araujo, M.A.; Skorupa, A.L.A. O material de origem do solo é o principal controle sobre as concentrações de metais pesados nas terras altas tropicais do Brasil. *Catena* **2020**, *185*, 104319. [CrossRef]
27. De Conti, L.; Ceretta, C.A.; Ferreira, P.A.; Lourenzi, C.R.; Girotto, E.; Lorensini, F.; Tiecher, T.L.; Marchezan, C.; Anchieta MGBrunetto, G. Soil solution concentrations and chemical species of copper and zinc in a soil with a history of pig slurry application and plant cultivation. *Agric. Ecosyst. Environ.* **2016**, *216*, 374–386. [CrossRef]
28. Tiecher, T.L.; Ceretta, C.A.; Comin, J.J.; Girotto, E.; Miotto, A.; Moraes, M.P.; Benedet, L.; Ferreira, P.A.A.; Lorenzi, C.R.; Couto, R.R.; et al. Forms and accumulation of copper and zinc in a sandy typic hapludalf soil after long-term application of pig slurry and deep litter. *Rev. Bras. Ciênc. Solo* **2013**, *37*, 812–824. [CrossRef]
29. Dresler, S.; Hanaka, A.; Bednarek, W.; Maksymiec, W. Accumulation of low molecular-weight organic acids in roots and leaf segments of Zea mays plants treated with cadmium and copper. *Acta Physiol. Plant* **2014**, *36*, 1565–1575. [CrossRef]
30. Trentin, E.; Facco, D.B.; Hammerschmitt, R.K.; Avelar Ferreira, P.A.; Morsch, L.; Belles, S.W.; Ricachenevsky, F.K.; Nicoloso, F.T.; Ceretta, C.A.; Tiecher, T.L.; et al. Potential of vermicompost and limestone in reducing copper toxicity in young grapevines grown in Cu-contaminated vineyard soil. *Chemosphere* **2019**, *226*, 421–430. [CrossRef] [PubMed]
31. Brunetto, G.; Comin, J.J.; Miotto, A.; de Moraes, M.P.; Sete, P.B.; Schmitt, D.E.; Gatiboni, L.C.; de Melo, G.W.B.; Morais, G.P. Copper and zinc accumulation, fractionation and migration in vineyard soils from Santa Catarina State, Brazil. *Bragantia* **2017**, *77*, 141–151. [CrossRef]
32. McBride, M.; Sauve, S.; Hendershot, W. Solubility control of Cu, Zn, Cd and Pb in contaminated soils. *Eur. J. Soil Sci.* **1997**, *48*, 337–346. [CrossRef]

33. Ouyang, W.; Wang, Y.; Lin, C.; He, M.; Hao, F.; Liu, H.; Zhu, W. Heavy metal loss from agricultural watershed to aquatic system: A scientometrics review. *Sci. Total Environ.* **2018**, *637*, 208–220. [CrossRef] [PubMed]
34. Tiecher, T.; Denardin, L.G.D.O.; Martins, A.P.; Pias, O.H.; Nunes, L.R.; Tiecher, T.L.; Alvarez, J.W.R.; Mallmann, F.J.K.; Brunetto, G.; Rheinheimer Dos Santos, D. Dynamics of sulfate and basic cations in soil solution as affected by gypsum fertilization in an Ultisol of Southern Brazil. *Arch. Agron. Soil Sci.* **2019**, *65*, 1272–1286. [CrossRef]
35. CETESB—Companhia de Tecnologia de Saneamento Ambiental. Decisão de Diretoria nº 256/2016. *Rev. Meio Ambiente* **2016**, *126*, 55–56.
36. USEPA—United States Environmental Protection Agency. *40 CFR-Standards for the Use and Disposal of Sewage Sludge*; USEPA: Washington, DC, USA, 1993; pp. 9387–9415.
37. ANZECC. Australian and New Zealand Guidelines for the Assessment and Management of Contaminated Sites [Electronic Resource]/Australian and New Zealand Environment and Conservation Council, National Health and Medical Research Council. Version Details—Trove. 1992. Available online: https://trove.nla.gov.au/work/18093032?q&versionId=21236483 (accessed on 15 February 2024).
38. Council Directive 86/278/EEC. Protection of the environment, and in particular of the soil, when sewage sludge is used in agriculture. *Off. J. Eur. Union* **1986**, *181*, 6–12.
39. Girotto, E.; Ceretta, C.A.; Brunetto, G.; Santos, D.R.; Silva, L.S.; Lourenzi, C.R.; Lorensini, F.; Vieira, R.C.B.; Schmatz, R. Acúmulo e formas de cobre e zinco no solo após aplicações sucessivas de dejeto líquido de suínos. *Rev. Bras. Ciênc. Solo* **2010**, *34*, 955–965. [CrossRef]